Mechanics of Fibrous Composites

CARL T. HERAKOVICH

University of Virginia

John Wiley & Sons, Inc.
New York • Chichester • Weinheim • Brisbane • Singapore • Toronto

ACQUISITIONS EDITOR Wayne Anderson
MARKETING MANAGER Harper Mooy
PRODUCTION EDITOR Tony VenGraitis
COVER DESIGNER Janice Noto-Helmers
DESIGN SUPERVISION Harry Nolan
ILLUSTRATION COORDINATOR Sigmund Malinowski

The book was set in Times Roman by Publication Services and printed and bound by Hamilton Printing
The cover was printed by Phoenix Color

This book is printed on acid-free paper. ♾

The paper in this book was manufactured by a mill whose forest management principals include sustained yield harvesting of its timberlands. Sustained yield harvesting principles ensure that the numbers of trees cut each year does not exceed the amount of new growth.

Library of Congress Cataloging in Publication Data:
Herakovich, Carl T.
Mechanics of fibrous composites / Carl T. Herakovich

p. cm.
Includes index.
ISBN 0-471-10636-4 (cloth : alk. paper)
1. Fibrous composites--Mechanical properties. I. Title.

TA418.9.C6H47 1998
620.1'183--dc21 97-28361
CIP

To Marlene,

wife, best friend, and the mother of our children . . .

Brad, Doug, Kris, and Russ

PREFACE

This text is intended for introductory and advanced courses on the study of fibrous composite materials from the mechanics perspective. The book is the result of teaching courses and conducting research on composites for more than a quarter-century. Although the book is self-contained, it is advisable to precede the material in this book with a fundamental course on mechanics of solids. The entire book can be used for a full academic year of study. For an introductory, one-semester course, coverage can be limited to Chapter 1; Chapter 2, Sections 2.1–2.12; Chapters 4, 5, and 6; Chapter 8, Sections 8.1–8.2; Chapter 9, Sections 9.1–9.3; and Chapter 11, Sections 11.1–11.3.

In writing this book the intent has been to provide fundamental information on the engineering properties and advantages of fibrous composites; the mathematical formulations of the governing equations of mechanics for anisotropic, laminated materials; descriptions of test methods for determining properties of composites; experimental results for the complete mechanical response of a variety of composites up to and including failure; correlation of actual material response in both the linear and nonlinear ranges with theoretical predictions, including the presentation of a theory for damage evolution in composites; in-depth coverage of micromechanics methods for predicting the elastic properties of composites based upon the properties of the fibers and the matrix; and in-depth coverage for the thermal-elastic response of laminated tubes. The theoretical aspects of the book never stray too far from the real world of the laboratory and actual structural applications. Sufficient data on actual materials have been included so that the book can serve as a resource on material properties and material response.

The author believes that certain topics are presented in a text on composites for the first time, including interlaminar stresses (Chapter 8), damage mechanics (Chapter 9), and laminated tubes (Chapter 10). A chapter is devoted to interlaminar stresses because they often play a vital role in determining the strength of laminated fibrous composites. Likewise, a damage theory for composites is presented in Chapter 9 because some composites exhibit extensive damage prior to ultimate failure. Finally, the chapter on laminated tubes is included because the tube is a fundamental structural member, because an analytical solution exists for many types of loadings, and because laminated tubes exhibit coupling phenomena that are common characteristics of composites. The chapter on micromechanics completes the range of coverage from micro to macro.

Both SI (Système international d'unités) and U.S. customary units are used throughout the book to facilitate use by readers around the world and also to reinforce the relationship between the two systems of units.

No project like this is completed without the help of many people. Thanks are gratefully extended to Professor Nicolae Taranu of Iasi University, Iasi, Rumania, for an extensive, in-depth editing of the entire book during a Fulbright year at the University of Virginia; to Professor James Simmonds, my colleague at the University of Virginia, for extensive editing of most of the chapters; to Professor Jacob Aboudi of Tel Aviv University for review of several chapters as well as many in-depth discussions on mechanics of composites during his three sabbatical years that we spent together at the University of Virginia (two) and Virginia Tech (one); to Professor Marek-Jerzy Pindera, my colleague at the University of Virginia and Virginia Tech, for numerous discussions on composites over a period of close to 15 years; to Dr. Cheryl Rose of NASA Langley Research Center for editing Chapter 8 on interlaminar stresses; to Professor Pierre Ladevèze of the Laboratoire de Mécanique et Technologie, ENS Cachan / CNRS / Université, Paris 6, and his colleague Dr. Alain Gasser, for an in-depth review and editing of the section on damage mechanics in Chapter 9; to Dr. Eric Le Dantec of Sociéte Bertin & Cie, Plaisir, France, for supplying data for many of the figures in the section on damage; to Professor Niall Horgan of the University of Virginia for review of several of the early chapters; to Mrs. Eman Siragy for an outstanding job in converting old figures to a form suitable for inclusion in this text; to Mrs. Penny Laferriere, my former secretary, and Mrs. Beverly Martin, my current secretary, for assistance with word processing and scanning figures; and to Dr. Farshad Mirzadeh and Mrs. Sydney Heasley for assistance in the preparation and presentation of figures. I would also like to thank all of my current and former graduate students whom I have had the opportunity to work with. Through them I have learned much about composites, and much of their work is included in this book, either directly or indirectly. Thanks are also due to all of the students who have taken courses in composites from me and enhanced my understanding of the subject through their questions and reviews of the manuscript while it was in preparation. Thanks also go to the University of Virginia and Virginia Polytechnic Institute and State University (Virginia Tech), the two universities that have provided me the opportunity to teach courses on mechanics of composites and supported me in this endeavor. Thanks to NASA Langley Research Center for continuous support of research on composites over more than a quarter-century, and, more recently, to the Air Force Office of Scientific Research for major support of research on high-temperature composites at the University of Virginia. And finally, a sincere thank you to all my colleagues and friends at the University of Virginia, Virginia Tech, and other universities around the world who have improved my understanding of composites through our numerous discussions.

This book has been in preparation for many years, and it is with a sense of relief that I write this preface.

Carl T. Herakovich
Charlottesville, Virginia
February 1997

"More important than any one new application is the new 'materials' concept itself. It marks a shift from concern with substances to concern with structures, a shift from artisan to scientist as man's artificer, a shift from chemistry to physics as the basic discipline, and a shift, above all, from the concrete experience of the workshop to abstract mathematics, a shift from starting with what nature provides to what man wants to accomplish."

Peter F. Drucker,
The Age of Discontinuity, 1969

"Where there is much desire to learn, there of necessity will be much arguing, much writing, many opinions: for opinion in good men is but knowledge in the making."

John Milton,
On Opinion and Knowledge
(Aeropagitica, 1644)

"Nothing happens unless first a dream."

Carl Sandburg

NOTATION

BASIC CONCEPTS

a_{ij} direction cosines

σ_{ij} components of stress tensor

u_1, u_2, u_3 displacements

θ angular orientation

m $\cos\theta$

n $\sin\theta$

ε_{ij} components of strain tensor

γ_{ij} engineering shear strains

$[\varepsilon]$ strain matrix

$[\sigma]$ stress matrix

τ_{ij} components of shear stress

C_{ijkl} fourth-order stiffness tensor

F_i components of body force per unit volume

T_i traction

$\sigma_{nn}, \sigma_{nt}, \sigma_n$ components of boundary stress

x_1, x_2, x_3 material principal coordinates

x, y, z global coordinates

V_i vector

W strain energy density

S surface

S_T traction surface

S_D displacement surface

V volume

Π potential energy

Π^* kinematically admissible Π

Π_c^o statically admissible Π

CONSTITUTIVE EQUATIONS

σ_i stresses (contracted notation)

ε_i strains (contracted notation)

C_{ij} stiffness coefficients (contracted notation)

S_{ij} compliance coefficients (contracted notation)

$\bar{C}_{ij}$ transformed stiffness

$\bar{S}_{ij}$ transformed compliance

Q_{ij} reduced stiffness

$\bar{Q}_{ij}$ transformed reduced stiffness

E Young's modulus—isotropic

G shear modulus—isotropic

ν Poisson's ratio—isotropic

E_1, E_2, E_3 axial moduli—principal material directions

ν_{12}, ν_{13} axial Poisson's ratios

ν_{23} transverse Poisson's ratio

G_{12}, G_{13} axial shear moduli

G_{23} transverse shear modulus

$\{\varepsilon^T\}$ free thermal strains

$\{\alpha\}_1$ coefficients of thermal expansion (CTE)—principal material directions

$\{\alpha\}_x$ coefficients of thermal expansion—global x-y directions

ΔT temperature change

$\{\varepsilon^\sigma\}$ strains associated with stress

$\{\varepsilon\}$ total strains

$\{\beta\}$ coefficient of hygroscopic expansion (CHE)

$\{\varepsilon^H\}$ hygroscopic strains

ΔM percent moisture change

E_x axial modulus

E_y transverse modulus

G_{xy} in-plane shear modulus

ν_{xy} in-plane Poisson's ratio

$\eta_{xy,x}$ coefficient of mutual influence—second kind

$\eta_{xy,y}$ coefficient of mutual influence—second kind

$\eta_{x,xy}$ coefficient of mutual influence—first kind

$\eta_{y,xy}$ coefficient of mutual influence—first kind

$[T_1]$ stress transformation matrix

$[T_2]$ strain transformation matrix—engineering shear

U_i reduced stiffness invariants

$\{M^H\}$ equivalent hygroscopic moments/length

$\{\beta\}$ laminate coefficient of hygroscopic expansion

$\{\beta\}^k$ *k*th-layer coefficient of hygroscopic expansion

ν_{xz}, ν_{yz} effective through-thickness Poisson's ratios

α_z through-thickness effective CTE

β_z through-thickness effective CHE

LAMINATE ANALYSIS

u, v, w displacement components

z_k coordinate to the *bottom* of *k*th layer

t_k thickness of *k*th layer

$2H$ laminate total thickness

$\{\kappa\}_x = (\kappa_x, \kappa_y, \kappa_{xy})$ laminate curvatures

$\{\varepsilon^o\}_x = (\varepsilon_x^o, \varepsilon_y^o, \gamma_{xy}^o)$ laminate midplane strains

$[\bar{Q}]^k$ *k*th-layer reduced stiffness matrix

$\{N\} = (N_x, N_y, N_{xy})$ in-plane forces/length

$\{M\} = (M_x, M_y, M_{xy})$ moments/length

$[A]$ laminate in-plane stiffness

$[B]$ laminate bending-stretching coupling

$[D]$ laminate bending stiffness

N or $2N$ number of layers in a laminate

$\{\bar{\sigma}\}$ laminate average stresses

$[a^*]$ laminate compliance

$\{N^T\}$ equivalent thermal force/length

$\{M^T\}$ equivalent thermal moment/length

$\{\alpha\}$ laminate coefficient of thermal expansion (CTE)

$\{\alpha\}^k$ *k*th-layer coefficient of thermal expansion

$\{N^H\}$ equivalent hygroscopic forces/length

INTERLAMINAR STRESSES

F_{yz}, F_{zx}, M_z interlaminar forces and moment

$\delta\nu_{xy}$ Poisson mismatch

$\delta\eta_{xy,x}$ coefficient of mutual influence mismatch

$U(y,z), V(y,z), W(y,z)$ displacement warping functions

$[\sigma_{ij}^{(k)}]_E$ equilibrium solution

$[\sigma_{ij}^{(k)}]_\eta$ η mismatch solution

$[\sigma_{ij}^{(k)}]_\upsilon$ ν mismatch solution

FAILURE

Micro-level Failure:

σ_c^{ult} composite ultimate average stress

σ_m' matrix stress at composite failure

σ_f' fiber stress at composite failure

σ_m^{ult} failure stress of matrix portion of composite

V_f^{min} minimum V_f for actual reinforcement

V_f^{crit} critical V_f for additional strength of bulk matrix

σ_f^{ult} fiber ultimate stress

σ_c^{fult} fiber-dominated composite ultimate stress

σ_c^{mult} failure stress of matrix volume at failure

Macroscopic Failure:

$X_T, X_C, Y_T, Y_C, Z_T, Z_C$ normal stress ultimates

Q, R, S shear stress ultimates

$\varepsilon_i^T, \varepsilon_i^C$ normal strain ultimates

Γ_{ij} shear strain ultimates

F_i, F_{ij} strength tensors

F_i', F_{ij}' transformed strength tensors

$f(\sigma_i)$ tensor polynomial function

DAMAGE

$\{\tilde{\sigma}\}$ effective stresses

d_1, d_2, d_6 damage parameters

E_1^0, E_2^0, G_{12}^0 undamaged moduli

E_1, E_2, G_{12} damaged moduli

Y_1, Y_2, Y_6 thermodynamic forces

E_D damaged strain energy density

ψ thermodynamic potential

$\underline{Y}_2, \underline{Y}_6$ maximum value of thermodynamic forces

$\hat{Y}$ combined thermodynamic force

$\hat{\underline{Y}}$ maximum value of $\hat{Y}$

Y_2^C brittle-damage threshold for transverse tension

Y_O, Y_C, Y_C', Y_O' damage constants

ε_{ij}^p inelastic (permanent) strain components

$\tilde{p}$ effective permanent strain

$\dot{\tilde{\varepsilon}}_{ij}^p$ effective inelastic strain rates

$f(\tilde{\sigma}_{ij}, \tilde{p})$ elastic domain function

Φ_p inelastic dissipation

$R(\tilde{p})$ hardening function

R_0 initial value of hardening function

$\dot{\lambda}_p$ hardening parameter

α, β material parameters for hardening function

a^2 coupling parameter in elastic domain function

$\dot{\varepsilon}_{ij}$ total strain rates

$\dot{\varepsilon}_{ij}^E$ elastic strain rates

$\dot{\varepsilon}_{ij}^p$ inelastic strain rates

σ_L applied axial stress on coupon

$\varepsilon_L, \varepsilon_T$ axial and transverse coupon strains

τ current time

t total elapsed time

TUBES

R_I, R_o inside and outside radii

u, v, w axial, tangential, and radial displacements

x, θ, r axial, tangential, and radial coordinates

P_x, T_x axial force and torque

p_I, p_o internal and external pressure

ϕ fiber orientation for tubes

$\bar{C}_{ij}, \bar{S}_{ij}$ stiffness and compliance in x, θ, r coordinates

ε_x^0 uniform axial strain

γ^0 angle of twist per unit length

$\sigma_x, \sigma_\theta, \sigma_r, \tau_{\theta r}, \tau_{xr}, \tau_{x\theta}$ stresses in x, θ, r coordinates

$\varepsilon_x, \varepsilon_\theta, \varepsilon_r$ normal strains in x, θ, r coordinates

$\gamma_{xr}, \gamma_{\theta r}, \gamma_{x\theta}$ shear strains in x, θ, r coordinates

$\varepsilon_i^T, \gamma_{ij}^T$ thermal strains

$\alpha_x, \alpha_\theta, \alpha_r, \alpha_{x\theta}$ coefficient of thermal expansion

$\Gamma, \Omega, \Psi, \Sigma$ defined material constants combinations

N number of layers

$\bar{E}_x, \bar{G}_{x\theta}$ tube effective axial moduli

$\bar{\alpha}_x$, $\bar{\alpha}_r$ tube effective CTE

J tube polar moment of inertia

L tube length

$\zeta_{P\gamma}$ axial force coupling coefficient

$\zeta_{T\varepsilon}$ torque coupling coefficient

$\zeta_{\Delta T}$ thermal coupling coefficient

ζ_{PI} internal pressure coupling coefficient

MICROMECHANICS

Strength of Materials:

V_f fiber volume fraction

V_m matrix volume fraction

V total volume

A cross-sectional area

A_f, A_m fiber and matrix cross-sectional area

C^*_{ij} effective stiffness

E^*_1, E^*_2, E^*_3 effective moduli

ν^*_{12}, ν^*_{13}, ν^*_{23} effective Poisson's ratios

$G^*_{12} = G^*_{13}$ effective axial shear moduli

E_f, E_m, V_f, V_m, ν_f, ν_m fiber and matrix properties

$\bar{\sigma}_1$, $\bar{\varepsilon}_1$ average axial stress and strain

Continuum Mechanics:

$\bar{\sigma}_{ij}$ composite average stress

$\sigma_{ij}(\underset{\sim}{x})$ local stress at position $\underset{\sim}{x}$

$\bar{\varepsilon}_{ij}$ composite average strain

C^*_{ijkl} effective stiffness

S^*_{ijkl} effective compliance

$\bar{\sigma}^{(f)}_{ij}$ average stress in fiber

$\bar{\sigma}^{(m)}_{ij}$ average stress in matrix

$\bar{\varepsilon}^{(f)}_{ij}$ average strain in fiber

$\bar{\varepsilon}^{(m)}_{ij}$ average strain in matrix

$T_i = \sigma^o_{ji} n_j$ boundary traction

σ^o_{ij} applied stress

ε^o_{ij} applied strain

$A_{ijkl}(\underset{\sim}{x})$ strain concentration factors

$B_{ijkl}(\underset{\sim}{x})$ stress concentration factors

$\bar{A}^{(\alpha)}_{ijkl}$ phase average strain concentration factors

$\bar{B}^{(\alpha)}_{ijkl}$ phase average stress concentration factors

I_{ijkl} fourth-order identity tensor

$\bar{A}^{(f)}_{ijkl}$, $\bar{A}^{(m)}_{ijkl}$ fiber and matrix average strain concentration factors

$\bar{B}^{(f)}_{ijkl} = \bar{B}^{(m)}_{ijkl}$ fiber and matrix average stress concentration factors

k^*_{23} plane strain bulk modulus

G^*_{23} effective shear modulus

K_m matrix bulk modulus

Method of Cells:

$\bar{\underset{\sim}{\sigma}}^{(\beta\gamma)}$ subcell average stress

$\underset{\sim}{A}^{(\beta\gamma)}$ subcell strain concentration factors

$\bar{\underset{\sim}{\varepsilon}}^{(\beta\gamma)}$ subcell average strain

$\underset{\sim}{C}^{(\beta\gamma)}$ subcell stiffness matrix

$\bar{\underset{\sim}{\varepsilon}}$ composite (cell) strains

$\phi^{(\beta\gamma)}_i(x_1)$, $\psi^{(\beta\gamma)}_i(x_1)$ microvariables

$w^{(\beta\gamma)}_i(x_1, x_2, x_3)$ displacement of subcell center

h_i, l_i subcell dimensions

$[A_G]$ geometric coefficient matrix

$[J]$ displacement continuity coefficient matrix

$[A_M]$ traction continuity coefficient matrix

$[\tilde{A}]$ combined coefficient matrix

$[A]$ matrix of Hill's strain concentrations

Mori-Tanaka:

$\underset{\sim}{T}$ Mori-Tanaka interaction tensor

$\underset{\sim}{S}$ Eshelby's tensor

$\underset{\sim}{C}_f$, $\underset{\sim}{C}_m$ fiber and matrix stiffness tensors

GENERAL ABBREVIATIONS

f fiber

m matrix

1, 2, 3 material principal coordinates

x, y, z global coordinates

$[\bar{\ }]$ transformed quantity

$\{\ \}_x$ global coordinates

$\{\ \}_1$ material principal coordinates

CTE coefficient of thermal expansion

Gr/Ep graphite/epoxy

B/Al boron/aluminum

Gl/Ep glass/epoxy

Prefix	Symbol	Factor
nano	n	10^{-9}
micro	μ	10^{-6}
milli	m	10^{-3}
centi	c	10^{-2}
kilo	k	10^{3}
mega	M	10^{6}
giga	G	10^{9}

UNITS CONVERSION

Quantity	SI to U.S. Customary	U.S. Customary to SI
Length	1.0 m = 3.281 ft = 39.7 in 1.0 μm = 1.0×10^{-6} m = 39.7×10^{-6} in	1.0 in = 2.54 cm = 25.4 mm
Stress	1.0 MPa = 145.0 psi[a] = 0.145 ksi	1.0 ksi = 6.8948 MPa
Modulus	1.0 GPa = 0.145 Msi	1.0 Msi = 6.8948 GPa
Force	1.0 N = 0.225 lbf[b]	1.0 lbf = 4.448 N
Force/length	1.0 N/m = 0.00571 lbf/in	1.0 lbf/in = 175.2 N/m
Moment	1.0 N · m = 8.850 lbf · in	1.0 lbf · in = 0.1130 N · m
Moment/length	$1.0 \frac{\text{N} \cdot \text{m}}{\text{m}} = 0.2248 \frac{\text{lbf} \cdot \text{in}}{\text{in}}$	$1.0 \frac{\text{lbf} \cdot \text{in}}{\text{in}} = 4.448 \frac{\text{N} \cdot \text{m}}{\text{m}}$
Temperature	T (°F) = 1.8T (°C) + 32 T (°F) = 1.8T (K) – 459.67 (Note: T (K) = T (°C) + 273.15)	T (°C) = [T (°F) – 32] / 1.8 T (K) = [T (°F) + 459.67] / 1.8 (Note: T (K) = T (°R) / 1.8)
Thermal expansion	1.0 με/°C = 0.556 με/°F	1.0 με/°F = 1.8 με/°C
Density	1.0 g/cm^3 = 0.0361 lb[c]/in^3	1.0 lb/in^3 = 27.68 g/cm^3

[a] psi = lbf/in^2

[b] lbf = pound-force

[c] lb = pound-mass

CHAPTER 1 THE WHAT AND THE WHY OF FIBROUS COMPOSITES 1

CHAPTER 2 CONCEPTS OF SOLID MECHANICS 29

CHAPTER 3 3-D CONSTITUTIVE EQUATIONS 47

CHAPTER 5 LAMINATION THEORY 112

CHAPTER 6 TEST METHODS 185

CHAPTER 7 MATERIAL RESPONSE 222

CHAPTER 8 INTERLAMINAR STRESSES 247

CHAPTER 11 MICROMECHANICS 402

CHAPTER 1 THE WHAT AND THE WHY OF FIBROUS COMPOSITES

"Oh, but a man's reach should exceed his grasp,
or what's a heaven for?"
Robert Browning

1.1 Introduction

The field of composite materials is both old and new. It is old in the sense that most natural objects, including the human body, plants, and animals, are composites. It is new in the sense that only since the early 1960s have engineers and scientists exploited seriously the vast potential of fabricated fibrous composite materials. Development of new composites and new applications of composites is now accelerating. Because of this, the material properties and applications cited in this introductory chapter should be considered only as typical of current examples. As better and more economical composites are introduced, and as more engineers become knowledgeable in the analysis, design, and fabrication of composites, the applications will expand even faster. Those mentioned here will serve as a peek through the crack in the door to a world of engineered materials and structures whose potential appears to be far-reaching. Future applications of composites will be limited only by the lack of ingenuity and the unwillingness of individuals and society to explore this vast unknown. There can be no question that in the worldwide competitive markets of the 21st century some individuals and some societies will meet the challenges and opportunities offered through the use of "engineered composites." The potential benefits are too great for this not to be true.

In order to provide a motivation for the use of composites, their engineering properties are compared with those of monolithic materials, and applications where composites have proven advantages over monolithics are presented in this chapter. However, the main focus of this book is the *mechanics of composites*. The fundamental laws of mechanics are timeless. What will change is the synthesis and fabrication of new materials, and the designs that can be accomplished with materials exhibiting superior properties, properties that can be tailored to meet the demands of the application.

1.2 Definition of a Composite

Very simply, a *composite* is a material which is composed of two or more distinct phases (Fig. 1.1). Thus a composite is *heterogeneous*. To a certain extent this definition depends upon the level of analysis, as all materials may be considered heterogeneous if the scale of interest is sufficiently small. As used in this book, *fibrous composites* are materials in which one phase acts as a

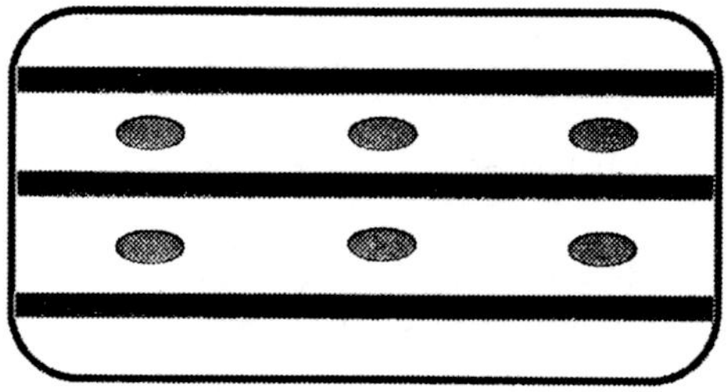

FIGURE 1.1 Multiphase Media

reinforcement of a second phase. The second phase is called the *matrix*. The challenge is to combine the fibers and the matrix to form the most efficient material for the intended application.

One of the most efficient structures known is the human body. The sketch of the muscles in the human body in Fig. 1.2 is a beautiful example of the efficient use of fibrous composites. The muscles are present in a layered system consisting of fibers at different angles and in different concentrations. This system results in a very strong, efficient, versatile, and adaptable structure which is capable of performing amazing feats. It is a challenge to engineers to design and fabricate structures which function in as versatile and efficient a manner as the human body. The book by Clayman (1995) includes numerous excellent views clearly showing the heterogeneous and fibrous nature of the human body. Other obvious examples of naturally occurring fibrous composites include the wings of a bird, the fins of a fish, trees, and grass. All of these structures are typified by two or more phases, one of which is stronger and stiffer than the others and serves as the primary load-carrying component. The book by Wainwright et al. (1976) is concerned with the mechanics of a wide variety of organisms that are fibrous composites. The book includes a discussion of fiber-wound cylinders treated as reinforced membrane systems and shows the fundamental role played by the fiber orientation.

Fibrous composite materials were used in Egypt as early as 4000 B.C. for making laminated writing materials from the papyrus plant. Cuttings from the papyrus plant were also used by early Egyptians to make boats, sails, baskets, and ropes. Papyrus parchments are still in use today. Around 1300 B.C., as indicated in the Book of Exodus ("do not give them straw for their bricks, make them find their own straw"), straw was used as reinforcement for mud bricks, a practice that is still in use today. According to Hartman et al. (1994), ancient Egyptians also made containers of coarse fibers drawn from heat-softened glass, and the French scientist Reaumur considered the potential of forming fine glass fibers as early as the 18th century. It was not until 1939 that continuous glass fibers were produced commercially (Knox, 1982). These fibers were produced mainly for high-temperature electrical applications. Two more decades passed before the so-called advanced fibers were produced: boron (late 1950s) and carbon (1960s). Thus, from the earliest applications of composites by the Egyptians to the introduction of advanced composites in the second half of the 20th century, roughly 6000 years have passed. Without doubt, the progress in the use of fibrous composites in the latter half of the 20th century has been much greater than that during the preceding 6000 years.

And where are we today? We are on the verge of an explosion in the use of these fibrous materials for structural applications. The applications over the past quarter-century have been primarily in specialty areas such as athletic equipment and aerospace structures. More recently we are seeing applications in the infrastructure, including plans for an all-composite bridge over an interstate highway and the anticipation of an all-composite automobile that will result in significant improvements in fuel mileage, improved safety, and reduced maintenance. It is clear that as more engineers

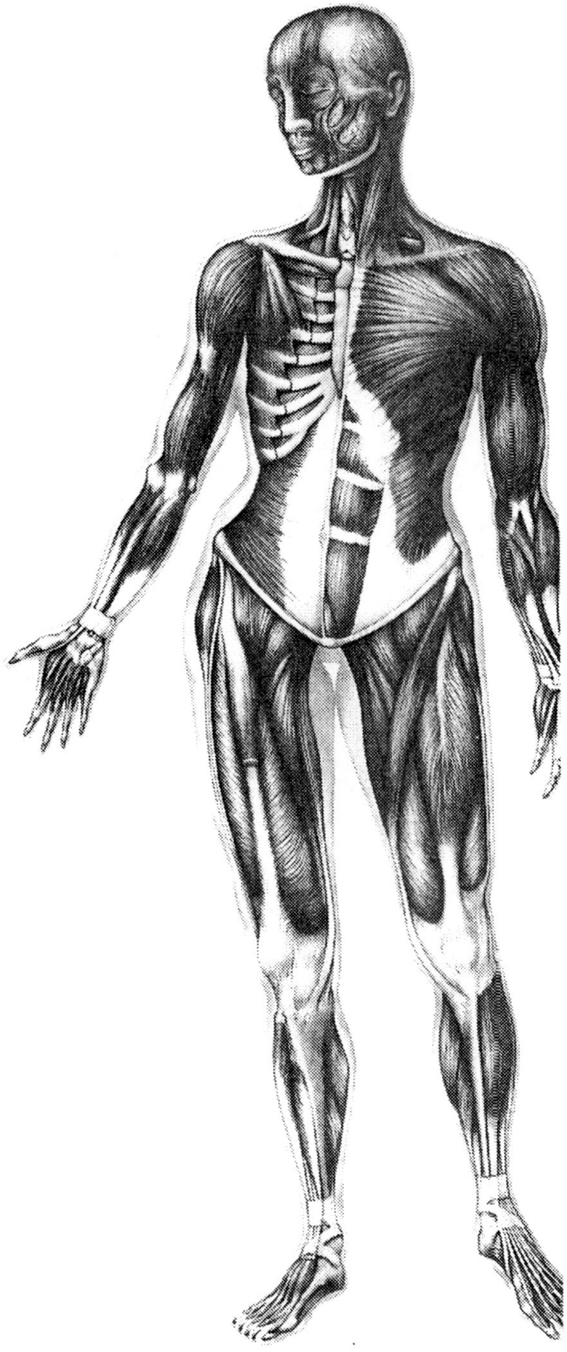

FIGURE 1.2 Muscles in the Human Body
(Photo courtesy of Dorling Kindersley Limited)

become familiar with the potential engineering and economic advantages, as well as the analysis and design procedures with these materials, the number and range of applications will grow even more rapidly. And as the volume of material produced increases, the cost per pound will decrease dramatically, thereby making composites more affordable for even more applications.

A wide variety of fibers and matrix materials are now available for use in advanced composites. The selection of the specific fiber and matrix to be used in a composite is not arbitrary. The two (or more) phases of a composite must be carefully chosen if the composite material is to be structurally efficient. The composite generally must be resistant to debonding at the fiber/matrix interface, and it must also be resistant to fiber breakage and matrix cracking. However, in applications where it is desired to dissipate energy during the failure process (such as in crashworthy or impact-resistant structures), progressive fiber failure and fiber/matrix debonding (damage development) are positive features because they dissipate energy. Thus, a major challenge for the mechanics and materials community is to understand the factors influencing damage development and to know how to design for it under severe environmental and mechanical loading conditions, including the fabrication phase as well as the in-service phase.

In the remainder of this chapter we will discuss a representative sample of fibrous composite materials—their properties, advantages, fabrication methods, and uses as structural composites.

1.3 Fibers

1.3.1 Fiber Types

A wide variety of fibers are available for use in composites, and the number is ever-increasing. Glass fibers have been in use since the 1930s; however, it was only in the late 1950s that fibers which exhibit significantly higher stiffness were developed for structural applications. These new high–*specific stiffness* (stiffness divided by density) and high–*specific strength* (strength divided by density) fibers are called *advanced fibers*. Composites made from them are called *advanced composites*. An in-depth discussion of fiber types and fabrication methods can be found in the book by Chawla (1987).

The first advanced fiber developed for applications in structural composites was the boron fiber introduced by Talley (1959). Other examples of advanced fibers include carbon (some forms of which are graphite), aramid (sold under the trade name Kevlar by Du Pont), silicon carbide, alumina, and sapphire. As mentioned, glass fibers generally are not considered to be advanced fibers because of their relatively low modulus compared with those of other advanced fibers. However, glass is an important engineering fiber because of its high specific strength and low cost. Thus, it will be included in this book when relative properties are discussed. The difference between carbon and graphite fibers is the carbon content. Carbon fibers typically have a carbon content of 80–95% whereas graphite fibers have a carbon content in excess of 99% (Lubin, 1982). Carbon fibers become graphitized by heat treatment at temperatures in excess of 1800°C (3272°F). For the remainder of this book, the general term *carbon* will be used for both carbon and graphite fibers.

As indicated in Table 1.1, the different fibers have different morphology, origin, size, and shape. Some fibers, such as glass, carbon, and alumina, are supplied in the form of tows (also called rovings or strands) consisting of many individual continuous fiber filaments. The size of the individual filaments ranges from 3 to 147 μm (0.1×10^{-3} to 5.8×10^{-3} in). The maximum use temperature of the fibers ranges from as low as 250°C (482°F) to as high as 2000°C (3632°F); however, in most applications, the use temperature of a composite is controlled by the use temperature of the matrix.

Fiber Type		Origin	Fabrication Method	Filament Size, μm (μin)	Filaments/ Tow
Glass	S-2 glass	Molten glass	Fiber-drawing	6–14 (230–550)	2000
Organic	Kevlar 49	Liquid crystal	Spinning	12 (472)	1000
Carbon	AS4	PAN	Heat and stress	8 (315)	12,000
	P-100S	Pitch	Heat	10 (393)	2000
	IM8	PAN	Heat and stress	5 (197)	12,000
Ceramic	Boron	Tungsten core	CVD	142 (5600)	1
	Nicalon (SiC)	Polymer precursor	Pyrolysis	15 (600)	500
	SCS-6 (SiC)	Carbon core	CVD	127 (5000)	1
	Alumina	Slurry mix	Spin and heat	20 (800)	1

TABLE 1.1 Typical Features of Fibers

Boron is a ceramic monofilament fiber manufactured by chemical vapor deposition (CVD) of boron on a (usually) tungsten core. Thus the fiber itself is a composite. It has a circular cross section and has been produced over a wide range of fiber diameters (33–400 μm) with the typical boron fiber diameter being approximately 140 μm. This is a relatively large fiber diameter and results in lower flexibility, in particular because boron is a very brittle material. The mismatch in the coefficient of thermal expansion of the tungsten core and the deposited boron results in residual stresses which develop during fabrication cool-down to room temperature.

Carbon filaments are made by controlled pyrolysis (chemical decomposition by heat) of a precursor material in fiber form such as polyacrylonitrile (PAN), rayon, or pitch by heat treatment at temperatures ranging from 1000 to 3000°C, with the fiber properties varying considerably with the fabrication temperature. The fabrication process and properties of carbon filaments were summarized by Riggs (1990). Individual carbon filaments have a diameter of 4–10 μm. Tows can consist of from 3000 to 30,000 filaments. These tows are generally what is referred to when carbon fibers are discussed. The small filament size and tow arrangement result in a very flexible fiber (tow) which can actually be tied into a knot without breaking the fiber. The modulus and strength of carbon fibers is controlled by the (proprietary) process, which consists of thermal decomposition of the organic precursor under well-controlled conditions of temperature and stress.

A second type of carbon fiber is made from a pitch precursor. The pitch fibers are made by spinning a petroleum-based product to form a pitch precursor. The cross section of carbon fibers is often noncircular. Indeed, many have the shape of a kidney bean. Carbon fibers have a heterogeneous microstructure consisting of numerous lamellar ribbons. The morphology is very dependent on the manufacturing process. PAN-based carbon fibers typically have an onionskin appearance with the basal planes in more or less circular arcs, whereas the morphology of pitch-based fibers is such that the basal planes lie along radial planes. Thus, the properties of carbon fibers are anisotropic.

Glass fibers are available in a variety of forms: E-glass and S-2 (Owens-Corning Fiberglas Corporation) are the most common for structural applications. E-glass is used where strength and high electrical resistivity are required, and S-2 glass is used in composite structural applications which require high strength, modulus, and stability under extreme temperature and corrosive environments. Glass fibers are produced by drawing molten glass through numerous tiny orifices in a gravity-fed tank to form continuous filaments which are gathered together in a strand or tow. A chemical sizing (a coating) that serves to both protect the filament surfaces as well as bind them together is applied to the filaments during production. The strands are wound onto a drum at very high speeds (up to 61 m/s, 200 ft/sec) at the end of the process. This fabrication method results in individual filaments that are small in diameter, isotropic, and very flexible.

Alumina fibers are ceramics fabricated by spinning a slurry mix of alumina particles and additives to form a yarn which is then subjected to controlled heating. The most important feature of alumina fibers is their strength retention at high temperature.

Aramid is an organic fiber which is melt-spun from a liquid polymer solution. The Du Pont Company developed these fibers and sells their product under the trade name Kevlar; four grades of Kevlar with varying engineering properties are available. The chemical and physical structure of aramid fibers as well as the fabrication process are described in considerable detail by Morgan and Allred (1990). The morphology of the fiber consists of radially arranged crystalline sheets resulting in anisotropic properties. The filaments are small in diameter (~12 μm) and, partially because of this, very flexible. Aramid fibers tend to have high tensile strength but only intermediate modulus. They also exhibit significantly lower strength in compression.

Silicon carbide (SiC) is a ceramic fiber made by one of two methods. The first method consists of chemical vapor deposition (CVD) of silicon and carbon onto a pyrolytic graphite-coated carbon core. This fiber (developed by AVCO Specialty Materials Co. in the United States and designated SCS-6) is very similar in size and microstructure to boron fiber. The SCS-6 fiber is relatively stiff in flexure, having a diameter of 140 μm (0.00556 in). The second method for producing silicon carbide fibers (developed in Japan under the trade name Nicalon) is controlled pyrolysis (chemical decomposition by heat) of a polymeric precursor. This method results in filaments which are similar to carbon filaments in terms of size (~14 μm, 0.00056 in) and microstructure. The diameter of a Nicalon filament is approximately one-tenth that of an SCS-6 fiber, and hence it is much more flexible. The Nicalon filaments are arranged in tows with 250 to 500 filaments per tow. Although CVD is more expensive than the pyrolysis process, it results in superior properties.

1.3.2 Fiber Properties

Typical engineering properties of specific fibers are compared with the properties of structural and matrix materials in Table 1.2 and Fig. 1.3. A wide variety of values for fiber properties can be found in the literature depending upon the manufacturer, the fabrication process, and the test method. High-end values are presented to demonstrate the best properties that can be attained at the present time. Experience has shown that these values continue to improve as new fibers are developed. The table includes density ρ, axial modulus E_L, axial Poisson's ratio ν_L, axial tensile strength σ_L^u, specific stiffness and specific strength normalized with respect to the values for aluminum, and axial coefficient of thermal expansion (CTE) α_L.

The modulus and strength values are for tensile loading along the axis of the fiber (longitudinal). Possibly the most important properties given in Table 1.2 are the specific stiffness, the specific strength, and the coefficient of thermal expansion. The specific stiffness and strength values are normalized with those of aluminum. The normalized values of specific stiffness and specific

Material	Density ρ, g/cm^3 (lb/in^3)	Modulus E_L, GPa (Msi)	Poisson's Ratio ν_L	Strength σ_L^u, MPa (ksi)	Specific Stiffness $(E/\rho)/(E/\rho)_{Al}$	Specific Strength $(\sigma^u/\rho)/(\sigma^u/\rho)_{Al}$	Thermal Expansion Coefficient α_L, μ/°C (μ/°F)
METALS							
Steel	7.8 (0.284)	200 (29)	0.32	1724 (250)	1.0	1.2	12.8 (7.1)
Aluminum	2.7 (0.097)	69 (10)	0.33	483 (70)	1.0	1.0	23.4 (13.0)
Titanium	4.5 (0.163)	91 (13.2)	0.36	758 (110)	0.95	1.2	8.8 (4.9)
FIBERS (Axial Properties)							
AS4	1.80 (0.065)	235 (34)	0.20	3599 (522)	5.1	11.1	–0.8 (–0.44)
T300	1.76 (0.064)	231 (33)	0.20	3654 (530)	5.1	11.5	–0.5 (–0.3)
P100S	2.15 (0.078)	724 (105)	0.20	2199 (319)	13.2	5.5	–1.4 (–0.78)
IM8	1.8 (0.065)	310 (45)	0.20	5171 (750)	6.7	16.1	—
Boron	2.6 (0.094)	385 (55.8)	0.21	3799 (551)	5.8	8.3	8.3 (4.6)
Kevlar 49	1.44 (0.052)	124 (18)	0.34	3620 (525)	3.6	13.9	–2.0 (–1.1)
SCS-6	3.3 (0.119)	400 (58.0)	0.25	3496 (507)	5.1	6.1	5.0 (2.77)
Nicalon	2.55 (0.092)	180 (28)	0.25	2000 (290)	2.8	4.4	4.0 (2.2)
Alumina	3.95 (0.143)	379 (55)	0.25	1585 (230)	3.7	1.9	7.5 (4.2)
S-2 Glass	2.46 (0.090)	86.8 (12.6)	0.23	4585 (665)	1.4	10.4	1.6 (0.9)
E-Glass	2.58 (0.093)	69 (10.0)	0.22	3450 (550)	1.05	7.5	5.4 (3.0)
Sapphire	3.97 (0.143)	435 (63)	0.28	3600 (522)	4.3	5.1	8.8 (4.9)
MATRIX MATERIALS							
Epoxy	1.38 (0.050)	4.6 (0.67)	0.36	58.6 (8.5)	0.08	0.4	63 (35)
Polyimide	1.46 (0.053)	3.5 (0.5)	0.35	103 (15)	0.03	0.4	36 (20)
Copper	8.9 (0.32)	117 (17)	0.33	400 (58)	0.5	0.3	17 (9.4)
Silicon carbide	3.2 (0.116)	400 (58)	0.25	310 (45)	4.9	0.5	4.8 (2.67)

TABLE 1.2 Properties of Engineering Materials, Fibers and Matrix

strength are compared graphically in Fig. 1.3, and the stress-strain responses of the fibers are compared in Fig. 1.4.

As indicated in Table 1.2 and Figs. 1.3 and 1.4, advanced fibers exhibit a broad range of properties. Indeed, the properties of carbon fibers can vary significantly depending upon the fabrication process. At the time of this writing carbon fibers can be fabricated to have high modulus (>700 GPa, 100 Msi) or high strength (>5 GPa, 725 ksi), but not both. A method to fabricate a carbon fiber with both high modulus and high strength is desired. The SCS-6, IM8, boron, and sapphire fibers offer the best combination of stiffness and strength, but three of these fibers have large diameters and thus have limited flexibility; the IM8 carbon fiber with its small diameter is the exception.

The values presented in the table show that the specific stiffness and the specific strength of the monolithic structural materials—steel, aluminum, and titanium—are all very similar. In contrast, the specific stiffness of a fiber can be more than 13 times that of structural metals, and the specific strength can be more than 16 times that of structural metals. Thus the potential for large weight savings is present through the use of advanced composites.

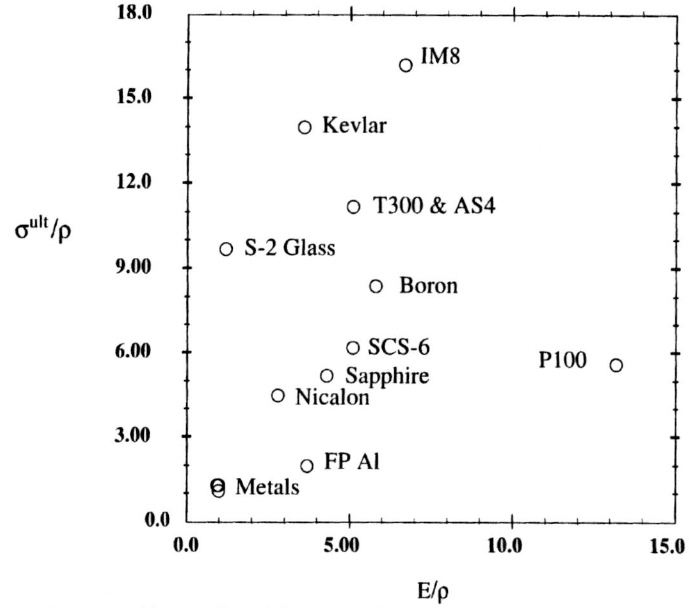

FIGURE 1.3 Normalized Specific Stiffness and Strength

The fiber data in Table 1.2 and Figs. 1.3 and 1.4 are for the fiber only, with the loading along the fiber axis. These properties are reduced significantly when the fiber is used with a matrix material to form a composite. The specific properties are reduced even further when the loading is in a direction other than along the fibers. Nevertheless, actual experience has shown that significant weight savings are possible in primary engineering structures through the use of advanced composites. As will be discussed later in this chapter, weight is not the only reason for choosing composites; indeed, for some applications composites are chosen when there is a weight penalty, but there are other advantages such as heat transfer characteristics or nonconductive properties which are more important. The values in Fig. 1.3 are normalized with respect to those of aluminum.

1.4 Matrix Materials

Polymers, metals, and ceramics are all used as matrix materials in continuous fiber composites. Polymeric matrix materials can be further subdivided into thermoplastics and thermosets. The thermoplastics soften upon heating and can be reshaped with heat and pressure. Thermoplastic polymers used for composites include polypropylene, polyvinyl chloride, nylon, polyurethane, poly-ether-ether-ketone (PEEK), polyphenylene sulfide (PPS), and polysulfone. The thermoplastic composites offer the potential for higher toughness and high-volume, low-cost processing. They have a useful temperature range upwards of 225°C (437°F).

Thermoset polymers become cross-linked during fabrication and do not soften upon reheating. The most common thermoset polymer matrix materials are polyesters, epoxies, and polyimides. *Polyesters* are used extensively with glass fibers. They are inexpensive, are lightweight, have a useful temperature range up to 100°C (212°F), and are somewhat resistant to environmental exposures. *Epoxies* are more expensive but have better moisture resistance and lower shrinkage on curing.

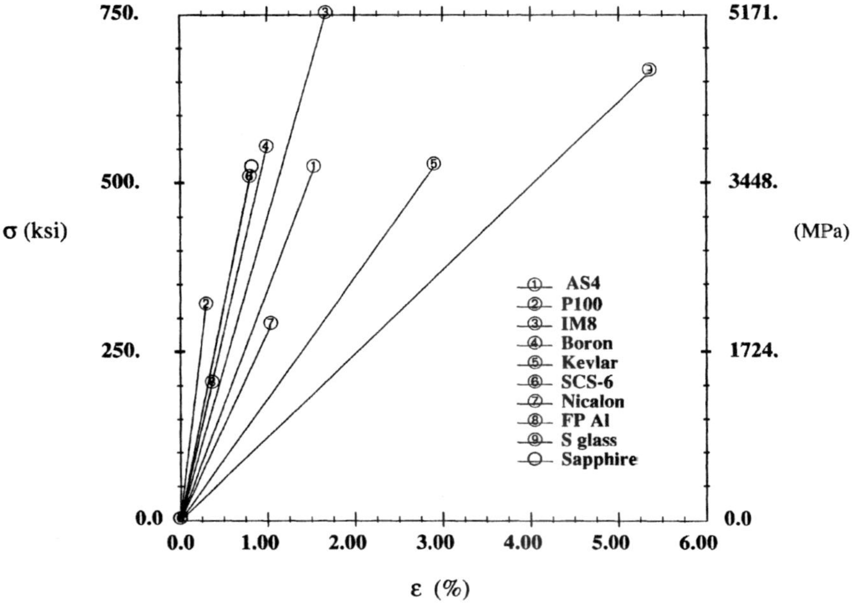

FIGURE 1.4 Stress-Strain Response of Advanced Fibers

Their maximum use temperature is in the vicinity of 175°C (347°F). *Polyimides* have a higher use temperature (300°C, 572°F) but are more difficult to fabricate.

Potential problems with the use of polymer matrix materials are the limited use temperature range; susceptibility to environmental degradation due to moisture, radiation, and atomic oxygen (in the space environment); low transverse strength; and high residual stresses due to the large mismatch in coefficients of thermal expansion between fiber and matrix. Typically, polymer matrix composites cannot be used near or above the glass transition temperature T_g, at which many physical properties change (degrade) abruptly.

The most common metals used as matrix materials are *aluminum, titanium,* and *copper*. Reasons for choosing a metal as the matrix material include higher use temperature range, higher transverse strength, toughness (as contrasted with the brittle behavior of polymers and ceramics), the absence of moisture effects, and high thermal conductivity (copper). On the negative side, metals are heavier and more susceptible to interfacial degradation at the fiber/matrix interface and to corrosion. Aluminum matrix composites have a use temperature upwards of 300°C (572°F), and titanium can be used at 800°C (1470°F). Essentially all materials exhibit degradation of properties at the highest temperatures.

The main reasons for choosing *ceramics* as the matrix include a very high use temperature range (>2000°C, 3600°F), high elastic modulus, and low density. The major disadvantage to ceramic matrix materials is their brittleness, which makes them susceptible to flaws. *Carbon, silicon carbide,* and *silicon nitride* are ceramics that have been used as matrix materials.

Carbon/carbon is a composite that consists of carbon fibers in a carbon matrix. The primary advantage of this material is that it can withstand temperatures in excess of 2200°C (4000°F). Indeed, carbon/carbon can be stronger at elevated temperature than at room temperature. The disadvantage of carbon/carbon composites is that their fabrication is an expensive, multistage process. Thus this material is used only where its high temperature capabilities are essential for the application. One application of carbon/carbon is on the leading edge of the space shuttle, where very high temperatures occur during reentry.

1.5 Composite Forms

1.5.1 Unidirectional Lamina

The unidirectional lamina (Fig. 1.5) is the basic form of continuous fiber composites. The lamina (or ply) may be composed of one or more layers of material, but all fibers are in the same direction. It may be fabricated in a variety of ways, including from prepreg tape, filament winding, pultrusion, or resin transfer molding (RTM). The stiffness and strength in the fiber direction are typically much greater than in the transverse directions, depending on the matrix material and the quality of the fiber/matrix bond. The properties of a unidirectional lamina are orthotropic, with different properties in the material principal directions (parallel and perpendicular to the fibers). For a sufficient number of filaments (or layers) in the thickness direction, the effective properties in the transverse plane (perpendicular to the fibers) may be isotropic. Such a material is called "transversely isotropic."

Figure 1.6 shows a photomicrograph of a typical cross section of unidirectional carbon/epoxy, and Fig. 1.7 shows a typical cross section of unidirectional silicon carbide/titanium. It is evident

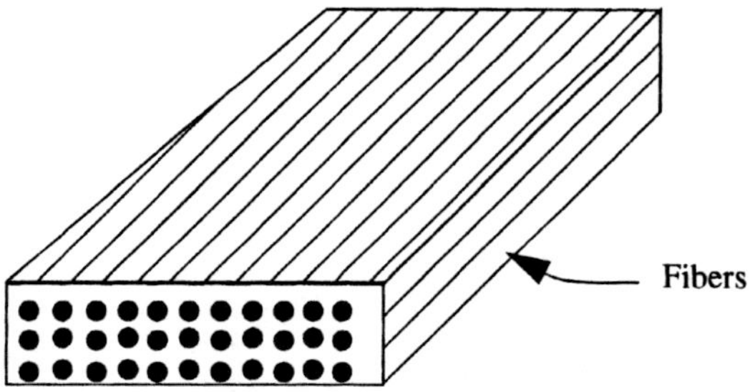

FIGURE 1.5 Unidirectional Lamina

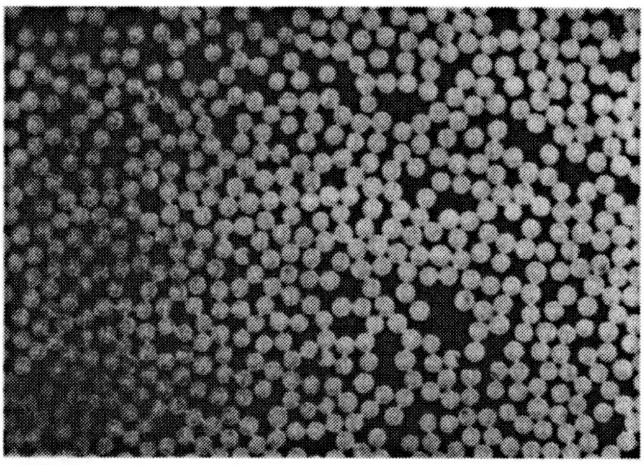

FIGURE 1.6 Carbon/Epoxy
(Photo courtesy of Joseph M. Kunze, University of Virginia)

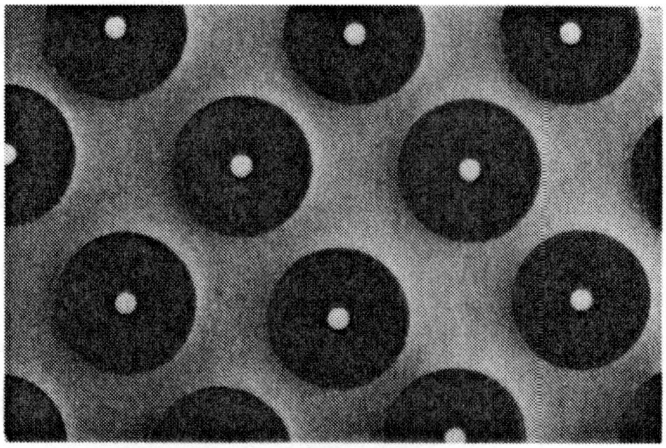

FIGURE 1.7 Silicon Carbide/Titanium
(Photo courtesy of Joseph M. Kunze, University of Virginia)

from these figures that the fiber volume is quite variable, with fibers actually touching in some locations, for the carbon/epoxy. The metal matrix composite exhibits much more uniform fiber spacing and, therefore, fiber volume fraction. The tungsten core is clearly visible in the silicon carbide fiber.

Figure 1.8 shows the fracture surface of a unidirectional (parallel-fiber) composite consisting of boron fibers in aluminum matrix. Again, the tungsten of the boron fiber is very evident. This fracture surface shows fiber breakage, matrix failure, fiber/matrix debonding, and fiber "pull-out."

1.5.2 Woven Fabrics

Woven fibers have been in use for centuries for products made from fibrous materials. Examples include cloth, baskets and other containers, and hats. The flexible fibers, such as glass, carbon, and aramid, can be woven into cloth fabric, which can then be impregnated with a matrix material. A wide variety of weave patterns are available. Two patterns are depicted in Fig. 1.9, a plain weave (every fiber over and under every other perpendicular fiber) and a five-harness satin weave (under only every fifth fiber). Woven fabrics naturally have better in-plane transverse effective properties than unidirectional lamina. They lay or "drape" better in structural configurations with substantial curvature and are more durable during handling.

1.5.3 Laminates

Laminates (Fig. 1.10) are made by stacking the unidirectional (or woven fabric) layers at different fiber orientations. The effective properties of the laminate vary with the orientation, thickness, and stacking sequence of the individual layers. A shorthand notation has been developed to describe the laminate stacking sequence and orientation of the individual layers. This notation will be described in detail in Chapter 5. It also will be demonstrated later that an important class of laminates are those which have a symmetric stacking sequence about the laminate midplane.

1.5.4 Hybrid Composites

Composites are used in a variety of applications where either more than one fiber type is used or where a composite is combined with another material such as a metal. Such combinations of

FIGURE 1.8 Fracture Surface of Boron/Aluminum
(Photo by Frank Wawner, University of Virginia)

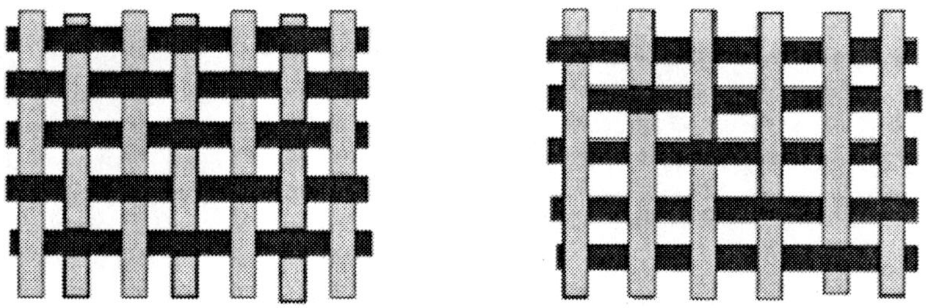

Plain Weave | Five-Harness Satin Weave

FIGURE 1.9 Woven Fabrics

different materials are called *hybrids*. Different fiber types are combined in the same material system when it is desired to take advantage of the best attributes of the different fibers while maintaining the lowest possible cost. An example is the combination of Kevlar and carbon fibers. Kevlar exhibits excellent tensile properties and is less expensive than carbon; however, the compressive properties of Kevlar are not as good as the tensile properties. Hence, some carbon fibers may be combined with the Kevlar to provide the desired compressive properties.

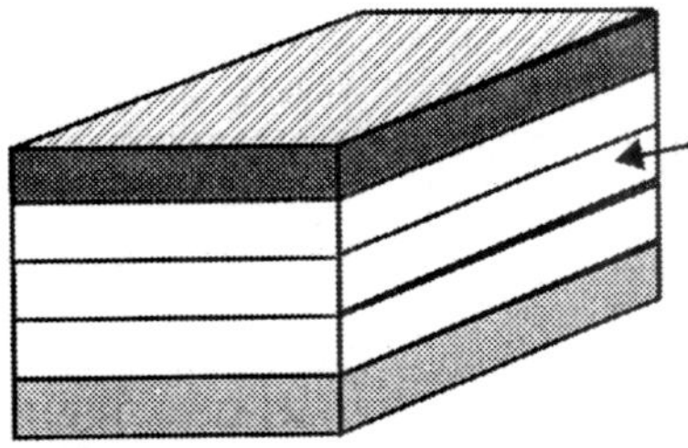

FIGURE 1.10 Laminate

Another example of a hybrid system is a laminate consisting of layers of composite sandwiched between layers of metal. Alcoa makes a material (known by the trade name ARALL) in which layers of aramid/epoxy are laminated with layers of aluminum. The advantages of the hybrid system as compared with the all-metal structure include better fatigue life and higher specific stiffness and specific strength. The hybrid can be machined using traditional metal techniques, and the outer aluminum layers provide a moisture barrier as well as lightning protection.

1.5.5 Chopped Fiber

Many fibers, such as carbon and glass, can be chopped into short lengths and then used in compression- and injection-molding compounds to produce industrial components such as machine parts, gears, and valves. The molding processes have greater flexibility for components with complex geometries. The finished products have many of the advantages of continuous fiber composites including resistance to creep, corrosion, and fatigue as well as high specific stiffness and strength.

1.6 Composite Properties

Table 1.3 presents typical average or effective properties for unidirectional composites. The designation of the different composites consists of the name of the fiber followed by the name of the matrix. As will be discussed in detail in later chapters, unidirectional fibrous composites exhibit different properties in different directions. This is reflected in Table 1.3 by the labels *axial* and *transverse*, which refer to properties in the direction of the fiber (axial) and the properties perpendicular to the fiber (transverse). The properties of a unidirectional composite are also a function of the volume fraction of fibers. Thus fiber volume fractions are also given in the table. The values presented in this table have been gleaned from a wide variety of sources, including personal research by the author and collaborators.

As indicated in the table, there can be a wide range of composite properties depending upon the choice of fiber and matrix and the direction of loading. The particular composites presented in the table were chosen to demonstrate the range of properties possible for polymeric and metallic matrix composites. The axial moduli range from a low of 43.5 GPa (6.31 Msi) for S glass/epoxy (S glass fibers in epoxy matrix) to a high of 227 GPa (32.9 Msi) for boron/aluminum. The transverse moduli are much smaller, ranging from 5.5 GPa (0.8 Msi) for Kevlar to 145 GPa (21 Msi) for SCS-6/Ti-15-3. The degree of orthotropy as measured by the ratio of axial to transverse modulus also varies over a wide range, from a low of 1.5 for a metal matrix composite, SCS-6/Ti-15-3, to a high of 14.8 for Kevlar. (Carbon/epoxy composites made with high-modulus fibers can exhibit a degree

Material	AS4/ 3501-6	T300/ 5208	Kevlar/ epoxy	Boron/ Al	SCS-6/ Ti-15-3	S-2 glass/ epoxy
Density, g/cm^3 (lb/in^3)	1.52 (0.055)	1.54 (0.056)	1.38 (0.05)	2.65 (0.096)	3.86 (0.14)	2.00 (0.072)
Axial modulus E_1, GPa (Msi)	148 (21.5)	132 (19.2)	76.8 (11.0)	227 (32.9)	221 (32)	43.5 (6.31)
Transverse modulus E_2, GPa (Msi)	10.50 (1.46)	10.8 (1.56)	5.5 (0.8)	139 (20.2)	145 (21)	11.5 (1.67)
Poisson's ratio ν_{12}	0.30	0.24	0.34	0.24	0.27	0.27
Poisson's ratio ν_{23}	0.59	0.59	0.37	0.36	0.40	0.40
Shear modulus G_{12}, GPa (Msi)	5.61 (0.81)	5.65 (0.82)	2.07 (0.3)	57.6 (8.35)	53.2 (7.78)	3.45 (0.50)
Shear modulus G_{23}, GPa (Msi)	3.17 (0.46)	3.38 (0.49)	1.4 (0.20)	49.1 (7.12)	51.7 (7.50)	4.12 (0.60)
Modulus ratio E_1/E_2	12.6	12.3	14.8	1.6	1.5	4.6
Axial tensile strength X_T, MPa (ksi)	2137 (310)	1513 (219.5)	1380 (200)	1290 (187)	1517 (220)	1724 (250)
Transverse tensile strength Y_T, MPa (ksi)	53.4 (7.75)	43.4 (6.3)	27.6 (4.0)	117 (17)	317 (46)	41.4 (6.0)
Strength ratio X_T / Y_T	27	35	50	11	4.8	42
Axial CTE α_1, µ/°C (µ/°F)	−0.8 (−0.44)	−0.77 (−0.43)	−4 (−2.2)	5.94 (3.3)	6.15 (3.4)	6.84 (3.8)
Transverse CTE α_2, µ/°C (µ/°F)	29 (16)	25 (13.6)	57 (32)	16.6 (9.2)	7.90 (4.4)	29 (16)
Fiber volume fraction V_f	0.62	0.62	0.55	0.46	0.39	0.60
Ply thickness, mm (in)	0.127 (0.005)	0.127 (0.005)	0.127 (0.005)	0.178 (0.007)	0.229 (0.009)	

TABLE 1.3 Typical Properties of Unidirectional Composites

of orthotropy as high as 40.) It is also evident from the table that the coefficients of thermal expansion range from small negative values in the fiber direction to large positive values in the transverse direction.

Additional properties of unidirectional and laminated composites, including nonlinear stress-strain response curves, are presented in Chapter 7.

1.7 Advantages of Composites

The initial development and application of advanced fibrous composites were pursued primarily because of the potential for lighter structures. The first applications in the early 1960s were in aerospace structures, where weight critically affects fuel consumption, performance, and payload, and in sports equipment, where lighter equipment often leads to improved performance. Today fibrous composites are often the material of choice of designers for a variety of reasons, including low weight, high stiffness, high strength, electrical conductivity (or nonconductivity), low thermal expansion, low or high rate of heat transfer, corrosion resistance, longer fatigue life, optimal design, reduced maintenance, fabrication to net shape, and retention of properties at high operating temperature. Some of these advantages are discussed in more detail in the following paragraphs.

1.7.1 Specific Stiffness and Specific Strength

Undoubtedly the most often cited advantage of fibrous composites is their high specific stiffness and high specific strength as compared with traditional engineering materials. These properties lead to improved performance and reduced energy consumption, both vitally important in the design of almost all engineering structures. The specific stiffness and specific strength of composites have been discussed in detail (see Figs. 1.3 and 1.4).

1.7.2 Tailored Design

Because composites are fabricated, they can be *engineered* to meet the specific demands of each particular application. Available design options include (but are not limited to) the choice of materials (fiber and matrix), the volume fraction of fiber and matrix, fabrication method, layer orientations, number of layers in a given direction, thickness of individual layers, type of layer (unidirectional or fabric), and the layer stacking sequence.

This vast array of design variables for composites contrasts sharply with more traditional engineering materials, where the choices are much more limited. The availability of a wide array of structured materials means that more efficient structures can be fabricated with less material waste. Unlike isotropic materials, composites are directionally dependent. Thus composites can be designed to have the desired properties in specified directions without overdesigning in other directions. The large number of variables available with composites also means that the use of computers, optimization, expert systems, and artificial intelligence should be considered in improving designs.

1.7.3 Fatigue Life

Figure 1.11 shows the fatigue lives of several composites compared with that of aluminum. These *S-N* curves indicate the number of load cycles N that the material can withstand under tensile stress S. Clearly, composites exhibit much better resistance to fatigue than does aluminum. This can be critical in structures such as aircraft, where fatigue life is often the most important design consideration. Improved fatigue life is one of the major reasons why there has been a shift to composites by the aircraft industry. Fatigue life is also important for many other structures that experience cyclic loading, such as transportation vehicles, bridges, industrial components, and structures exposed to variable wind or water loading.

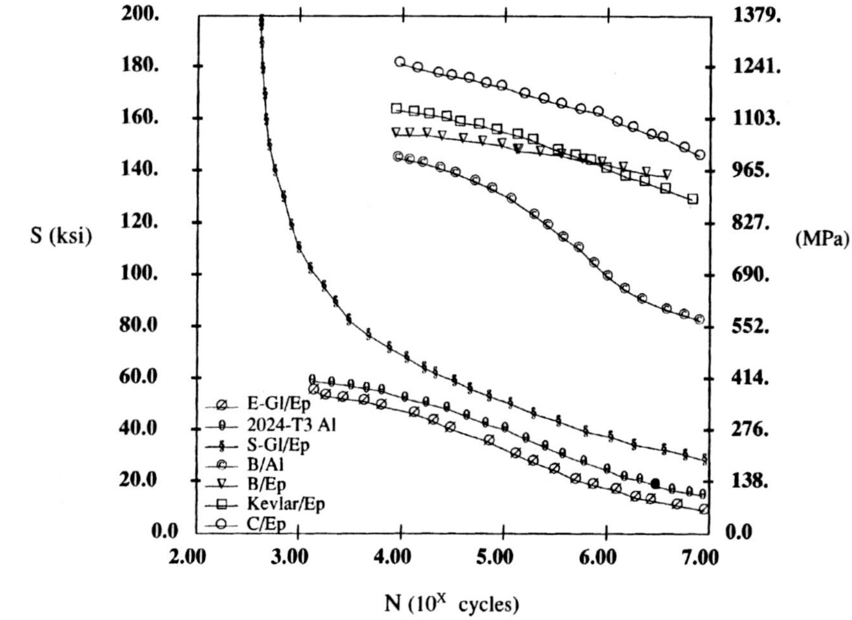

FIGURE 1.11 Fatigue Life Comparisons (Courtesy of Hexel Composites)

1.7.4 Dimensional Stability

Nearly all structures are exposed to temperature changes during their lifetimes. The strains associated with temperature change can result in changes in size or shape, increased friction and wear, and thermal stresses. In some applications these thermal effects can be critical. Increased friction between moving parts (in an engine, for example) can result in failure because of overheating. Small changes in the shape of a space antenna can render it totally inoperative for the intended use. Cyclic thermal loading, as experienced by some space structures, can result in thermal fatigue. Highways buckle due to thermal expansion, and roof systems develop leaks and, at times, fail due to thermal expansion and contraction. Thus, there are many applications where a zero- or near-zero-CTE material can result in significant benefits. Through proper design, it is possible to have zero-CTE composites or to design the CTE of the composite to match that of other components to minimize thermal mismatch and the resulting thermal stresses. Figure 1.12 shows a carbon/epoxy optical bench for the Space Telescope. The bench was designed to have a zero coefficient of thermal expansion over a wide range of temperatures.

1.7.5 Corrosion Resistance

Polymer and ceramic matrix materials can be selected to make composites resistant to corrosion from moisture and other chemicals. Current applications of glass fiber composites that have been driven by corrosion considerations include filament-wound underground storage tanks, structural members for offshore drilling platforms and chemical plants, sucker rod used in pumping oil from wells, pipe, and domestic applications including doors, window frames, and deck furniture in coastal regions where saltwater corrosion is a major problem.

FIGURE 1.12 Optical Bench
(Photo courtesy of Alliant Techsystems)

Polymeric and ceramic matrix composites can often be made to be essentially maintenance free compared with traditional engineering materials. This is true primarily because of the corrosion resistance. Reduced maintenance can represent substantial savings and should be considered in all total cost evaluations. Unfortunately, all too often, cost decisions are based primarily on the initial capital expenditure without regard for the total lifetime cost of maintaining the structure. Corrosion resistance results in longer life of a structure and hence reduced (or delayed) replacement costs.

1.7.6 Cost-Effective Fabrication

Composite structures can be fabricated efficiently through the use of automated methods such as filament winding, pultrusion, and tape laying. Efficiencies in fabrication can also be achieved because composites can be fabricated with very little material waste. In many cases, composite components can be fabricated exactly to size specifications with no material waste. This is in stark contrast to the use of metals, where it is often necessary to "hog out" large portions of a material to arrive at the final configuration.

Fabrication costs also are directly related to the number of parts in a structure. The use of composites can substantially reduce this number because of the ability to fabricate to net shape and because of the use of bonded rather than riveted joints. As an example, two sections of a fuselage were made by (a) riveting aluminum components and (b) adhesively bonding composite components. The number of parts in the aluminum structure was ~11,000, whereas the composite structure had only ~1000. This tenfold reduction represents a significant savings in both the cost of components and the cost of assembly.

1.7.7 Conductivity

It is desirable that many engineering structures be electrically nonconducting. Excellent examples are the glass/polyester ladders and booms which have replaced steel and aluminum in order to reduce the possibility of electrocution. Nonconducting components are also important for applications in the electronics industry, whether it be a computer chip or the entire building in which the chips are fabricated. In contrast, copper matrix composites are now under consideration for high-temperature applications because of the high thermal conductivity of copper. Copper matrix composites can serve as radiators in regions where it is necessary to maintain lower temperatures. It is noteworthy that the fiberglass ladders and the copper matrix composites are chosen even though there is a weight penalty.

1.7.8 Overall Cost Considerations

In evaluating the cost competitiveness of structures made from composite materials the total lifetime cost should be included. Per pound, composites are usually more expensive than traditional materials; however, many other factors must be included in a meaningful cost comparison. First, fewer pounds of composite material are required because of the higher specific stiffness and strength. Second, it is possible that fabrication costs can be lower. Third, transportation and erection costs are generally lower for composite structures. Finally, the composite structure will generally last much longer than the traditional material and will require much less maintenance during its life. Composite materials have been shown to be cost competitive in a wide variety of aerospace, automotive, industrial, domestic, oil drilling, and electronic applications, among others.

1.8 Applications

It would be impossible in this introductory chapter to cover the full range of composite applications, even at this relatively early state of the art. It is accurate to say that there is a continuing explosion of new applications. A few examples in a wide variety of fields will be given in order to demonstrate that these applications do indeed cover a wide range, for a multitude of reasons.

1.8.1 Aerospace

Aircraft, spacecraft, satellites, space telescopes, the space shuttle, the space station, missiles, booster rockets, and helicopters are all examples where composite materials have been (or will be) used to advantage. The primary reasons for using composites in aircraft include specific stiffness and specific strength, design tailorability, and fatigue resistance. As noted previously, dimensional stability also is often a major consideration for space applications. Some aircraft would not have been possible without the availability of advanced composites. The all-composite Voyager airplane (Fig. 1.13) flew nonstop around the world without refueling, and the forward-swept X-29

FIGURE 1.13 Voyager Composite Aircraft
(Photo courtesy of Alliant Techsystems)

was possible only because of the special features of composites. Many aerospace structures include components that are made with different composites depending upon the type of loading experienced by the individual component. Possibly the most exotic example of the advantages which can be gained through the use of composites are the stealth aircraft which were used so effectively in the early 1990s. The stealth feature of the planes would not have been possible were it not for the availability of advanced composites.

Shown in Fig. 1.14 is the Perseus high-altitude (24,300 m, 80,000 ft) all-composite, remote-controlled aircraft used for weather and other atmospheric studies (manufactured by Aurora Flight Sciences, Inc.). The primary reason for choosing composites for this plane is weight. The space shuttle (Fig. 1.15) has many structural components made from composites. Carbon/epoxy, Kevlar, and carbon/carbon are used on different components. Carbon/carbon is used on the leading edges and nose cone of the shuttle, where temperatures upwards of 1500°C (2732°F) occur during re-entry. Another example of the use of composites in airplanes is depicted in Fig. 1.16, where an exploded view of the B2 bomber shows the variety of composites used for different structural applications in this plane. Both fiberglass and graphite fibers are used with epoxy matrix and polyimide matrix.

1.8.2 Composite Railway Carrier

A most interesting example that includes a relatively large number of factors that tipped the scales in favor of composites over structural metals is a composite railway auto carrier (Fig. 1.17) designed and built by Alcoa/Goldsworthy Engineering for the Union Pacific Railroad in 1990.

FIGURE 1.14 Perseus All-Composite High-Altitude Airplane
(Photo courtesy of Aurora Flight Sciences Corporation)

FIGURE 1.15 United States Space Shuttle

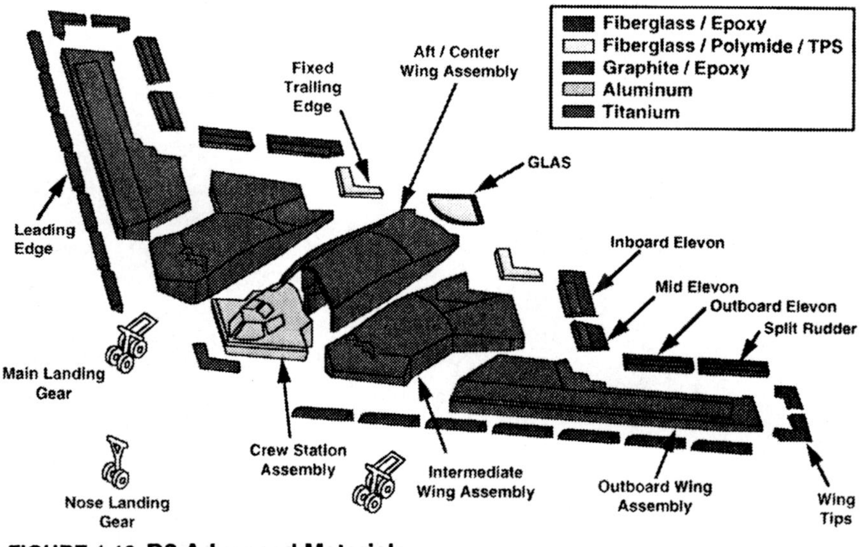

FIGURE 1.16 B2 Advanced Materials
(Photo courtesy of Northrop Grumman)

Automobiles can now be transported in stackable polymer matrix composite modules that are 53 ft (16.1 m) long and 8.5 ft (2.6 m) wide. At the time, the modules were the largest single units ever assembled from pultruded structural components. The composite carriers weigh from 5 to 10% less than the traditional "car racks"; are designed to eliminate shock and vibration damage to auto components at speeds up to 80 mph (129 kph), compared to the maximum of 50 mph (80 kph) for standard car racks; are completely enclosed to protect against vandalism, theft, and corrosion; and have reduced maintenance costs, higher reliability, and better utilization (modules come in two different heights and can be stacked up to three levels high).

FIGURE 1.17 Composite Railway Carrier
(Photo courtesy of W. Brandt Goldsworthy and Assocs. Inc.)

1.8.3 Athletic and Recreational Equipment

Examples of athletic and recreational equipment made from composites include tennis rackets, golf clubs, baseball bats, helmets, skis, hockey sticks, fishing rods, boat hulls, windsurfing boards, water skis, sails, canoes and racing shells, paddles, yachting rope, speed boats, scuba diving tanks, and race cars. Advanced composites are used in athletic equipment primarily to improve performance through lighter weight and improved tailoring. In the recreation area low weight is often desirable, but reduced maintenance and corrosion resistance are also major factors resulting in the selection of composites. A notable example of the use of composites in sailboating is the success of the composite boats and composite sails in the America's Cup races. Several examples of the use of composites in athletic and recreational applications are shown in Fig. 1.18.

1.8.4 Automotive

The reasons for choosing composites in automotive applications include lower weight and greater durability (improved corrosion resistance, fatigue life, wear and impact resistance). Applications include drive shafts, fan blades and shrouds, springs, bumpers, interior panels, tires, brake shoes, clutch plates, gaskets, hoses, belts, and engine parts. An interesting example is the use of a hybrid composite drive shaft for trucks manufactured by pultrusion (by Morrison Molded Fiberglass). Carbon and glass fiber composites are pultruded over an aluminum cylinder to create a drive shaft that is significantly lighter and less expensive. One reason the composite shaft is less expensive is that it is fabricated as a single component, whereas the metal shaft was fabricated in two parts and required a connection device.

A most interesting and exciting article that discusses the tremendous potential for more fuel-efficient cars through the use of advanced composites and "hybrid-electric drive" is that by Lovins and Lovins (1995). In this article the authors argue that saving one pound of structural weight may save as much as five pounds of total weight, partly because power steering, power brakes, engine

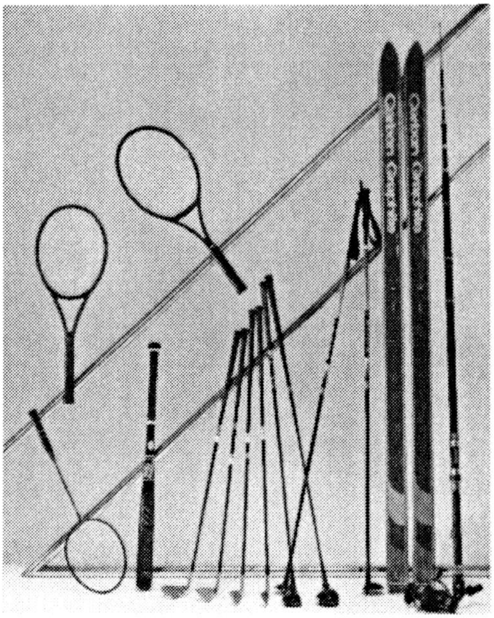

FIGURE 1.18 Athletic and Recreational Uses of Composites
(Photo courtesy of Toray Industries, Inc.)

cooling, and other components and systems will no longer be required. They indicate that the most efficient cars, called "hypercars," can be obtained through a combination of ultralight structures made of advanced composites and wheels driven by special electric motors that also operate as electronic brakes, thus converting unwanted motion back into useful electricity. Indeed, the braking energy can be stored in a carbon-fiber "superflywheel." Lovins and Lovins suggest that very early in the 21st century "hypercars" may be available that will carry a family from the East Coast to the West Coast of the United States, more safely and more comfortably, on one tank of liquid fuel. The liquid fuel would be used in a small on-board engine to generate electricity for the wheel motors and other systems. Liquid fuel would be used because it is a hundred times more efficient per pound than batteries.

1.8.5 Infrastructure Structures

The use of advanced composites in structures such as bridges and buildings has lagged behind applications in other areas. One major reason for this is that weight is not as important a consideration in static structures. However, as the benefits of reduced maintenance and erection costs combined with architectural enhancements are recognized, the application of composites in these structures will follow. Composites are now often used in structures where corrosion is a dominant design consideration, such as in the chemical industry and on off-shore oil platforms. A dramatic example is the use of Kevlar 49 to reinforce the 5600 m^2 (60,000 ft^2) retractable roof of the Montreal Olympic stadium. The roof is light yet able to withstand snow loads and high winds. Composites are also in use in lightweight overhead walkways, as well as lighting and communications poles, to name just a few other applications.

1.8.6 Industrial

Composites are used throughout industry to reduce costs by prolonging lifetimes as well as by improving performance. Examples of longer-life applications include drive and conveyor belts, hoses, tear- and puncture-resistant fabrics, rotor vanes, mandrels, ropes, and cables. As mentioned previously, many ladders as well as truck booms are now being made with glass fibers and resin matrices because they are electrically nonconducting materials. This has significantly reduced the possibility of injury. The fiberglass sucker rod used in the oil industry has resulted in longer life and increased pumping efficiency. Composite flywheels have the potential for performance not attainable with traditional materials because of the higher speeds possible.

1.8.7 Medical

Composites are strong candidates for a variety of medical applications. They have been used to reduce weight and extend durability. Lightweight wheelchairs and crutches offer an obvious advantage to the user. Medical implant devices should be light, durable, and biocompatible. One application of considerable interest is the use of composites as a replacement hip joint. Also, as pointed out earlier in this chapter, the human body is a fibrous composite. Thus, the mechanics of the body parts must obey the same fundamental laws, with appropriate modification of the constitutive equations.

1.8.8 Electronic

The integrated circuits or "chips" used in electronic computing devices are laminated hybrid systems composed of a number of layers (materials) which serve different functions. An important design consideration from a mechanics point of view is environmental durability. The chip must have good heat transfer properties so as not to overheat and it must be able to withstand induced thermal stresses without delaminating. The laminated, heterogeneous nature of the chip requires that it be treated as a composite.

1.8.9 Military

Helmets, bulletproof vests, impact-resistant vehicles, lighter and less detectable ships, portable bridges, and lighter weapons are but a few examples of military applications which are useful for protection of the individual as much as for increased performance. However, any good student of military history knows the value of reduced fuel consumption, which can be gained through lighter vehicles. The stealth capability of composites was discussed in the section on aerospace applications.

1.9 Fabrication Methods

1.9.1 Autoclave Curing

An autoclave is a closed vessel for controlling temperature and pressure and is used for curing polymeric matrix composites. The composite to be cured is prepared either through hand layup or machine placement of individual laminae in the form of fibrous tape which has been impregnated with resin. The component is then placed in an autoclave and subjected to a controlled cycle of temperature and pressure. After curing, the composite is "solidified" and exhibits the corresponding

FIGURE 1.19 Autoclave for Composite Curing
(Photo courtesy of Alliant Techsystems)

engineering properties. Some autoclaves are large enough to cure major structural components. Figure 1.19 shows such a large autoclave.

1.9.2 Filament Winding

Filament winding is an automated process used in the fabrication of components or structures made with flexible fibers. Individual fiber tows wetted with liquid resin are "wound" on a mandrel which has the shape of the final structure. The wound component is then cured in an oven or autoclave. Sophisticated winding machines are now available with as many as six independent axes of movement. Examples of filament-wound structures include pressure bottles, cylinders, and rocket motor casings (Fig. 1.20). Filament winding is a relatively inexpensive process.

1.9.3 Pultrusion

Pultrusion is a continuous process in which composites in the form of fibers and fabrics are pulled through a bath of liquid resin, passed through formers that closely align the components to the desired configuration, and then passed through a heated die that has the exact final shape of the product. The method is very cost-effective for fabrication of prismatic structures or components, for which the pultruded product is simply cut to the desired length. Components can be pultruded at a rate as high as 30 cm (1 ft) per minute. Standard structural shapes such as I-beams, T-beams, channels, angles, and a variety of thin-walled tubes are now commonly available as pultruded glass fiber products (see Fig. 1.21).

1.9.4 Braiding

Braiding is a cost-effective, automated technique for interlacing fibers into complex shapes. Rope, tubes, narrow flat strips, contoured shapes, and solid three-dimensional shapes including I-beams and T-beams can be produced through braiding. Braiding of a complex shape is demonstrated in Fig. 1.22.

FIGURE 1.20 Filament Winding of Rocket Motor Case
(Photo courtesy of Alliant Techsystems)

FIGURE 1.21 Pultruded Structural Shapes
(Photo courtesy of Strongwell)

FIGURE 1.22 Composite Braider
(Photo courtesy of Frank Ko, Drexel University)

1.9.5 Resin Transfer Molding

Resin transfer molding (RTM) consists of arranging fibers or cloth fabric to the final desired configuration in a preform, placing the preform in a mold, and then injecting resin into the mold. Advantages of RTM include efficiency, suitability for complex shapes, superior surface definition, better reproducibility, relatively low clamping pressure, and ability to include inserts.

1.9.6 Metal Matrix Composites

Fibrous metal matrix composites (MMCs) are fabricated using a variety of methods that can be broadly classified as either solid-state fabrication or liquid-state fabrication. In solid-state fabrication, monolayers of unidirectional fibers and matrix are consolidated (diffusion-bonded) under pressure and elevated temperature. The monolayers may be a row of matrix-coated fibers, a row of fibers on a thin metal foil, or unidirectional layers of composite that have been fabricated by plasma spray or vapor deposition. During the consolidation phase the matrix flows around the fibers, fills the voids, and bonds to adjacent layers. Figure 1.23 shows a schematic representation of the diffusion-bonding consolidation process.

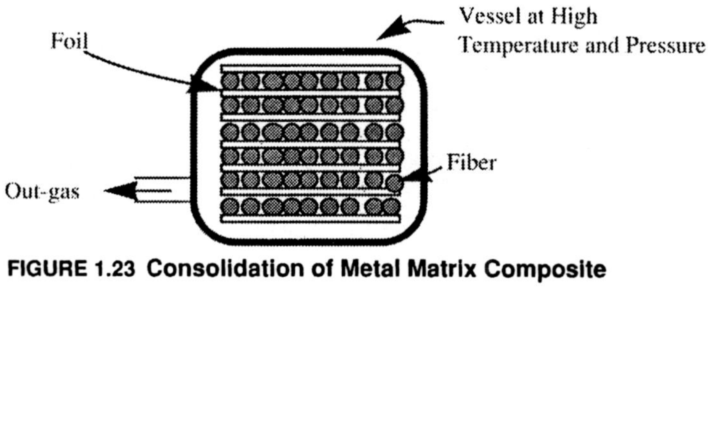

FIGURE 1.23 Consolidation of Metal Matrix Composite

In liquid-state fabrication, liquid metal is infiltrated between fibers or fiber preforms. Squeeze casting is such a method that involves the infiltration of molten metal between fibers which have been placed in a die. High pressure is usually required.

The fibers in MMCs are often coated prior to consolidation. The coating serves several purposes, including protection from fiber damage during processing, providing a barrier which reduces undesirable interaction between the fiber and matrix, and reduction of residual thermal stresses.

1.10 Summary

This introductory chapter has demonstrated that composites have the potential to be the material of choice in many applications for a wide variety of reasons, including their high specific stiffness and strength. The most severe limitation to the application of composites appears to be the lack of engineers with the knowledge and experience to design with these materials.

References

"Characteristics and Uses of Kevlar 49 Aramid High Modulus Organic Fiber," Du Pont Bulletin K-5, Wilmington, DE.

Chawla, K. K. (1987), *Composite Materials,* Springer-Verlag, New York.

Clayman, C. B. (editor-in-chief) (1995), *The Human Body,* Dorling Kindersley Publishing, New York.

Hartman, D. R., Greenwood, M. E., and Miller, D. M. (1994), "High Strength Glass Fibers," Owens-Corning Fiberglas Corporation Technical Paper, Toledo, OH.

"Hercules Carbon Fiber Type IM8," Product Data Sheet 871-1, Hercules, Inc., Wilmington, DE.

Knox, C. E. (1982), "Fiberglass Reinforcement," in *Handbook of Composites* (G. Lubin, ed.), Van Nostrand Reinhold, New York.

Lovins, A. B., and Lovins, L. H. (1995), "Reinventing the Wheels," *Atlantic Monthly,* January, pp. 75–86.

Lubin, G. (1982), *Handbook of Composites,* Van Nostrand Reinhold, New York.

"Magnamite Graphite Fiber Type AS4," Product Data Sheet 847-3, Hercules, Inc., Wilmington, DE.

Morgan, R. J., and Allred, R. E. (1989), "Aramid Fiber Composites," in *International Encyclopedia of Composites,* vol. 1 (S. M. Lee, ed.), VCH Publishers, New York, pp. 37–56.

"NICALON Ceramic Fiber," Dow Corning Corporation Information Sheet, Midland, MI.

Riggs, J. P. (1989), "Carbon Fibers," in *International Encyclopedia of Composites,* vol. 1 (S. M. Lee, ed.), VCH Publishers, New York, pp. 197–225.

Talley, C. P. (1959), "Mechanical Properties of Glassy Boron," *J. Appl. Phys.*, vol. 30, p. 1114.

"Typical Properties Thornel Pitch Based Fibers," Amoco Thornel Product Information Sheet P-100S, Ridgefield, CT.

Wainwright, S. A., Biggs, W. D., Currey, J. D., and Gosline, J. M. (1976), *Mechanical Design in Organisms,* John Wiley & Sons, New York.

Additional Reading Selections

Lee, Stuart M., ed. (1989), *International Encyclopedia of Composites,* vols. 1–6, VCH Publishers, New York.

Scientific American, vol. 255, no. 4, October 1986.

Exercises

1.1 List five examples of fibrous composite structures not given in this chapter.

1.2 List five possible advantages of composite structures not given in this chapter.

1.3 List five actual or potential applications of composite materials observed during your normal routine over the next 48 hours.

1.4 List five applications of composite materials discussed in the media (press, radio, TV, engineering magazines) during the past six months.

1.5 Determine the maximum possible fiber volume fraction of a composite consisting of circular fibers of radius R in (a) a square array and (b) a hexagonal array.

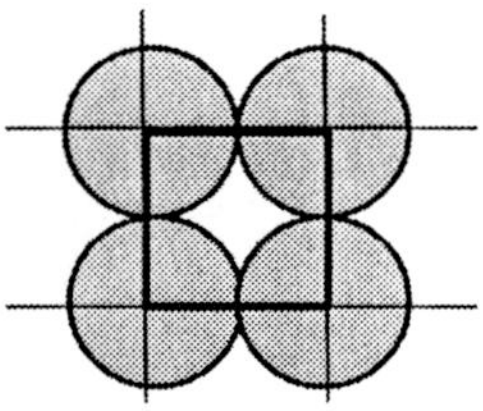

a) SQUARE ARRAY

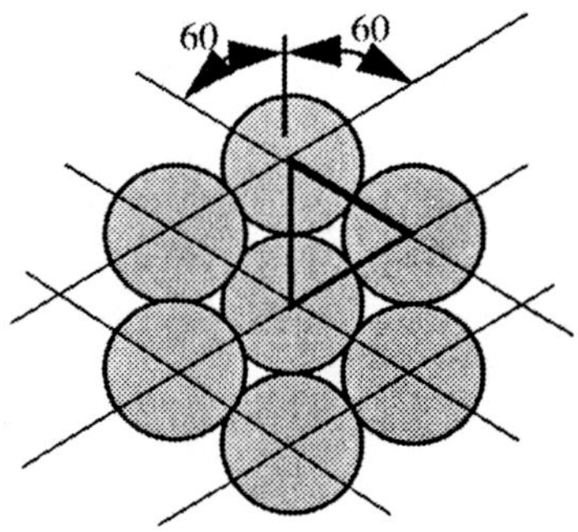

b) HEXAGONAL ARRAY

1.6 The three-hinged arch shown is subjected to a uniform vertical loading, which results in a horizontal thrust in the tie rod (A–B) of 28.5 Kips. If the tie rod is 64 ft in length and nine arches are required for the building, determine the total weight of tie rods required for the building if the tie rods are (a) steel with a tensile yield strength of 58 ksi, (b) Kevlar fibers with a tensile strength of 525 ksi, (c) IM8 carbon fibers with a tensile strength of 750 ksi.

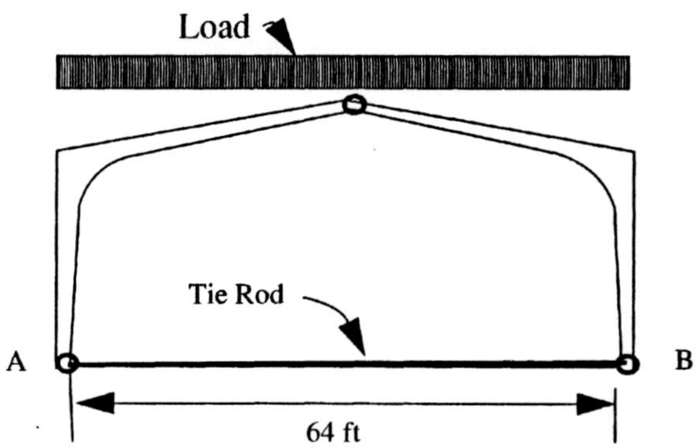

CHAPTER 2 CONCEPTS OF SOLID MECHANICS

"Perhaps the most incomprehensible thing about the world is that it is comprehensible."
Albert Einstein

2.1 Introduction

This chapter is a concise review of some of the fundamental concepts of solid mechanics in which individual phases of a material are considered to be *continua* and where, in the absence of debonding or other types of damage, the composite is also considered to be a continuum. A continuum is a material considered on a scale sufficiently large that the finest structure of the material can be ignored. Thus the individual fibers and the matrix of a composite are continua. Likewise, the heterogeneous composite is a multiphase continuum.

This chapter also will serve to establish much of the notation that will be used throughout the book. The reader is referred to the references at the end of the chapter for in-depth treatments of the topics covered in this chapter.

2.2 Tensors

Tensors are mathematical representations of physical quantities. They have components that change from one coordinate system to another (e.g., unprimed coordinates to primed coordinates in Fig. 2.1) according to so-called transformation equations. We will be concerned only with rectangular Cartesian coordinates and hence *Cartesian tensors*. Tensors usually are written using index notation, with the order of a tensor indicated by the number of live (nonrepeated) subscripts. If a subscript is repeated, summation over the range of that subscript is implied, unless otherwise indicated. For rectangular Cartesian coordinates the subscripts have a range of 1, 2, 3 in three dimensions and a range of 1, 2 in two dimensions. Examples of the types of tensor quantities used in this book are presented in Table 2.1.

The equations for transforming a tensor quantity from one coordinate system to another are written in terms of the direction cosines a_{ij} $(i, j = 1, 2, 3)$ of the angles measured from the unprimed axes, x_i, to the primed axes, x'_i (e.g., $a_{13} = \cos \theta_{13}$ in Fig. 2.1). *In this book, we shall use the convention that the first subscript (i) of a_{ij} corresponds to the initial, unprimed axes and the second subscript (j) corresponds to the final, primed axes.* (The reader is forewarned that some authors use the opposite convention for a_{ij}; i.e., the first subscript corresponds to the final, primed axes and the second subscript corresponds to the initial, unprimed axes.)

For the fully three-dimensional case, the direction cosines for coordinate transformation can be depicted conveniently in the tabular form indicated in Table 2.2.

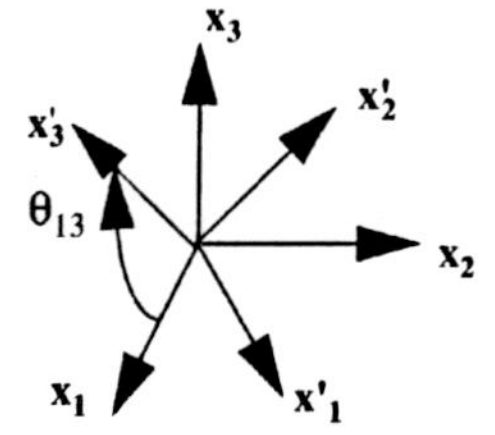

FIGURE 2.1 Rectangular Cartesian Coordinates

Example	Quantity	Live Subscripts
f	Scalar, zeroth-order tensor	0
σ_{ii}	Scalar: $(\sigma_{11} + \sigma_{22} + \sigma_{33})$	0
u_i	Vector: first-order tensor	1
σ_{ij}	Second-order tensor	2
C_{ijkl}	Fourth-order tensor	4

TABLE 2.1 Examples of Tensor Quantities

		x'_2	x'_3
x_1	a_{11}	a_{12}	a_{13}
x_2	a_{21}	a_{22}	a_{23}
x_3	a_{31}	a_{32}	a_{33}

TABLE 2.2 Direction Cosine Correspondence

The a_{ij} for transformation in a plane (coordinate rotation through an angle θ about an axis normal to the plane) are depicted in Fig. 2.2. From this figure it is apparent that $\angle\,(x_1$ to $x'_2) \neq \angle\,(x'_1$ to $x_2)$, and hence it is important that the convention for a_{ij} be known and understood. The individual a_{ij} for this planar rotation of coordinate systems about the x_3 axis are listed in Fig. 2.2. All components of a_{ij} are also given in terms of the sine or cosine of the angle, $\theta = \theta_{11}$, for this special case of rotation about an axis.

In matrix form, the transformation coefficients a_{ij} for rotation about the 3-axis are

$$a_{ij} = \begin{bmatrix} \cos\theta & -\sin\theta & 0 \\ \sin\theta & \cos\theta & 0 \\ 0 & 0 & 1 \end{bmatrix} \tag{2.1}$$

It is common practice to define $m = \cos\theta$ and $n = \sin\theta$, and thus (2.1) can be written

$$a_{ij} = \begin{bmatrix} m & -n & 0 \\ n & m & 0 \\ 0 & 0 & 1 \end{bmatrix} \tag{2.2}$$

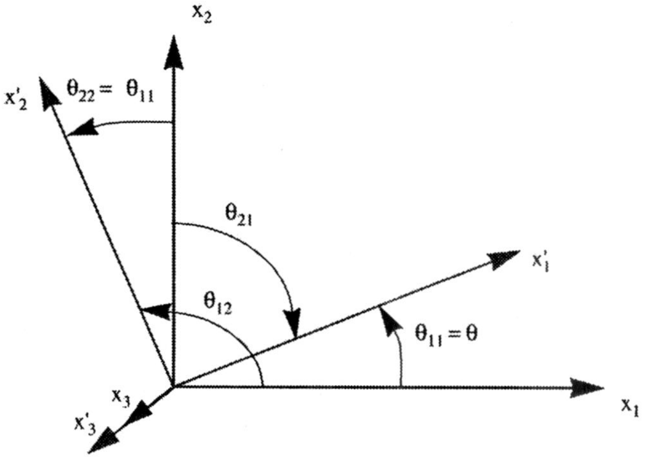

$a_{11} = \cos\theta_{11} = \cos\theta$
$a_{12} = \cos\theta_{12} = -\sin\theta$
$a_{13} = \cos 90 = 0$
$a_{21} = \cos\theta_{21} = \sin\theta$
$a_{22} = \cos\theta_{22} = \cos\theta$
$a_{23} = \cos 90 = 0$
$a_{31} = \cos 90 = 0$
$a_{32} = \cos 90 = 0$
$a_{33} = \cos 0 = 1$

FIGURE 2.2 Transformation about an Axis

Using the definitions of a_{ij} in the previous paragraphs, the transformation equation for a vector V_i' (in a rotated coordinate system) in terms of the components V_i in the original, unrotated coordinate system is

$$V_i' = a_{ji}V_j \tag{2.3}$$

Expanding (2.3) for the repeated subscript, j, over the range 1, 2, 3 gives

$$V_i' = a_{1i}V_1 + a_{2i}V_2 + a_{3i}V_3 \tag{2.4}$$

And writing the transformation equations explicitly in terms of direction cosines, we have

$$V_i' = \cos\theta_{1i}V_1 + \cos\theta_{2i}V_2 + \cos\theta_{3i}V_3 \tag{2.5}$$

Equation (2.3) can be inverted and written

$$V_j = a_{ji}^{-1}V_i' \tag{2.6}$$

However, for the special case of a transformation about an axis, we have from the definition of a_{ij} in (2.1)

$$a_{ji}^{-1} = a_{ij} = (a_{ji})^T \tag{2.7}$$

i.e., the inverse of the matrix is equal to the transform of the matrix. Thus, we can write (2.6) as

$$V_j = a_{ij}V_i' \tag{2.8}$$

The transformation for a first-order tensor u_i $(i, j = 1, 2)$ in two-dimensional space is

$$u_i' = a_{ji}u_j = a_{1i}u_1 + a_{2i}u_2 = \cos\theta_{1i}u_1 + \cos\theta_{2i}u_2 \tag{2.9}$$

and for a second-order tensor σ_{ij} in two-dimensional space, we have

$$\sigma'_{ij} = a_{ki}a_{lj}\sigma_{kl}$$

$$\sigma'_{ij} = a_{1i}a_{1j}\sigma_{11} + a_{1i}a_{2j}\sigma_{12} + a_{2i}a_{1j}\sigma_{21} + a_{2i}a_{2j}\sigma_{22} \tag{2.10}$$

$$\sigma'_{ij} = \cos\theta_{1i}\cos\theta_{1j}\sigma_{11} + \cos\theta_{1i}\cos\theta_{2j}\sigma_{12} + \cos\theta_{2i}\cos\theta_{1j}\sigma_{21} + \cos\theta_{2i}\cos\theta_{2j}\sigma_{22}$$

The following gives the expanded transformation expression for a second-order tensor σ_{ij} in three-dimensional space ($i, j = 1, 2, 3$) for a rotational transformation about an axis:

$$\begin{aligned} \sigma'_{ij} &= a_{ki}a_{lj}\sigma_{kl} \\ &= a_{1i}a_{1j}\sigma_{11} + a_{1i}a_{2j}\sigma_{12} + a_{1i}a_{3j}\sigma_{13} + a_{2i}a_{1j}\sigma_{21} + a_{2i}a_{2j}\sigma_{22} \\ &\quad + a_{2i}a_{3j}\sigma_{23} + a_{3i}a_{1j}\sigma_{31} + a_{3i}a_{2j}\sigma_{32} + a_{3i}a_{3j}\sigma_{33} \end{aligned} \tag{2.11}$$

It follows from the definition (2.1) that

$$a_{ik}a_{jk} = \delta_{ij} = \begin{cases} 1 & (i = j) \\ 0 & (i \neq j) \end{cases} \tag{2.12}$$

The term δ_{ij}, called the Kronecker delta, has the value 1 on the diagonal ($i = j$) and 0 on the off-diagonal ($i \neq j$).

The transformation equations for a fourth-order tensor C_{ijkl} can be written

$$C'_{ijkl} = a_{mi}a_{nj}a_{rk}a_{sl}C_{mnrs} \tag{2.13}$$

2.3 Deformation

Under the action of forces, a body S may translate and rotate as a rigid body as well as deform to occupy a new region S' as indicated in Fig. 2.3. We define the *displacements* u_i of any point P in the body to its new location P' in terms of the three components of the vector u_i (in a rectangular Cartesian coordinate system) as $u_i = (u_1, u_2, u_3)$. An equivalent notation for displacements that will be used is $u_i = (u, v, w)$. Displacement is a vector or first-order tensor quantity.

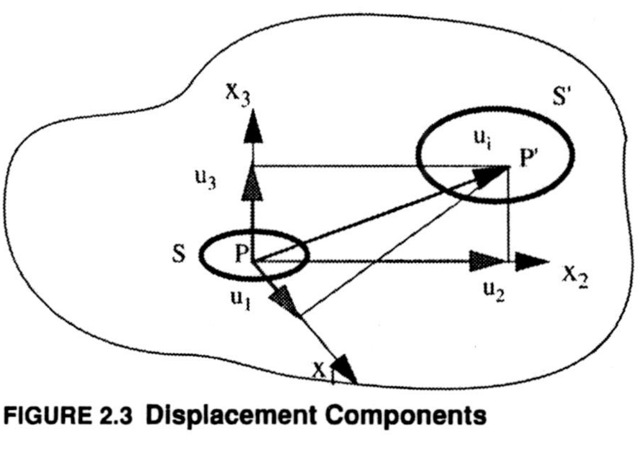

FIGURE 2.3 Displacement Components

2.4 Strain

If the gradients of the displacements are so small that products of partial derivatives of u_i are negligible compared with linear (first-order) derivative terms, then the (infinitesimal) *strain tensor* ε_{ij} is given by

$$\varepsilon_{ij} = \frac{1}{2}(u_{i,j} + u_{j,i}) \qquad (i, j = 1, 2, 3) \tag{2.14}$$

where the notation $u_{i,j}$ is defined

$$u_{i,j} = \frac{\partial u_i}{\partial x_j} \tag{2.15}$$

From the definition (2.14), strain is a second-order, symmetric tensor (i.e., $\varepsilon_{ij} = \varepsilon_{ji}$). In expanded form the strains are

$$\begin{aligned}
\varepsilon_{11} &= \frac{\partial u_1}{\partial x_1} & \varepsilon_{12} &= \varepsilon_{21} = \frac{1}{2}\left(\frac{\partial u_1}{\partial x_2} + \frac{\partial u_2}{\partial x_1}\right) \\
\varepsilon_{22} &= \frac{\partial u_2}{\partial x_2} & \varepsilon_{13} &= \varepsilon_{31} = \frac{1}{2}\left(\frac{\partial u_1}{\partial x_3} + \frac{\partial u_3}{\partial x_1}\right) \\
\varepsilon_{33} &= \frac{\partial u_3}{\partial x_3} & \varepsilon_{23} &= \varepsilon_{32} = \frac{1}{2}\left(\frac{\partial u_2}{\partial x_3} + \frac{\partial u_3}{\partial x_2}\right)
\end{aligned} \tag{2.16}$$

The normal components of strain (the change in length per unit length, Fig. 2.4a) correspond to the terms $i = j$, and the shear components (one-half the change in an original right angle, Fig. 2.4b) correspond to the terms $i \neq j$. We also note that it is common in the study of mechanics of materials to make use of the *engineering shear strain* $\gamma_{ij} = 2\varepsilon_{ij}$, for $i \neq j$. The engineering shear strain is introduced because of its relationship with the shear modulus. The engineering shear strains are written

$$\begin{aligned}
\gamma_{12} &= \gamma_{21} = \frac{\partial u_1}{\partial x_2} + \frac{\partial u_2}{\partial x_1} \\
\gamma_{13} &= \gamma_{31} = \frac{\partial u_1}{\partial x_3} + \frac{\partial u_3}{\partial x_1} \\
\gamma_{23} &= \gamma_{32} = \frac{\partial u_2}{\partial x_3} + \frac{\partial u_3}{\partial x_2}
\end{aligned} \tag{2.17}$$

In matrix notation, the second-order strain tensor is written

$$[\varepsilon] = \begin{bmatrix} \varepsilon_{11} & \varepsilon_{12} & \varepsilon_{13} \\ \varepsilon_{12} & \varepsilon_{22} & \varepsilon_{23} \\ \varepsilon_{13} & \varepsilon_{23} & \varepsilon_{33} \end{bmatrix} \tag{2.18}$$

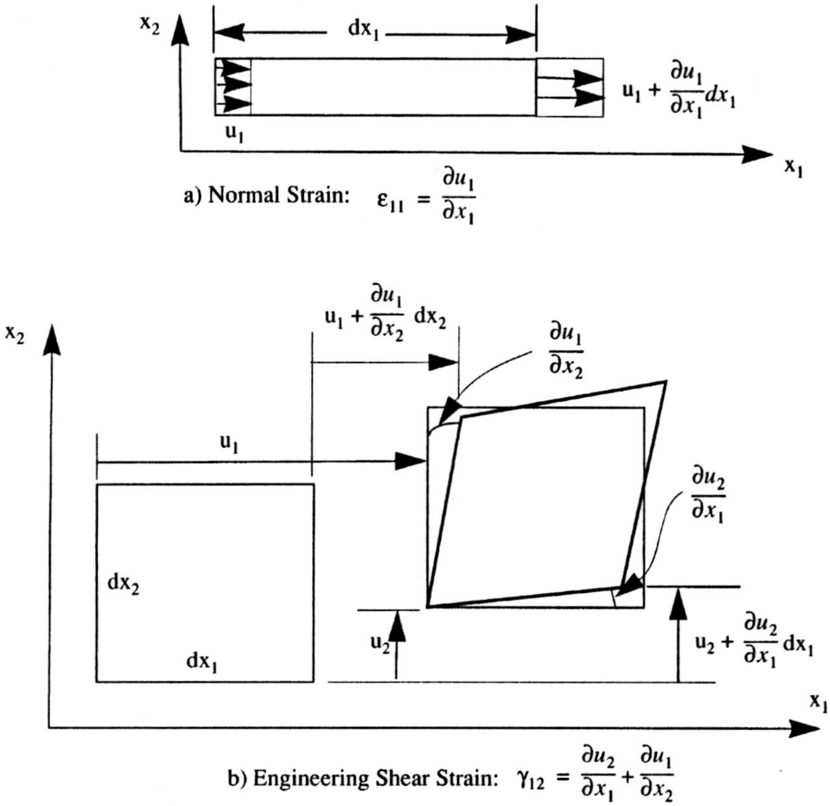

FIGURE 2.4 Components of Strain

or, equivalently, using the engineering shear strains,

$$[\varepsilon] = \begin{bmatrix} \varepsilon_{11} & \frac{\gamma_{12}}{2} & \frac{\gamma_{13}}{2} \\ \frac{\gamma_{12}}{2} & \varepsilon_{22} & \frac{\gamma_{23}}{2} \\ \frac{\gamma_{13}}{2} & \frac{\gamma_{23}}{2} & \varepsilon_{33} \end{bmatrix} \tag{2.19}$$

2.5 Stress

The components of stress at a point are the forces per unit area (in the limit) acting on planes passing through the point. The second-order stress tensor can be expressed in terms of the components acting on three mutually perpendicular planes aligned with the orthogonal coordinate directions as indicated in Fig. 2.5. The tensor notation for *stress* is σ_{ij} $(i, j = 1, 2, 3)$, where the first subscript

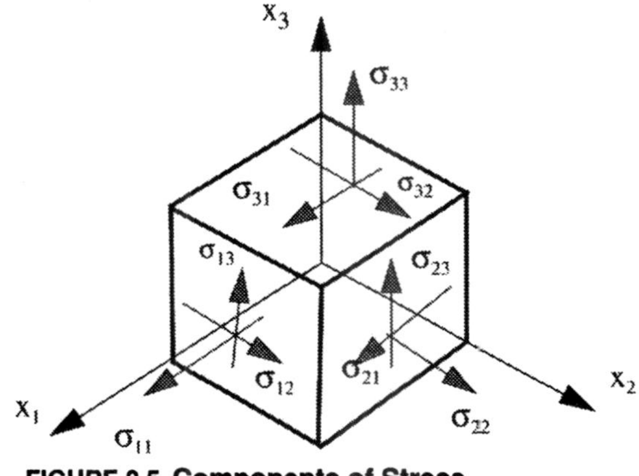

FIGURE 2.5 Components of Stress

corresponds to the direction of the normal to the plane of interest and the second subscript corresponds to the direction of the stress. Tensile normal stresses $(i = j)$ are positive, and shear stresses $(i \neq j)$ are defined to be positive when the normal to the plane and the stress component directions are either both positive or both negative. All components of stress depicted in Fig. 2.5 have a positive sense.

In matrix notation, the second-order stress tensor is written

$$[\sigma] = \begin{bmatrix} \sigma_{11} & \sigma_{12} & \sigma_{13} \\ \sigma_{12} & \sigma_{22} & \sigma_{23} \\ \sigma_{13} & \sigma_{23} & \sigma_{33} \end{bmatrix} \tag{2.20}$$

where we have used the fact that force and moment equilibrium of the element in Fig. 2.5 requires that the stress tensor be symmetric (i.e., $\sigma_{ij} = \sigma_{ji}$). The proof is left as an exercise.

It is common practice to express quantities in terms of a global x-y-z coordinate system, and it is also common to use the notation τ_{ij} for shear stresses. Thus the stress tensor will also be written

$$[\sigma] = \begin{bmatrix} \sigma_{xx} & \tau_{xy} & \tau_{xz} \\ \tau_{xy} & \sigma_{yy} & \tau_{yz} \\ \tau_{xz} & \tau_{yz} & \sigma_{zz} \end{bmatrix} \tag{2.21}$$

2.6 Equilibrium

For a body in *static equilibrium*, the equations of equilibrium at a point are written in tensor notation as

$$\sigma_{ij,j} + F_i = 0 \tag{2.22}$$

where F_i is the body force per unit volume. Body forces will be considered to be negligible for the applications in this book and, hence, the equilibrium equations take the simple form

$$\sigma_{ij,j} = 0 \tag{2.23}$$

The expanded form of the equilibrium equations, when written in terms of a global x-y-z coordinate system, is

$$\begin{aligned}\frac{\partial\sigma_{xx}}{\partial x} + \frac{\partial\sigma_{xy}}{\partial y} + \frac{\partial\sigma_{xz}}{\partial z} &= 0\\ \frac{\partial\sigma_{xy}}{\partial x} + \frac{\partial\sigma_{yy}}{\partial y} + \frac{\partial\sigma_{yz}}{dz} &= 0\\ \frac{\partial\sigma_{xz}}{\partial x} + \frac{\partial\sigma_{yz}}{\partial y} + \frac{\partial\sigma_{zz}}{\partial z} &= 0\end{aligned} \tag{2.24}$$

2.7 Boundary Conditions

2.7.1 Tractions

The solution of problems in solid mechanics requires that boundary conditions be specified. The boundary conditions may be specified in terms of components of displacement or stress, or a combination of displacements and stresses. For any point on an arbitrary surface we define the *traction* T_i to be the vector consisting of the three components of stress acting *on* the surface at the point of interest. As indicated in Fig. 2.6, the traction vector consists of one component of normal stress, σ_{nn}, and two components of shear stress, σ_{nt} and σ_{ns}. The traction vector can be written

$$T_i = \sigma_{ji}n_j \tag{2.25}$$

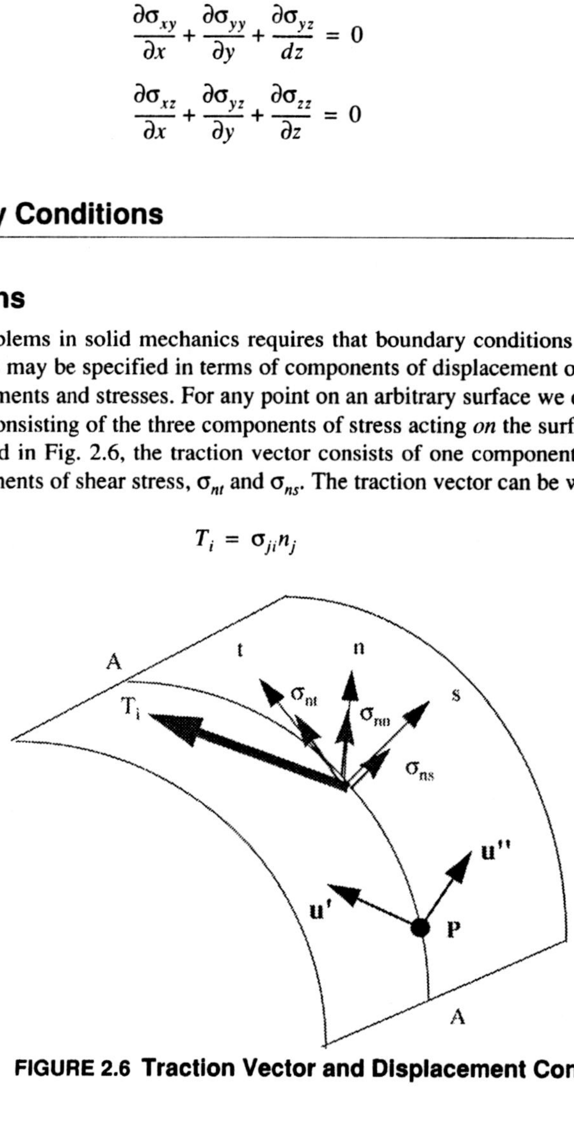

FIGURE 2.6 Traction Vector and Displacement Continuity

where n_j is the unit normal to the surface at the point under consideration. For a plane perpendicular to the x_1 axis (Fig. 2.5), $n_i = (1, 0, 0)$ and the components of the traction are

$$\begin{aligned} T_1 &= \sigma_{11} \\ T_2 &= \sigma_{12} \\ T_3 &= \sigma_{13} \end{aligned} \tag{2.26}$$

2.7.2 Free Surface Boundary Conditions

The condition that a surface be free of stress is equivalent to all components of traction (Fig. 2.6) being zero, i.e.,

$$\begin{aligned} T_n &= \sigma_{nn} = 0 \\ T_t &= \sigma_{nt} = 0 \\ T_s &= \sigma_{ns} = 0 \end{aligned} \tag{2.27}$$

It is possible, of course, that only selected components of the traction be zero while others are nonzero. For example, pure pressure loading would correspond to nonzero normal stress and zero shear stresses.

2.8 Continuity Conditions

2.8.1 Displacement Continuity

Certain conditions on displacements must be satisfied along any surface in a perfectly bonded continuum. Consider for example the line *A-A* shown in Fig. 2.6. The displacements associated with the material from either side of the line at *any* point **P** must be identical. Thus $u_i' = u_i''$ at **P** and the components of displacements must satisfy

$$\begin{aligned} u_1' &= u_1'' \\ u_2' &= u_2'' \\ u_3' &= u_3'' \end{aligned} \tag{2.28}$$

The continuity condition (2.28) must be satisfied at every point in a perfectly bonded continuum. However, continuity is not required in the presence of debonding or sliding between regions or phases of a material.

2.8.2 Traction Continuity

Equilibrium (action and reaction) requires that the tractions T_i must be continuous across any surface. Mathematically this is stated

$$T_i' = T_i'' \tag{2.29}$$

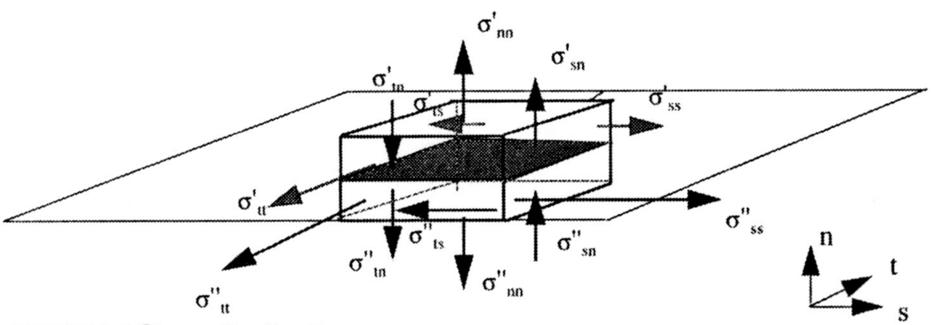

FIGURE 2.7 Stress Continuity

or, using (2.25),

$$\sigma'_{ji} n'_j = \sigma''_{ji} n''_j \tag{2.30}$$

In terms of individual stress components, this is written

$$\begin{aligned} \sigma'_{nn} &= \sigma''_{nn} \\ \sigma'_{nt} &= \sigma''_{nt} \\ \sigma'_{ns} &= \sigma''_{ns} \end{aligned} \tag{2.31}$$

Thus, the normal and shear components of stress acting *on* a surface (Fig. 2.7) must be continuous across that surface. We emphasize that there are no continuity requirements on the other three components of stress. In fact, they can be discontinuous. Thus it is permissible that

$$\begin{aligned} \sigma'_{tt} &\neq \sigma''_{tt} \\ \sigma'_{ss} &\neq \sigma''_{ss} \\ \sigma'_{ts} &\neq \sigma''_{ts} \end{aligned} \tag{2.32}$$

For heterogeneous materials (including laminated composites), the lack of continuity of the two normal and one shear components of stress indicated in (2.32) is very common. This is true because the material properties are discontinuous across phase (or layer) boundaries.

2.9 Compatibility

The strain displacement equations (2.14) provide six equations for only three unknown displacements u_i. Thus, integration of the equations to determine the unknown displacements will not have a singled-valued solution unless the strains ε_{ij} satisfy certain conditions. Arbitrary specification of the ε_{ij} could result in a situation in which there were discontinuities (gaps) in the material or overlapping regions of material.

The conditions that are necessary to provide single-valued displacements are called the *compatibility conditions*. They are conditions to be satisfied by the strains ε_{ij} and are obtained by eliminating u_i from (2.14). Differentiation of (2.14) gives

$$\varepsilon_{ij,kl} = \frac{1}{2}(u_{i,jkl} + u_{j,ikl}) \tag{2.33}$$

This expression can be rewritten by interchanging subscripts with the results

$$\begin{aligned} \varepsilon_{kl,ij} &= \frac{1}{2}(u_{k,lij} + u_{l,kij}) \\ \varepsilon_{jl,ik} &= \frac{1}{2}(u_{j,lik} + u_{l,jik}) \\ \varepsilon_{ik,jl} &= \frac{1}{2}(u_{i,kjl} + u_{k,ijl}) \end{aligned} \tag{2.34}$$

From (2.33) and (2.34), it can be verified that

$$\varepsilon_{ij,kl} + \varepsilon_{kl,ij} - \varepsilon_{ik,jl} - \varepsilon_{jl,ik} = 0 \tag{2.35}$$

Equations (2.35) are the *compatibility equations* (St. Venant, 1864).

Only 6 of the 81 equations represented by the equations of compatibility provide useful information. The remaining equations are either identities or repetitions of previous equations (when the symmetry of ε_{ij} is considered).

For rectangular Cartesian coordinates the six equations are written explicitly in the form

$$\begin{aligned} \varepsilon_{xx,zy} &= -\varepsilon_{yz,xx} + \varepsilon_{zx,yx} + \varepsilon_{xy,zx} \\ \varepsilon_{yy,xz} &= -\varepsilon_{zx,yy} + \varepsilon_{xy,zy} + \varepsilon_{yz,xy} \\ \varepsilon_{zz,yx} &= -\varepsilon_{xy,zz} + \varepsilon_{yz,xz} + \varepsilon_{zx,yz} \\ 2\varepsilon_{xy,yx} &= \varepsilon_{xx,yy} + \varepsilon_{yy,xx} \\ 2\varepsilon_{yz,zy} &= \varepsilon_{yy,zz} + \varepsilon_{zz,yy} \\ 2\varepsilon_{zx,xz} &= \varepsilon_{zz,xx} + \varepsilon_{xx,zz} \end{aligned} \tag{2.36}$$

2.10 Constitutive Equations

Constitutive equations provide the relationship between stress and strain. We shall be primarily concerned with linear elastic material response in this book. Thus the relationships between stress and strain are linear, with the most general form being

$$\sigma_{ij} = C_{ijkl}\varepsilon_{kl} \tag{2.37}$$

where the C_{ijkl} are elastic constants called *elastic moduli* or *stiffnesses*. This relation is called the generalized Hooke's law (Hooke, 1678; Love, 1892). We will be concerned with a variety of material types, distinguished by their degree of anisotropy. The number of independent constants ranges from 2 to 21 depending upon the degree of anisotropy. This topic is discussed in detail in Chapter 3.

Inverting (2.37) gives the strains in terms of the stresses in the form

$$\varepsilon_{ij} = C_{ijkl}^{-1}\sigma_{kl} = S_{ijkl}\sigma_{kl} \tag{2.38}$$

where we have defined the *compliance* S_{ijkl} as the inverse of the stiffness.

As will be shown in Chapter 3, the constitutive equations for an orthotropic material (a material with three mutually perpendicular planes of material symmetry) can be written in terms of nine independent stiffness coefficients in the form

$$\begin{aligned}
\varepsilon_{11} &= S_{1111}\sigma_{11} + S_{1122}\sigma_{22} + S_{1133}\sigma_{33} \\
\varepsilon_{22} &= S_{2211}\sigma_{11} + S_{2222}\sigma_{22} + S_{2233}\sigma_{33} \\
\varepsilon_{33} &= S_{3311}\sigma_{11} + S_{3322}\sigma_{22} + S_{3333}\sigma_{33} \\
\varepsilon_{23} &= S_{2323}\sigma_{23} \\
\varepsilon_{31} &= S_{3131}\sigma_{31} \\
\varepsilon_{12} &= S_{1212}\sigma_{12}
\end{aligned} \tag{2.39}$$

2.11 Plane Stress

Plane stress corresponds to a condition in which all three out-of-plane components of stress are identically zero throughout the region. For the plane stress case shown in Fig. 2.8, the normal stress σ_{zz} and both out-of-plane shear components σ_{zx} and σ_{zy} are zero for plane stress in the x-y plane. It is noted that the z-components of strain are not necessarily zero for plane stress in the x-y plane.

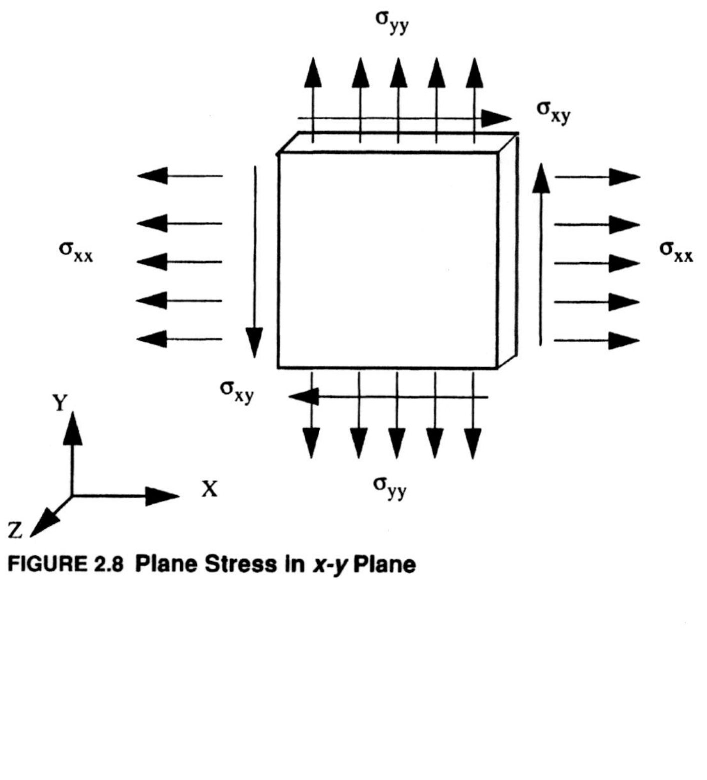

FIGURE 2.8 Plane Stress In *x-y* Plane

The magnitudes of the z-components of strain depend on the constitutive equation as expressed in (2.38). The out-of-plane normal strain ε_{zz} is nonzero for materials having a nonzero Poisson's ratio.

For the case of plane stress in the x-y plane, the equilibrium equations (2.24) reduce to the simpler form

$$\begin{aligned} \frac{\partial \sigma_{xx}}{\partial x} + \frac{\partial \sigma_{xy}}{\partial y} &= 0 \\ \frac{\partial \sigma_{xy}}{\partial x} + \frac{\partial \sigma_{yy}}{\partial y} &= 0 \end{aligned} \tag{2.40}$$

and the compatibility equations (2.36) reduce to a single equation:

$$2\varepsilon_{xy,\,yx} = \varepsilon_{xx,\,yy} + \varepsilon_{yy,\,xx} \tag{2.41}$$

2.12 Plane Strain

Plane strain corresponds to a condition in which all three out-of-plane components of strain (ε_{zz}, ε_{zx}, ε_{zy}) are zero (see Fig. 2.9) and the stresses are, at most, functions of x and y. As for the plane stress case, the out-of-plane components of stress depend upon the form of the constitutive equations, and the normal stress σ_{zz} is generally nonzero. Equilibrium (2.24) reduces to

$$\begin{aligned} \frac{\partial \sigma_{xx}}{\partial x} + \frac{\partial \sigma_{xy}}{\partial y} &= 0 \\ \frac{\partial \sigma_{xy}}{\partial x} + \frac{\partial \sigma_{yy}}{\partial y} &= 0 \\ \sigma_{zz} &= f(x, y) \end{aligned} \tag{2.42}$$

With the out-of-plane strain components (ε_{zz}, ε_{zx}, ε_{zy}) zero and strains independent of z through (2.38) (since the stresses are independent of z), compatibility (2.36) reduces to the single equation

$$2\varepsilon_{xy,\,yx} = \varepsilon_{xx,\,yy} + \varepsilon_{yy,\,xx} \tag{2.43}$$

Thus we see that the compatibility equation is the same for plane stress and plane strain.

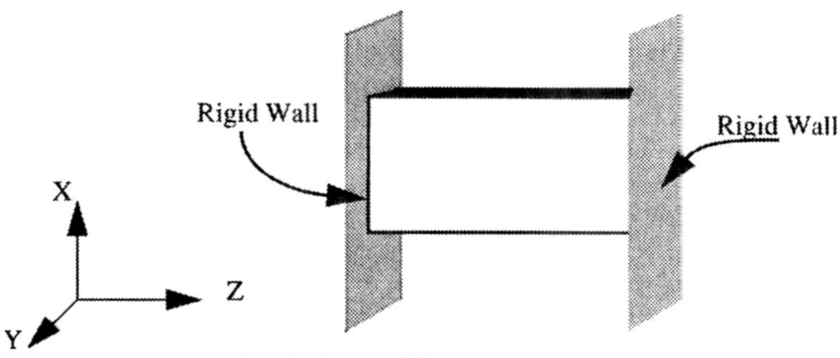

FIGURE 2.9 Plane Strain in *x*-*y* Plane

2.13 Generalized Plane Problems

Generalized plane problems are those in which all six components of stress and strain are (may be) nonzero, but stresses and strains do not vary along a prescribed direction (say the z-direction). The state of stress and strain can then be represented by that in a generic plane (Lekhnitskii, 1963). The compatibility equations then reduce to the single equation (2.43), and equilibrium (2.24) reduces to

$$\begin{aligned} \frac{\partial \sigma_{xx}}{\partial x} + \frac{\partial \sigma_{xy}}{\partial y} &= 0 \\ \frac{\partial \sigma_{xy}}{\partial x} + \frac{\partial \sigma_{yy}}{\partial y} &= 0 \\ \frac{\partial \sigma_{xz}}{\partial x} + \frac{\partial \sigma_{yz}}{\partial y} &= 0 \end{aligned} \tag{2.44}$$

2.14 Strain Energy Density

The strain energy density is the stored internal energy per unit volume in a body upon which work has been done. If the internal energy is zero in the unstressed state, the strain energy density in the stressed state is

$$W = \frac{1}{2}\sigma_{ij}\varepsilon_{ij} \tag{2.45}$$

Expansion of (2.45) in a global x-y-z coordinate system, invoking the symmetry of stress σ_{ij} and strain ε_{ij}, gives the expression

$$W = \frac{1}{2}(\sigma_{xx}\varepsilon_{xx} + 2\sigma_{xy}\varepsilon_{xy} + 2\sigma_{yz}\varepsilon_{yz} + 2\sigma_{xz}\varepsilon_{xz} + \sigma_{yy}\varepsilon_{yy} + \sigma_{zz}\varepsilon_{zz}) \tag{2.46}$$

It is noted that the strain energy density is a scalar function. Integration of the strain energy density over a given volume provides the strain energy of the body.

2.15 Minimum Principles

Minimum principles, which have their foundation in the calculus of variations, have proven to be very effective for obtaining approximate solutions to problems in solid mechanics. The principles are based upon the concept that a specific function, *chosen from a class of admissible functions that satisfy some, but not all, of the requirements of the complete solution,* minimizes a certain functional expression, *defined for this class of functions,* when the specific admissible function *satisfies the remaining requirements for the complete solution.*

For the general statement of these principles it is convenient to consider the boundary of the body under consideration to be composed of a displacement portion, S_D, and a remaining traction portion, S_T. *A portion of the boundary is S_D if, in each of three independent directions, either the component of displacement is given or the corresponding component of traction vanishes. A por-*

tion of the boundary is S_T if, in each of three independent directions, either the component of traction is given or the corresponding component of displacement vanishes (Hodge, 1958).

The minimum principles are expressed in terms of *statically admissible stress states* σ_{ij}^{o} which satisfy the equations of equilibrium and the traction boundary condition, or *kinematically admissible displacements* u_i^* which are single-valued continuous displacement fields that satisfy the displacement boundary conditions. Designating the strains associated (through the constitutive equations) with the statically admissible stresses as ε_{ij}^{o}, the stresses associated with the kinematically admissible strain as σ_{ij}^*, and the material elastic compliance as S_{ijkl}, two minimum principles in the theory of elasticity are the principle of minimum potential energy and the principle of minimum complementary energy. They are defined in the following sections.

2.15.1 Minimum Potential Energy

The principle of minimum potential energy states that among all kinematically admissible displacements fields u_i^*, the actual solution to the problem minimizes the functional Π^* defined as

$$\Pi^* = \frac{1}{2}\int_V S_{ijkl}\sigma_{ij}^*\sigma_{kl}^* dV - \int_{S_T} T_i u_i^* dS \tag{2.47}$$

The proof proceeds along the following lines. For a given boundary value problem with tractions T_i specified on S_T, let:

- u_i be the actual displacement field which is the exact solution to the problem
- σ_{ij} be the actual stresses
- u_i^* be any kinematically admissible displacement field
- σ_{ij}^* be the stresses associated with the kinematically admissible displacements
- Π be the actual potential energy
- Π^* be the potential energy associated with the admissible displacement field

Now define $\Delta\Pi$ as the difference between the admissible and actual potential energies.

$$\Delta\Pi = \Pi^* - \Pi \tag{2.48}$$

In order to prove the theorem, we must show that

$$\Delta\Pi \geq 0 \tag{2.49}$$

for all admissible displacement fields *with the equality holding if and only if* $u_i^* = u_i$.

Substituting the actual and admissible quantities in (2.47) and (2.48), we can write

$$\Delta\Pi = \frac{1}{2}\int_V S_{ijkl}(\sigma_{ij}^*\sigma_{kl}^* - \sigma_{ij}\sigma_{kl})dV - \int_{S_T} T_i(u_i^* - u_i)dS \tag{2.50}$$

Now we can write

$$\int_{S_T} T_i(u_i^* - u_i)dS = \int_S T_i(u_i^* - u_i)dS \tag{2.51}$$

since $S = S_T + S_D$ and $u_i{}^* = u_i$ on S_D, giving

$$\int_{S_D} T_i(u_i^* - u_i)dS = 0 \tag{2.52}$$

Using the divergence theorem and the definition (2.25) for tractions in terms of stresses, we can write the surface integral as a volume integral.

$$\int_S T_i(u_i^* - u_i)dS = \int_V \frac{\partial[\sigma_{ij}(u_i^* - u_i)]}{\partial x_j} dV \tag{2.53}$$

Now, performing the partial differentiation indicated in (2.53), and using equilibrium (2.23), strain displacement (2.14), and constitutive equations (2.38), we can write

$$\int_S T_i(u_i^* - u_i)dS = \int_V S_{ijkl}\sigma_{ij}(\sigma_{kl}^* - \sigma_{kl})dV \tag{2.54}$$

Combining (2.50) and (2.54) and using the symmetry of S_{ijkl}, we write

$$\Delta\Pi = \frac{1}{2}\int_V S_{ijkl}(\sigma_{ij}^* - \sigma_{ij})(\sigma_{kl}^* - \sigma_{kl})dV \tag{2.55}$$

The integrand in (2.55) is positive definite for all nonzero $(\sigma_{ij}^* - \sigma_{ij})$ since S_{ijkl} is symmetric and the strain energy density is assumed to be positive definite (see Kreyszig, 1967). Thus

$$\Delta\Pi \geq 0 \tag{2.56}$$

with $\Delta\Pi = 0$ *if and only if* $\sigma_{ij}^* = \sigma_{ij}$. This implies that the displacements satisfy $u_i^* = u_i$ (to within a rigid body motion). Hence the theorem is proven.

2.15.2 Minimum Complementary Energy

The principle of minimum complementary energy states that among all statically admissible stress fields σ_{ij}^o the actual solution to the problem minimizes the admissible complementary energy, Π_c^o, defined as

$$\Pi_c^o = \frac{1}{2}\int_V S_{ijkl}\sigma_{ij}^o\sigma_{kl}^o dV - \int_{S_D} T_i^o u_i dS \tag{2.57}$$

In order to prove the theorem, it is necessary to show that

$$\Pi_c^o - \Pi_c \geq 0 \tag{2.58}$$

where Π_c is the complementary energy of the exact solution.

The proof is similar to that for potential energy. It is left as an exercise.

2.15.3 Bounds and Uniqueness

From the definitions (2.47) and (2.57) the sum of the potential and complementary energies for the actual solution must satisfy the principle of virtual work, i.e.,

$$\Pi + \Pi_c = 0 \tag{2.59}$$

For a proof of the principle of virtual work, see Fung (1965). It can be combined with (2.48), (2.49), and (2.58) with the result

$$-\Pi_c^o \leq -\Pi_c = \Pi \leq \Pi^* \tag{2.60}$$

Equation (2.60) shows that the complementary energy and potential energy of admissible states can be used to provide bounds on the exact solution.

Uniqueness of the solution is proved by considering two distinct complete solutions whose potential energies are $\Pi^{(1)}$ and $\Pi^{(2)}$. By definition, complete solutions must also be kinematically admissible. Therefore, considering $\Pi^{(1)}$ as the complete solution and $\Pi^{(2)}$ as the kinematically admissible solution, we can write

$$\Pi^{(2)} \geq \Pi^{(1)} \tag{2.61}$$

Reversing the procedure and taking $\Pi^{(2)}$ as the complete solution and $\Pi^{(1)}$ as the kinematically admissible solution, we write

$$\Pi^{(1)} \geq \Pi^{(2)} \tag{2.62}$$

The only way for both equations to be satisfied is for

$$\Pi^{(1)} = \Pi^{(2)} \tag{2.63}$$

Hence the solution is unique. This is a most important principle of elasticity theory. If a solution that satisfies all the requirements of an elasticity solution is determined, it is the *only* solution.

References

Fung, Y. C. (1965), *Foundations of Solid Mechanics,* Prentice-Hall, Inc., Englewood Cliffs, NJ, p. 285.

Hodge, P. G., Jr. (1958), "The Mathematical Theory of Plasticity," in *Elasticity and Plasticity* (J. N. Goodier and P. G. Hodge, Jr., eds.), John Wiley & Sons, Inc., New York, p. 81.

Hooke, R. (1678), *De Potentia Restitutiva,* London.

Kreyszig, E. (1967), *Advanced Engineering Mathematics,* 2nd ed., John Wiley & Sons, Inc., New York, p. 416.

Lekhnitskii, S. G. (1963), *Theory of Elasticity of an Anisotropic Body,* Holden-Day, Inc., San Francisco (translation of *Teoriia Uprugosti Anisotropnovo Tela,* Government Publishing House for Technical-Theoretical Works, Moscow, 1950; also Mir Publishers, Moscow, 1981).

Lemaitre, J., and Chaboche, J. L. (1985), *Mecanique des materiaux solides* (English translation 1990, *Mechanics of Solid Materials*), Cambridge University Press, Cambridge, England.

Love, A. E. H. (1892), *A Treatise on the Mathematical Theory of Elasticity,* Cambridge University Press, London (also 4th ed., 1944, Dover Publications, Inc., New York).

Malvern, L. E. (1969), *Introduction to the Mechanics of a Continuous Medium,* Prentice-Hall, Inc., Englewood Cliffs, NJ.

Sokolnikoff, I. S. (1956), *Mathematical Theory of Elasticity* (1st ed. 1946), McGraw-Hill, New York.

St. Venant, B. de (1864), in his edition of Navier's *Résumé des Leçons sur l'application de la Mécanique,* app. 3, Carilian-Goeury, Paris.

Exercises

2.1 Consider a transformation about the x_3 axis and show that $a_{ij}^{-1} = (a_{ij})^T = a_{ji}$.

2.2 Consider a transformation about the x_3 axis and show that $a_{ij}(-\theta) = a_{ji}(\theta)$.

2.3 Show that (2.12) is true, i.e., $a_{ik}a_{jk} = \delta_{ij}$.

2.4 Use force and moment equilibrium of an infinitesimal element to show that the stress tensor is symmetric.

2.5 Write the equations of equilibrium for a state of plane stress.

2.6 Write the compatibility equations for a state of plane strain.

2.7 Prove the principle of minimum complementary energy as expressed by (2.58).

2.8 Prove the principle of virtual work as expressed by (2.59).

CHAPTER 3 3-D CONSTITUTIVE EQUATIONS

"I like the dreams of the future better than the history of the past."
Thomas Jefferson

3.1 Effective Property Concept

In this chapter we consider ways to write the linear elastic constitutive equations for the effective (or average) response of fibrous composite materials. Our goal is to develop equations for predicting the elastic constants required for the average stress-strain relationship. This relationship is usually referred to as Hooke's law. We will consider three-dimensional constitutive equations in this chapter and two-dimensional (plane stress) equations in the following chapter.

We want to write constitutive equations for the effective or smeared properties of the composite because an analysis of the exact configuration of the individual fibers and matrix in a composite structure would be impractical for structural analysis. To do so, we model the properties of the composite as the properties of a repeating *representative volume element (RVE)*, which is small enough to be representative of local material response yet large enough to represent "average" material response. In a fibrous composite the length scale over which the properties can be averaged in a meaningful manner must be several times that of the mean fiber spacing. If this dimension is small compared with the characteristic dimension of the structure, the material can be modeled as *effectively homogeneous*. Figure 3.1 shows an example representative volume element as well as volume elements that are too small to be representative of the composite. See Christensen (1979) for a more in-depth discussion of equivalent homogeneity, and Chapter 11 of this book for methods to determine the effective properties from knowledge of the fiber and matrix properties and their arrangement.

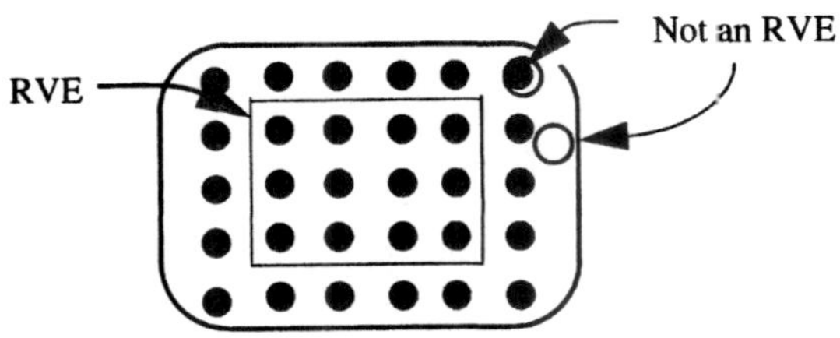

FIGURE 3.1 Representative Volume Element

3.2 Generalized Hooke's Law

As indicated in Chapter 2, the most general linear relationship between stresses and strains such that the stresses vanish when the strains are zero is

$$\sigma_{ij} = C_{ijkl}\varepsilon_{kl} \tag{3.1}$$

where C_{ijkl} is a fourth-order tensor with 81 (3^4) elastic "constants." This linear elastic stress-strain constitutive relationship is called the *Generalized Hooke's law.*

Invoking symmetry of the stress tensor gives $C_{ijkl} = C_{jikl}$. Similarly, symmetry of the strain tensor gives $C_{ijkl} = C_{ijlk}$. Thus the number of independent constants is reduced from 81 to 36 when the stresses and strains are symmetric.

In view of this reduced number of constants, Hooke's law can be written in contracted notation as

$$\sigma_i = C_{ij}\varepsilon_j \qquad (i, j = 1, 2, \ldots, 6) \tag{3.2}$$

where the contracted notation for stress and strain is defined by the following equivalences:

$$\begin{array}{ll}
\sigma_1 \equiv \sigma_{11} & \varepsilon_1 \equiv \varepsilon_{11} \\
\sigma_2 \equiv \sigma_{22} & \varepsilon_2 \equiv \varepsilon_{22} \\
\sigma_3 \equiv \sigma_{33} & \varepsilon_3 \equiv \varepsilon_{33} \\
\sigma_4 \equiv \sigma_{23} & \varepsilon_4 \equiv 2\varepsilon_{23} \\
\sigma_5 \equiv \sigma_{31} & \varepsilon_5 \equiv 2\varepsilon_{31} \\
\sigma_6 \equiv \sigma_{12} & \varepsilon_6 \equiv 2\varepsilon_{12}
\end{array} \tag{3.3}$$

We note here that, in the contracted notation, the shear strain terms are the engineering shear strains, e.g., $\varepsilon_4 = 2\varepsilon_{23} = \gamma_{23}$ (see Section 2.4).

The C_{ij} are referred to by a variety of names, including *elastic constants, moduli,* and *stiffness coefficients.* In order for (3.2) to be solvable for the strains in terms of the stresses, the determinant of the stiffness matrix must be nonzero, i.e., $|C_{ij}| \neq 0$.

The number of independent elastic constants can be reduced further if there exists a *strain energy density function* W such that

$$W = \frac{1}{2}C_{ij}\varepsilon_i\varepsilon_j \tag{3.4}$$

with the property

$$\sigma_i = \frac{\partial W}{\partial \varepsilon_i} \tag{3.5}$$

If the quadratic form (3.4) exists such that (3.5) is true, then C_{ij} is symmetric and the number of constants is reduced accordingly, i.e.,

$$C_{ij} = C_{ji} \tag{3.6}$$

The proof is left as an exercise.

The symmetric stiffness, C_{ij}, can be written in matrix notation as

$$[C_{ij}] = \begin{bmatrix} C_{11} & C_{12} & C_{13} & C_{14} & C_{15} & C_{16} \\ C_{12} & C_{22} & C_{23} & C_{24} & C_{25} & C_{26} \\ C_{13} & C_{23} & C_{33} & C_{34} & C_{35} & C_{36} \\ C_{14} & C_{24} & C_{34} & C_{44} & C_{45} & C_{46} \\ C_{15} & C_{25} & C_{35} & C_{45} & C_{55} & C_{56} \\ C_{16} & C_{26} & C_{36} & C_{46} & C_{56} & C_{66} \end{bmatrix} \tag{3.7}$$

This symmetric matrix has 21 independent constants; i.e., counting the independent C_{ij} by row gives 6 + 5 + 4 + 3 + 2 + 1 = 21.

The existence of the function W is based upon the first and second laws of thermodynamics. It was first proposed by Green (1839). An in-depth discussion of the existence and positive definiteness of the strain energy function is given in Fung (1965). Experimental results are consistent with this theory for the elastic response of materials.

The inverted form of Hooke's law (3.2) can be written

$$\varepsilon_i = S_{ij}\sigma_j \tag{3.8}$$

The coefficients S_{ij} in (3.8) are called *compliance coefficients*. They will be discussed in more detail in the following sections. From (3.2) and (3.8) it is evident that the compliance matrix is the inverse of the stiffness matrix; thus,

$$S_{ij} = C_{ij}^{-1} \tag{3.9}$$

We note here that since the stiffness matrix is symmetric, the compliance matrix also is symmetric.

A material with 21 independent constants is called *anisotropic*. The maximum number of independent constants for an anisotropic material is limited by the symmetry of the stress and strain tensors and the assumption of the existence of the strain energy density function. Materials which exhibit *material symmetry* about planes passing through the material have fewer independent constants. Different degrees of material symmetry are discussed in the following section.

3.3 Material Symmetry

3.3.1 Monoclinic Material

Consider a material which is symmetric about the x_1-x_2 plane (Fig. 3.2); that is, the elastic constants in the primed and the unprimed coordinate systems are identical $C_{ij} = C'_{ij}$. The implications of this symmetry can be deduced by considering Hooke's law in each coordinate system. In the unprimed coordinates,

$$\sigma_i = C_{ij}\varepsilon_j \tag{3.10}$$

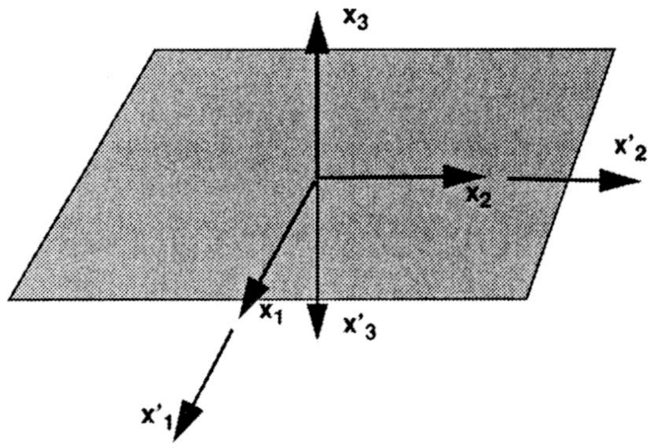

FIGURE 3.2 Symmetry about x_1-x_2 Plane

and in the primed coordinates,

$$\sigma'_i = C'_{ij}\varepsilon'_j \tag{3.11}$$

From the definition of the coordinate systems, it is apparent that

$$\begin{aligned} \sigma'_1 &= \sigma_1 & \sigma'_4 &= -\sigma_4 \\ \sigma'_2 &= \sigma_2 & \sigma'_5 &= -\sigma_5 \\ \sigma'_3 &= \sigma_3 & \sigma'_6 &= \sigma_6 \end{aligned} \tag{3.12}$$

$$\begin{aligned} \varepsilon'_1 &= \varepsilon_1 & \varepsilon'_4 &= -\varepsilon_4 \\ \varepsilon'_2 &= \varepsilon_2 & \varepsilon'_5 &= -\varepsilon_5 \\ \varepsilon'_3 &= \varepsilon_3 & \varepsilon'_6 &= \varepsilon_6 \end{aligned} \tag{3.13}$$

Expressing the first stress equality $\sigma'_1 = \sigma_1$ in (3.12) in terms of strains and stiffnesses using the constitutive equations (3.10) and (3.11) gives

$$C'_{11}\varepsilon'_1 + C'_{12}\varepsilon'_2 + \cdots + C'_{16}\varepsilon'_6 = C_{11}\varepsilon_1 + C_{12}\varepsilon_2 + \cdots + C_{16}\varepsilon_6 \tag{3.14}$$

Now substituting for the ε'_i in terms of the ε_i from (3.13) and comparing like terms in (3.14) (noting that $C_{ij} = C'_{ij}$), we obtain

$$C_{14} = C_{15} = 0 \tag{3.15}$$

In a similar fashion, the remaining equations in (3.12) lead to the results

$$C_{24} = C_{25} = C_{34} = C_{35} = C_{46} = C_{56} = 0 \tag{3.16}$$

The proof of (3.16) is left as an exercise. Thus, for material symmetry about the (x_1-x_2) plane, the stiffness matrix reduces to

$$C_{ij} = \begin{bmatrix} C_{11} & C_{12} & C_{13} & 0 & 0 & C_{16} \\ C_{12} & C_{22} & C_{23} & 0 & 0 & C_{26} \\ C_{13} & C_{23} & C_{33} & 0 & 0 & C_{36} \\ 0 & 0 & 0 & C_{44} & C_{45} & 0 \\ 0 & 0 & 0 & C_{45} & C_{55} & 0 \\ C_{16} & C_{26} & C_{36} & 0 & 0 & C_{66} \end{bmatrix} \tag{3.17}$$

As indicated in (3.17), a material with one plane of material symmetry has *13 independent elastic constants*. Such a material is called *monoclinic*.

The effective properties of a unidirectional fibrous composite with the fibers oriented off-axis can be modeled as a homogeneous anisotropic material. As indicated in Fig. 3.3, two Cartesian coordinate systems are identified: the 1-2-3 system and the *x*-*y*-*z* system. The 1-direction is the fiber direction, with the 2- and 3-directions perpendicular to the fiber direction. The *x*-*y*-*z* coordinates are obtained by a rotation about the 3-axis. The material exhibits symmetry with respect to the *x*-*y* plane, but not with respect to the *x*-*z* or *y*-*z* plane. Such a material with one plane of symmetry is called a monoclinic material. It has a stiffness matrix of the form (3.17) when referred to the global *x*-*y*-*z* coordinate system. The symmetry that such a material possesses when viewed in the 1-2-3 coordinate system which is parallel and perpendicular to the fiber direction will be discussed in the following section.

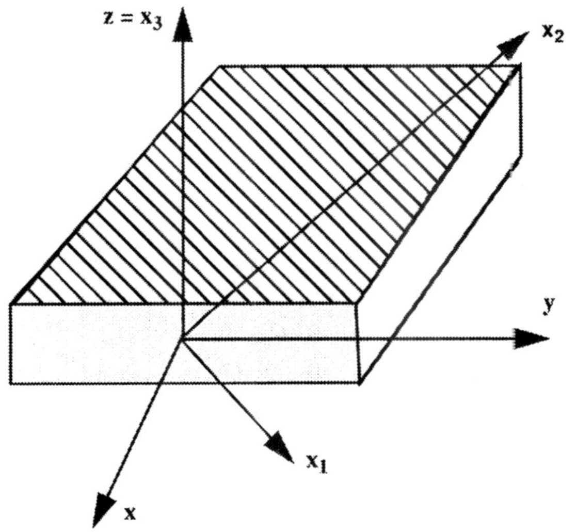

FIGURE 3.3 Off-Axis Unidirectional Lamina

3.3.2 Orthotropic Material

If the material under consideration has a second plane of material symmetry, say the x_2-x_3 plane (Fig. 3.4), the σ_1, σ_2, σ_3, and σ_5 stress components must be equal to the corresponding stresses in

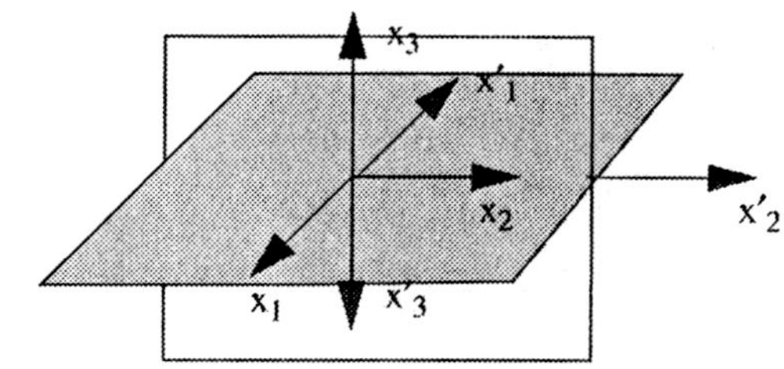

FIGURE 3.4 Symmetry about x_1-x_2 and x_2-x_3 Planes

the primed coordinate system. Following the same procedure as in the previous section gives

$$C_{16} = C_{26} = C_{36} = C_{45} = 0 \tag{3.18}$$

The stiffness matrix then has the form of an *orthotropic* material with nine independent constants, i.e.,

$$[C] = \begin{bmatrix} C_{11} & C_{12} & C_{13} & 0 & 0 & 0 \\ C_{12} & C_{22} & C_{23} & 0 & 0 & 0 \\ C_{13} & C_{23} & C_{33} & 0 & 0 & 0 \\ 0 & 0 & 0 & C_{44} & 0 & 0 \\ 0 & 0 & 0 & 0 & C_{55} & 0 \\ 0 & 0 & 0 & 0 & 0 & C_{66} \end{bmatrix} \tag{3.19}$$

For the stiffness matrix (3.19), the constitutive equation (3.2) can be written in matrix form as

$$\begin{Bmatrix} \sigma_1 \\ \sigma_2 \\ \sigma_3 \\ \tau_{23} \\ \tau_{13} \\ \tau_{12} \end{Bmatrix} = \begin{bmatrix} C_{11} & C_{12} & C_{13} & 0 & 0 & 0 \\ C_{12} & C_{22} & C_{23} & 0 & 0 & 0 \\ C_{13} & C_{23} & C_{33} & 0 & 0 & 0 \\ 0 & 0 & 0 & C_{44} & 0 & 0 \\ 0 & 0 & 0 & 0 & C_{55} & 0 \\ 0 & 0 & 0 & 0 & 0 & C_{66} \end{bmatrix} \begin{Bmatrix} \varepsilon_1 \\ \varepsilon_2 \\ \varepsilon_3 \\ \gamma_{23} \\ \gamma_{13} \\ \gamma_{12} \end{Bmatrix} \tag{3.20}$$

It is left as an exercise to show that a material with two perpendicular planes of material symmetry also exhibits symmetry about the third, mutually perpendicular plane. Thus, an orthotropic material has three mutually perpendicular planes of symmetry. An orthotropic fibrous material is depicted in Fig. 3.5, where the cross section of the fibers is oval.

3.3.3 Transversely Isotropic Material

A *transversely isotropic* material is defined to be a material whose effective properties are isotropic (independent of direction) in one of its planes. A unidirectional fibrous composite consisting of a random array of fibers such as that shown in Fig. 1.6 exhibits isotropic properties in the plane trans-

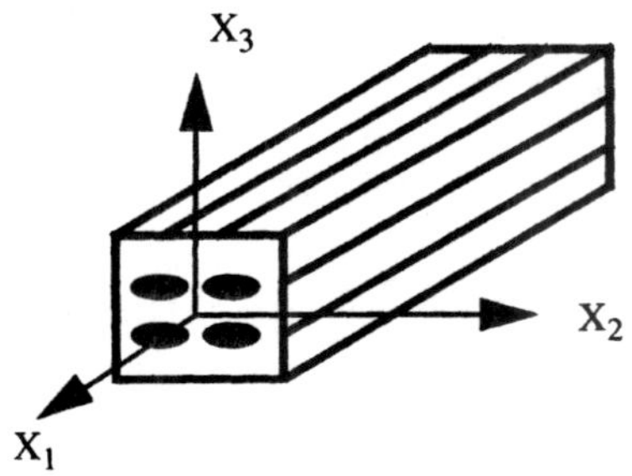

FIGURE 3.5 Orthotropic Composite

verse to the fibers, the 2-3 plane in Fig. 3.6. Transverse isotropy results in the following relations for stiffness coefficients:

$$\begin{gathered} C_{22} = C_{33} \qquad C_{12} = C_{13} \qquad C_{55} = C_{66} \\ C_{44} = (C_{22} - C_{23})/2 \end{gathered} \tag{3.21}$$

The relations (3.21) can be deduced directly from the equality of properties in the x_2 and x_3 directions and the E, G, ν relationship ($G = E/2(1+\nu)$) in the plane of isotopy (see, for example, Popov, 1990). The stiffness matrix for a transversely isotropic material then has the form

$$[C] = \begin{bmatrix} C_{11} & C_{12} & C_{12} & 0 & 0 & 0 \\ C_{12} & C_{22} & C_{23} & 0 & 0 & 0 \\ C_{12} & C_{23} & C_{22} & 0 & 0 & 0 \\ 0 & 0 & 0 & \dfrac{C_{22} - C_{23}}{2} & 0 & 0 \\ 0 & 0 & 0 & 0 & C_{66} & 0 \\ 0 & 0 & 0 & 0 & 0 & C_{66} \end{bmatrix} \tag{3.22}$$

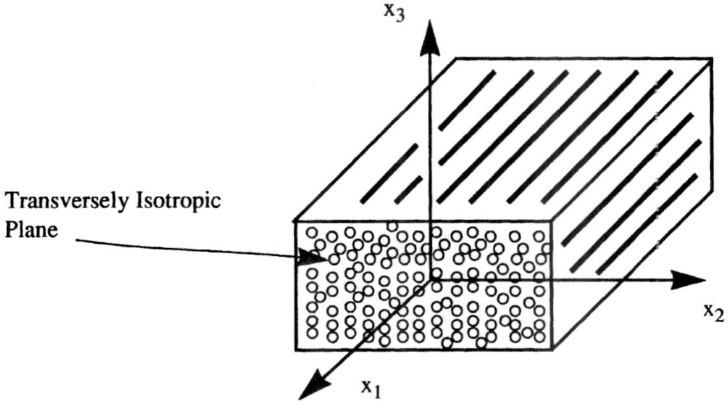

FIGURE 3.6 Transversely Isotropic Material

It is evident from (3.22) that a *transversely isotropic* material has only *five independent elastic constants.*

3.3.4 Isotropic Material

An *isotropic* material is one whose properties are independent of direction and whose planes are all planes of symmetry. In this case we have the following additional relations between stiffness coefficients:

$$\begin{aligned} C_{11} &= C_{22} \qquad C_{12} = C_{23} \\ C_{66} &= (C_{22} - C_{23})/2 = (C_{11} - C_{12})/2 \end{aligned} \tag{3.23}$$

Thus an *isotropic material* has only *two independent elastic constants,* and the stiffness matrix can be written

$$[C] = \begin{bmatrix} C_{11} & C_{12} & C_{12} & 0 & 0 & 0 \\ C_{12} & C_{11} & C_{12} & 0 & 0 & 0 \\ C_{12} & C_{12} & C_{11} & 0 & 0 & 0 \\ 0 & 0 & 0 & \frac{C_{11}-C_{12}}{2} & 0 & 0 \\ 0 & 0 & 0 & 0 & \frac{C_{11}-C_{12}}{2} & 0 \\ 0 & 0 & 0 & 0 & 0 & \frac{C_{11}-C_{12}}{2} \end{bmatrix} \tag{3.24}$$

3.4 Engineering Constants

The simplest form of the constitutive equations is obtained when they are written in terms of stiffness coefficients C_{ij} and compliance coefficients S_{ij}. However, in general, these coefficients are not measured directly in the laboratory. The constants that are measured in the laboratory are called *engineering constants.* In order to establish the relationship between the engineering constants and the compliance coefficients, we consider an orthotropic material in the *principal material coordinates,* i.e., for a unidirectional fibrous composite, the axes aligned parallel and perpendicular to the fiber direction (Fig. 3.6). The compliance coefficients can then be expressed in terms of the related engineering constants through superposition of the strains associated with the individual components of stress. For an orthotropic material subjected to a three-dimensional state of stress, the equations for strains in terms of stresses and engineering constants take the form

$$\varepsilon_1 = \frac{\sigma_1}{E_1} - \frac{\nu_{21}}{E_2}\sigma_2 - \frac{\nu_{31}}{E_3}\sigma_3 \tag{3.25}$$

$$\varepsilon_2 = -\frac{\nu_{12}}{E_1}\sigma_1 + \frac{\sigma_2}{E_2} - \frac{\nu_{32}}{E_3}\sigma_3 \tag{3.26}$$

$$\varepsilon_3 = -\frac{\nu_{13}}{E_1}\sigma_1 - \frac{\nu_{23}}{E_2}\sigma_2 + \frac{\sigma_3}{E_3} \tag{3.27}$$

$$\gamma_{23} = \frac{\tau_{23}}{G_{23}} \tag{3.28}$$

$$\gamma_{13} = \frac{\tau_{13}}{G_{13}} \tag{3.29}$$

$$\gamma_{12} = \frac{\tau_{12}}{G_{12}} \tag{3.30}$$

In the preceding equations, the convention for the subscripts on the engineering constants is as follows. E_1, E_2, and E_3 are the Young's moduli in the three coordinate directions. Thus E_1 is the axial (fiber direction) modulus, and E_2 and E_3 are the transverse moduli. For Poisson's ratios, ν_{ij}, the first subscript, i, refers to the direction of the applied stress, and the second subscript corresponds to the direction of associated lateral strain. Thus, for $\sigma_i \neq 0$, $i \neq j$, and $i, j = 1, 2, 3$,

$$\nu_{ij} = \frac{-\varepsilon_j}{\varepsilon_i} \tag{3.31}$$

It is important to note that, in general, $\nu_{ij} \neq \nu_{ji}$. The shear moduli G_{ij} are defined for shear stress loading in the i-j plane. Thus, for a unidirectional fibrous composite, G_{12} and G_{13} are axial shear moduli in the two orthogonal planes that contain the fibers, and G_{23} is the transverse (out-of-plane) shear modulus. The shear modulus corresponds to the shear stiffness of the associated plane, and hence $G_{ij} = G_{ji}$.

The relationships between engineering constants and applied stress states are depicted in Fig. 3.7 for a transversely isotropic composite.

The constitutive equation (3.8) for an orthotropic material can be written in matrix form as

$$\begin{Bmatrix} \varepsilon_1 \\ \varepsilon_2 \\ \varepsilon_3 \\ \gamma_{23} \\ \gamma_{13} \\ \gamma_{12} \end{Bmatrix} = \begin{bmatrix} S_{11} & S_{12} & S_{13} & 0 & 0 & 0 \\ S_{12} & S_{22} & S_{23} & 0 & 0 & 0 \\ S_{13} & S_{23} & S_{33} & 0 & 0 & 0 \\ 0 & 0 & 0 & S_{44} & 0 & 0 \\ 0 & 0 & 0 & 0 & S_{55} & 0 \\ 0 & 0 & 0 & 0 & 0 & S_{66} \end{bmatrix} \begin{Bmatrix} \sigma_1 \\ \sigma_2 \\ \sigma_3 \\ \tau_{23} \\ \tau_{13} \\ \tau_{12} \end{Bmatrix} \tag{3.32}$$

where we have used the engineering shear strains $\gamma_{ij} = 2\varepsilon_{ij}$ and denoted shear stresses by τ_{ij}.

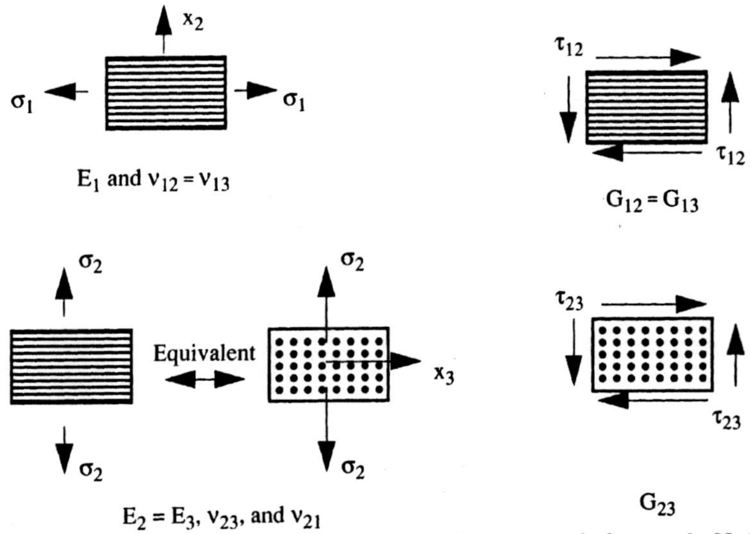

FIGURE 3.7 Engineering Constants for Transversely Isotropic Material

Equating equivalent terms in (3.25)–(3.30) with those in (3.32) gives the individual compliance coefficients explicitly in terms of the engineering constants:

$$\begin{aligned}
S_{11} &= \frac{1}{E_1}; & S_{12} &= -\frac{\nu_{21}}{E_2}; & S_{13} &= -\frac{\nu_{31}}{E_3} \\
S_{21} &= -\frac{\nu_{12}}{E_1}; & S_{22} &= \frac{1}{E_2}; & S_{23} &= -\frac{\nu_{32}}{E_3} \\
S_{31} &= -\frac{\nu_{13}}{E_1}; & S_{32} &= -\frac{\nu_{23}}{E_2}; & S_{33} &= \frac{1}{E_3} \\
S_{44} &= \frac{1}{G_{23}}; & S_{55} &= \frac{1}{G_{13}}; & S_{66} &= \frac{1}{G_{12}}
\end{aligned} \tag{3.33}$$

Using the symmetry of the compliance matrix, i.e., $S_{ij} = S_{ji}$, gives

$$\frac{\nu_{21}}{E_2} = \frac{\nu_{12}}{E_1}; \qquad \frac{\nu_{31}}{E_3} = \frac{\nu_{13}}{E_1}; \qquad \frac{\nu_{23}}{E_2} = \frac{\nu_{32}}{E_3} \tag{3.34}$$

which are referred to as the reciprocal relations. They are often written in the form

$$E_1\nu_{21} = E_2\nu_{12}; \qquad E_1\nu_{31} = E_3\nu_{13}; \qquad E_3\nu_{23} = E_2\nu_{32} \tag{3.35}$$

As noted previously, in general, $\nu_{ij} \neq \nu_{ji}$. For example, from (3.35),

$$\nu_{21} = \frac{E_2}{E_1}\nu_{12} \tag{3.36}$$

and

$$\nu_{32} = \frac{E_3}{E_2}\nu_{23} \tag{3.37}$$

Thus, for a transversely isotropic material with $E_1 >> E_2$ and $E_2 = E_3$, ν_{21} is much smaller than ν_{12}, but $\nu_{23} = \nu_{32}$.

Substitution of the definitions (3.33) into (3.32) gives the compliance form of the constitutive equations for an orthotropic material explicitly in terms of engineering constants:

$$\begin{Bmatrix} \varepsilon_1 \\ \varepsilon_2 \\ \varepsilon_3 \\ \gamma_{23} \\ \gamma_{13} \\ \gamma_{12} \end{Bmatrix} = \begin{bmatrix} \frac{1}{E_1} & \frac{-\nu_{21}}{E_2} & \frac{-\nu_{31}}{E_3} & 0 & 0 & 0 \\ \frac{-\nu_{12}}{E_1} & \frac{1}{E_2} & \frac{-\nu_{32}}{E_3} & 0 & 0 & 0 \\ \frac{-\nu_{13}}{E_1} & \frac{-\nu_{23}}{E_2} & \frac{1}{E_3} & 0 & 0 & 0 \\ 0 & 0 & 0 & \frac{1}{G_{23}} & 0 & 0 \\ 0 & 0 & 0 & 0 & \frac{1}{G_{13}} & 0 \\ 0 & 0 & 0 & 0 & 0 & \frac{1}{G_{12}} \end{bmatrix} \begin{Bmatrix} \sigma_1 \\ \sigma_2 \\ \sigma_3 \\ \tau_{23} \\ \tau_{13} \\ \tau_{12} \end{Bmatrix} \tag{3.38}$$

Inversion of (3.38) gives the stiffness form of the constitutive equations explicitly in terms of engineering constants

$$\begin{Bmatrix} \sigma_1 \\ \sigma_2 \\ \sigma_3 \\ \tau_{23} \\ \tau_{13} \\ \tau_{12} \end{Bmatrix} = \begin{bmatrix} \frac{1-\nu_{23}\nu_{32}}{E_2E_3\Delta} & \frac{\nu_{21}+\nu_{23}\nu_{31}}{E_2E_3\Delta} & \frac{\nu_{31}+\nu_{21}\nu_{32}}{E_2E_3\Delta} & 0 & 0 & 0 \\ \frac{\nu_{21}+\nu_{23}\nu_{31}}{E_2E_3\Delta} & \frac{1-\nu_{13}\nu_{31}}{E_1E_3\Delta} & \frac{\nu_{32}+\nu_{12}\nu_{31}}{E_1E_3\Delta} & 0 & 0 & 0 \\ \frac{\nu_{31}+\nu_{21}\nu_{32}}{E_2E_3\Delta} & \frac{\nu_{32}+\nu_{12}\nu_{31}}{E_1E_3\Delta} & \frac{1-\nu_{12}\nu_{21}}{E_1E_2\Delta} & 0 & 0 & 0 \\ 0 & 0 & 0 & G_{23} & 0 & 0 \\ 0 & 0 & 0 & 0 & G_{13} & 0 \\ 0 & 0 & 0 & 0 & 0 & G_{12} \end{bmatrix} \begin{Bmatrix} \varepsilon_1 \\ \varepsilon_2 \\ \varepsilon_3 \\ \gamma_{23} \\ \gamma_{13} \\ \gamma_{12} \end{Bmatrix} \tag{3.39}$$

where

$$\Delta = (1-\nu_{12}\nu_{21} - \nu_{23}\nu_{32} - \nu_{13}\nu_{31} - 2\nu_{21}\nu_{32}\nu_{13})/(E_1E_2E_3) \tag{3.40}$$

is the determinant of the matrix in (3.38).

For a transversely isotropic material with $E_2 = E_3$, $\nu_{12} = \nu_{13}$, $G_{12} = G_{13}$, and $G_{23} = E_2/2(1 + \nu_{23})$ and using the E-ν relations of (3.34), (3.38) simplifies to

$$\begin{Bmatrix} \varepsilon_1 \\ \varepsilon_2 \\ \varepsilon_3 \\ \gamma_{23} \\ \gamma_{13} \\ \gamma_{12} \end{Bmatrix} = \begin{bmatrix} \frac{1}{E_1} & \frac{-\nu_{12}}{E_1} & \frac{-\nu_{12}}{E_1} & 0 & 0 & 0 \\ \frac{-\nu_{12}}{E_1} & \frac{1}{E_2} & \frac{-\nu_{23}}{E_2} & 0 & 0 & 0 \\ \frac{-\nu_{12}}{E_1} & \frac{-\nu_{23}}{E_2} & \frac{1}{E_2} & 0 & 0 & 0 \\ 0 & 0 & 0 & \frac{2(1+\nu_{23})}{E_2} & 0 & 0 \\ 0 & 0 & 0 & 0 & \frac{1}{G_{12}} & 0 \\ 0 & 0 & 0 & 0 & 0 & \frac{1}{G_{12}} \end{bmatrix} \begin{Bmatrix} \sigma_1 \\ \sigma_2 \\ \sigma_3 \\ \tau_{23} \\ \tau_{13} \\ \tau_{12} \end{Bmatrix} \tag{3.41}$$

Example 3.1 Stiffness and Compliance for Transversely Isotropic Material

Determine the compliance and stiffness coefficients for T300/5208 carbon/epoxy considering it to be a transversely isotropic material.

Solution

From Table 1.3, the material properties in material principal coordinates (in U.S. customary units) are: $E_1 = 19.2$ Msi; $E_2 = E_3 = 1.56$ Msi; $\nu_{12} = \nu_{13} = 0.24$; $\nu_{23} = 0.59$; $G_{12} = G_{13} = 0.82$ Msi; $G_{23} = 0.49$ Msi

From (3.33), the compliance coefficients are (in 1/Msi):

$$S_{11} = \frac{1}{E_1} = 0.05208 \qquad S_{12} = S_{13} = -\frac{\nu_{21}}{E_2} = -0.0125$$

$$S_{22} = S_{33} = \frac{1}{E_2} = 0.641 \qquad S_{23} = -\frac{\nu_{32}}{E_3} = -0.3782$$

$$S_{44} = \frac{1}{G_{23}} = 2.038 \qquad S_{55} = S_{66} = \frac{1}{G_{13}} = 1.220$$

The corresponding stiffness coefficients are, by inversion (in Msi),

$$C_{11} = 19.65 \qquad C_{12} = C_{13} = 0.9345$$

$$C_{22} = C_{33} = 2.437 \qquad C_{23} = 1.456$$

$$C_{44} = 0.4906 \qquad C_{55} = C_{66} = 0.820$$

3.5 Coordinate Transformations

As noted in Chapter 2, stress and strain are second-order tensors whereas stiffness and compliance are fourth-order tensors. Therefore, these quantities obey the tensor transformation laws which relate the components in one coordinate system to those in a rotated coordinate system. In the following development we use matrix notation to develop expressions for the transformed stiffness

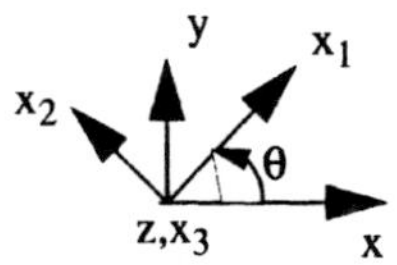

FIGURE 3.8 Transformation about an Axis

and compliance. Stress and strain matrices in the principal material (1-2-3) coordinate system are denoted $\{\ \}_1$ and quantities in the global x-y-z coordinate system are denoted $\{\ \}_x$. We consider rotations through an angle θ about the common z- (3) axis as indicated in Fig. 3.8. The angle θ is measured positive counterclockwise from the x- (unprimed) axis to the x_1- (primed) axis as indicated in the figure.

The appropriate direction cosines for this transformation are all expressed in terms of the angle θ in Table 3.1.

As an example, transformation of a vector $V(V_x, V_y, V_z)$ in the global x-y-z coordinates about the x_3 axis to the 1-2-3 coordinates is, from (2.4),

$$\begin{aligned} V_1 &= a_{11}V_x + a_{21}V_y + a_{31}V_z \\ V_2 &= a_{12}V_x + a_{22}V_y + a_{32}V_z \\ V_3 &= a_{13}V_x + a_{23}V_y + a_{33}V_z \end{aligned} \tag{3.42}$$

Using the values in Table 3.1, we have

$$\begin{aligned} V_1 &= (\cos\theta)V_x + (\sin\theta)V_y \\ V_2 &= (-\sin\theta)V_x + \cos\theta(V_y) \\ V_3 &= V_z \end{aligned} \tag{3.43}$$

3.5.1 Stress and Strain Transformation about an Axis

Stress and strain are second-order tensors and thus both follow the same transformation law (2.11). However, in transforming strain according to the tensor transformation law, tensor shear strains must be employed. The use of tensor shear strains results in unsymmetric transformed stiffness and compliance matrices and thus is undesirable. This problem can be overcome through the use of engineering shear strains, with the result that the transformed stiffness and compliance matrices are symmetric. (It is left as a homework exercise to show that in the final analysis the two approaches, i.e., using tensor or engineer shear strains, are equivalent.) In order to transform using engineering shear strain, it is necessary to define a second transformation matrix. Thus we define the transformation matrix $[T_1]$ (the standard second-order tensor transformation matrix) for transformation of stress, and transformation matrix $[T_2]$ for transformation of strain using engineering shear strain.

		To		
		x_1	x_2	x_3
From	x	$a_{11} = \cos\theta$	$a_{12} = -\sin\theta$	$a_{13} = 0$
	y	$a_{21} = \sin\theta$	$a_{22} = \cos\theta$	$a_{23} = 0$
	z	$a_{31} = 0$	$a_{32} = 0$	$a_{33} = 1$

TABLE 3.1 Direction Cosines for Rotation about an Axis

Expanding the stress transformation equations (2.11) (in condensed notation) and using the direction cosines in Table 3.1 gives a relationship between stresses in the principal material and global coordinates:

$$\{\sigma\}_1 = [T_1]\{\sigma\}_x \tag{3.44}$$

where the transformation matrix $[T_1]$ is

$$[T_1] = \begin{bmatrix} m^2 & n^2 & 0 & 0 & 0 & 2mn \\ n^2 & m^2 & 0 & 0 & 0 & -2mn \\ 0 & 0 & 1 & 0 & 0 & 0 \\ 0 & 0 & 0 & m & -n & 0 \\ 0 & 0 & 0 & n & m & 0 \\ -mn & mn & 0 & 0 & 0 & m^2-n^2 \end{bmatrix} \tag{3.45}$$

and we have used $m = \cos\theta$ and $n = \sin\theta$. We note that $[T_1]$ is asymmetric.

Likewise, the strain transformation equations using engineering shear strain are

$$\{\varepsilon\}_1 = [T_2]\{\varepsilon\}_x \tag{3.46}$$

where

$$[T_2] = \begin{bmatrix} m^2 & n^2 & 0 & 0 & 0 & mn \\ n^2 & m^2 & 0 & 0 & 0 & -mn \\ 0 & 0 & 1 & 0 & 0 & 0 \\ 0 & 0 & 0 & m & -n & 0 \\ 0 & 0 & 0 & n & m & 0 \\ -2mn & 2mn & 0 & 0 & 0 & m^2-n^2 \end{bmatrix} \tag{3.47}$$

We note that the transformation matrices $[T_1]$ and $[T_2]$ differ by factors of 2 in two terms.

3.5.2 Stiffness Transformation

An expression for the *transformed stiffness matrix,* $[\bar{C}]$, is determined by recalling the constitutive equation (3.7) in principal material coordinates and employing the stress and strain transformations (3.44) and (3.46).

$$\begin{aligned} \{\sigma\}_1 &= [C]\{\varepsilon\}_1 \\ [T_1]\{\sigma\}_x &= [C][T_2]\{\varepsilon\}_x \end{aligned} \tag{3.48}$$

$$\{\sigma\}_x = [T_1]^{-1}[C][T_2]\{\varepsilon\}_x \tag{3.49}$$

Equation (3.49) can be written

$$\{\sigma\}_x = [\bar{C}]\{\varepsilon\}_x \tag{3.50}$$

where we define the *transformed stiffness matrix* $[\bar{C}]$ as

$$[\bar{C}] \equiv [T_1]^{-1}[C][T_2] \tag{3.51}$$

It is noted here that $[T_i(\theta)]^{-1} = [T_i(-\theta)]$ $(i = 1, 2)$.

For transformation through an arbitrary angle θ about the 3-axis, the transformed stiffness matrix has the form of a monoclinic material (3.17), i.e.,

$$[\bar{C}] = \begin{bmatrix} \bar{C}_{11} & \bar{C}_{12} & \bar{C}_{13} & 0 & 0 & \bar{C}_{16} \\ \bar{C}_{12} & \bar{C}_{22} & \bar{C}_{23} & 0 & 0 & \bar{C}_{26} \\ \bar{C}_{13} & \bar{C}_{23} & \bar{C}_{33} & 0 & 0 & \bar{C}_{36} \\ 0 & 0 & 0 & \bar{C}_{44} & \bar{C}_{45} & 0 \\ 0 & 0 & 0 & \bar{C}_{45} & \bar{C}_{55} & 0 \\ \bar{C}_{16} & \bar{C}_{26} & \bar{C}_{36} & 0 & 0 & \bar{C}_{66} \end{bmatrix} \tag{3.52}$$

The individual $\bar{C}_{ij}$ terms of this matrix are determined from (3.45), (3.47), and (3.51) to be

$$\bar{C}_{11} = m^4C_{11} + 2m^2n^2(C_{12} + 2C_{66}) + n^4C_{22} \tag{3.53}$$

$$\bar{C}_{12} = n^2m^2(C_{11} + C_{22} - 4C_{66}) + (n^4 + m^4)C_{12} \tag{3.54}$$

$$\bar{C}_{13} = m^2C_{13} + n^2C_{23} \tag{3.55}$$

$$\bar{C}_{16} = nm[m^2(C_{11} - C_{12} - 2C_{66}) + n^2(C_{12} - C_{22} + 2C_{66})] \tag{3.56}$$

$$\bar{C}_{22} = n^4C_{11} + 2m^2n^2(C_{12} + 2C_{66}) + m^4C_{22} \tag{3.57}$$

$$\bar{C}_{23} = n^2C_{13} + m^2C_{23} \tag{3.58}$$

$$\bar{C}_{26} = nm[n^2(C_{11} - C_{12} - 2C_{66}) + m^2(C_{12} - C_{22} + 2C_{66})] \tag{3.59}$$

$$\bar{C}_{33} = C_{33} \tag{3.60}$$

$$\bar{C}_{36} = mn(C_{13} - C_{23}) \tag{3.61}$$

$$\bar{C}_{44} = m^2C_{44} + n^2C_{55} \tag{3.62}$$

$$\bar{C}_{45} = mn(C_{55} - C_{44}) \tag{3.63}$$

$$\bar{C}_{55} = n^2C_{44} + m^2C_{55} \tag{3.64}$$

$$\bar{C}_{66} = n^2m^2(C_{11} - 2C_{12} + C_{22}) + (n^2 - m^2)^2C_{66} \tag{3.65}$$

We see from these equations that the transformed quantities are fourth order in the sine and cosine functions. Thus it is very important to use high precision when calculating these values for very large and very small angles. We also note that $\bar{C}_{16}$, $\bar{C}_{26}$, $\bar{C}_{36}$, and $\bar{C}_{45}$ are identically zero for $\theta = 0°$ or $90°$ because of the common multiplying factor mn in these terms.

3.5.3 Compliance Transformation

Inverting the stiffness form of the constitutive equation (3.50) give the compliance form

$$\{\varepsilon\}_x = [\bar{S}]\{\sigma\}_x \tag{3.66}$$

where the compliance $[\bar{S}]$ is the inverse of stiffness and

$$\begin{aligned}
[\bar{S}] &= [\bar{C}]^{-1} = ([T_1]^{-1}[C][T_2])^{-1} \\
[\bar{S}] &= [T_2]^{-1}[C]^{-1}([T_1]^{-1})^{-1} \\
[\bar{S}] &= [T_2]^{-1}[S][T_1]
\end{aligned} \tag{3.67}$$

The transformed compliance matrix then has the same symmetric form as the transformed stiffness matrix:

$$[\bar{S}] = \begin{bmatrix} \bar{S}_{11} & \bar{S}_{12} & \bar{S}_{13} & 0 & 0 & \bar{S}_{16} \\ \bar{S}_{12} & \bar{S}_{22} & \bar{S}_{23} & 0 & 0 & \bar{S}_{26} \\ \bar{S}_{13} & \bar{S}_{23} & \bar{S}_{33} & 0 & 0 & \bar{S}_{36} \\ 0 & 0 & 0 & \bar{S}_{44} & \bar{S}_{45} & 0 \\ 0 & 0 & 0 & \bar{S}_{45} & \bar{S}_{55} & 0 \\ \bar{S}_{16} & \bar{S}_{26} & \bar{S}_{36} & 0 & 0 & \bar{S}_{66} \end{bmatrix} \tag{3.68}$$

The individual terms of the transformed compliance matrix exhibit forms very similar to those of the stiffness matrix, but with small differences (factors of 2) because of the differences between engineering and tensor shear strains. In expanded form, the equations are

$$\bar{S}_{11} = m^4 S_{11} + m^2 n^2 (2S_{12} + S_{66}) + n^4 S_{22} \tag{3.69}$$

$$\bar{S}_{12} = n^2 m^2 (S_{11} + S_{22} - S_{66}) + (n^4 + m^4) S_{12} \tag{3.70}$$

$$\bar{S}_{13} = m^2 S_{13} + n^2 S_{23} \tag{3.71}$$

$$\bar{S}_{16} = nm[m^2(2S_{11} - 2S_{12} - S_{66}) + n^2(2S_{12} - 2S_{22} + S_{66})] \tag{3.72}$$

$$\bar{S}_{22} = n^4 S_{11} + m^2 n^2 (2S_{12} + S_{66}) + m^4 S_{22} \tag{3.73}$$

$$\bar{S}_{23} = n^2 S_{13} + m^2 S_{23} \tag{3.74}$$

$$\bar{S}_{26} = nm[n^2(2S_{11} - 2S_{12} - S_{66}) + m^2(2S_{12} - 2S_{22} + S_{66})] \tag{3.75}$$

$$\bar{S}_{33} = S_{33} \tag{3.76}$$

$$\bar{S}_{36} = 2mn(S_{13} - S_{23}) \tag{3.77}$$

$$\bar{S}_{44} = m^2 S_{44} + n^2 S_{55} \tag{3.78}$$

$$\bar{S}_{45} = mn(S_{55} - S_{44}) \tag{3.79}$$

$$\bar{S}_{55} = n^2 S_{44} + m^2 S_{55} \tag{3.80}$$

$$\bar{S}_{66} = 4n^2 m^2 (S_{11} - 2S_{12} + S_{22}) + (n^2 - m^2)^2 S_{66} \tag{3.81}$$

It is noted that compliance transformation is fourth order in the sine and cosine of the fiber angle, as was the stiffness transformation.

The constitutive equation (3.66) can now be written in full as

$$\begin{Bmatrix} \varepsilon_x \\ \varepsilon_y \\ \varepsilon_z \\ \gamma_{yz} \\ \gamma_{zx} \\ \gamma_{xy} \end{Bmatrix} = \begin{bmatrix} \bar{S}_{11} & \bar{S}_{12} & \bar{S}_{13} & 0 & 0 & \bar{S}_{16} \\ \bar{S}_{12} & \bar{S}_{22} & \bar{S}_{23} & 0 & 0 & \bar{S}_{26} \\ \bar{S}_{13} & \bar{S}_{23} & \bar{S}_{33} & 0 & 0 & \bar{S}_{36} \\ 0 & 0 & 0 & \bar{S}_{44} & \bar{S}_{45} & 0 \\ 0 & 0 & 0 & \bar{S}_{45} & \bar{S}_{55} & 0 \\ \bar{S}_{16} & \bar{S}_{26} & \bar{S}_{36} & 0 & 0 & \bar{S}_{66} \end{bmatrix} \begin{Bmatrix} \sigma_x \\ \sigma_y \\ \sigma_z \\ \tau_{yz} \\ \tau_{zx} \\ \tau_{xy} \end{Bmatrix} \tag{3.82}$$

3.5.4 Transversely Isotropic Material

The preceding equations for transformation of stiffness and compliance are given for an orthotropic material. In the special case of a transversely isotropic material (Section 3.3.3), the equations can be simplified by using the equalities

$$C_{22} = C_{33}; \quad C_{12} = C_{13}; \quad C_{55} = C_{66}; \quad C_{44} = (C_{22} - C_{23})/2 \tag{3.83}$$

in the stiffness relations and

$$S_{22} = S_{33}; \quad S_{12} = S_{13}; \quad S_{55} = S_{66}; \quad S_{44} = 2(S_{22} - S_{23}) \tag{3.84}$$

in the compliance relations.

3.5.5 θ Dependence of Transformed Stiffness

The θ dependence of the transformed stiffness coefficients is shown in Figs. 3.9–3.11 for a transversely isotropic material (T300/5208 carbon/epoxy, Table 1.3). (Note that $\bar{C}_{33}$ is independent of θ and $\bar{C}_{66} = \bar{C}_{55}$.) As indicated by equations (3.53)–(3.65), all diagonal terms are expressed in even powers of the trigonometric functions. As a result, the diagonal terms are positive, as indicated in Fig. 3.9. The normal stiffnesses $\bar{C}_{11}$ and $\bar{C}_{22}$ are out of phase by 90° but otherwise identical. In contrast, the shear stiffnesses $\bar{C}_{44}$ and $\bar{C}_{66}$ exhibit different θ dependence. The in-plane shear stiffness $\bar{C}_{66}$ exhibits a maximum at 45° and minima at 0 and 90°; the out-of-plane shear stiffness $\bar{C}_{44}$ is much lower in magnitude, varies little over the full range of angles, and exhibits a maximum at 90° and a minimum at 0°. This figure clearly shows that the maximum in-plane normal stiffness can be several times the maximum shear stiffness, and that the in-plane shear stiffness is maximum at ±45°.

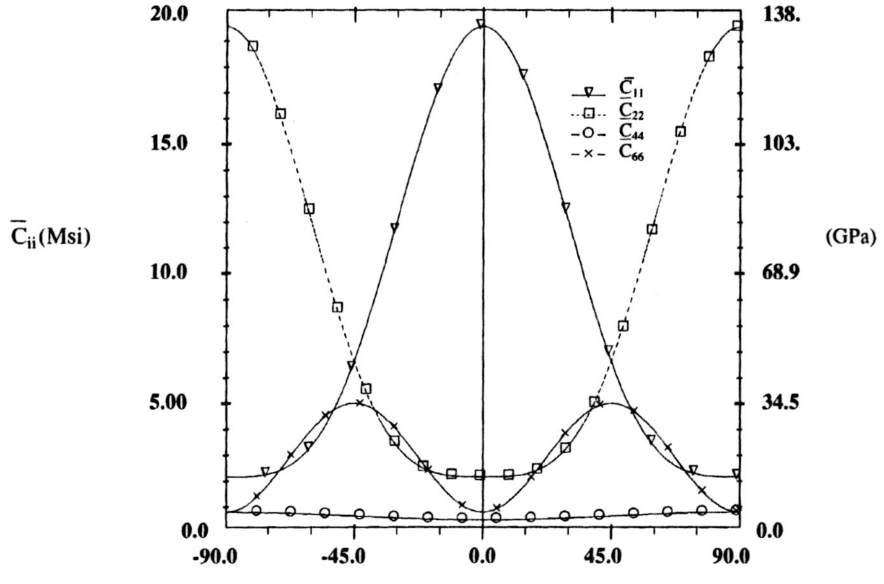

FIGURE 3.9 Diagonal Stiffness Terms: T300/5208

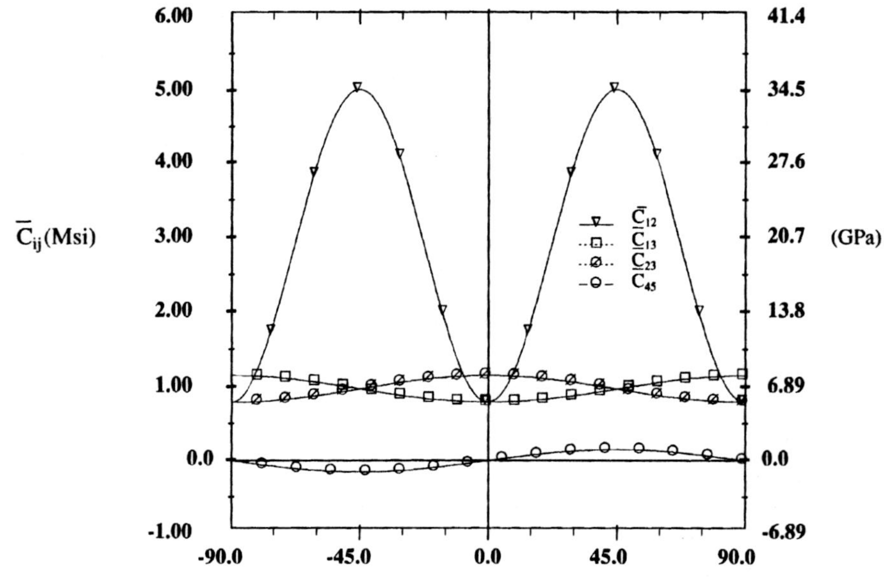

FIGURE 3.10 $\bar{C}_{12}$, $\bar{C}_{13}$, $\bar{C}_{23}$, and $\bar{C}_{45}$ Stiffness Terms: T300/5208

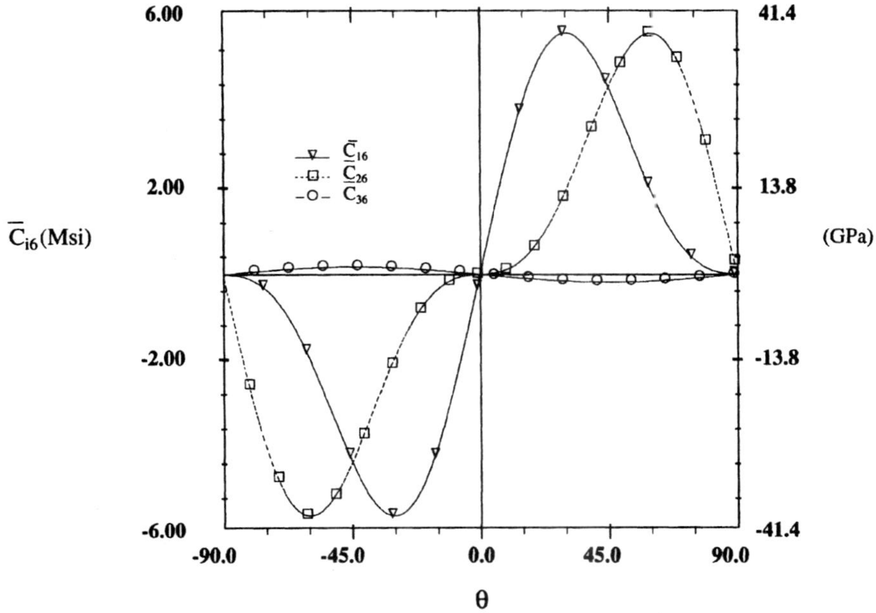

FIGURE 3.11 $\bar{C}_{16}$, $\bar{C}_{26}$, and $\bar{C}_{36}$ Stiffness Terms: T300/5208

The θ dependence of off-diagonal terms $\bar{C}_{12}$, $\bar{C}_{13}$, $\bar{C}_{23}$, and $\bar{C}_{45}$ is shown in Fig. 3.10. All terms except $\bar{C}_{45}$ are positive. $\bar{C}_{12}$ is the largest of the four terms, with maximum values at ±45°—five (or more) times greater than those for the remaining terms. The θ dependence of the $\bar{C}_{i6}$ (i = 1, 2, 3) terms is shown in Fig. 3.11. Here we see that the in-plane $\bar{C}_{16}$ and $\bar{C}_{26}$ are many times larger than the out-of-plane $\bar{C}_{36}$. All three terms are odd functions of θ with maximum and minimum values at differing angles.

Equations for the θ dependence of the compliance $\bar{S}_{ij}$ are given in (3.69) to (3.81). The results of such transformations are shown in Figs. 3.12–3.14. Since the compliance is the inverse of the stiffness, the curves follow accordingly.

3.6 Summary of 3-D Constitutive Equations

Expressions have been developed for the three-dimensional constitutive equations in terms of the stiffness matrix for a variety of materials ranging from fully anisotropic to isotropic. Also, it has been shown that the compliance and stiffness matrices are the inverse of one another. Finally, the compliance and stiffness coefficients for an orthotropic material have been expressed in terms of the engineering constants measured in the laboratory.

Equation (3.50) is the set of fundamental stress-strain equations describing the elastic response of a monoclinic material in terms of the stiffness matrix $[\bar{C}]$. The equations can be written explicitly

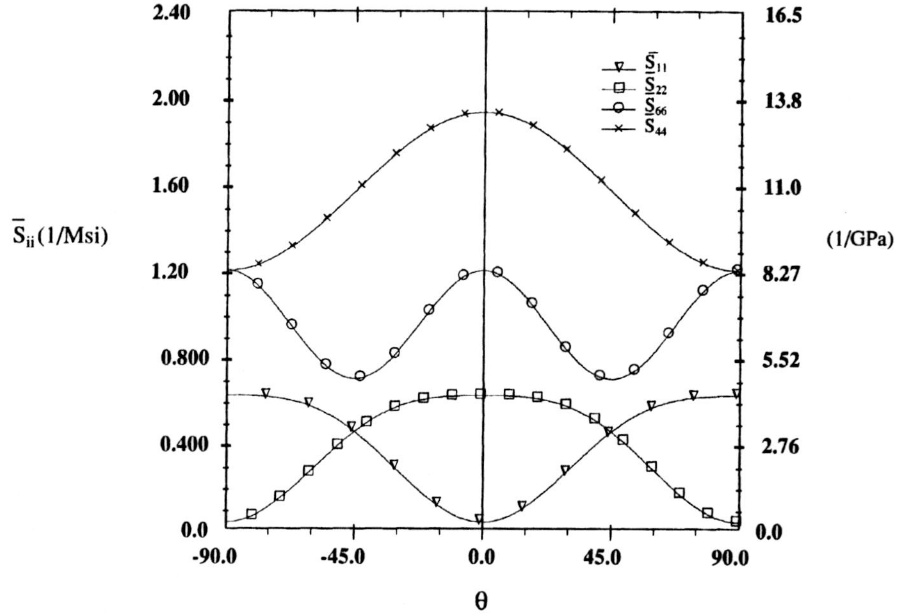

FIGURE 3.12 Diagonal Compliance Terms: T300/5208

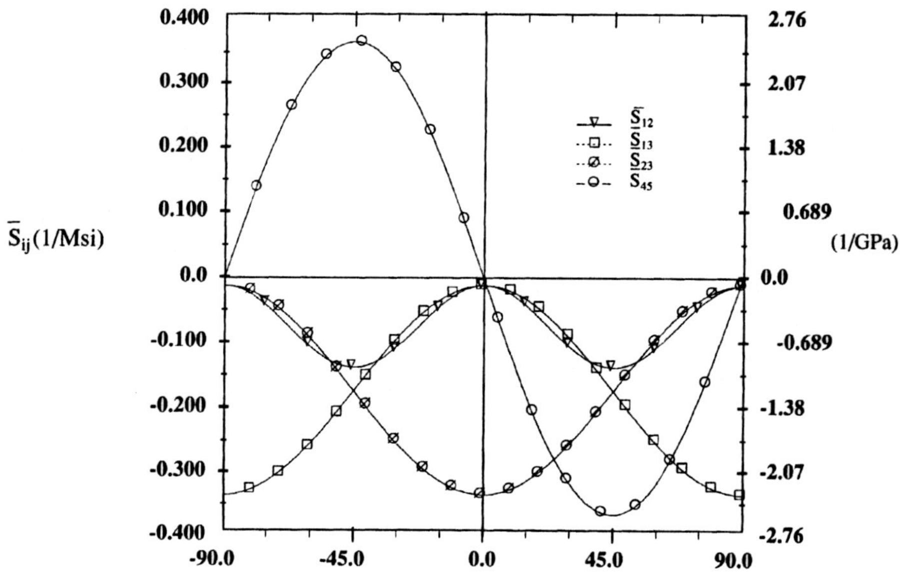

FIGURE 3.13 S_{12}, S_{13}, S_{23}, and S_{45} Compliance Terms: T300/5208

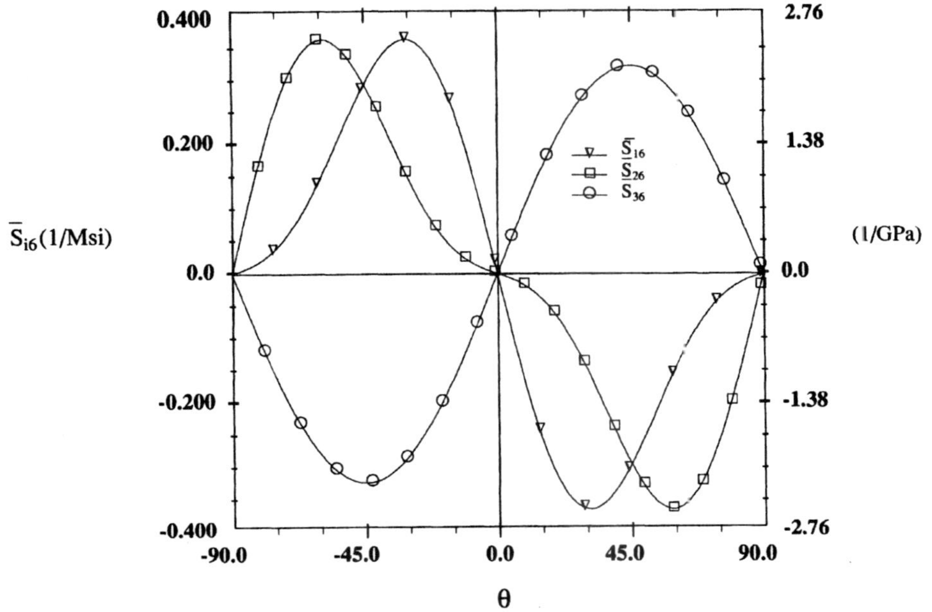

FIGURE 3.14 $\bar{S}_{16}$, $\bar{S}_{26}$, and $\bar{S}_{36}$ Compliance Terms: T300/5208

in expanded form as

$$\begin{aligned}
\sigma_x &= \bar{C}_{11}\varepsilon_x + \bar{C}_{12}\varepsilon_y + \bar{C}_{13}\varepsilon_z + \bar{C}_{16}\gamma_{xy} \\
\sigma_y &= \bar{C}_{12}\varepsilon_x + \bar{C}_{22}\varepsilon_y + \bar{C}_{23}\varepsilon_z + \bar{C}_{26}\gamma_{xy} \\
\sigma_z &= \bar{C}_{13}\varepsilon_x + \bar{C}_{23}\varepsilon_y + \bar{C}_{33}\varepsilon_z + \bar{C}_{36}\gamma_{xy} \\
\tau_{yz} &= \bar{C}_{44}\gamma_{yz} + \bar{C}_{45}\gamma_{zx} \\
\tau_{zx} &= \bar{C}_{45}\gamma_{yz} + \bar{C}_{55}\gamma_{zx} \\
\tau_{xy} &= \bar{C}_{16}\varepsilon_x + \bar{C}_{26}\varepsilon_y + \bar{C}_{36}\varepsilon_z + \bar{C}_{66}\gamma_{xy}
\end{aligned} \tag{3.85}$$

Likewise, the compliance form of the constitutive equation (3.82) can be expanded and written in the explicit form

$$\begin{aligned}
\varepsilon_x &= \bar{S}_{11}\sigma_x + \bar{S}_{12}\sigma_y + \bar{S}_{13}\sigma_z + \bar{S}_{16}\tau_{xy} \\
\varepsilon_y &= \bar{S}_{12}\sigma_x + \bar{S}_{22}\sigma_y + \bar{S}_{23}\sigma_z + \bar{S}_{26}\tau_{xy} \\
\varepsilon_z &= \bar{S}_{13}\sigma_x + \bar{S}_{23}\sigma_y + \bar{S}_{33}\sigma_z + \bar{S}_{36}\tau_{xy} \\
\gamma_{yz} &= \bar{S}_{44}\tau_{yz} + \bar{S}_{45}\tau_{zx} \\
\gamma_{zx} &= \bar{S}_{45}\tau_{yz} + \bar{S}_{55}\tau_{zx} \\
\gamma_{xy} &= \bar{S}_{16}\sigma_x + \bar{S}_{26}\sigma_y + \bar{S}_{36}\sigma_z + \bar{S}_{66}\tau_{xy}
\end{aligned} \tag{3.86}$$

The constitutive equations (3.85) and (3.86) will be referred to extensively throughout the remainder of this book. Written in matrix notation, they have the form

$$\begin{Bmatrix} \sigma_x \\ \sigma_y \\ \sigma_z \\ \tau_{yz} \\ \tau_{zx} \\ \tau_{xy} \end{Bmatrix} = \begin{bmatrix} \bar{C}_{11} & \bar{C}_{12} & \bar{C}_{13} & 0 & 0 & \bar{C}_{16} \\ \bar{C}_{12} & \bar{C}_{22} & \bar{C}_{23} & 0 & 0 & \bar{C}_{26} \\ \bar{C}_{13} & \bar{C}_{23} & \bar{C}_{33} & 0 & 0 & \bar{C}_{36} \\ 0 & 0 & 0 & \bar{C}_{44} & \bar{C}_{45} & 0 \\ 0 & 0 & 0 & \bar{C}_{45} & \bar{C}_{55} & 0 \\ \bar{C}_{16} & \bar{C}_{26} & \bar{C}_{36} & 0 & 0 & \bar{C}_{66} \end{bmatrix} \begin{Bmatrix} \varepsilon_x \\ \varepsilon_y \\ \varepsilon_z \\ \gamma_{yz} \\ \gamma_{zx} \\ \gamma_{xy} \end{Bmatrix} \tag{3.87}$$

$$\begin{Bmatrix} \varepsilon_x \\ \varepsilon_y \\ \varepsilon_z \\ \gamma_{yz} \\ \gamma_{zx} \\ \gamma_{xy} \end{Bmatrix} = \begin{bmatrix} \bar{S}_{11} & \bar{S}_{12} & \bar{S}_{13} & 0 & 0 & \bar{S}_{16} \\ \bar{S}_{12} & \bar{S}_{22} & \bar{S}_{23} & 0 & 0 & \bar{S}_{26} \\ \bar{S}_{13} & \bar{S}_{23} & \bar{S}_{33} & 0 & 0 & \bar{S}_{36} \\ 0 & 0 & 0 & \bar{S}_{44} & \bar{S}_{45} & 0 \\ 0 & 0 & 0 & \bar{S}_{45} & \bar{S}_{55} & 0 \\ \bar{S}_{16} & \bar{S}_{26} & \bar{S}_{36} & 0 & 0 & \bar{S}_{66} \end{bmatrix} \begin{Bmatrix} \sigma_x \\ \sigma_y \\ \sigma_z \\ \tau_{yz} \\ \tau_{zx} \\ \tau_{xy} \end{Bmatrix} \tag{3.88}$$

The presence of nonzero 16, 26, 36, and 45 terms in the monoclinic constitutive equations (3.85)–(3.88) shows that for such materials there is normal-shear coupling (e.g., normal strain is a function of shear stress and vice versa) and shear-shear coupling for the out-of-plane shears. Such coupling is not present in isotropic, transversely isotropic, or orthotropic materials when viewed in the principal material coordinates. The presence of coupling can have far-reaching consequences in the application of fibrous composites.

3.7 Material Dependence of Transformed Stiffness

The θ dependence of the diagonal stiffness terms $\bar{C}_{ij}$ $(i = j)$ is shown in Fig. 3.9 for a carbon/epoxy composite. In order to demonstrate the strong influence that the degree of material orthotropy has on the stiffness coefficients, the equivalent to Fig. 3.9 for a metal matrix composite (SCS-6/Ti-15-3, Table 1.3) is shown in Fig. 3.15. Comparison of these two figures clearly shows that the metal matrix composite exhibits significantly higher stiffness values for all θ and a much smaller range as a function of θ for all diagonal terms in the stiffness matrix. The higher values for the diagonal stiffness terms of a metal matrix composite are a direct consequence of their higher Young's moduli and shear modulus (Table 1.3). The smaller range of values as a function of θ is due to the smaller degree of orthotropy of the metal matrix composite ($E_1/E_2 = 1.5$) as compared with the orthotropy of the polymeric matrix composite ($E_1/E_2 = 12.6$). It is most interesting that $\bar{C}_{44}$ is independent of θ. This is a direct consequence of the fact that G_{23} and G_{13} are nearly identical for SCS-6/Ti-15-3 and the relation $\sin^2\theta + \cos^2\theta = 1$ in (3.62).

Likewise, the metal matrix equivalent to Fig. 3.11, which shows the θ dependence for the $\bar{C}_{i6}$ $(i = 1, 2, 3)$ stiffness terms of a polymeric matrix composite, is shown in Fig. 3.16 for SCS-6/Ti-15-3. Comparison of Figs. 3.11 and 3.16 shows that the in-plane coefficients $\bar{C}_{16}$ and $\bar{C}_{26}$ are significantly *smaller* for the metal matrix composite, but the out-of-plane coefficient $\bar{C}_{36}$

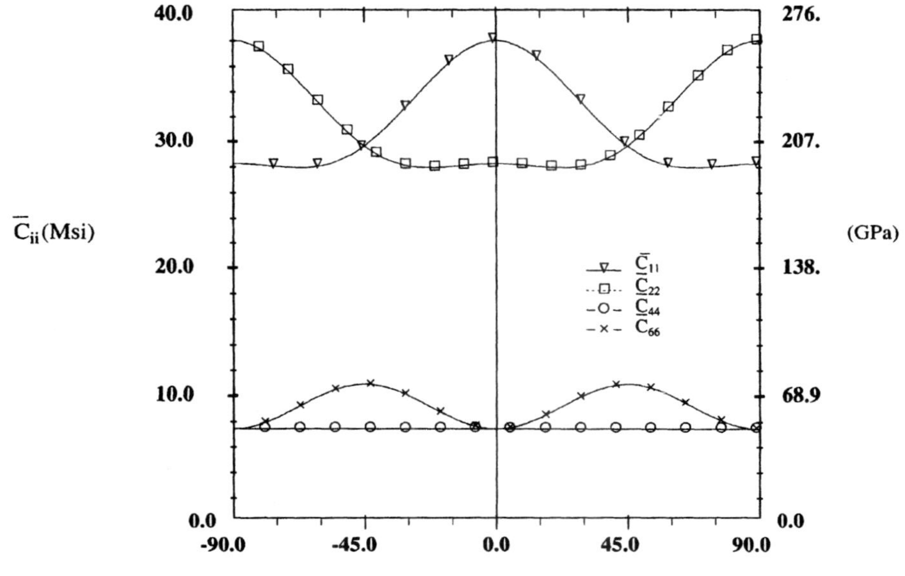

FIGURE 3.15 θ Dependence of Diagonal Stiffness Terms: SCS-6/Ti-15-3

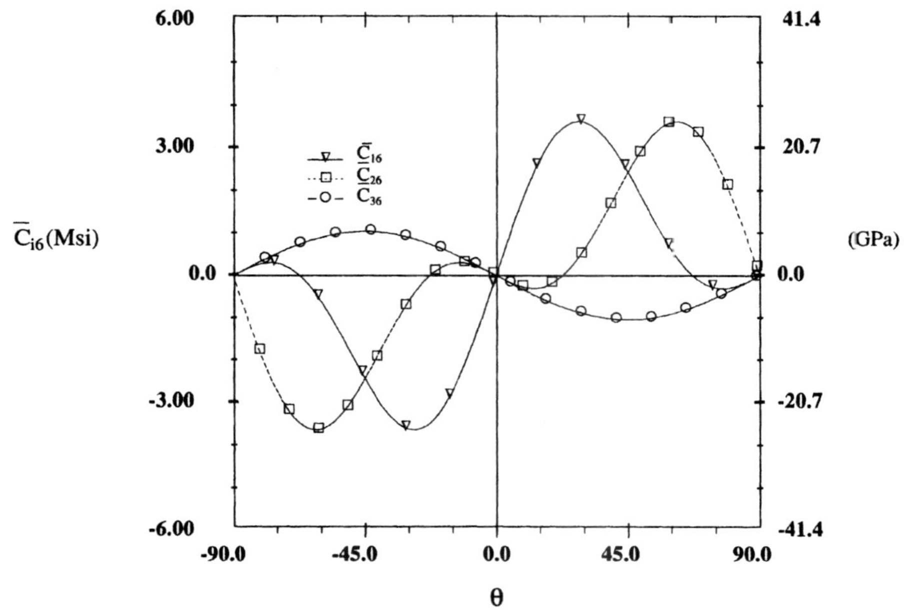

FIGURE 3.16 θ Dependence of $\bar{C}_{16}$, $\bar{C}_{26}$, and $\bar{C}_{36}$ Compliance: SCS-6/Ti-15-3

is significantly *larger* for the metal matrix composite. The fact that C_{36} is significantly *larger* for the metal matrix composite indicates that there is much stronger coupling between the out-of-plane normal stress σ_z and the in-plane shear strain γ_{xy} for a metal matrix composite than there is for a polymeric matrix composite, (3.87).

3.8 Thermal Effects

Thermal stresses play a very important role in the study of composite materials for a variety of reasons. On the one hand, composite materials have significant residual thermal stresses from the fabrication process. A second major reason for considering thermal effects in composites is that these materials often behave well at elevated temperature. The constituents can be chosen such that composites retain their properties at elevated temperature; they can also be chosen such that the coefficient of thermal expansion is a specified value, possibly zero. We will first develop the equations for thermal stress analysis of an orthotropic material in principal material coordinates and then show the transformation to arbitrary coordinates.

3.8.1 Thermal Strains and Coefficient of Thermal Expansion

When materials are exposed to a temperature change they exhibit strains (called *thermal strains* $\{\varepsilon^T\}$) which are proportional to the *temperature change* ΔT. The constant of proportionality is the *coefficient of thermal expansion* (CTE), denoted by $\{\alpha\}$. The thermal strains in material principal coordinates are then written

$$\{\varepsilon^T\}_1 = \{\alpha\}_1 \Delta T \tag{3.89}$$

These thermal strains are not accompanied by stresses when a homogeneous material undergoes a temperature change in an unconstrained manner. For this reason, thermal strains are often referred to as *free thermal strains*. It is emphasized here that thermal strains are present (for nonzero CTE materials) whenever a material changes temperature. However, when the material is constrained, as are the individual layers in a laminate or the fibers and matrix in a unidirectional composite, the thermal strains cannot develop "freely"; thermal stresses develop.

For a homogeneous, orthotropic material the directionally dependent coefficients of thermal expansion for three-dimensional problems can be expressed in matrix form (in reduced notation) as

$$\{\alpha\}_1 = \begin{Bmatrix} \alpha_1 \\ \alpha_2 \\ \alpha_3 \\ 0 \\ 0 \\ 0 \end{Bmatrix} \tag{3.90}$$

The shear terms in (3.90) are zero because an orthotropic material does not exhibit shear strains (in the principal material coordinates) if its temperature changes. However, thermal shear strains are present if we consider the strains in a global coordinate system that is obtained by a rotation θ about the 3-axis. This fact is demonstrated through consideration of the strain transformation equations (3.46):

$$\{\varepsilon\}_x = [T_2]^{-1}\{\varepsilon\}_1 \tag{3.91}$$

Combining (3.89)–(3.91) gives

$$\{\varepsilon^T\}_x = [T_2]^{-1}\{\alpha\}_1 \Delta T \tag{3.92}$$

From the preceding equation it is apparent that we can write the thermal strains in a global coordinate system as

$$\{\varepsilon^T\}_x = \{\alpha\}_x \Delta T \tag{3.93}$$

where $\{\alpha\}_x$ is

$$\{\alpha\}_x = \left[T_2\right]^{-1}\{\alpha\}_1 = \begin{Bmatrix} \alpha_x \\ \alpha_y \\ \alpha_z \\ 0 \\ 0 \\ \alpha_{xy} \end{Bmatrix} \tag{3.94}$$

The explicit forms for the nonzero coefficients of thermal expansion in a global coordinate system rotated about the common z (3) axis are (using engineering shear strain)

$$\begin{aligned} \alpha_x &= m^2\alpha_1 + n^2\alpha_2 \\ \alpha_y &= n^2\alpha_1 + m^2\alpha_2 \\ \alpha_z &= \alpha_3 \\ \alpha_{xy} &= 2mn(\alpha_1 - \alpha_2) \end{aligned} \tag{3.95}$$

Finally, combining (3.93) through (3.95), we can write the engineering thermal strains in global coordinates in terms of coefficients of thermal expansion in the principal material directions and the angle of rotation as

$$\begin{Bmatrix} \varepsilon_x^T \\ \varepsilon_y^T \\ \varepsilon_z^T \\ \gamma_{yz}^T \\ \gamma_{zx}^T \\ \gamma_{xy}^T \end{Bmatrix} = \begin{Bmatrix} m^2\alpha_1 + n^2\alpha_2 \\ n^2\alpha_1 + m^2\alpha_2 \\ \alpha_3 \\ 0 \\ 0 \\ 2mn(\alpha_1 - \alpha_2) \end{Bmatrix} \Delta T \tag{3.96}$$

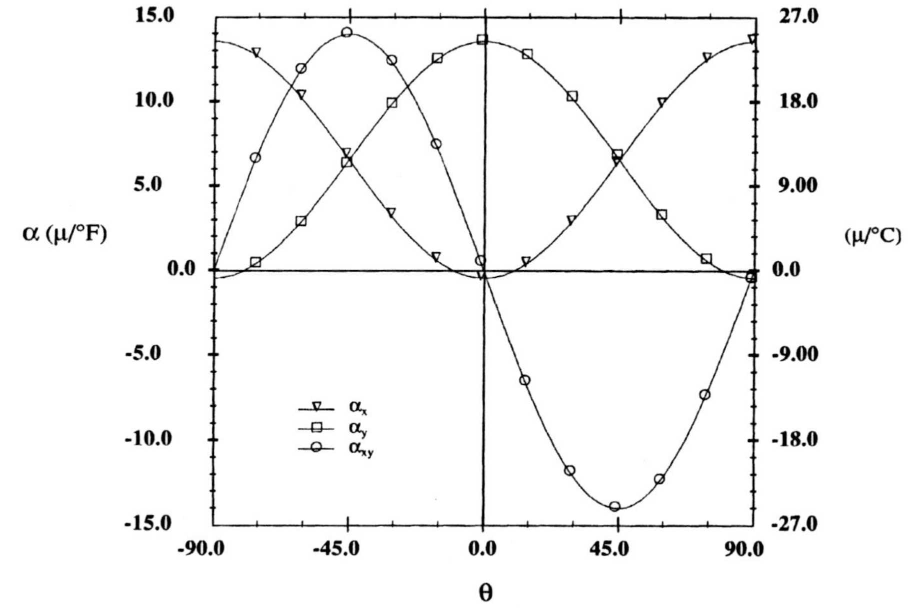

FIGURE 3.17 Coefficient of Thermal Expansion θ Dependence: T300/5208

Several important conclusions can be drawn from the preceding equations and a plot showing the variation of CTE values with fiber orientation. The variations of the coefficients of thermal expansion for T300/5208 carbon/epoxy, over the range –90° to +90°, are shown in Fig. 3.17. Only the in-plane components vary with fiber orientation. The through-thickness coefficient, α_z, retains its principal material coordinate value independent of θ. The in-plane CTE values range from a small negative number for expansion in the fiber direction to a large positive value in the transverse direction. Expansion in the x- and y-directions differs by a 90° phase shift.

For an orthotropic material ($\alpha_1 \neq \alpha_2$) and for fiber orientations other than 0° and 90°, α_{xy} is nonzero. The extreme values of the thermal shear strain occur at fiber orientations of +45° and –45°. Thus, temperature change introduces shear strain in other than principal material coordinates. There is no thermal shear strain in isotropic materials or in the principal material coordinate directions of an orthotropic material.

The mechanics of thermal shear strains in orthotropic materials can be explained through consideration of the thermal expansion of an isotropic material and an orthotropic material as shown in Fig. 3.18 (where it is assumed that $\alpha_2 >> \alpha_1$ in the orthotropic material). Shear strain is evident in the global coordinates of the orthotropic material by the change in the original right angle formed by the diagonals. In contrast, the diagonals remain orthogonal in the isotropic material, and, further, no shear is evident in the principal material coordinate directions of the orthotropic material.

3.8.2 Constitutive Equation

There is one major, fundamental assumption required for thermo-elastic stress analysis: the *total strain*, $\{\varepsilon\}$, is a superposition of the "free" thermal strain, $\{\varepsilon^T\}$, and the strain resulting from the

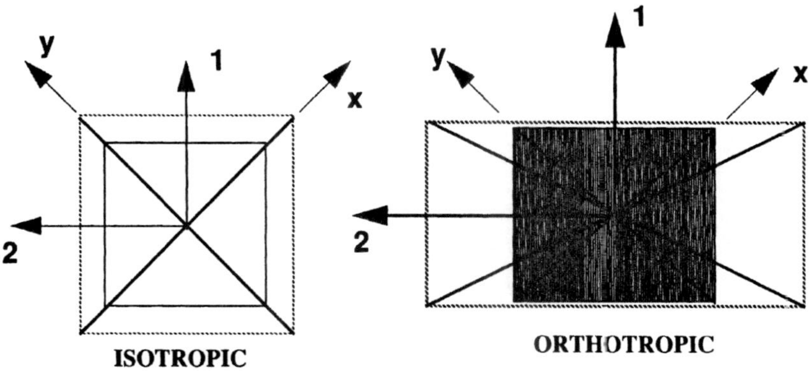

FIGURE 3.18 Thermal Expansion in Isotropic and Orthotropic Materials

stresses (i.e., the *mechanical strains,* denoted $\{\varepsilon^\sigma\}$).

$$\{\varepsilon\} = \{\varepsilon^\sigma\} + \{\varepsilon^T\} \tag{3.97}$$

With the previous definition of thermal strain and the assumption of strain superposition, thermal stress analysis can be carried out in a very straightforward manner. Our goal is to develop an expression for the stresses as a function of thermal and mechanical strains. Recalling the linear elastic constitutive equation, i.e., Hooke's law in matrix notation (3.8), relating the mechanical strains to the stresses,

$$\{\varepsilon^\sigma\} = [S]\{\sigma\} \tag{3.98}$$

Combining (3.98) with the strain superposition equation (3.97) gives

$$\{\varepsilon\} = [S]\{\sigma\} + \{\varepsilon^T\} \tag{3.99}$$

Solving for the stresses (using $[C] = [S]^{-1}$), we have

$$\{\sigma\} = [C](\{\varepsilon\} - \{\varepsilon^T\}) \tag{3.100}$$

Equation (3.100) is the basic constitutive equation for thermo-elastic stress analysis. It states that the stresses are proportional to the difference between the total strains and the free thermal strains. For a global x-y-z coordinate system this fundamental equation can be written in terms of the transformed stresses, strains, and stiffness as

$$\{\sigma\}_x = [\bar{C}](\{\varepsilon\}_x - \{\varepsilon^T\}_x) \tag{3.101}$$

It is noted that the subscripts x in (3.101) are not necessary because the bar on the stiffness matrix clearly identifies the equation as being in the global coordinate system. If the fiber orientation is 0°, the transformed equation remains valid. Thus for the remainder of this text, the subscript x on stress and strain quantities will be omitted unless explicitly required. In expanded matrix form, the thermo-elastic constitutive equation (3.101) is

$$\begin{Bmatrix} \sigma_x \\ \sigma_y \\ \sigma_z \\ \tau_{yz} \\ \tau_{zx} \\ \tau_{xy} \end{Bmatrix} = \begin{bmatrix} \bar{C}_{11} & \bar{C}_{12} & \bar{C}_{13} & 0 & 0 & \bar{C}_{16} \\ \bar{C}_{12} & \bar{C}_{22} & \bar{C}_{23} & 0 & 0 & \bar{C}_{26} \\ \bar{C}_{13} & \bar{C}_{23} & \bar{C}_{33} & 0 & 0 & \bar{C}_{36} \\ 0 & 0 & 0 & \bar{C}_{44} & \bar{C}_{45} & 0 \\ 0 & 0 & 0 & \bar{C}_{45} & \bar{C}_{55} & 0 \\ \bar{C}_{16} & \bar{C}_{26} & \bar{C}_{36} & 0 & 0 & \bar{C}_{66} \end{bmatrix} \begin{Bmatrix} \varepsilon_x - \varepsilon_x^T \\ \varepsilon_y - \varepsilon_y^T \\ \varepsilon_z - \varepsilon_z^T \\ \gamma_{yz} \\ \gamma_{zx} \\ \gamma_{xy} - \gamma_{xy}^T \end{Bmatrix} \tag{3.102}$$

Writing out equations (3.102) explicitly, we have the thermo-elastic form of (3.85):

$$\begin{aligned} \sigma_x &= \bar{C}_{11}(\varepsilon_x - \varepsilon_x^T) + \bar{C}_{12}(\varepsilon_y - \varepsilon_y^T) + \bar{C}_{13}(\varepsilon_z - \varepsilon_z^T) + \bar{C}_{16}(\gamma_{xy} - \gamma_{xy}^T) \\ \sigma_y &= \bar{C}_{12}(\varepsilon_x - \varepsilon_x^T) + \bar{C}_{22}(\varepsilon_y - \varepsilon_y^T) + \bar{C}_{23}(\varepsilon_z - \varepsilon_z^T) + \bar{C}_{26}(\gamma_{xy} - \gamma_{xy}^T) \\ \sigma_z &= \bar{C}_{13}(\varepsilon_x - \varepsilon_x^T) + \bar{C}_{23}(\varepsilon_y - \varepsilon_y^T) + \bar{C}_{33}(\varepsilon_z - \varepsilon_z^T) + \bar{C}_{36}(\gamma_{xy} - \gamma_{xy}^T) \\ \tau_{yz} &= \bar{C}_{44}\gamma_{yz} + \bar{C}_{45}\gamma_{zx} \\ \tau_{zx} &= \bar{C}_{45}\gamma_{yz} + \bar{C}_{55}\gamma_{zx} \\ \tau_{xy} &= \bar{C}_{16}(\varepsilon_x - \varepsilon_x^T) + \bar{C}_{26}(\varepsilon_y - \varepsilon_y^T) + \bar{C}_{36}(\varepsilon_z - \varepsilon_z^T) + \bar{C}_{66}(\gamma_{xy} - \gamma_{xy}^T) \end{aligned} \tag{3.103}$$

Equations (3.101)–(3.103) can, of course, be inverted to provide the total strains in terms of the mechanical (stress-induced) and free thermal strains. The results are

$$\{\varepsilon\}_x = [\bar{S}]\{\sigma\}_x + \{\varepsilon^T\}_x \tag{3.104}$$

$$\begin{Bmatrix} \varepsilon_x \\ \varepsilon_y \\ \varepsilon_z \\ \gamma_{yz} \\ \gamma_{zx} \\ \gamma_{xy} \end{Bmatrix} = \begin{bmatrix} \bar{S}_{11} & \bar{S}_{12} & \bar{S}_{13} & 0 & 0 & \bar{S}_{16} \\ \bar{S}_{12} & \bar{S}_{22} & \bar{S}_{23} & 0 & 0 & \bar{S}_{26} \\ \bar{S}_{13} & \bar{S}_{23} & \bar{S}_{33} & 0 & 0 & \bar{S}_{36} \\ 0 & 0 & 0 & \bar{S}_{44} & \bar{S}_{45} & 0 \\ 0 & 0 & 0 & \bar{S}_{45} & \bar{S}_{55} & 0 \\ \bar{S}_{16} & \bar{S}_{26} & \bar{S}_{36} & 0 & 0 & \bar{S}_{66} \end{bmatrix} \begin{Bmatrix} \sigma_x \\ \sigma_y \\ \sigma_z \\ \tau_{yz} \\ \tau_{zx} \\ \tau_{xy} \end{Bmatrix} + \begin{Bmatrix} \varepsilon_x^T \\ \varepsilon_y^T \\ \varepsilon_z^T \\ 0 \\ 0 \\ \gamma_{xy}^T \end{Bmatrix} \tag{3.105}$$

$$\begin{aligned} \varepsilon_x &= \bar{S}_{11}\sigma_x + \bar{S}_{12}\sigma_y + \bar{S}_{13}\sigma_z + \bar{S}_{16}\tau_{xy} + \varepsilon_x^T \\ \varepsilon_y &= \bar{S}_{12}\sigma_x + \bar{S}_{22}\sigma_y + \bar{S}_{23}\sigma_z + \bar{S}_{26}\tau_{xy} + \varepsilon_y^T \\ \varepsilon_z &= \bar{S}_{13}\sigma_x + \bar{S}_{23}\sigma_y + \bar{S}_{33}\sigma_z + \bar{S}_{36}\tau_{xy} + \varepsilon_z^T \\ \gamma_{yz} &= \bar{S}_{44}\tau_{yz} + \bar{S}_{45}\tau_{zx} \\ \gamma_{zx} &= \bar{S}_{45}\tau_{yz} + \bar{S}_{55}\tau_{zx} \\ \gamma_{xy} &= \bar{S}_{16}\sigma_x + \bar{S}_{26}\sigma_y + \bar{S}_{36}\sigma_z + \bar{S}_{66}\tau_{xy} + \gamma_{xy}^T \end{aligned} \tag{3.106}$$

3.9 Moisture Effects

3.9.1 Hygroscopic Expansion

Polymer matrix composites, such as carbon/epoxy, absorb moisture, which results in hygroscopic (moisture) expansion. The hygroscopic expansion problem is treated exactly in the same manner as is thermal expansion. Hygroscopic strains $\{\varepsilon^H\}$ are assumed to be proportional to the *percent moisture (by weight) absorbed,* ΔM. The constant of proportionality $\{\beta\}$ is the *coefficient of hygroscopic (or moisture) expansion.* The hygroscopic strains are then

$$\{\varepsilon^H\}_1 = \{\beta\}_1 \Delta M \tag{3.107}$$

Following the thermal problem, in material principal coordinates we have

$$\{\beta\}_1 = \begin{Bmatrix} \beta_1 \\ \beta_2 \\ \beta_3 \\ 0 \\ 0 \\ 0 \end{Bmatrix} \tag{3.108}$$

and in arbitrary x-y-z coordinates,

$$\{\varepsilon^H\}_x = \{\beta\}_x \Delta M \tag{3.109}$$

$$\{\beta\}_x = \left[T_2\right]^{-1}\{\beta\}_1 = \begin{Bmatrix} \beta_x \\ \beta_y \\ \beta_z \\ 0 \\ 0 \\ \beta_{xy} \end{Bmatrix} \tag{3.110}$$

The coefficients of hygroscopic expansion will vary like the coefficients of thermal expansion (Fig. 3.17). Typical values of β_i for polymer matrix composites range from 0.0 in the fiber direction to 0.005/%M in the transverse direction.

3.9.2 Hygro-Thermo-Elastic Constitutive Equation

The hygro-thermo-elastic constitutive equation including mechanical, thermal, and moisture effects is developed by assuming that the total strain is the superposition of all three types of strains. Thus

$$\{\varepsilon\} = \{\varepsilon^{\sigma}\} + \{\varepsilon^T\} + \{\varepsilon^H\} \tag{3.111}$$

Using Hooke's law for the mechanical strain (as we did in the thermal case) and solving for stress, the hygro-thermo-elastic constitutive equation, in global coordinates, is

$$\{\sigma\} = [\bar{C}](\{\varepsilon^{\sigma}\} - \{\varepsilon^{T}\} - \{\varepsilon^{H}\}) \tag{3.112}$$

3.10 Summary

Three-dimensional constitutive equations have been presented for a variety of materials with symmetry ranging from fully anisotropic to isotropic. The number of elastic constants for such materials ranges from 21 for full anisotropy to 2 for isotropy The constitutive equations have been expressed in terms of both the compliance coefficients S_{ij} and the stiffness coefficients C_{ij}. Further, these coefficients have been expressed in terms of the engineering constants normally measured in the laboratory. It has been shown how these constants can be determined for a coordinate system rotated through an arbitrary angle θ about the 3 (z) material principal axis. Finally, the thermal and hygroscopic strains have been defined and the constitutive equations modified to include hygrothermal effects.

References

Boley, B. A., and Weiner, J. H. (1960), *Theory of Thermal Stresses,* John Wiley & Sons, Inc., New York.

Christensen, R. M. (1979), *Mechanics of Composite Materials,* John Wiley & Sons, Inc., New York.

Fung, Y. C. (1965), *Foundations of Solid Mechanics,* Prentice-Hall, Inc., Englewood Cliffs, NJ, p. 285.

Green, G. (1839), "On the Laws of Reflexion and Refraction of Light at the Common Surface of Two Noncrystallized Media," *Trans. Camb. Phil. Soc.*, vol. 7, p. 121.

Hearmon, R. F. S. (1961), *An Introduction to Applied Anisotropic Elasticity,* Oxford University Press, Oxford, England.

Lekhnitskii, S. G. (1950), *Teoriia Uprugosti Anisotropnovo Tela,* Government Publishing House for Technical-Theoretical Works, Moscow. (Translated as *Theory of Elasticity of an Anisotropic Body,* Holden-Day, Inc., San Francisco, 1963.)

Popov, E. P. (1990), *Engineering Mechanics of Solids,* Prentice-Hall, Englewood Cliffs, NJ.

Exercises

3.1 Show that the stiffness matrix C_{ij} is symmetric, i.e., $C_{ij} = C_{ji}$ (Eq. (3.6)), by taking the second derivatives of the strain energy density, W, with respect to ε_i and ε_j in reverse order, and using the fact that the final result is independent of the order of differentiation.

3.2 Show that Eqs. (3.16) are true.

3.3 Show that Eqs. (3.18) are true.

3.4 Show that a material that exhibits symmetry about two mutually perpendicular planes of symmetry also exhibits symmetry about the third mutually perpendicular plane.

3.5 Show that Eqs. (3.21) are true for a transversely isotropic material.

3.6 Show that Eqs. (3.23) are true for an isotropic material.

3.7 Show that for the transformation matrices $[T_1]$ and $[T_2]$ given by (3.45) and (3.47), respectively, $[T_1(\theta)]^{-1} = [T_1(-\theta)]$ and $[T_2(\theta)]^{-1} = [T_2(-\theta)]$.

3.8 Show that, when using tensor shear strains and the standard second-order tensor transformation matrix $[T_1]$ for rotation about the 3-axis, the transformed stiffness matrix $[\bar{C}]$ can be written as

$$[\bar{C}] = [T_1]^{-1}\begin{bmatrix} C_{11} & C_{12} & C_{13} & 0 & 0 & 0 \\ C_{12} & C_{22} & C_{23} & 0 & 0 & 0 \\ C_{13} & C_{23} & C_{33} & 0 & 0 & 0 \\ 0 & 0 & 0 & 2C_{44} & 0 & 0 \\ 0 & 0 & 0 & 0 & 2C_{55} & 0 \\ 0 & 0 & 0 & 0 & 0 & 2C_{66} \end{bmatrix}[T_1] = \begin{bmatrix} \bar{C}_{11} & \bar{C}_{12} & \bar{C}_{13} & 0 & 0 & 2\bar{C}_{16} \\ \bar{C}_{12} & \bar{C}_{22} & \bar{C}_{23} & 0 & 0 & 2\bar{C}_{26} \\ \bar{C}_{13} & \bar{C}_{23} & \bar{C}_{33} & 0 & 0 & 2\bar{C}_{36} \\ 0 & 0 & 0 & 2\bar{C}_{44} & 2\bar{C}_{45} & 0 \\ 0 & 0 & 0 & 2\bar{C}_{45} & 2\bar{C}_{55} & 0 \\ \bar{C}_{16} & \bar{C}_{26} & \bar{C}_{36} & 0 & 0 & 2\bar{C}_{66} \end{bmatrix}$$

3.9 Write a computer program to read in the five engineering constants (E_1, E_2, ν_{12}, G_{12}, and ν_{23}) of a transversely isotropic material, and calculate the individual terms of the compliance matrix $[S]$ and the stiffness matrix $[C]$. Calculate and print (in matrix form) the values of $[S]$ and $[C]$ for T300/5208 carbon/epoxy.

3.10 Modify your program to develop curves for the stiffness $[\bar{C}]$ of SCS-6/Ti-15-3 as a function of fiber orientation and compare your results with those for T300/5208 (Figs. 3.9 to 3.11).

3.11 Modify your program to develop curves for the compliance $[\bar{S}]$ of SCS-6/Ti-15-3 as a function of fiber orientation, and compare your results with those for T300/5208 (Figs. 3.12 to 3.14).

3.12 Modify your program to develop curves for the CTE of SCS-6/Ti-15-3 as a function of fiber orientation, and compare your results with those for T300/5208 (Fig. 3.17).

CHAPTER 4

PLANE STRESS CONSTITUTIVE EQUATIONS

"Everything should be made as simple as possible, but not simpler."

—Albert Einstein

4.1 Plane Stress Stiffness and Compliance

It is often the case in the analysis of composites that a condition of plane stress actually exists or is a very good approximation. Thus we need to develop constitutive equations for plane stress. We start with the 3-D constitutive equation (3.88) for a single layer (lamina) of a unidirectional composite with a fiber orientation θ relative to the global coordinates (Fig. 4.1).

$$\begin{Bmatrix} \varepsilon_x \\ \varepsilon_y \\ \varepsilon_z \\ \gamma_{yz} \\ \gamma_{zx} \\ \gamma_{xy} \end{Bmatrix} = \begin{bmatrix} \bar{S}_{11} & \bar{S}_{12} & \bar{S}_{13} & 0 & 0 & \bar{S}_{16} \\ \bar{S}_{12} & \bar{S}_{22} & \bar{S}_{23} & 0 & 0 & \bar{S}_{26} \\ \bar{S}_{13} & \bar{S}_{23} & \bar{S}_{33} & 0 & 0 & \bar{S}_{36} \\ 0 & 0 & 0 & \bar{S}_{44} & \bar{S}_{45} & 0 \\ 0 & 0 & 0 & \bar{S}_{45} & \bar{S}_{55} & 0 \\ \bar{S}_{16} & \bar{S}_{26} & \bar{S}_{36} & 0 & 0 & \bar{S}_{66} \end{bmatrix} \begin{Bmatrix} \sigma_x \\ \sigma_y \\ \sigma_z \\ \tau_{yz} \\ \tau_{zx} \\ \tau_{xy} \end{Bmatrix} \tag{4.1}$$

For a state of plane stress with $\sigma_z = \tau_{yz} = \tau_{zx} = 0$, (4.1) indicates that the two out-of-plane shear strains are identically zero, i.e.,

$$\begin{aligned} \gamma_{yz} &= \bar{S}_{44}\tau_{yz} + \bar{S}_{45}\tau_{zx} = 0 \\ \gamma_{zx} &= \bar{S}_{45}\tau_{yz} + \bar{S}_{55}\tau_{zx} = 0 \end{aligned} \tag{4.2}$$

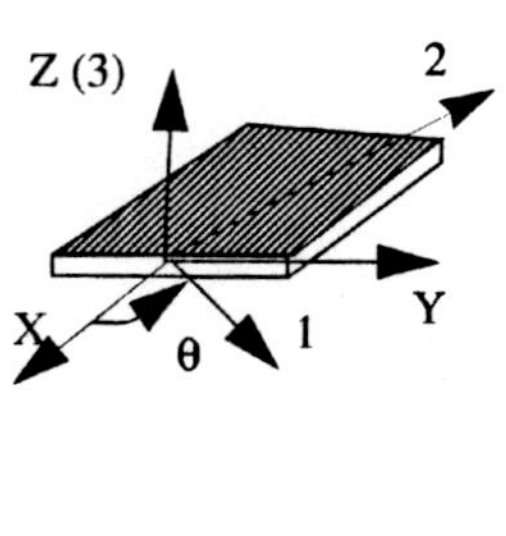

FIGURE 4.1 Global and Principal Material Coordinates

We also have that the out-of-plane normal strain, ε_z, can be expressed in terms of the in-plane components of stress using the fact that $\sigma_z = 0$.

$$\varepsilon_z = \bar{S}_{13}\sigma_x + \bar{S}_{23}\sigma_y + \bar{S}_{36}\tau_{xy} \tag{4.3}$$

The in-plane components of strains for a plane stress state can be written in matrix form from (4.1) as

$$\begin{Bmatrix} \varepsilon_x \\ \varepsilon_y \\ \gamma_{xy} \end{Bmatrix} = \begin{bmatrix} \bar{S}_{11} & \bar{S}_{12} & \bar{S}_{16} \\ \bar{S}_{12} & \bar{S}_{22} & \bar{S}_{26} \\ \bar{S}_{16} & \bar{S}_{26} & \bar{S}_{66} \end{bmatrix} \begin{Bmatrix} \sigma_x \\ \sigma_y \\ \tau_{xy} \end{Bmatrix} \tag{4.4}$$

Now from the stiffness formulation of the 3-D constitutive equation (3.85), we have

$$\sigma_z = \bar{C}_{13}\varepsilon_x + \bar{C}_{23}\varepsilon_y + \bar{C}_{33}\varepsilon_z + \bar{C}_{36}\gamma_{xy} = 0 \tag{4.5}$$

And this can be solved for ε_z:

$$\varepsilon_z = \frac{-(\bar{C}_{13}\varepsilon_x + \bar{C}_{23}\varepsilon_y + \bar{C}_{36}\gamma_{xy})}{\bar{C}_{33}} \tag{4.6}$$

We now have the out-of-plane normal strain expressed in terms of in-plane strains and known stiffness coefficients. Thus once the in-plane problem is solved, all out-of-plane strains (ε_z, γ_{yz}, and γ_{zx}) are also known.

4.1.1 Reduced Stiffness Matrix

We now write the equations for the in-plane components of stress in terms of the transformed stiffness coefficients. From (3.85),

$$\begin{aligned} \sigma_x &= \bar{C}_{11}\varepsilon_x + \bar{C}_{12}\varepsilon_y + \bar{C}_{13}\varepsilon_z + \bar{C}_{16}\gamma_{xy} \\ \sigma_y &= \bar{C}_{12}\varepsilon_x + \bar{C}_{22}\varepsilon_y + \bar{C}_{23}\varepsilon_z + \bar{C}_{26}\gamma_{xy} \\ \tau_{xy} &= \bar{C}_{16}\varepsilon_x + \bar{C}_{26}\varepsilon_y + \bar{C}_{36}\varepsilon_z + \bar{C}_{66}\gamma_{xy} \end{aligned} \tag{4.7}$$

Substituting ε_z from (4.6) into (4.7) allows us to write the plane stress constitutive equation as

$$\begin{aligned} \sigma_x &= \left(\bar{C}_{11} - \frac{\bar{C}_{13}\bar{C}_{13}}{\bar{C}_{33}}\right)\varepsilon_x + \left(\bar{C}_{12} - \frac{\bar{C}_{13}\bar{C}_{23}}{\bar{C}_{33}}\right)\varepsilon_y + \left(\bar{C}_{16} - \frac{\bar{C}_{13}\bar{C}_{36}}{\bar{C}_{33}}\right)\gamma_{xy} \\ \sigma_y &= \left(\bar{C}_{12} - \frac{\bar{C}_{13}\bar{C}_{23}}{\bar{C}_{33}}\right)\varepsilon_x + \left(\bar{C}_{22} - \frac{\bar{C}_{23}\bar{C}_{23}}{\bar{C}_{33}}\right)\varepsilon_y + \left(\bar{C}_{26} - \frac{\bar{C}_{23}\bar{C}_{36}}{\bar{C}_{33}}\right)\gamma_{xy} \\ \tau_{xy} &= \left(\bar{C}_{16} - \frac{\bar{C}_{13}\bar{C}_{36}}{\bar{C}_{33}}\right)\varepsilon_x + \left(\bar{C}_{26} - \frac{\bar{C}_{23}\bar{C}_{36}}{\bar{C}_{33}}\right)\varepsilon_y + \left(\bar{C}_{66} - \frac{\bar{C}_{63}\bar{C}_{36}}{\bar{C}_{33}}\right)\gamma_{xy} \end{aligned} \tag{4.8}$$

or in matrix form as

$$\begin{Bmatrix} \sigma_x \\ \sigma_y \\ \tau_{xy} \end{Bmatrix} = \begin{bmatrix} \bar{Q}_{11} & \bar{Q}_{12} & \bar{Q}_{16} \\ \bar{Q}_{12} & \bar{Q}_{22} & \bar{Q}_{26} \\ \bar{Q}_{16} & \bar{Q}_{26} & \bar{Q}_{66} \end{bmatrix} \begin{Bmatrix} \varepsilon_x \\ \varepsilon_y \\ \gamma_{xy} \end{Bmatrix} \tag{4.9}$$

where the *transformed reduced stiffness coefficients,* $\bar{Q}_{ij}$, are defined as

$$\bar{Q}_{ij} = \bar{C}_{ij} - \frac{\bar{C}_{i3}\bar{C}_{3j}}{\bar{C}_{33}} \qquad (i, j = 1, 2, 6) \tag{4.10}$$

In the two preceding equations, we have introduced the standard notation in composites that the indices range in the order 1, 2, 6 in the plane stress problem. This notation recalls the 3-D heritage of these terms; it will be used throughout this book. We have also implicitly noted that the transformed reduced stiffness matrix is symmetric. *It is very important to note that the transformed reduced stiffness terms for plane stress,* $\bar{Q}_{ij}$, *are not simply the corresponding terms,* $\bar{C}_{ij}$, *taken from the 3-D stiffness matrix.*

4.1.2 Reduced Stiffness and Compliance for Orthotropic Material

In principal material coordinates, the stiffness terms C_{i6} $(i \neq 6)$ are zero (see (3.19)) and hence, from (4.10), $Q_{16} = Q_{26} = 0$. Thus, in principal material coordinates of an orthotropic material, the plane stress constitutive equations (4.9) have the simplified form

$$\begin{Bmatrix} \sigma_1 \\ \sigma_2 \\ \tau_{12} \end{Bmatrix} = \begin{bmatrix} Q_{11} & Q_{12} & 0 \\ Q_{12} & Q_{22} & 0 \\ 0 & 0 & Q_{66} \end{bmatrix} \begin{Bmatrix} \varepsilon_1 \\ \varepsilon_2 \\ \gamma_{12} \end{Bmatrix} \tag{4.11}$$

Inverting this equation, we obtain

$$\begin{Bmatrix} \varepsilon_1 \\ \varepsilon_2 \\ \gamma_{12} \end{Bmatrix} = \begin{bmatrix} S_{11} & S_{12} & 0 \\ S_{12} & S_{22} & 0 \\ 0 & 0 & S_{66} \end{bmatrix} \begin{Bmatrix} \sigma_1 \\ \sigma_2 \\ \tau_{12} \end{Bmatrix} \tag{4.12}$$

where the compliance is the inverse of the stiffness, i.e., $S_{ij} = Q_{ij}^{-1}$. Comparison of (3.32) and (4.12) shows that (unlike the stiffness terms) the compliance terms of the constitutive equations are identical for three-dimensional and plane stress analysis.

Using the fact that $Q_{ij} = S_{ij}^{-1}$, we can write expressions for the individual Q_{ij} terms as functions of the S_{ij}:

$$\begin{aligned} Q_{11} &= \frac{S_{22}}{S_{11}S_{22} - S_{12}^2} & \qquad Q_{22} &= \frac{S_{11}}{S_{11}S_{22} - S_{12}^2} \\ Q_{12} &= \frac{-S_{12}}{S_{11}S_{22} - S_{12}^2} & \qquad Q_{66} &= \frac{1}{S_{66}} \end{aligned} \tag{4.13}$$

4.1.3 Compliance and Stiffness in Terms of Engineering Constants

Since the 3-D compliance terms (3.33) carry over directly to the 2-D problem, the compliance terms of interest for plane stress are

$$
\begin{aligned}
S_{11} &= \frac{1}{E_1} \qquad S_{12} = \frac{-\nu_{21}}{E_2} \\
S_{21} &= \frac{-\nu_{12}}{E_1} \qquad S_{22} = \frac{1}{E_2} \\
S_{66} &= \frac{1}{G_{12}}
\end{aligned}
\tag{4.14}
$$

Since the compliance matrix is symmetric, $S_{12} = S_{21}$. Hence, we have the reciprocal relationship

$$
\frac{\nu_{21}}{E_2} = \frac{\nu_{12}}{E_1} \tag{4.15}
$$

From the preceding equations we see that *only four of the five material constants for plane stress of an orthotropic material are independent.* The constants that can be measured most accurately in the laboratory are E_1, E_2, ν_{12}, and G_{12}. Accurate measurement of Poisson's ratio ν_{21} is often very difficult because it is very small for many composites.

Equations (4.13) and (4.14) can be combined to yield explicit expressions for the Q_{ij} in terms of engineering constants:

$$
\begin{aligned}
Q_{11} &= \frac{E_1}{1-\nu_{12}\nu_{21}} \qquad Q_{12} = \frac{\nu_{12}E_2}{1-\nu_{12}\nu_{21}} = \frac{\nu_{21}E_1}{1-\nu_{12}\nu_{21}} \\
Q_{22} &= \frac{E_2}{1-\nu_{12}\nu_{21}} \qquad Q_{66} = G_{12}
\end{aligned}
\tag{4.16}
$$

We note here that for a transversely isotropic material there is no reduction of the number of independent constants for the plane stress problem.

Example 4.1 Reduced Stiffness and Compliance

Determine the reduced stiffness coefficients Q_{ij} and compliance coefficients S_{ij} for plane stress of T300/5208 carbon/epoxy.

Solution

From Table 1.3, the material properties in material principal coordinates (in U.S. customary units) are $E_1 = 19.2\,\text{Msi}$, $E_2 = 1.56\,\text{Msi}$, $\nu_{12} = 0.24$, $G_{12} = 0.82\,\text{Msi}$.

From (4.14), the compliance coefficients are (in 1/Msi)

$$
\begin{aligned}
S_{11} &= \frac{1}{E_1} = 0.05208 \qquad S_{12} = -\frac{\nu_{21}}{E_2} = -0.0125 \\
S_{22} &= \frac{1}{E_2} = 0.641 \qquad S_{66} = \frac{1}{G_{12}} = 1.2195 \\
S_{16} &= S_{26} = 0
\end{aligned}
$$

The corresponding stiffness coefficients are, by inversion (in Msi),

$$Q_{11} = 19.29 \qquad Q_{12} = 0.376$$
$$Q_{22} = 1.567 \qquad Q_{66} = 0.820$$
$$Q_{16} = Q_{26} = 0$$

4.2 Constitutive Equations in Material Coordinates

The constitutive equations (4.11) and (4.12) can be combined with (4.14) and (4.16) to give the constitutive equations in principal material coordinates in terms of engineering constants. The results for strain in terms of stress are

$$\varepsilon_1 = \frac{\sigma_1}{E_1} - \frac{\nu_{21}}{E_2}\sigma_2 \tag{4.17}$$

$$\varepsilon_2 = -\frac{\nu_{12}}{E_1}\sigma_1 + \frac{\sigma_2}{E_2} \tag{4.18}$$

$$\gamma_{12} = \frac{\tau_{12}}{G_{12}} \tag{4.19}$$

And for stress in terms of strain,

$$\sigma_1 = \frac{E_1}{1-\nu_{12}\nu_{21}}\varepsilon_1 + \frac{\nu_{12}E_2}{1-\nu_{12}\nu_{21}}\varepsilon_2 \tag{4.20}$$

$$\sigma_2 = \frac{\nu_{12}E_2}{1-\nu_{12}\nu_{21}}\varepsilon_1 + \frac{E_2}{1-\nu_{12}\nu_{21}}\varepsilon_2 \tag{4.21}$$

$$\tau_{12} = G_{12}\gamma_{12} \tag{4.22}$$

We see from (4.17)–(4.22) that there clearly is no coupling between the normal and shear responses of an orthotropic material in the principal material coordinates.

4.3 2-D Transformations about an Axis

The 2-D transformation equations for rotation about the 3 (z) axis (Fig. 4.1) are straightforward simplifications of the 3-D equations (3.44) and (3.46). We use $[T_1]$ as the transformation matrix for stress (a second-order tensor) and $[T_2]$ for transformation of strain, where we are using engineering shear strain rather than tensor shear strain. Thus,

$$\begin{Bmatrix} \sigma_1 \\ \sigma_2 \\ \tau_{12} \end{Bmatrix} = [T_1]\begin{Bmatrix} \sigma_x \\ \sigma_y \\ \tau_{xy} \end{Bmatrix} \tag{4.23}$$

$$\begin{Bmatrix} \varepsilon_1 \\ \varepsilon_2 \\ \gamma_{12} \end{Bmatrix} = [T_2] \begin{Bmatrix} \varepsilon_x \\ \varepsilon_y \\ \gamma_{xy} \end{Bmatrix} \tag{4.24}$$

where for plane stress problems with $m = \cos\theta$ and $n = \sin\theta$

$$[T_1] = \begin{bmatrix} m^2 & n^2 & 2mn \\ n^2 & m^2 & -2mn \\ -mn & mn & m^2 - n^2 \end{bmatrix} \tag{4.25}$$

$$[T_2] = \begin{bmatrix} m^2 & n^2 & mn \\ n^2 & m^2 & -mn \\ -2mn & 2mn & m^2 - n^2 \end{bmatrix} \tag{4.26}$$

Note that, as in the 3-D case of Chapter 3, $[T_1(\theta)]^{-1} = [T_1(-\theta)]$ and $[T_2(\theta)]^{-1} = [T_2(-\theta)]$. The proof is left as an exercise.

Combining (4.23) and (4.25), the stresses in principal material coordinates are

$$\begin{aligned} \sigma_1 &= m^2\sigma_x + n^2\sigma_y + 2mn\tau_{xy} \\ \sigma_2 &= n^2\sigma_x + m^2\sigma_y - 2mn\tau_{xy} \\ \tau_{12} &= -mn\sigma_x + mn\sigma_y + (m^2 - n^2)\tau_{xy} \end{aligned} \tag{4.27}$$

Example 4.2 Stress Transformation

The same results for stress transformation are obtained by considering equilibrium of a unit-thickness wedge of material at a point where the state of stress is $(\sigma_x, \sigma_y, \tau_{xy})$, as indicated in Fig. 4.2. If the area of the surface at the angle θ (perpendicular to the fiber direction) is A, equilibrium in the 1- and 2-directions gives

$$\begin{aligned} \Sigma F_1 &= 0 \\ \sigma_1 A &= \sigma_x A m^2 + \sigma_y A n^2 + \tau_{xy} A mn \\ \sigma_1 &= \sigma_x m^2 + \sigma_y n^2 + \tau_{xy} 2mn \\ \Sigma F_2 &= 0 \\ \tau_{12} A &= -\sigma_x A mn + \sigma_y A mn + \tau_{xy} A(m^2 - n^2) \\ \tau_{12} &= -\sigma_x mn + \sigma_y mn + \tau_{xy}(m^2 - n^2) \end{aligned}$$

Likewise, equilibrium in the 2-direction for a wedge angle parallel to the fiber direction gives

$$\sigma_2 = \sigma_x n^2 + \sigma_y m^2 - \tau_{xy} 2mn$$

The proof is left to the reader.

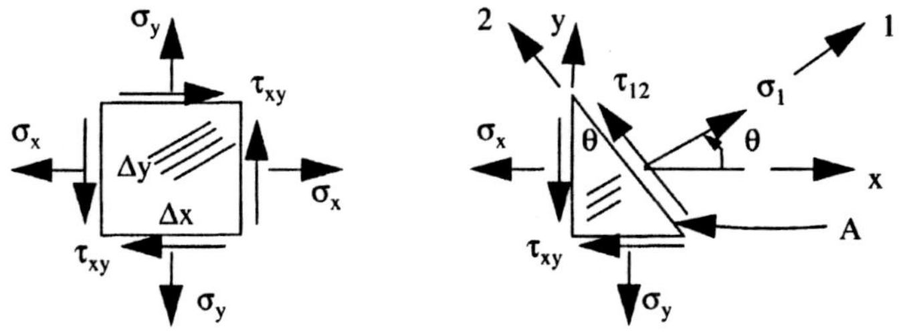

FIGURE 4.2 Stress Transformation

4.3.1 Lamina Stress-Strain Relations in Global Coordinates

The plane stress constitutive equation in principal material coordinates (4.11) is

$$\{\sigma\}_1 = [Q]\{\varepsilon\}_1 \tag{4.28}$$

Combining this with the transformation equations (4.23) and (4.24), we have

$$\{\sigma\}_x = [T_1]^{-1}[Q][T_2]\{\varepsilon\}_x \tag{4.29}$$

or

$$\begin{Bmatrix} \sigma_x \\ \sigma_y \\ \tau_{xy} \end{Bmatrix} = [T_1]^{-1}\begin{bmatrix} Q_{11} & Q_{12} & 0 \\ Q_{12} & Q_{22} & 0 \\ 0 & 0 & Q_{66} \end{bmatrix}[T_2]\begin{Bmatrix} \varepsilon_x \\ \varepsilon_y \\ \gamma_{xy} \end{Bmatrix} \tag{4.30}$$

We now define the *plane stress transformed reduced stiffness matrix* $[\bar{Q}]$:

$$[\bar{Q}] = [T_1]^{-1}[Q][T_2] \tag{4.31}$$

The plane stress constitutive equation in an arbitrary *x-y* coordinate system is then written

$$\{\sigma\}_x = [\bar{Q}]\{\varepsilon\}_x \tag{4.32}$$

or

$$\begin{Bmatrix} \sigma_x \\ \sigma_y \\ \tau_{xy} \end{Bmatrix} = \begin{bmatrix} \bar{Q}_{11} & \bar{Q}_{12} & \bar{Q}_{16} \\ \bar{Q}_{12} & \bar{Q}_{22} & \bar{Q}_{26} \\ \bar{Q}_{16} & \bar{Q}_{26} & \bar{Q}_{66} \end{bmatrix}\begin{Bmatrix} \varepsilon_x \\ \varepsilon_y \\ \gamma_{xy} \end{Bmatrix} \tag{4.33}$$

The individual Q_{ij} terms of the transformed reduced stiffness matrix (4.31) are

$$\begin{aligned}
\bar{Q}_{11} &= Q_{11}m^4 + 2(Q_{12} + 2Q_{66})m^2n^2 + Q_{22}n^4 \\
\bar{Q}_{12} &= (Q_{11} + Q_{22} - 4Q_{66})m^2n^2 + Q_{12}(n^4 + m^4) \\
\bar{Q}_{22} &= Q_{11}n^4 + 2(Q_{12} + 2Q_{66})m^2n^2 + Q_{22}m^4 \\
\bar{Q}_{16} &= (Q_{11} - Q_{12} - 2Q_{66})m^3n + (Q_{12} - Q_{22} + 2Q_{66})n^3m \\
\bar{Q}_{26} &= (Q_{11} - Q_{12} - 2Q_{66})mn^3 + (Q_{12} - Q_{22} + 2Q_{66})nm^3 \\
\bar{Q}_{66} &= (Q_{11} + Q_{22} - 2Q_{12} - 2Q_{66})m^2n^2 + Q_{66}(n^4 + m^4)
\end{aligned} \tag{4.34}$$

As in the three-dimensional case, the reduced stiffness coefficients are fourth order in the sine and cosine functions. We note that $[\bar{Q}]$ is symmetric and, in general, fully populated with nonzero $\bar{Q}_{16}$ and $\bar{Q}_{26}$ coefficients. The $\bar{Q}_{16}$ and $\bar{Q}_{26}$ coefficients are very important in that they define the coupling between the in-plane normal and shear responses. These two coefficients are identically zero for isotropic materials and for orthotropic materials in the principal material coordinates. Hence, there is no coupling between shear and normal responses in these cases.

The presence or absence of normal-shear coupling is demonstrated in Fig. 4.3. This figure shows the undeformed and deformed shapes of a rectangular element subjected to pure tensile stress. For isotropic and orthotropic materials loaded in principal material directions, there is no distortion of the original right angle (i.e., no normal-shear coupling); however, for the unidirectional off-axis lamina (a monoclinic material in the global coordinate system), coupling is clearly demonstrated through the distortion of the original right angle.

The transformed plane stress constitutive equation (4.32) can be inverted to give

$$\{\varepsilon\}_x = [\bar{S}]\{\sigma\}_x \tag{4.35}$$

where the transformed compliance, $[\bar{S}]$, is the inverse of the transformed reduced stiffness, i.e.,

$$[\bar{S}] = [\bar{Q}]^{-1} \tag{4.36}$$

In expanded form (using (4.31)), this can be written

$$[\bar{S}] = [T_2]^{-1}[Q]^{-1}[T_1] \tag{4.37}$$

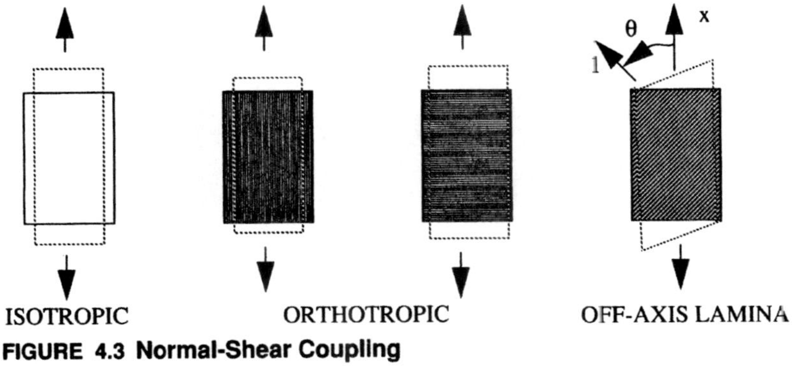

FIGURE 4.3 Normal-Shear Coupling

The explicit expressions for the $\bar{S}_{ij}$ terms as functions of S_{ij} and the fiber orientation θ are the same as those given in (3.69) to (3.81) for $i, j = 1, 2,$ and 6. They are rewritten here for convenience:

$$\bar{S}_{11} = m^4 S_{11} + m^2 n^2 (2S_{12} + S_{66}) + n^4 S_{22} \tag{4.38}$$

$$\bar{S}_{12} = n^2 m^2 (S_{11} + S_{22} - S_{66}) + (n^4 + m^4) S_{12} \tag{4.39}$$

$$\bar{S}_{16} = nm[m^2(2S_{11} - 2S_{12} - S_{66}) + n^2(2S_{12} - 2S_{22} + S_{66})] \tag{4.40}$$

$$\bar{S}_{22} = n^4 S_{11} + m^2 n^2 (2S_{12} + S_{66}) + m^4 S_{22} \tag{4.41}$$

$$\bar{S}_{26} = nm[n^2(2S_{11} - 2S_{12} - S_{66}) + m^2(2S_{12} - 2S_{22} + S_{66})] \tag{4.42}$$

$$\bar{S}_{66} = 4n^2 m^2 (S_{11} - 2S_{12} + S_{22}) + (n^2 - m^2)^2 S_{66} \tag{4.43}$$

These equations can also be expressed in terms of the engineering constants of the material using (4.14) and (4.15). The results are

$$\bar{S}_{11} = \left[m^4 + m^2 n^2 \left(-2\nu_{12} + \frac{E_1}{G_{12}}\right) + n^4 \frac{E_1}{E_2}\right]\left(\frac{1}{E_1}\right) \tag{4.44}$$

$$\bar{S}_{12} = \left[n^2 m^2 \left(1 + \frac{E_1}{E_2} - \frac{E_1}{G_{12}}\right) - (n^4 + m^4)\nu_{12}\right]\left(\frac{1}{E_1}\right) \tag{4.45}$$

$$\bar{S}_{16} = nm\left[m^2\left(2 + 2\nu_{12} - \frac{E_1}{G_{12}}\right) + n^2\left(-2\nu_{12} - 2\frac{E_1}{E_2} + \frac{E_1}{G_{12}}\right)\right]\left(\frac{1}{E_1}\right) \tag{4.46}$$

$$\bar{S}_{22} = \left[n^4 + m^2 n^2 \left(-2\nu_{12} + \frac{E_1}{G_{12}}\right) + m^4 \frac{E_1}{E_2}\right]\left(\frac{1}{E_1}\right) \tag{4.47}$$

$$\bar{S}_{26} = nm\left[n^2\left(2 + 2\nu_{12} - \frac{E_1}{G_{12}}\right) + m^2\left(-2\nu_{12} - 2\frac{E_1}{E_2} + \frac{E_1}{G_{12}}\right)\right]\left(\frac{1}{E_1}\right) \tag{4.48}$$

$$\bar{S}_{66} = \left[4n^2 m^2 \left(1 + 2\nu_{12} + \frac{E_1}{E_2}\right) + (n^2 - m^2)^2 \frac{E_1}{G_{12}}\right]\left(\frac{1}{E_1}\right) \tag{4.49}$$

In summary, the 2-D constitutive equations for an orthotropic material in an arbitrary set of orthogonal Cartesian coordinates can be written in the forms

$$\begin{Bmatrix} \sigma_x \\ \sigma_y \\ \tau_{xy} \end{Bmatrix} = \begin{bmatrix} \bar{Q}_{11} & \bar{Q}_{12} & \bar{Q}_{16} \\ \bar{Q}_{12} & \bar{Q}_{22} & \bar{Q}_{26} \\ \bar{Q}_{16} & \bar{Q}_{26} & \bar{Q}_{66} \end{bmatrix} \begin{Bmatrix} \varepsilon_x \\ \varepsilon_y \\ \gamma_{xy} \end{Bmatrix} \tag{4.50}$$

or

$$\begin{aligned} \sigma_x &= \bar{Q}_{11}\varepsilon_x + \bar{Q}_{12}\varepsilon_y + \bar{Q}_{16}\gamma_{xy} \\ \sigma_y &= \bar{Q}_{12}\varepsilon_x + \bar{Q}_{22}\varepsilon_y + \bar{Q}_{26}\gamma_{xy} \\ \tau_{xy} &= \bar{Q}_{16}\varepsilon_x + \bar{Q}_{26}\varepsilon_y + \bar{Q}_{66}\gamma_{xy} \end{aligned} \tag{4.51}$$

and

$$\begin{Bmatrix} \varepsilon_x \\ \varepsilon_y \\ \gamma_{xy} \end{Bmatrix} = \begin{bmatrix} \bar{S}_{11} & \bar{S}_{12} & \bar{S}_{16} \\ \bar{S}_{12} & \bar{S}_{22} & \bar{S}_{26} \\ \bar{S}_{16} & \bar{S}_{26} & \bar{S}_{66} \end{bmatrix} \begin{Bmatrix} \sigma_x \\ \sigma_y \\ \tau_{xy} \end{Bmatrix} \tag{4.52}$$

or

$$\begin{aligned} \varepsilon_x &= \bar{S}_{11}\sigma_x + \bar{S}_{12}\sigma_y + \bar{S}_{16}\tau_{xy} \\ \varepsilon_y &= \bar{S}_{12}\sigma_x + \bar{S}_{22}\sigma_y + \bar{S}_{26}\tau_{xy} \\ \gamma_{xy} &= \bar{S}_{16}\sigma_x + \bar{S}_{26}\sigma_y + \bar{S}_{66}\tau_{xy} \end{aligned} \tag{4.53}$$

Equations (4.51) and (4.53) clearly show the coupling between normal and shear responses by the presence of nonzero 16 and 26 stiffness and compliance coefficients.

Before leaving this subject, we note that the same notation is used for compliances S_{ij} and $\bar{S}_{ij}$ in 2-D and 3-D. This is because the corresponding terms are identical, unlike the stiffness coefficients, which are different in 2-D and 3-D. The stiffness values differ because the inverse of the 3 × 3 compliance matrix for plane stress is different from the inverse of the 6 × 6 matrix for 3-D analysis.

Example 4.3 Transformed Reduced Stiffness and Compliance

Determine the transformed reduced stiffness coefficients $\bar{Q}_{ij}$ and compliance coefficients $\bar{S}_{ij}$ for plane stress of a T300/5208 carbon/epoxy lamina at 30° to the global x-axis.

Solution

From Table 1.3, the material properties in material principal coordinates (in U.S. customary units) are $E_1 = 19.2\,\text{Msi}$; $E_2 = 1.56\,\text{Msi}$; $\nu_{12} = 0.24$; $G_{12} = 0.82\,\text{Msi}$.

From (4.44) to (4.49), the compliance coefficients are (in 1/Msi)

$$\begin{aligned} &\bar{S}_{11} = 0.2933 \quad \bar{S}_{12} = -0.1065 \quad \bar{S}_{16} = -0.3636 \\ &\bar{S}_{22} = 0.5878 \quad \bar{S}_{26} = -1.465 \\ &\bar{S}_{66} = 0.8434 \end{aligned}$$

The corresponding stiffness coefficients are, by inversion (in Msi),

$$\begin{aligned} &\bar{Q}_{11} = 11.705 \quad \bar{Q}_{12} = 3.531 \quad \bar{Q}_{16} = 5.658 \\ &\bar{Q}_{22} = 2.843 \quad \bar{Q}_{26} = 2.017 \\ &\bar{Q}_{66} = 3.975 \end{aligned}$$

4.4 Stiffness and Compliance θ Dependence

For fibrous composites modeled as homogeneous orthotropic materials, it is instructive to study the variation of the stiffness and compliance coefficients with fiber orientation. Plots showing the variation of these terms for T300/5208 carbon/epoxy with fiber orientations ranging from –90° to +90° are presented in Figs. 4.4–4.7.

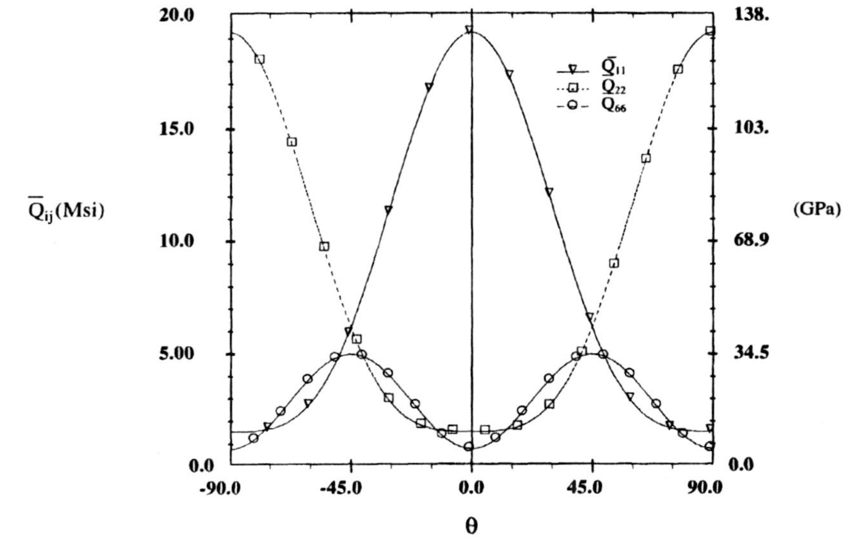

FIGURE 4.4 θ **Dependence of $\bar{Q}_{11}$, $\bar{Q}_{22}$, and $\bar{Q}_{66}$: T300/5208**

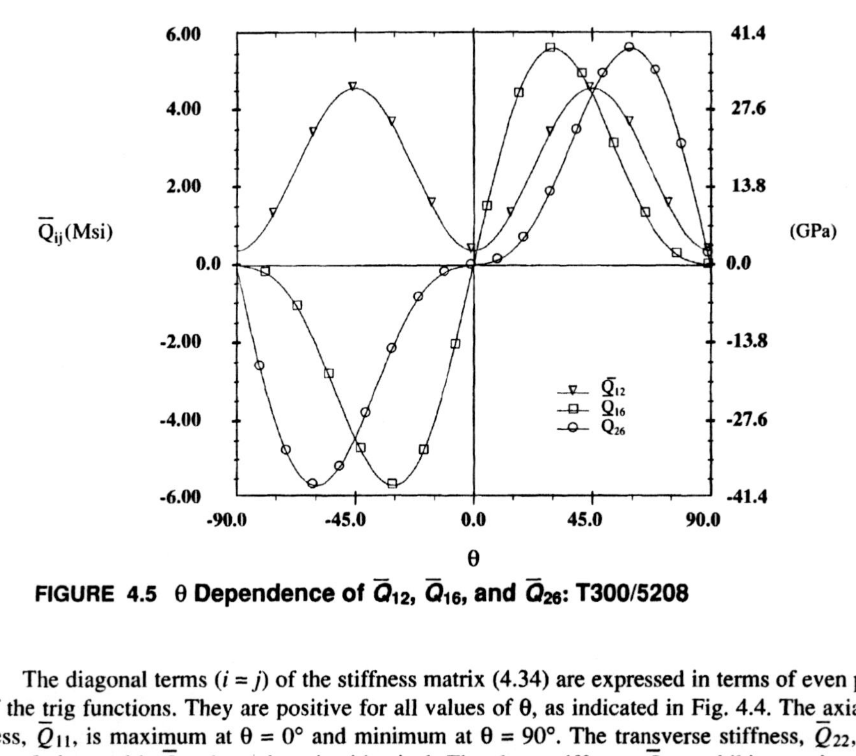

FIGURE 4.5 θ **Dependence of $\bar{Q}_{12}$, $\bar{Q}_{16}$, and $\bar{Q}_{26}$: T300/5208**

The diagonal terms ($i = j$) of the stiffness matrix (4.34) are expressed in terms of even powers of the trig functions. They are positive for all values of θ, as indicated in Fig. 4.4. The axial stiffness, $\bar{Q}_{11}$, is maximum at θ = 0° and minimum at θ = 90°. The transverse stiffness, $\bar{Q}_{22}$, is 90° out of phase with $\bar{Q}_{11}$, but otherwise identical. The shear stiffness, $\bar{Q}_{66}$, exhibits maxima at θ = +45° and –45° and minima at 0° and 90°.

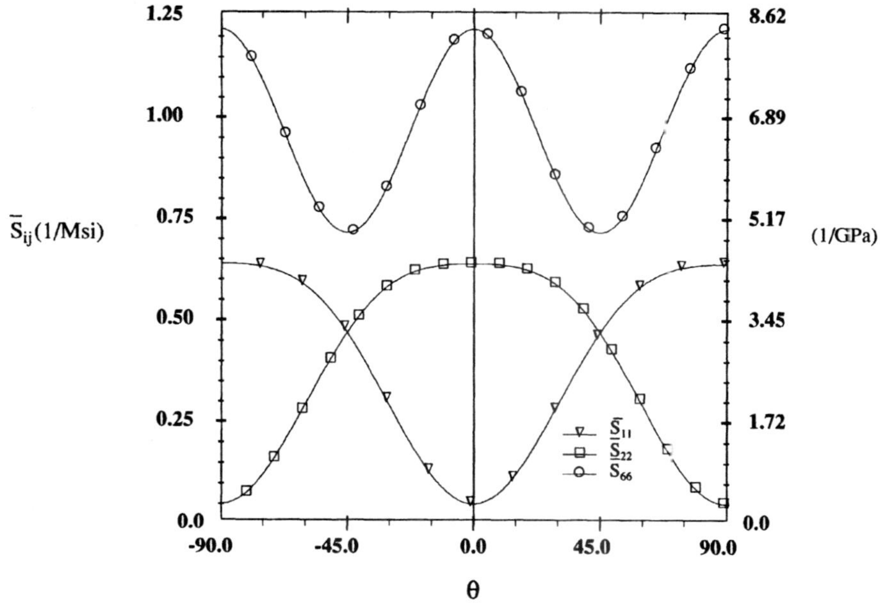

FIGURE 4.6 θ Dependence of $\bar{S}_{11}$, $\bar{S}_{22}$, and $\bar{S}_{66}$: T300/5208

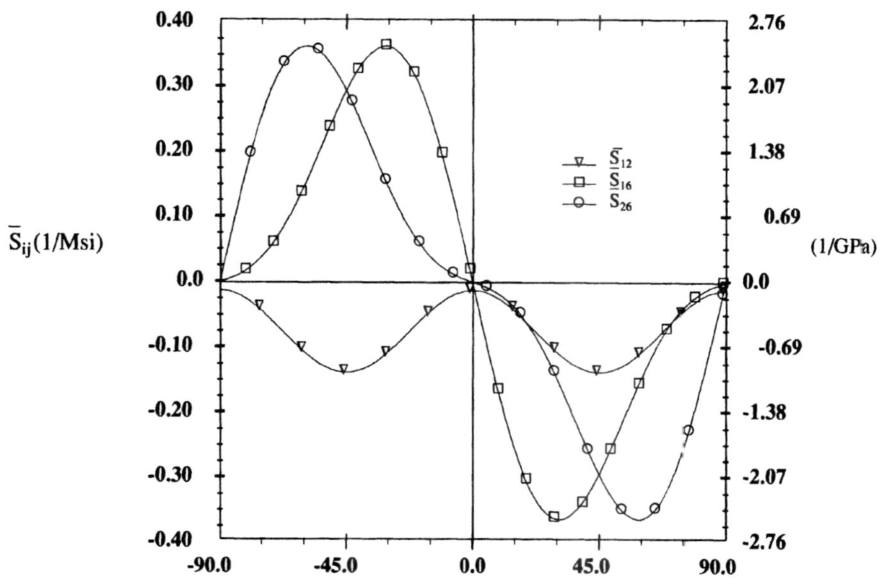

FIGURE 4.7 θ Dependence of $\bar{S}_{12}$, $\bar{S}_{16}$, and $\bar{S}_{26}$: T300/5208

The variations of the off-diagonal terms ($i \neq j$) are presented in Fig. 4.5. The $\bar{Q}_{12}$ coefficient is always positive and varies in the same manner as $\bar{Q}_{66}$, but with different extreme values, as prescribed by (4.34). The coupling coefficients $\bar{Q}_{16}$ and $\bar{Q}_{26}$ are the only terms of all six $\bar{Q}_{ij}$ whose sign varies with the sign of the fiber orientation. The coupling coefficients are zero for $\theta = -90°, 0°$, and $+90°$, the principal material orientations, and exhibit maxima and minima at ±30° ($\bar{Q}_{16}$) and ±60° ($\bar{Q}_{26}$).

The variations of the compliance coefficients are presented in Figs. 4.6 (diagonal terms, $i = j$) and 4.7 (off-diagonal terms, $i \neq j$). All coefficients are reflections of the corresponding stiffness coefficient, with the one exception being the shear compliance $\bar{S}_{66}$, which also differs in magnitude.

4.5 Lamina Engineering Constants

It is possible to develop expressions for the engineering constants of an off-axis lamina in terms of the known material properties and the fiber orientation angle, θ, by considering a series of one-dimensional stress states applied to the lamina. This procedure is sometimes referred to as conducting *thought experiments*. We start by rewriting the constitutive equation (4.52) for plane stress in a global x-y coordinate system, i.e.,

$$\begin{Bmatrix} \varepsilon_x \\ \varepsilon_y \\ \gamma_{xy} \end{Bmatrix} = \begin{bmatrix} \bar{S}_{11} & \bar{S}_{12} & \bar{S}_{16} \\ \bar{S}_{12} & \bar{S}_{22} & \bar{S}_{26} \\ \bar{S}_{16} & \bar{S}_{26} & \bar{S}_{66} \end{bmatrix} \begin{Bmatrix} \sigma_x \\ \sigma_y \\ \tau_{xy} \end{Bmatrix} \tag{4.54}$$

For a given state of stress (σ_x, σ_y, τ_{xy}), the strains can be determined from (4.54), and the engineering constants can be determined as described in the following sections.

4.5.1 Axial Modulus

Consider the unidirectional off-axis lamina under the loading $\sigma_x \neq 0$ with $\sigma_y = \tau_{xy} = 0$, as depicted in Fig. 4.8.

The axial strain from (4.54) is

$$\varepsilon_x = \bar{S}_{11}\sigma_x \tag{4.55}$$

and from the definition of Young's modulus,

$$\varepsilon_x = \frac{1}{E_x}\sigma_x \tag{4.56}$$

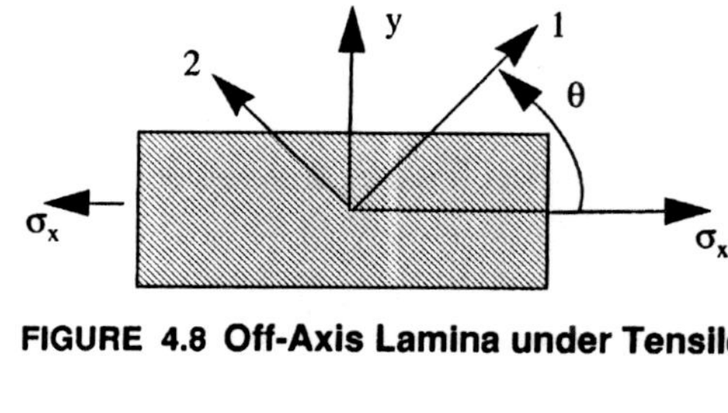

FIGURE 4.8 Off-Axis Lamina under Tensile Stress

Combining (4.55) and (4.56) gives the desired expression for the *axial modulus*, E_x, in terms of the transformed compliance coefficient:

$$E_x = \frac{1}{\bar{S}_{11}} \tag{4.57}$$

This modulus can be expressed in terms of the compliances in principal material coordinates and the fiber orientation angle using (3.69):

$$E_x = \frac{1}{\bar{S}_{11}} = \frac{1}{m^4 S_{11} + m^2 n^2 (2S_{12} + S_{66}) + n^4 S_{22}} \tag{4.58}$$

or, in terms of engineering properties in principal material coordinates using (4.44),

$$E_x = \frac{E_1}{\left[m^4 + m^2 n^2\left(-2\nu_{12} + \frac{E_1}{G_{12}}\right) + n^4 \frac{E_1}{E_2}\right]} \tag{4.59}$$

Results for the predicted axial modulus of unidirectional off-axis laminae of T300/5208 carbon/epoxy and SCS-6/Ti-15-3 silicon carbide/titanium are shown in Fig. 4.9. As the fiber orientation is rotated off-axis, the modulus decreases sharply and then gradually approaches the transverse modulus. The variation with fiber orientation is identical for positive and negative values of θ. The modulus of the metal matrix composite is significantly higher than that of the carbon/epoxy and exhibits an interesting characteristic in that the decrease in modulus with increasing fiber orientation is not monotonic; the lowest modulus occurs at $\theta = 58°$, not 90° as for the polymer matrix composite. The θ dependence of E_x is evident from (4.59), which clearly shows $E_x = E_1$ at $\theta = 0°$ and $E_x = E_2$ at $\theta = 90°$.

4.5.2 Poisson's Ratio

Poisson's ratio, ν_{xy}, is defined as the negative ratio of the lateral strain ε_y to the axial strain ε_x, associated with an applied stress, σ_x. Thus, we consider the same loading as in the previous case (Fig. 4.8). Using (4.54) for the strains, we can write

$$\nu_{xy} = \frac{-\varepsilon_y}{\varepsilon_x} = \frac{-\bar{S}_{12}}{\bar{S}_{11}} \tag{4.60}$$

Combining (4.60) with (4.44) and (4.55), we have

$$\nu_{xy} = \frac{-\left[n^2 m^2\left(1 + \frac{E_1}{E_2} - \frac{E_1}{G_{12}}\right) - (n^4 + m^4)\nu_{12}\right]}{\left[m^4 + m^2 n^2\left(-2\nu_{12} + \frac{E_1}{G_{12}}\right) + n^4 \frac{E_1}{E_2}\right]} \tag{4.61}$$

Poisson's ratio is defined with the minus sign because most materials exhibit negative lateral strains when subjected to tensile stress. This results in a positive value for Poisson's ratio for such materials. However, as we shall see later, not all materials exhibit positive Poisson's ratios.

The variation of ν_{xy} with fiber orientation for T300/5208 carbon/epoxy and SCS-6/Ti-15-3 silicon carbide/titanium is shown in Fig. 4.10. For both materials, as the fiber orientation is rotated

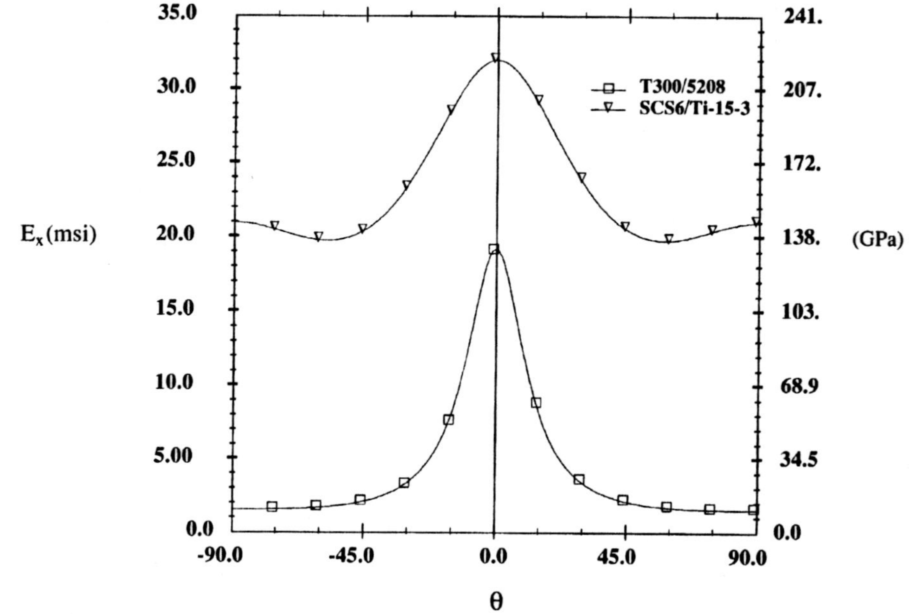

FIGURE 4.9 Lamina Axial Modulus E_x versus θ

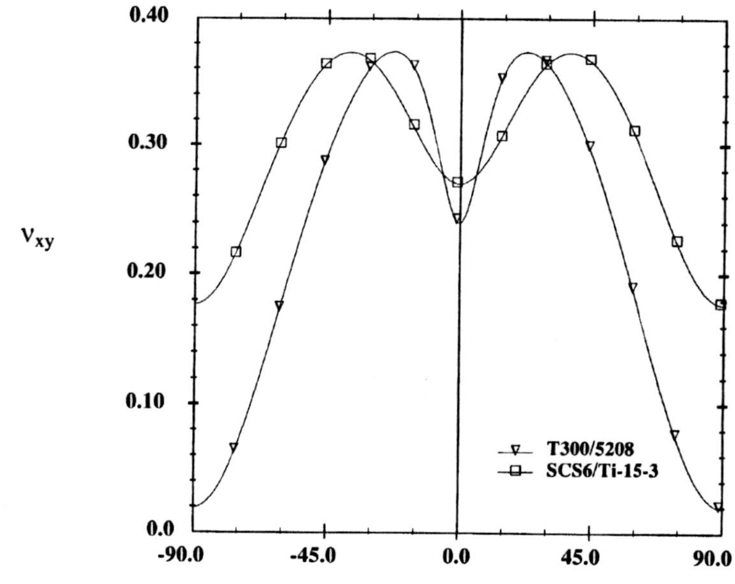

FIGURE 4.10 Lamina Poisson's Ratio ν_{xy} versus θ

off-axis, Poisson's ratio increases to a maximum value and then decreases to the minimum value at $\theta = \pm 90°$. For carbon/epoxy the maximum value is at $\theta = \pm 22°$, whereas for the metal matrix composite the maximum value is at $\theta = \pm 38°$. The range of values is smaller for the metal matrix composite because it has a lower degree of orthotropy. It is most interesting that both materials exhibit nearly identical maximum values of Poisson's ratio. These results follow from the θ dependence of the transformed compliances $\bar{S}_{ij}$ as depicted in Figs. 4.6 and 4.7 for carbon/epoxy.

4.5.3 Transverse Modulus

The expression for the *transverse modulus*, E_y, is determined from the stress state $\sigma_y \neq 0$, $\sigma_x = \tau_{xy} = 0$ applied to the lamina in Fig. 4.8. Using this stress state in (4.54) gives the strain ε_y, and Hooke's law then gives the transverse modulus:

$$E_y = \frac{\sigma_y}{\varepsilon_y} = \frac{1}{\bar{S}_{22}} \tag{4.62}$$

Now using (4.47) in (4.62), we have

$$E_y = \frac{E_1}{\left[n^4 + m^2 n^2\left(-2\nu_{12} + \frac{E_1}{G_{12}}\right) + m^4 \frac{E_1}{E_2}\right]} \tag{4.63}$$

The θ dependence of the transverse modulus E_y for T300/5208 carbon/epoxy and SCS-6/Ti-15-3 silicon carbide/titanium is shown in Fig. 4.11. As expected, the curves are identical to those for E_x, but shifted 90°.

4.5.4 Shear Modulus

The *shear modulus*, G_{xy}, is determined for the stress state $\tau_{xy} \neq 0$, $\sigma_y = \sigma_x = 0$ in (4.54) and Hooke's law for shear, with the result

$$G_{xy} = \frac{\tau_{xy}}{\gamma_{xy}} = \frac{1}{\bar{S}_{66}} \tag{4.64}$$

Combining (4.64) with (4.49) gives

$$G_{xy} = \frac{E_1}{\left[4n^2 m^2\left(1 + 2\nu_{12} + \frac{E_1}{E_2}\right) + (n^2 - m^2)^2 \frac{E_1}{G_{12}}\right]} \tag{4.65}$$

The θ dependence of the shear modulus of T300/5208 carbon/epoxy and SCS-6/Ti-15-3 silicon carbide/titanium as a function of fiber orientation is shown in Fig. 4.12. In contrast to the previous properties discussed, the in-plane shear modulus, G_{xy}, exhibits a larger range of values for metal matrix composites than it does for polymer matrix composite. The shear modulus is a maximum at $\pm 45°$ and minimum for loading in the principal material directions, 0° and $\pm$ 90°, for both materials.

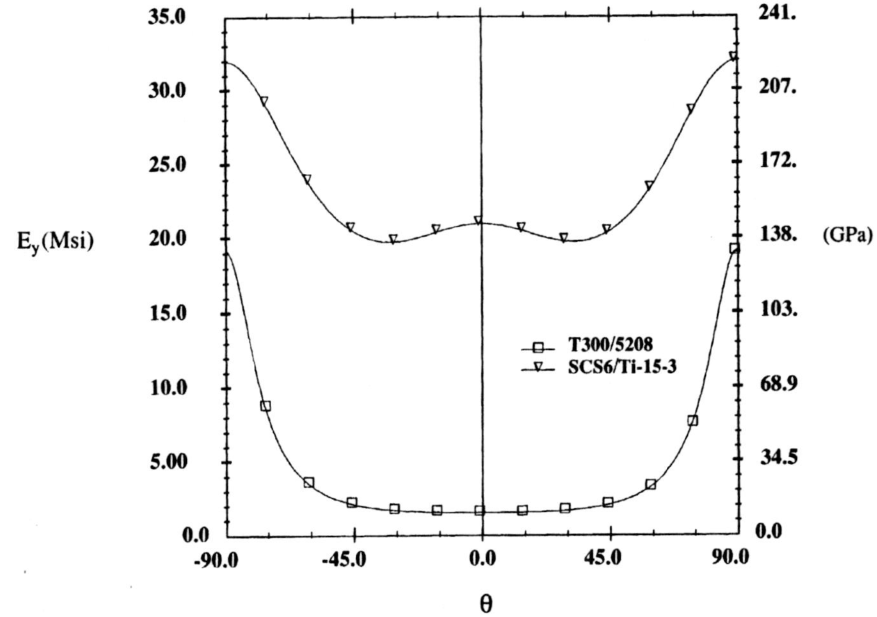

FIGURE 4.11 Lamina Transverse Modulus E_y versus θ

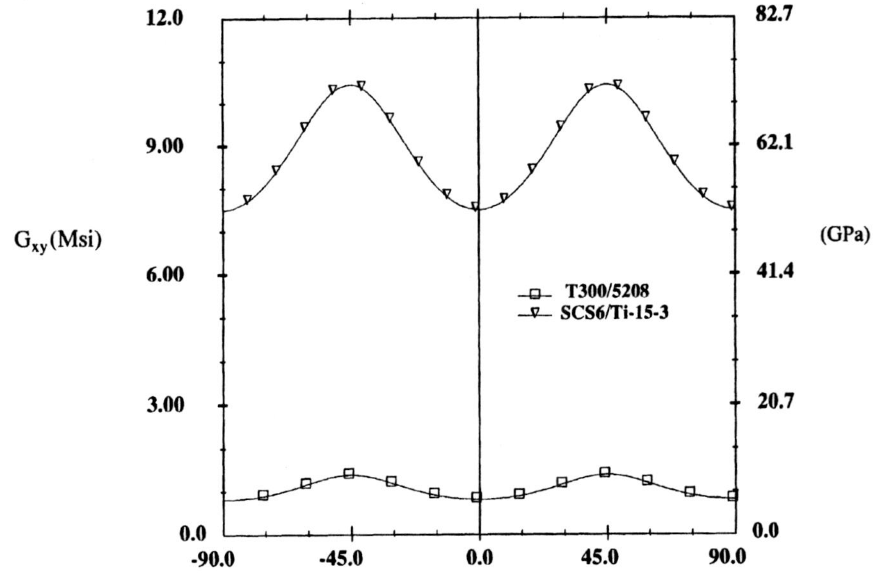

FIGURE 4.12 Lamina Shear Modulus G_{xy} versus θ

From (4.65) we see that at $\theta = \pm 45°$, the shear modulus is

$$G_{xy} = \frac{E_1}{\left(1 + 2\nu_{12} + \frac{E_1}{E_2}\right)} \tag{4.66}$$

The well-known E, G, ν relationship, $G = E/2(1+\nu)$, for isotropic materials is recovered from (4.66) for $E_1 = E_2$.

The maximum shear modulus at ±45° can be understood physically through consideration of the stress state and the fiber orientation for these cases. As shown in Fig. 4.13, pure shear loading is equivalent to tension and compression along the 45° diagonal lines of a square element. When the fibers are oriented at ±45°, either tensile normal stress (positive shear stress case) or compressive normal stress (negative shear stress case) is aligned with the fibers, thus providing the high shear stiffness at ±45° exhibited in Fig. 4.12.

4.5.5 Coefficients of Mutual Influence

As has been demonstrated in Section 4.1, the response of orthotropic materials exhibits normal-shear coupling when the material is loaded in other than principal material coordinates. Another way of saying this is that there can be shear strain associated with normal stress and, likewise, there can be normal strain associated with shear stress. It is convenient to define material properties which quantify the normal-shear coupling. The normal-shear coupling properties are called *coefficients of mutual influence*.

Two types of coefficients of mutual influence must be defined. The *coefficient of mutual influence of the first kind*, $\eta_{i,ij}$, is defined for applied shear stress, and the *coefficient of mutual influence of the second kind*, $\eta_{ij,i}$, is defined for applied normal stress (Lekhnitskii, 1950). Both coefficients are defined as the ratio of an associated strain to the applied strain for the given state of stress. In this sense they are similar to Poisson's ratios and, likewise, dimensionless.

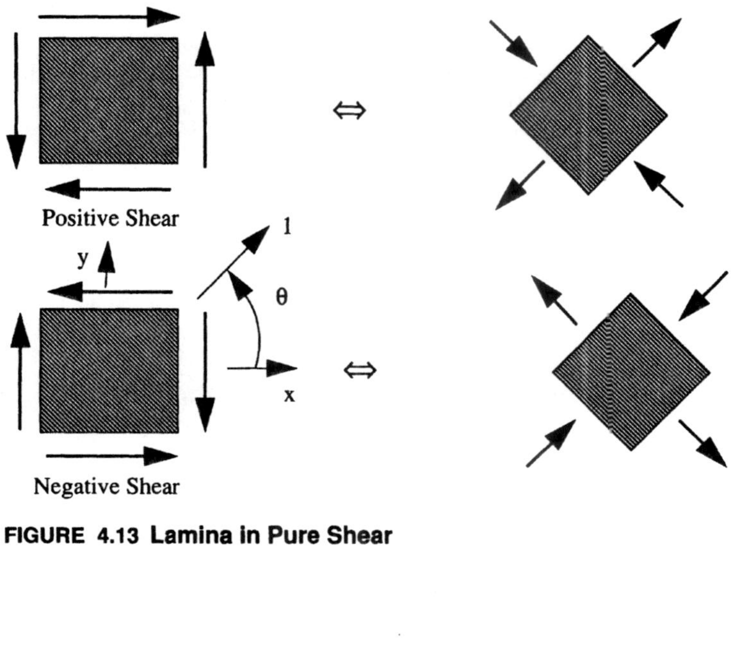

FIGURE 4.13 Lamina in Pure Shear

More specifically, the coefficient of mutual influence of the first kind, $\eta_{i,ij}$, is defined as the ratio of normal strain to shear strain for an applied shear stress $\tau_{xy} \neq 0$, with $\sigma_x = \sigma_y = 0$, i.e.,

$$\eta_{i,\,ij} = \frac{\varepsilon_i}{\gamma_{ij}} \tag{4.67}$$

The coefficient of mutual influence of the second kind, $\eta_{ij,i}$, is defined as the ratio of shear strain to normal strain for an applied normal stress $\sigma_x \neq 0$, with $\sigma_y = \tau_{xy} = 0$, i.e.,

$$\eta_{ij,\,i} = \frac{\gamma_{ij}}{\varepsilon_i} \tag{4.68}$$

Thus, the notation convention is that the first subscript corresponds to the associated (or induced) strain and the second subscript corresponds to the applied strain. It is noted that this is opposite to the convention used for Poisson's ratio. (C'est la vie!)

For the loading $\sigma_x \neq 0$ with $\sigma_y = \tau_{xy} = 0$, the coefficient of mutual influence of the second kind, $\eta_{xy,x}$, is obtained using (4.54) and (4.68):

$$\eta_{xy,\,x} = \frac{\gamma_{xy}}{\varepsilon_x} = \frac{\bar{S}_{16}}{\bar{S}_{11}} \tag{4.69}$$

Combining (4.44), (4.46), and (4.69), we have

$$\eta_{xy,\,x} = \frac{nm\left[m^2\left(2 + 2\nu_{12} - \frac{E_1}{G_{12}}\right) + n^2\left(-2\nu_{12} - 2\frac{E_1}{E_2} + \frac{E_1}{G_{12}}\right)\right]}{\left[m^4 + m^2n^2\left(-2\nu_{12} + \frac{E_1}{G_{12}}\right) + n^4\frac{E_1}{E_2}\right]} \tag{4.70}$$

Predictions for the variation of $\eta_{xy,x}$ in T300/5208 carbon/epoxy and SCS-6/Ti-15-3 silicon carbide/titanium are shown in Fig. 4.14. Differences between polymer matrix and metal matrix composites are very evident in this figure. For both materials, $\eta_{xy,x}$ is an odd function of θ, being negative for positive θ and positive for negative θ, with an extremum at $\theta = \pm 12°$ for the carbon/epoxy and at $\theta = \pm 22°$ for the metal matrix composite. The extreme values are greater than 2.0 for carbon/epoxy, indicating that normal-shear coupling can be much stronger than the Poisson effect where the maximum ratio of transverse to axial strain is less than 0.4 for T300/5208 (Fig. 4.10). In contrast, $\eta_{xy,x}$ for the metal matrix composite is much smaller in magnitude, with a maximum of 0.40.

The strong normal-shear coupling possible in polymer matrix composites will be shown later to have a dominant effect on response and failure in some situations. Because of the lack of experience with anisotropic materials by the uninitiated, normal-shear coupling is not normally part of one's intuition. It is extremely important that this phenomenon be understood and considered in working with polymer matrix composites. Not to do so is to invite disaster.

4.5.6 Additional Lamina Engineering Constants

Following similar steps as outlined in Sections 4.5.1–4.5.5, with the one-dimensional stress states $\sigma_y \neq 0$, $\sigma_x = \tau_{xy} = 0$ and then $\tau_{xy} \neq 0$, $\sigma_x = \sigma_y = 0$, leads to the following relationships for other engineering lamina constants.

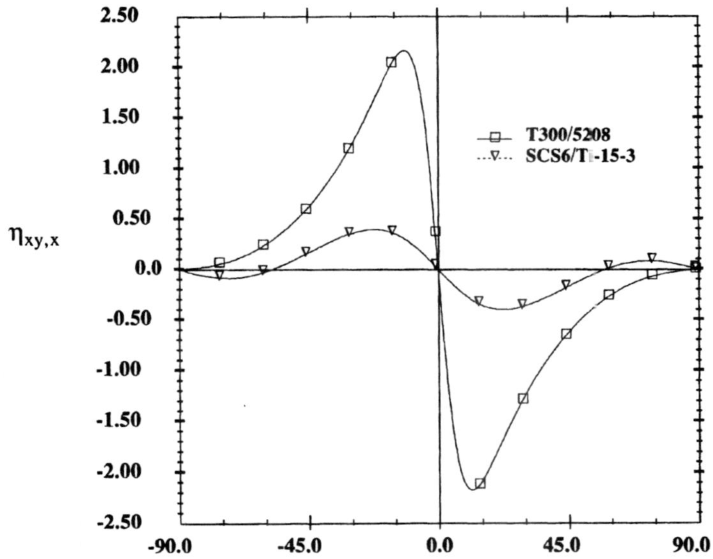

FIGURE 4.14 Lamina Coefficient of Mutual Influence $\eta_{xy,x}$ versus θ

For $\sigma_y \neq 0$, $\sigma_y = \tau_{xy} = 0$, Poisson's ratio ν_{yx} is

$$\nu_{yx} = \frac{-\varepsilon_x}{\varepsilon_y} = \frac{-\bar{S}_{12}}{\bar{S}_{22}} = \frac{-\left[n^2 m^2\left(1 + \frac{E_1}{E_2} - \frac{E_1}{G_{12}}\right) - (n^4 + m^4)\nu_{12}\right]}{\left[n^4 + m^2 n^2\left(-2\nu_{12} + \frac{E_1}{G_{12}}\right) + m^4 \frac{E_1}{E_2}\right]} \tag{4.71}$$

and the coefficient of mutual influence $\eta_{xy,x}$ is

$$\eta_{xy,y} = \frac{\gamma_{xy}}{\varepsilon_y} = \frac{\bar{S}_{26}}{\bar{S}_{22}} = \frac{nm\left[n^2\left(2 + 2\nu_{12} - \frac{E_1}{G_{12}}\right) + m^2\left(-2\nu_{12} - 2\frac{E_1}{E_2} + \frac{E_1}{G_{12}}\right)\right]}{\left[n^4 + m^2 n^2\left(-2\nu_{12} + \frac{E_1}{G_{12}}\right) + m^4 \frac{E_1}{E_2}\right]} \tag{4.72}$$

For $\tau_{xy} \neq 0$, $\sigma_y = \tau_{xy} = 0$, the coefficient of mutual influence $\eta_{x,xy}$ is

$$\eta_{x,xy} = \frac{\varepsilon_x}{\gamma_{xy}} = \frac{\bar{S}_{16}}{\bar{S}_{66}} = \frac{nm\left[m^2\left(2 + 2\nu_{12} - \frac{E_1}{G_{12}}\right) + n^2\left(-2\nu_{12} - 2\frac{E_1}{E_2} + \frac{E_1}{G_{12}}\right)\right]}{\left[4n^2 m^2\left(1 + 2\nu_{12} + \frac{E_1}{E_2}\right) + (n^2 - m^2)^2 \frac{E_1}{G_{12}}\right]} \tag{4.73}$$

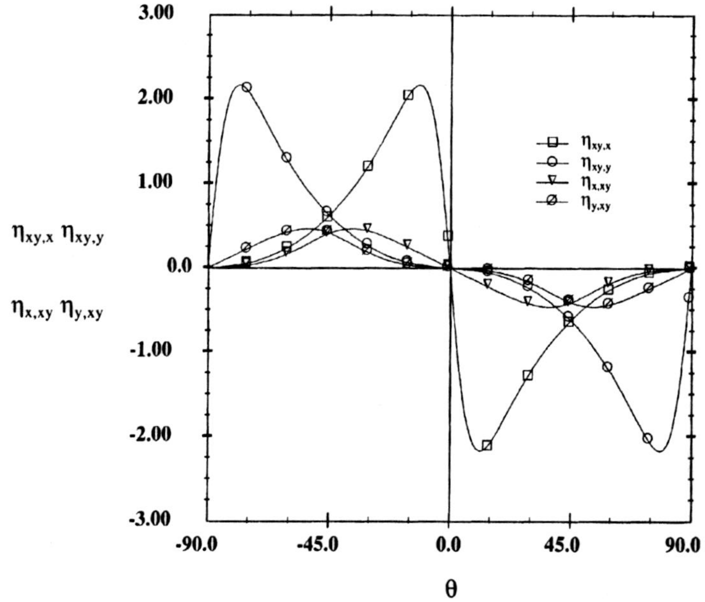

FIGURE 4.15 Coefficients of Mutual Influence θ Dependence

and the coefficient of mutual influence $\eta_{y,xy}$ is

$$\eta_{y,xy} = \frac{\varepsilon_y}{\gamma_{xy}} = \frac{\bar{S}_{26}}{\bar{S}_{66}} = \frac{nm\left[n^2\left(2 + 2\nu_{12} - \frac{E_1}{G_{12}}\right) + m^2\left(-2\nu_{12} - 2\frac{E_1}{E_2} + \frac{E_1}{G_{12}}\right)\right]}{\left[4n^2m^2\left(1 + 2\nu_{12} + \frac{E_1}{E_2}\right) + (n^2 - m^2)^2\frac{E_1}{G_{12}}\right]} \tag{4.74}$$

Figure 4.15 shows a comparison of all four coefficients of mutual influence for T300/5208 carbon/epoxy. It is apparent from the figure that the coefficients associated with applied axial strain (the second kind) are larger by a factor of 4 than those associated with applied shear strain (the first kind). All four coefficients are odd functions of the fiber orientation angle, and the critical angle for extremum values varies depending upon the coefficient.

4.5.7 Comparison of Theory and Experiment

Comparisons of experimental results and theoretical predictions for the axial modulus and Poisson's ratio of unidirectional off-axis aramid/epoxy are shown in Figs. 4.16 and 4.17. The experimental results in these figures are from Pindera et al. (1989). As the figures indicate, there is excellent correlation for the axial modulus and very good correlation for the Poisson's ratio. The slightly lower degree of correlation for the Poisson's ratio is due to limitation in the accuracy of measurements for small strain rather than to any limitation of the theory.

Additional comparisons between theory and experiment for elastic modulus and Poisson's ratio of unidirectional off-axis lamina for other materials can be found in Chapter 7.

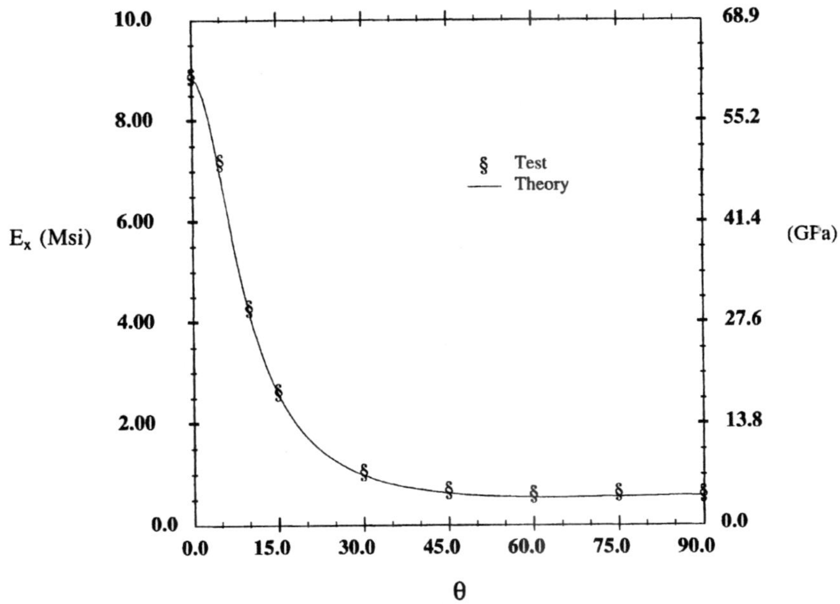

FIGURE 4.16 Aramid/Epoxy Off-Axis Modulus: Theory and Experiment

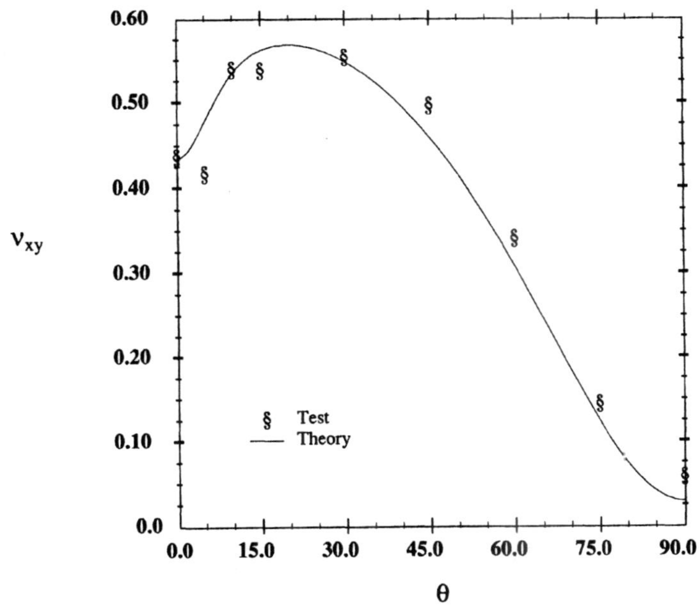

FIGURE 4.17 Aramid/Epoxy Off-Axis Poisson's Ratio: Theory and Experiment

4.6 Measurement of Material Engineering Constants

Details of the test methods used to determine the four engineering constants needed for plane stress analysis (E_1, E_2, ν_{12}, and G_{12}) are discussed in Chapter 6. However, we can show here that, in principle, all required constants can be determined from one 0°, one 90°, and one off-axis tensile test. For a uniform far-field axial stress, σ_x, applied to a unidirectional off-axis coupon of fiber orientation θ (Fig. 4.8), the stress transformation equations (4.23) (with $\sigma_y = \tau_{xy} = 0$) give the principal material coordinate stresses in terms of sin θ (n) and cos θ (m) as

$$\begin{aligned} \sigma_1 &= m^2\sigma_x \\ \sigma_2 &= n^2\sigma_x \\ \tau_{12} &= -mn\sigma_x \end{aligned} \tag{4.75}$$

Likewise, the strains in principal material coordinates are determined from the strains ε_x, ε_y, and γ_{xy} measured with a strain gage rosette (see Section 6.4.2) and the strain transformation equations (4.24):

$$\begin{aligned} \varepsilon_1 &= m^2\varepsilon_x + n^2\varepsilon_y + mn\gamma_{xy} \\ \varepsilon_2 &= n^2\varepsilon_x + m^2\varepsilon_y - mn\gamma_{xy} \\ \gamma_{12} &= -2mn\varepsilon_x + 2mn\varepsilon_y + (m^2 - n^2)\gamma_{xy} \end{aligned} \tag{4.76}$$

for $\theta = 0°$, $\sigma_1 = \sigma_x$, and $\sigma_2 = \tau_{12} = 0$. The axial modulus E_1 and Poisson's ratio ν_{12} are obtained using the measured stress σ_x and the measured strains $\varepsilon_1 = \varepsilon_x$ and $\varepsilon_2 = \varepsilon_y$ in (4.17), with the result

$$E_1 = \frac{\sigma_x}{\varepsilon_x} = \frac{\sigma_1}{\varepsilon_1} \tag{4.77}$$

$$\nu_{12} = \frac{-\varepsilon_y}{\varepsilon_x} = \frac{-\varepsilon_2}{\varepsilon_1} \tag{4.78}$$

Likewise, for $\theta = 90°$, $\sigma_2 = \sigma_x$, $\sigma_1 = \tau_{12} = 0$, and $\varepsilon_2 = \varepsilon_x$,

$$E_2 = \frac{\sigma_x}{\varepsilon_x} = \frac{\sigma_2}{\varepsilon_2} \tag{4.79}$$

For any angle other than 0° or 90°, the shear stress τ_{12} is given by the last equation of (4.75). From (4.19) or (4.22), the shear modulus G_{12}, which is independent of σ_1 and σ_2, is given in terms of the applied axial stress σ_x and the shear strain γ_{12} from the last equation of (4.76):

$$G_{12} = \frac{\tau_{12}}{\gamma_{12}} = \frac{-mn\sigma_x}{-2mn\varepsilon_x + 2mn\varepsilon_y + (m^2 - n^2)\gamma_{xy}} \tag{4.80}$$

For the special case $\theta = 45°$, (4.80) simplifies to

$$G_{12} = \frac{\tau_{12}}{\gamma_{12}} = \frac{\sigma_x}{2(\varepsilon_x - \varepsilon_y)} \tag{4.81}$$

Thus, we see that, in principle, all four engineering constants required for plane stress analysis of unidirectional composites can be determined from tension tests on three specimens. As will be demonstrated in Chapter 6, unless very special loading conditions are present, an off-axis coupon is subjected to a combined state of far-field stress when its ends are displaced as in the standard tensile test. This is true because of the axial-shear coupling that exists for off-axis fiber orientations.

Before we leave this discussion of the off-axis tension test, it is instructive to note the variation of the stresses in principal material coordinates as a function of fiber orientation. We can rewrite (4.75) as ratios of principal material stresses to applied axial stress:

$$\frac{\sigma_1}{\sigma_x} = m^2$$
$$\frac{\sigma_2}{\sigma_x} = n^2 \qquad (4.82)$$
$$\frac{\tau_{12}}{\sigma_x} = -mn$$

The stresses (Fig. 4.18) are statically determinate and independent of material properties. For $\theta = \pm 45°$, all three components of stress are of equal magnitude. The normal components of stress are tensile for all fiber orientations, whereas the sign of the shear stress is opposite that of the fiber orientation. Varying the fiber orientation alters the stress ratios. Thus different off-axis specimens potentially can be used to study material response under different biaxial stress states.

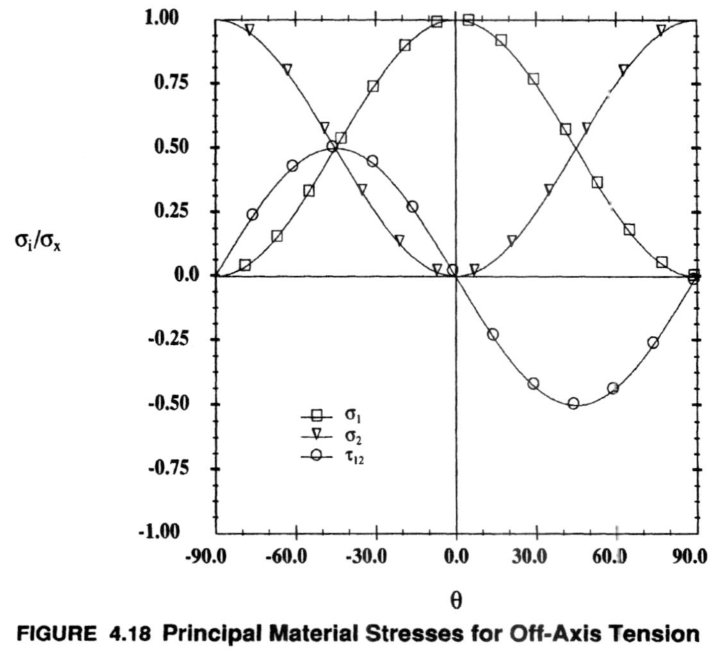

FIGURE 4.18 Principal Material Stresses for Off-Axis Tension

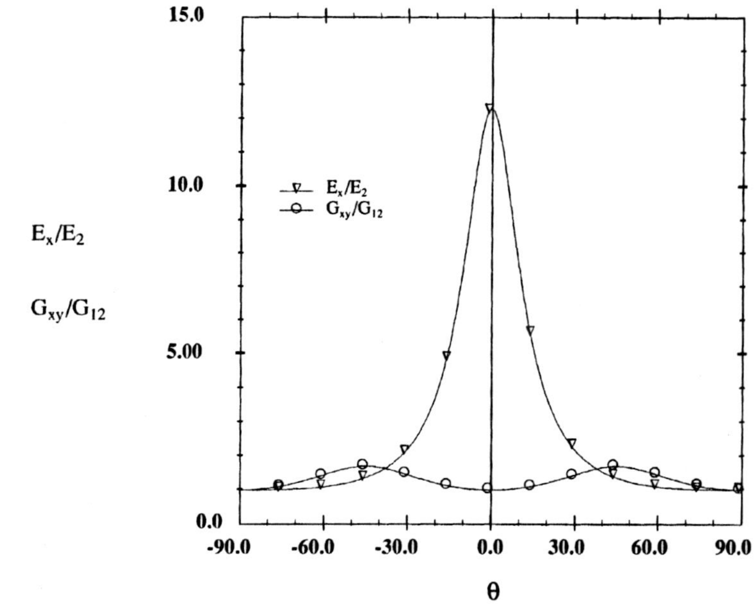

FIGURE 4.19 **Axial and Shear Modulus θ Dependence**

4.6.1 Axial and Shear Moduli

It is instructive to compare the θ dependence of the axial modulus E_x with that of the shear modulus G_{xy}. Figure 4.19 compares these two moduli, normalized with respect to principal material values, for carbon/epoxy. Clearly, the variation of axial modulus is much greater than that of the shear modulus. Indeed, even though the shear modulus varies by as much as 75% over the range of fiber angles, this is quite small in comparison to the variation in axial modulus, which can vary by a factor of more than 10.

4.7 Invariant Properties of Reduced Stiffness

The transformed reduced stiffness matrix $[\bar{Q}]$ of (4.34) can be written in terms of invariant quantities U_i and functions of the fiber orientation angle θ (Tsai and Pagano, 1968). The expressions are

$$\begin{aligned}
\bar{Q}_{11} &= U_1 + U_2\cos 2\theta + U_3\cos 4\theta \\
\bar{Q}_{12} &= U_4 - U_3\cos 4\theta \\
\bar{Q}_{22} &= U_1 - U_2\cos 2\theta + U_3\cos 4\theta \\
\bar{Q}_{16} &= \frac{1}{2}U_2\sin 2\theta + U_3\sin 4\theta \\
\bar{Q}_{26} &= \frac{1}{2}U_2\sin 2\theta - U_3\sin 4\theta \\
\bar{Q}_{66} &= U_5 - U_3\cos 4\theta
\end{aligned} \tag{4.83}$$

where the U_i are defined as follows:

$$
\begin{aligned}
U_1 &= \frac{3Q_{11}+3Q_{22}+2Q_{12}+4Q_{66}}{8} \\
U_2 &= \frac{Q_{11}-Q_{22}}{2} \\
U_3 &= \frac{Q_{11}+Q_{22}-2Q_{12}-4Q_{66}}{8} \\
U_4 &= \frac{Q_{11}+Q_{22}+6Q_{12}-4Q_{66}}{8} \\
U_5 &= \frac{Q_{11}+Q_{22}-2Q_{12}+4Q_{66}}{8}
\end{aligned}
\tag{4.84}
$$

Use of the invariants can simplify the calculation of stiffness terms and also is helpful in examining the influence of fiber orientation on the stiffness properties of an orthotropic material. These invariants will prove to be most helpful in Chapter 5 for proving that the in-plane properties of quasi-isotropic laminates are indeed independent of direction.

4.8 Thermal Effects

4.8.1 Coefficient of Thermal Expansion

The three-dimensional thermal problem was discussed in detail in Section 3.8. The plane stress condition results in the following simplifications. The thermal strains in principal material coordinates are proportional to the temperature change ΔT, i.e.,

$$
\{\varepsilon^T\}_1 = \{\alpha\}_1 \Delta T \tag{4.85}
$$

where the coefficients of thermal expansion (CTEs), α_i, for two-dimensional problems are

$$
\{\alpha\}_1 = \begin{Bmatrix} \alpha_1 \\ \alpha_2 \\ 0 \end{Bmatrix} \tag{4.86}
$$

The shear term, α_{12}, is zero because an orthotropic material does not exhibit shear strains (in the principal material coordinates) when undergoing a uniform temperature change. Transformation of the thermal strains $\{\varepsilon^T\}_1$ to the strains $\{\varepsilon^T\}_x$ in global coordinates gives

$$
\begin{Bmatrix} \varepsilon_x^T \\ \varepsilon_y^T \\ \gamma_{xy}^T \end{Bmatrix} = [T_2]^{-1} \begin{Bmatrix} \varepsilon_1^T \\ \varepsilon_2^T \\ \gamma_{12}^T \end{Bmatrix} \tag{4.87}
$$

Combining (4.85)–(4.87) to express the global thermal strains in terms of material principal CTE and the temperature change gives

$$\begin{Bmatrix} \varepsilon_x^T \\ \varepsilon_y^T \\ \gamma_{xy}^T \end{Bmatrix} = [T_2]^{-1} \begin{Bmatrix} \alpha_1 \\ \alpha_2 \\ 0 \end{Bmatrix} \Delta T \tag{4.88}$$

From the preceding equation, we can write the thermal strains in a global coordinate system as

$$\begin{Bmatrix} \varepsilon_x^T \\ \varepsilon_y^T \\ \gamma_{xy}^T \end{Bmatrix} = \begin{Bmatrix} \alpha_x \\ \alpha_y \\ \alpha_{xy} \end{Bmatrix} \Delta T \tag{4.89}$$

where, with α_{xy} being the engineering shear thermal CTE, we have

$$\{\alpha\}_x \equiv \begin{Bmatrix} \alpha_x \\ \alpha_y \\ \alpha_{xy} \end{Bmatrix} = [T_2]^{-1}\{\alpha\}_1 = \begin{bmatrix} m^2\alpha_1 + n^2\alpha_2 \\ n^2\alpha_1 + m^2\alpha_2 \\ 2mn(\alpha_1 - \alpha_2) \end{bmatrix} \tag{4.90}$$

The explicit forms for the individual coefficients of thermal expansion in global *x-y* coordinates are then

$$\begin{aligned} \alpha_x &= m^2\alpha_1 + n^2\alpha_2 \\ \alpha_y &= n^2\alpha_1 + m^2\alpha_2 \\ \alpha_{xy} &= 2mn(\alpha_1 - \alpha_2) \end{aligned} \tag{4.91}$$

These equations are identical to the corresponding equations in (3.95) for three-dimensional analysis, and thus Figs. 3.17 and 3.18 and the associated discussion apply directly. The figures and discussion are reproduced here for convenience. The plane stress thermal strains in global coordinates can be summarized as

$$\{\varepsilon^T\}_x \equiv \begin{Bmatrix} \varepsilon_x^T \\ \varepsilon_y^T \\ \gamma_{xy}^T \end{Bmatrix} = \begin{Bmatrix} \alpha_x \\ \alpha_y \\ \alpha_{xy} \end{Bmatrix} \Delta T \tag{4.92}$$

or, more simply,

$$\{\varepsilon^T\}_x = \{\alpha\}_x(\Delta T) \tag{4.93}$$

Several important conclusions can be drawn from the preceding equations and a plot showing the variation of in-plane CTE values with fiber orientation (Fig. 4.20). The CTE values for T300/5208 carbon/epoxy range from a small negative number for expansion in the fiber direction to a large positive value in the transverse direction. Expansion in the x- and y-directions differs by a 90° phase shift.

For an orthotropic material ($\alpha_1 \neq \alpha_2$) and for fiber orientations other than 0° and 90°, α_{xy} is nonzero. Thus, temperature change introduces shear strain in other than principal material coordinates. The extreme values of the thermal shear strain occur at fiber orientations of +45° and −45°. There is no thermal shear strain in isotropic materials or in the principal material coordinate directions of an orthotropic material.

The mechanics of thermal shear strains in orthotropic materials can be explained through consideration of the thermal expansion of an isotropic material and an orthotropic material as shown in Fig. 4.21 (where it is assumed that $\alpha_2 >> \alpha_1$ in the orthotropic material). Shear strain is evident in the global coordinates for the orthotropic material by the change in the original right angle formed by the diagonals. In contrast, the diagonals remain orthogonal in the isotropic material, and, further, no shear is evident in the principal material coordinate directions of the orthotropic material.

4.8.2 Thermo-Elastic Constitutive Equation

As in the three-dimensional case, we assume that the total strain $\{\varepsilon\}$ is a superposition of the "free" thermal strain $\{\varepsilon^T\}$ and the mechanical strain $\{\varepsilon^\sigma\}$. In principal material coordinates we have

$$\{\varepsilon\} = \{\varepsilon^\sigma\} + \{\varepsilon^T\} \tag{4.94}$$

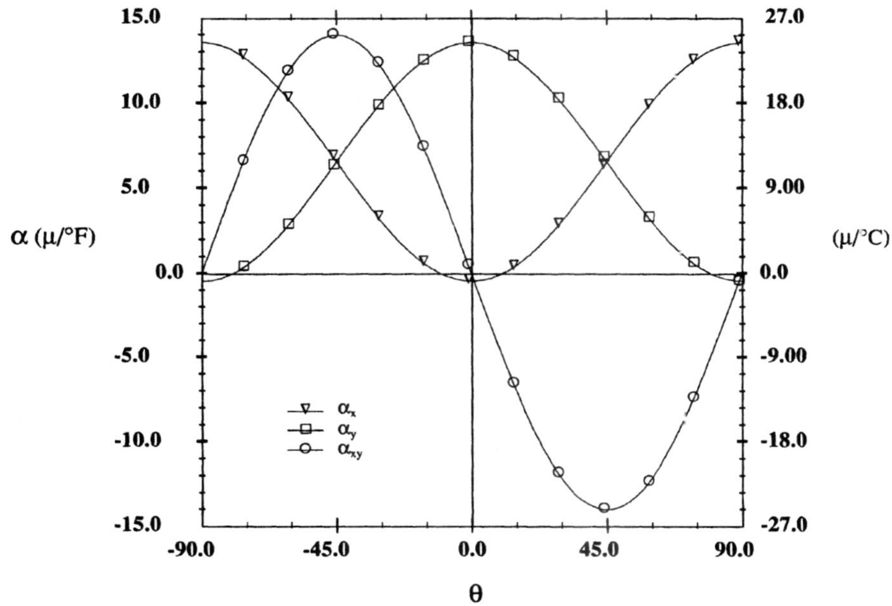

FIGURE 4.20 Coefficients of Thermal Expansion θ Dependence: T300/5208

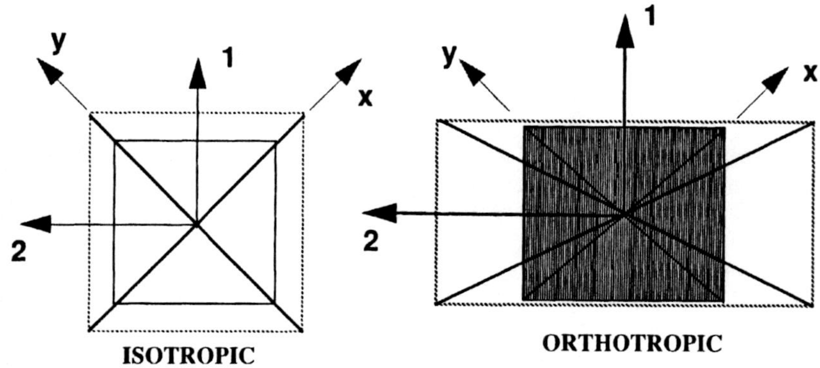

FIGURE 4.21 Thermal Expansion in Isotropic and Orthotropic Material

Recalling that Hooke's law (4.12) relates the mechanical strains and stresses, we have

$$\{\varepsilon^\sigma\} = [S]\{\sigma\} \tag{4.95}$$

Substitution of Hooke's law (4.95) into the strain superposition equation (4.94) gives

$$\{\varepsilon\} = [S]\{\sigma\} + \{\varepsilon^T\} \tag{4.96}$$

Solving the above equations for the stresses (using $[Q] = [S]^{-1}$), we have

$$\{\sigma\} = [Q](\{\varepsilon\} - \{\varepsilon^T\}) \tag{4.97}$$

This is the fundamental constitutive equation for plane, thermo-elastic stress analysis. For global coordinates this equation can be written in terms of the transformed stresses, strains, and stiffness as

$$\{\sigma\}_x = [\bar{Q}](\{\varepsilon\}_x - \{\varepsilon^T\}_x) \tag{4.98}$$

This thermo-elastic constitutive equation can be written in terms of the coefficients of thermal expansion and the temperature change using (4.93), with the result

$$\{\sigma\}_x = [\bar{Q}](\{\varepsilon\}_x - \{\alpha\}_x \Delta T) \tag{4.99}$$

Comparison of (3.101) and (4.98) shows that the only significant difference between the plane stress and three-dimensional thermo-elastic constitutive equations is the different forms of the stiffness matrix.

Example 4.4 Thermal Stress Analysis

Determine the stresses and strains in a unidirectional off-axis lamina that is constrained in the x-direction between smooth, frictionless walls, as shown in the figure, and subjected to a temperature change ΔT. Consider the specific case of T300/5208 carbon/epoxy at 45° and subjected to a temperature change ΔT = 180°F.

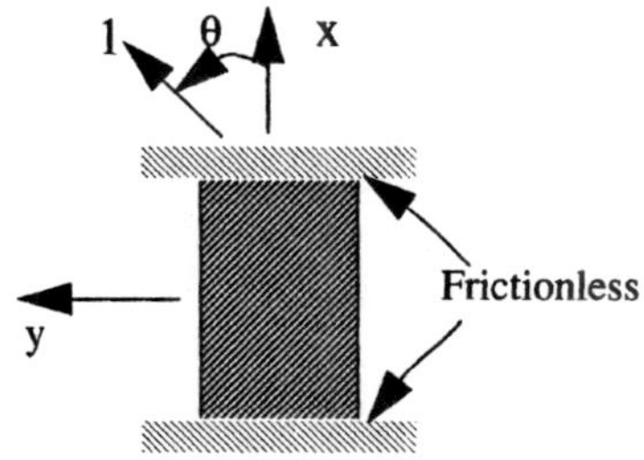

Solution

The boundary conditions for this problem are

$$\varepsilon_x = 0 \qquad \sigma_y = 0 \qquad \tau_{xy} = 0$$

From (4.99) the constitutive equations are

$$\begin{Bmatrix} \sigma_x \\ \sigma_y \\ \tau_{xy} \end{Bmatrix} = \begin{bmatrix} \bar{Q}_{11} & \bar{Q}_{12} & \bar{Q}_{16} \\ \bar{Q}_{12} & \bar{Q}_{22} & \bar{Q}_{26} \\ \bar{Q}_{16} & \bar{Q}_{26} & \bar{Q}_{66} \end{bmatrix} \left(\begin{Bmatrix} \varepsilon_x \\ \varepsilon_y \\ \gamma_{xy} \end{Bmatrix} - \begin{Bmatrix} \alpha_x \\ \alpha_y \\ \alpha_{xy} \end{Bmatrix} \Delta T \right)$$

This represents three equations in the three unknowns: σ_x, ε_y, γ_{xy}. For T300/5208 at $\theta = 45°$, the transformed reduced stiffness, (4.31), and coefficients of thermal expansion, (4.90), are

$$[\bar{Q}] = \begin{bmatrix} 6.222 & 4.582 & 4.431 \\ 4.582 & 6.222 & 4.431 \\ 4.431 & 4.431 & 5.026 \end{bmatrix} 10^6 \text{ (psi)} \qquad \{\alpha\}_x = \begin{Bmatrix} 6.565 \\ 6.565 \\ -13.990 \end{Bmatrix} 10^{-6} \; (1/°\text{F})$$

Thus, with $\Delta T = 180°$F the three simultaneous equations are

$$\sigma_x = -1608.9426 + 4.582 \times 10^6 \varepsilon_y + 4.431 \times 10^6 \gamma_{xy}$$
$$0 = -1608.9426 + 6.222 \times 10^6 \varepsilon_y + 4.431 \times 10^6 \gamma_{xy}$$
$$0 = 2184.2478 + 4.431 \times 10^6 \varepsilon_y + 5.026 \times 10^6 \gamma_{xy}$$

Solving the three simultaneous equations gives

$$\sigma_x = -2.50339 \text{ (ksi)}$$
$$\varepsilon_y = 0.153 \; (\%)$$
$$\gamma_{xy} = -0.178 \; (\%)$$

4.9 Moisture Effects

4.9.1 Hygroscopic Expansion

The plane stress constitutive equation including hygroscopic effects follows a similar development. The hygroscopic strains in principal material coordinates, $\{\varepsilon^H\}_1$, are

$$\{\varepsilon^H\}_1 = \{\beta\}_1 \Delta M \tag{4.100}$$

where ΔM is the percent change in moisture content by weight, and the coefficients of hygroscopic expansion, $\{\beta\}_1$, in principal material coordinates, are

$$\{\beta\}_1 = \begin{Bmatrix} \beta_1 \\ \beta_2 \\ 0 \end{Bmatrix} \tag{4.101}$$

In global coordinates, the coefficients of hygroscopic expansion are

$$\{\varepsilon^H\}_x = \{\beta\}_x \Delta M \tag{4.102}$$

where the strain transformation equations (4.24) give

$$\{\beta\}_x = [T_2]^{-1}\{\beta\}_1 = \begin{Bmatrix} \beta_x \\ \beta_y \\ \beta_{xy} \end{Bmatrix} \tag{4.103}$$

or

$$\begin{aligned} \beta_x &= m^2\beta_1 + n^2\beta_2 \\ \beta_y &= n^2\beta_1 + m^2\beta_2 \\ \beta_{xy} &= 2mn(\beta_1 - \beta_2) \end{aligned} \tag{4.104}$$

In view of (4.91) and (4.104), the θ dependence of the coefficients of hygroscopic expansion is identical to that of the coefficients of thermal expansion.

4.9.2 Hygrothermal Constitutive Equations

The constitutive equation including mechanical, thermal, and moisture effects is developed by assuming that the total strain is the superposition of all three types of strains. Thus

$$\{\varepsilon\} = \{\varepsilon^\sigma\} + \{\varepsilon^T\} + \{\varepsilon^H\} \tag{4.105}$$

Using Hooke's law for the mechanical strain (as we did in the thermal case) and solving for stress, we obtain the hygrothermal constitutive equation:

$$\{\sigma\}_x = [\bar{Q}](\{\varepsilon\}_x - \{\varepsilon^T\}_x - \{\varepsilon^H\}_x) \tag{4.106}$$

or

$$\{\sigma\}_x = [\bar{Q}](\{\varepsilon\}_x - \{\alpha\}_x \Delta T - \{\beta\}_x \Delta M) \tag{4.107}$$

4.10 Summary

The three-dimensional constitutive equations have been reduced for the case of plane stress. It has been shown that the corresponding compliance coefficients are identical to those in the 3-D case, but that the stiffness coefficients for plane stress take a different form from those for the 3-D case.

Equations for the engineering constants of a unidirectional lamina, oriented at an arbitrary angle θ to the global x-y axes, have been developed. Material properties, called coefficients of mutual influence, have been introduced. These new coefficients represent the coupling between normal and shear responses for unidirectional lamina.

Finally, the plane stress constitutive equations have been expanded to include thermal and moisture effects.

References

Ashton, J. E., Halpin, J. C., and Petit, P. H. (1969), *Primer on Composite Materials: Analysis,* Technomic Publishing Co., Stamford, CT.

Dong, S. B., Pister, K. S., and Taylor, R. L. (1962), "On the Theory of Laminated Anisotropic Shells and Plates," *J. Aerosp. Sci.*, vol. 29, pp. 969–975.

Hearmon, R. F. S. (1961), *An Introduction to Applied Anisotropic Elasticity,* Oxford University Press, Oxford, England.

Lekhnitskii, S. G. (1950), *Teoriia Uprugosti Anisotropnogo Tela,* Government Publishing House for Technical-Theoretical Works, Moscow. (Translated as *Theory of Elasticity of an Anisotropic Body* (1963), Holden-Day, Inc., San Francisco.)

Pindera, M.-J., Gurdal, Z., Hidde, J. S., and Herakovich, C. T. (1989), "Mechanical Response of Aramid/Epoxy under Tensile, Compressive and Shear Loading," *J. Reinf. Plast. Compos.*, vol. 8, pp. 410–420.

Pister, K. S., and Dong, S. B. (1959), "Elastic Bending of Layered Plates," *J. Eng. Mech. Div.*, EM 4, October, pp. 1–10.

Reissner, E., and Stavsky, Y. (1961), "Bending and Stretching of Certain Types of Heterogeneous Aeolotropic Elastic Plates," *J. Appl. Mech.*, vol. 28, pp. 402–408.

Tsai, S. W. (1966), "Mechanics of Composite Materials," part II, Technical Report AFML-TR-66-149, Wright-Patterson Air Force Base, Ohio.

Tsai, S. W., and Pagano, N. J. (1968), "Invariant Properties of Composite Materials," in *Composite Materials Workshop* (S. W. Tsai and J. C. Halpin, eds.), Technomic Publishing Co., Stamford, CT.

Exercises

Use the material properties given in Table 1.3 as needed for the following problems.

4.1 Show that for the transformation matrices $[T_1]$ and $[T_2]$ given by (4.25) and (4.26), respectively, $[T_1(\theta)]^{-1} = [T_1(-\theta)]$ and $[T_2(\theta)]^{-1} = [T_2(-\theta)]$.

4.2 Determine the transformed stiffness matrix $[\bar{Q}]$ for a 90° layer in terms of the principal material stiffnesses Q_{ij}.

4.3 Determine the transformed stiffness matrix $[\bar{Q}]$ for a 45° layer in terms of the principal material stiffnesses Q_{ij}.

4.4 Determine the transformed stiffness matrix $[\bar{Q}]$ for a –45° layer in terms of the principal material stiffnesses Q_{ij}.

4.5 Determine the transformed stiffness matrix $[\bar{Q}]$ for a 30° layer in terms of the principal material stiffnesses Q_{ij}.

4.6 Determine the transformed stiffness matrix $[\bar{Q}]$ for a –30° layer in terms of the principal material stiffnesses Q_{ij}.

4.7 Show that $\bar{Q}_{16}$ and $\bar{Q}_{26}$ are identically zero for an isotropic material.

4.8 Show that, with the use of tensor shear strains, the transformed stiffness matrix for plane stress is

$$[\bar{Q}]^{(T)} = [T_1]^{-1}\begin{bmatrix} Q_{11} & Q_{12} & 0 \\ Q_{12} & Q_{22} & 0 \\ 0 & 0 & 2Q_{66} \end{bmatrix}[T_1]$$

4.9 Write a computer program to read the engineering constants E_1, E_2, G_{12}, and ν_{12}, and then compute and print out the 3 × 3 plane stress matrices $[S]$, $[\bar{S}]$, $[Q]$, and $[\bar{Q}]$ for any material. Run your program for the properties of T300/5208 carbon/epoxy. Print out values for the fiber orientations 0°, 90°, +45°, –45°, +30°, and –30°. Use your program to generate the required values, and then plot $[\bar{Q}_{11}]$, $[\bar{Q}_{12}]$, $[\bar{Q}_{22}]$, $[\bar{Q}_{16}]$, $[\bar{Q}_{26}]$, and $[\bar{Q}_{66}]$ for values of θ ranging from –90° to 90° in 1° increments.

4.10 Graphically compare the θ dependences of $[\bar{Q}_{ij}]$ over the range –90° to +90° for T300/5208, aramid/epoxy, and SCS-6/Ti-15-3.

4.11 Graphically compare the θ dependences of $[\bar{S}_{ij}]$ over the range –90° to +90° for T300/5208, aramid/epoxy, and SCS-6/Ti-15-3.

4.12 Given the strain vector $\{\varepsilon\}_1 = \begin{Bmatrix} 0.001 \\ 0.002 \\ 0.0005 \end{Bmatrix}$, compute the corresponding stress vector $\{\sigma\}_1$ for T300/5208 carbon/epoxy. Use this result to compute $\{\varepsilon\}_1$, which should be your original vector.

4.13 For $\theta = 45°$ and given the stress vector (in psi) $\{\sigma\}_x = \begin{Bmatrix} 1000 \\ 3000 \\ 500 \end{Bmatrix}$, compute the corresponding strain vector $\{\varepsilon\}_x$. Use this result to compute $\{\sigma\}_x$, which should be the original stress vector.

4.14 Determine expressions for the four coefficients of mutual influence in equations (4.70) and (4.72) to (4.74) for $\theta = \pm 45°$ in terms of engineering constants in the principal material directions.

4.15 Give your program the capability to compute the off-axis engineering constants E_x, E_y, G_{xy}, ν_{xy}, $\eta_{xy,x}$, $\eta_{xy,y}$, α_x, α_y, and α_{xy} as functions of lamina properties and fiber orientation. Use this capability to plot these constants for T300/5208 over the range –90° to 90° in 1° increments.

4.16 Develop an expression for the coefficients of thermal expansion $\{\alpha\}_x$ for $\theta = 90°$ in terms of the CTEs α_1 and α_2 in principal material coordinates.

4.17 Develop an expression for the coefficients of thermal expansion $\{\alpha\}_x$ for $\theta = 45°$ and $\theta = -45°$ in terms of the CTEs α_1 and α_2 in principal material coordinates.

4.18 Verify the plots for engineering constants E_x, G_{xy}, ν_{xy}, and $\eta_{xy,x}$ for T300/5208 and SCS-6/Ti-15-3 given in Figs. 4.9–4.12 and 4.14.

4.19 Modify your program to calculate the stresses $\{\sigma\}_x$ and total strains $\{\varepsilon\}_x$ as functions of given mechanical strains $\{\varepsilon^\sigma\}_x$ and uniform temperature change ΔT. As an example, calculate the stresses associated with applied mechanical strains $\{\varepsilon^\sigma\}_x = \begin{Bmatrix} 0.001 \\ 0.002 \\ 0.0005 \end{Bmatrix}$ and a temperature change $\Delta T = -180°\text{F}$ for T300/5208 carbon/epoxy for angles ranging from –90° to 90° in 1° increments.

4.20 Repeat Exercise 4.19 for the case where the total strains are given as $\{\varepsilon\}_x = \begin{Bmatrix} 0.001 \\ 0.002 \\ 0.0005 \end{Bmatrix}$, rather than the mechanical strains being given.

CHAPTER 5 LAMINATION THEORY

"Never trust an experimental result until it has been confirmed by theory."
Sir Arthur Eddington

5.1 Introduction

This chapter is devoted to the development of the equations which describe the linear elastic response of a laminated composite subjected to in-plane loads and bending moments. Individual layers are assumed to be homogeneous, orthotropic, or transversely isotropic and in a state of plane stress. A typical laminate with local lamina and global laminate coordinate systems is depicted in Fig. 5.1.

The theory presented here follows the original works of Pister and Dong (1959), Reissner and Stavsky (1961), and Dong et al. (1962).

5.2 Notation

For the laminate in Fig. 5.1, we take a global *x-y-z* coordinate system with *z* perpendicular to the plane of the laminate and *positive downward.* The origin of the coordinate system is located on the laminate midplane, centered between the top and bottom surfaces. The laminate has *N* layers *num-*

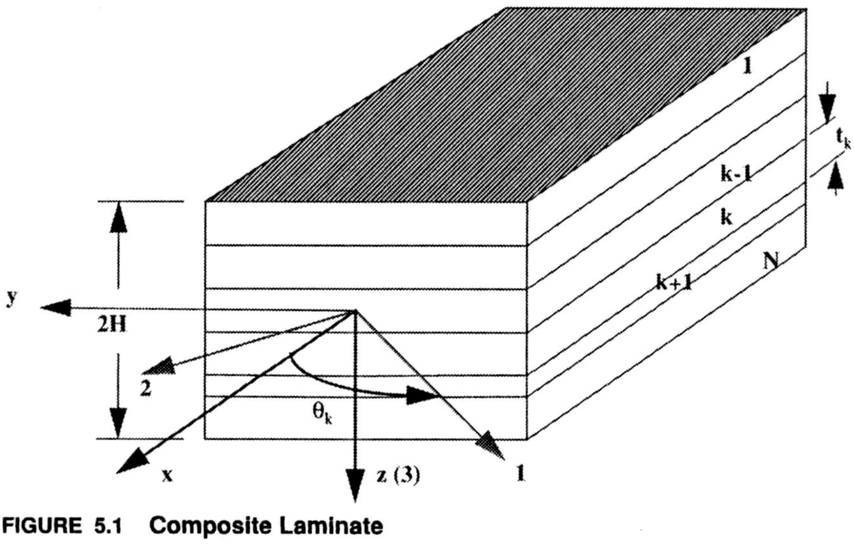

FIGURE 5.1 Composite Laminate

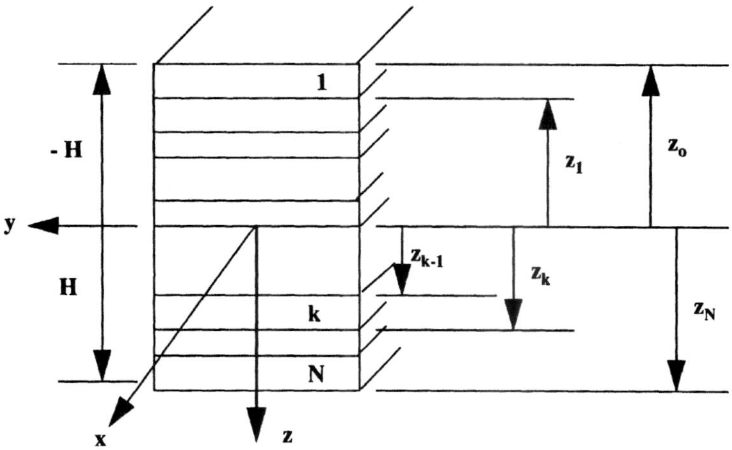

FIGURE 5.2 Laminate Coordinates

bered from top to bottom. Each layer has a distinct fiber orientation denoted θ_k. As indicated in Fig. 5.2, the z-coordinate of the *bottom* of the kth layer is designated z_k with the top of the layer being z_{k-1}. The thickness t_k of any layer is then $t_k = z_k - z_{k-1}$. The top surface of the laminate is denoted z_0 and the total thickness is $2H$.

A system has been devised to describe the stacking sequence of the layers in a laminate through the use of brackets, parentheses, slashes, and subscripts. The most common approach is to enclose the sequence within brackets; designate the fiber orientations starting from the top of the laminate ($k = 1$); separate distinct layers or groups of layers with a slash (/); designate repeated groups by a subscript n, where n is the number of times the group is repeated; designate laminates which are symmetric about the midplane by a subscript s on the entire bracket; and, on occasion, place a subscript T on the entire bracket to designate that the total stacking sequence is given.

Examples of laminate notation are given in Table 5.1.

Layer Sequence	Laminate	Layers
8 layers @ 0°	$[0_8]$	8
2 @ +45, 2 @ −45, symmetric	$[45_2/-45_2]_s$	8
+45/−45/+45/−45, symmetric	$[(\pm 45)_2]_s$	8
+45/−45/0/90, symmetric	$[\pm 45/0/90]_s$	8
50 groups of [±45/0/90], symmetric	$[(\pm 45/0/90)_{50}]_s$	400
+θ, −θ, symmetric	$[\pm\theta]_s$	4

TABLE 5.1 Example Laminate Notations

5.3 Assumptions of Lamination Theory

The following assumptions are fundamental to lamination theory:

1. The laminate consists of perfectly bonded layers (laminae).
2. Each layer is a homogeneous material with known effective properties.
3. Individual layer properties can be isotropic, orthotropic, or transversely isotropic.
4. Each layer is in a state of plane stress.
5. The laminate deforms according to the following Kirchhoff (1850) (also know as Kirchhoff-Love (Love, 1892)) assumptions for bending and stretching of thin plates:
   - Normals to the midplane remain straight and normal to the deformed midplane after deformation.
   - Normals to the midplane do not change length.

5.4 Strain-Displacement Relationships

The Kirchhoff assumption that normals remain straight and normal to the deformed midplane requires that the shear strains γ_{zx} and γ_{zy} be identically zero. The assumption that normals to the midplane do not change length requires that the z-displacement of the midplane be a function of x and y coordinates only, i.e., $w = w(x,y)$. The strain-displacement equation (2.16) then requires that the normal strain $\varepsilon_{33} = \varepsilon_z = 0$.

A schematic of the x-z plane of a deformed middle surface is shown in Fig. 5.3. If we further assume that the displacements are small, the slope, α, of the deformed middle surface also will be

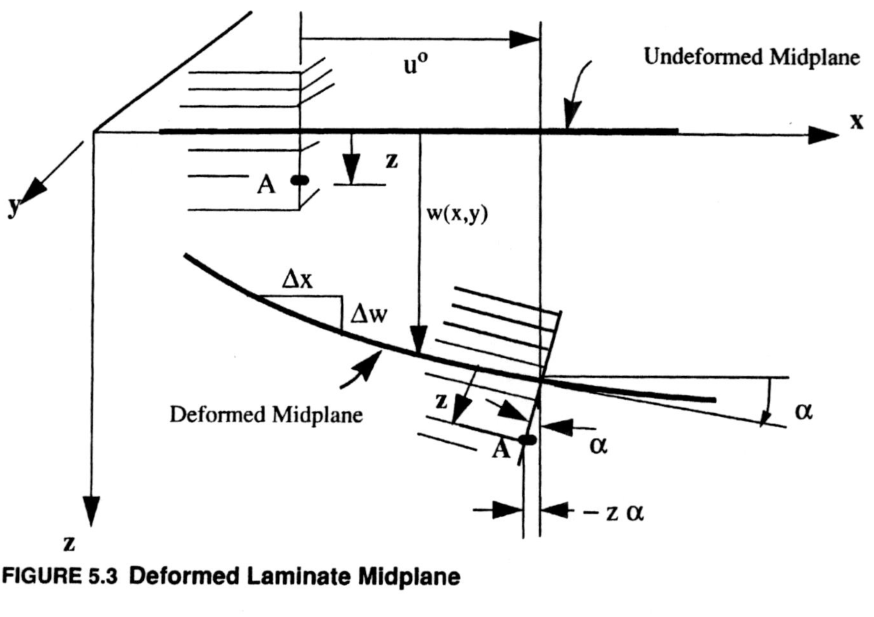

FIGURE 5.3 Deformed Laminate Midplane

small, and for small angles the tangent of the angle is approximately equal to the angle; thus

$$\tan\alpha = \frac{\partial w}{\partial x} \cong \alpha \tag{5.1}$$

The total x-displacement, u, of a generic point A in Fig. 5.3 can be written as the sum of the *midplane displacement*, u^o, plus the displacement due to the *rotation*, α, of the normal to the midplane. Thus,

$$u = u^o - z\tan\alpha = u^o - z\alpha = u^o - z\frac{\partial w}{\partial x} \tag{5.2}$$

Likewise, consideration of the y-displacements, v, in a y-z plane are expressed in terms of midplane y-displacement, v^o, and the slope, $\frac{\partial w}{\partial y}$:

$$v = v^o - z\frac{\partial w}{\partial y} \tag{5.3}$$

Since the normals do not change length, the plate deflection, w, is independent of z and we can write

$$w(x, y) = w^o(x, y) \tag{5.4}$$

where the superscript o refers to the midplane.

The displacements (5.2)–(5.4) can be combined with the strain-displacement equations (2.16) (written in x-y-z coordinates) to give the planar strains:

$$\begin{aligned}
\varepsilon_x &= \frac{\partial u}{\partial x} = \frac{\partial u^o}{\partial x} - z\frac{\partial^2 w}{\partial x^2} = \varepsilon_x^o + z\kappa_x \\
\varepsilon_y &= \frac{\partial v}{\partial y} = \frac{\partial v^o}{\partial y} - z\frac{\partial^2 w}{\partial y^2} = \varepsilon_y^o + z\kappa_y \\
\gamma_{xy} &= \left(\frac{\partial u}{\partial y} + \frac{\partial v}{\partial x}\right) = \frac{\partial u^o}{\partial y} - 2z\frac{\partial^2 w}{\partial x \partial y} + \frac{\partial v^o}{\partial x} = \gamma_{xy}^o + z\kappa_{xy}
\end{aligned} \tag{5.5}$$

where the curvatures $\{\kappa\}$ are defined

$$\begin{aligned}
\kappa_x &= -\frac{\partial^2 w}{\partial x^2} \\
\kappa_y &= -\frac{\partial^2 w}{\partial y^2} \\
\kappa_{xy} &= -2\frac{\partial^2 w}{\partial x \partial y}
\end{aligned} \tag{5.6}$$

Combining (5.5) and (5.6) in matrix form, we have

$$\begin{Bmatrix} \varepsilon_x \\ \varepsilon_y \\ \gamma_{xy} \end{Bmatrix} = \begin{Bmatrix} \varepsilon_x^o \\ \varepsilon_y^o \\ \gamma_{xy}^o \end{Bmatrix} + z \begin{Bmatrix} \kappa_x \\ \kappa_y \\ \kappa_{xy} \end{Bmatrix} \tag{5.7}$$

or more simply

$$\{\varepsilon\}_x = \{\varepsilon^o\}_x + z\{\kappa\}_x \tag{5.8}$$

The last equation expresses the total strains, $\{\varepsilon\}_x$, at any z-location in the laminate in terms of the *midplane strains*, $\{\varepsilon^o\}_x$, and the *curvatures*, $\{\kappa\}_x$; it is a fundamental equation of lamination theory. The total strains are the sum of the midplane strains and the strain associated with curvature. It is important to note that we have arrived at this result without specification of the type of material or any statement as to whether or not the plate is composed of more than one layer. It is the result of the Kirchhoff assumptions on displacements, independent of material considerations.

5.5 Stresses

The stresses at any z-location can now be determined simply by a substitution of the strains (5.8) into the plane stress constitutive equation (4.32), rewritten here as (5.9):

$$\{\sigma\}_x = [\bar{Q}]^k\{\varepsilon\}_x \tag{5.9}$$

where $[\bar{Q}]^k$ is the transformed reduced stiffness of the kth layer corresponding to the z-location. A very important point in using the laminate constitutive equation is the fact that the transformed reduced stiffness matrix $[\bar{Q}]^k$ varies with the fiber orientation of each layer. Combining (5.8) and (5.9) gives a general expression for the z-location stresses in the kth layer of a laminate:

$$\{\sigma\}^k = [\bar{Q}]^k\{\varepsilon^o\} + [\bar{Q}]^k z\{\kappa\} \tag{5.10}$$

The first term in (5.10) corresponds to the stresses associated with midplane strains, and the second term corresponds to the stresses associated with bending strains. The subscript x has been dropped since presence of the transformed reduced stiffness $[\bar{Q}]^k$ indicates that the equation is valid for any arbitrary coordinate system. *It is noted that* $\{\varepsilon^o\}$ *and* $\{\kappa\}$ *are associated with the laminate and are independent of z.*

5.6 In-Plane Forces per Unit Length

The *in-plane forces per unit length* $\{N_x, N_y, N_{xy}\}$ (depicted in Fig. 5.4) are defined as the through-thickness integrals of the planar stresses in the laminate. Thus

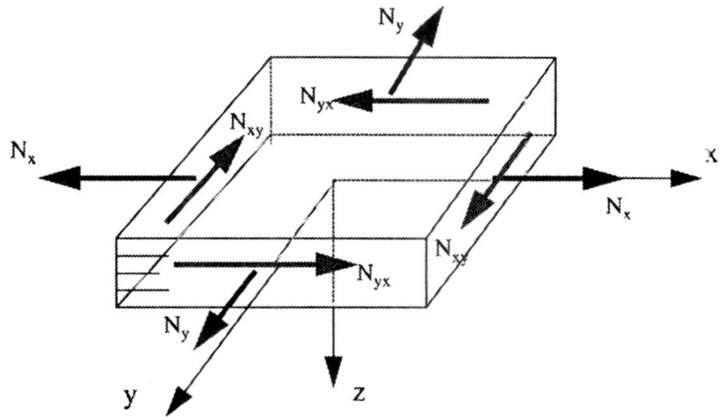

FIGURE 5.4 In-Plane Forces per Unit Length

$$\begin{aligned} N_x &= \int_{-H}^{H} \sigma_x \, dz \\ N_y &= \int_{-H}^{H} \sigma_y \, dz \\ N_{xy} &= \int_{-H}^{H} \tau_{xy} \, dz \end{aligned} \tag{5.11}$$

These three equations can be written in the condensed form

$$\{N\} = \int_{-H}^{H} \{\sigma\} dz \tag{5.12}$$

Combining equations (5.10) and (5.12) gives

$$\{N\} = \int_{-H}^{H} [\bar{Q}]^k \{\varepsilon^o\} dz + \int_{-H}^{H} [\bar{Q}]^k \{\kappa\} z \, dz \tag{5.13}$$

Recalling that $\{\varepsilon^o\}$ and $\{\kappa\}$ are independent of z, the integral over the laminate thickness can be replaced by a summation of integrals over the thicknesses of the individual layers. Thus

$$\{N\} = \sum_{k=1}^{N} \left(\int_{z_{k-1}}^{z_k} [\bar{Q}]^k dz \right) \{\varepsilon^o\} + \sum_{k=1}^{N} \left(\int_{z_{k-1}}^{z_k} [\bar{Q}]^k z \, dz \right) \{\kappa\} \tag{5.14}$$

This equation for the in-plane forces per unit length can be written

$$\{N\} = [A]\{\varepsilon^o\} + [B]\{\kappa\} \tag{5.15}$$

where the $[A]$ and $[B]$ matrices are defined as summations over all N layers,

$$[A] = \sum_{k=1}^{N} [\bar{Q}]^k (z_k - z_{k-1}) \tag{5.16}$$

$$[B] = \frac{1}{2}\sum_{k=1}^{N} [\bar{Q}]^k (z_k^2 - z_{k-1}^2) \tag{5.17}$$

and we have used the fact that the stiffness $[\bar{Q}]^k$ is constant within each layer.

The fundamental relationship (5.15) relates the in-plane forces per unit length to the midplane strains $\{\varepsilon^o\}$ and the laminate curvatures $\{\kappa\}$ through the $[A]$ and $[B]$ matrices. The $[A]$ matrix represents the in-plane stiffness and the $[B]$ matrix defines the bending-stretching coupling. From (5.16) it is evident that $[A]$ is a function of the layer thicknesses, $t_k = (z_k - z_{k-1})$, but is independent of the stacking sequence of the layers. In contrast, from (5.17) $[B]$ is dependent on the stacking sequence of the individual layers through the term $(z_k^2 - z_{k-1}^2)$. This point will be discussed more fully in subsequent sections. It is also evident from the definitions (5.16) and (5.17) that both $[A]$ and $[B]$ are symmetric matrices since $[\bar{Q}]^k$ is symmetric.

5.7 Moments per Unit Length

We define the *moments per unit length* $\{M_x, M_y, M_{xy}\}$ (Fig. 5.5) as the integrals of the differential forces, $\sigma_i\, dz$, times the moment arm, z, integrated over the laminate thickness. Thus

$$\begin{aligned} M_x &= \int_{-H}^{H} \sigma_x z\, dz \\ M_y &= \int_{-H}^{H} \sigma_y z\, dz \\ M_{xy} &= \int_{-H}^{H} \tau_{xy} z\, dz \end{aligned} \tag{5.18}$$

In condensed form, we write

$$\{M\} = \int_{-H}^{H} \{\sigma\} z\, dz \tag{5.19}$$

Substituting for the stresses from (5.10) into (5.19) gives the desired expression for the moments per unit length:

$$\{M\} = [B]\{\varepsilon^o\} + [D]\{\kappa\} \tag{5.20}$$

where the bending stiffness matrix, $[D]$, is defined as

$$[D] = \frac{1}{3}\sum_{k=1}^{N} [\bar{Q}]^k (z_k^3 - z_{k-1}^3) \tag{5.21}$$

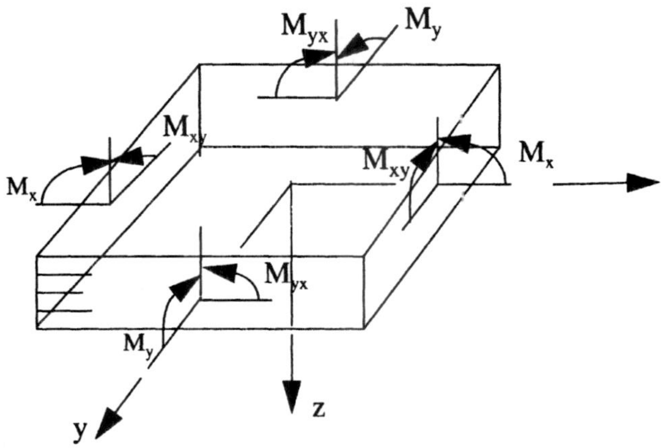

FIGURE 5.5 Moments per Unit Length

From (5.21) we see that the $[D]$ matrix is dependent on the stacking sequence of the individual layers through the $(z_k^3 - z_{k-1}^3)$ term. It is also evident from the definition (5.21) that the $[D]$ matrix is a symmetric matrix. From (5.20) we see that the moments per unit length are related to the mid-plane strains $\{\varepsilon^o\}$ and the laminate curvatures $\{\kappa\}$ through the $[B]$ and $[D]$ matrices, both of which are symmetric and functions of the laminate stacking sequence.

5.8 Laminate Constitutive Relations

Equations (5.15) and (5.20) can be combined to give the fundamental equation of lamination theory:

$$\begin{Bmatrix} N \\ M \end{Bmatrix} = \begin{bmatrix} A & B \\ B & D \end{bmatrix} \begin{Bmatrix} \varepsilon^o \\ \kappa \end{Bmatrix} \tag{5.22}$$

This equation can be written in expanded form as

$$\begin{Bmatrix} N_x \\ N_y \\ N_{xy} \\ M_x \\ M_y \\ M_{xy} \end{Bmatrix} = \begin{bmatrix} A_{11} & A_{12} & A_{16} & B_{11} & B_{12} & B_{16} \\ A_{12} & A_{22} & A_{26} & B_{12} & B_{22} & B_{26} \\ A_{16} & A_{26} & A_{66} & B_{16} & B_{26} & B_{66} \\ B_{11} & B_{12} & B_{16} & D_{11} & D_{12} & D_{16} \\ B_{12} & B_{22} & B_{26} & D_{12} & D_{22} & D_{26} \\ B_{16} & B_{26} & B_{66} & D_{16} & D_{26} & D_{66} \end{bmatrix} \begin{Bmatrix} \varepsilon_x^o \\ \varepsilon_y^o \\ \gamma_{xy}^o \\ \kappa_x \\ \kappa_y \\ \kappa_{xy} \end{Bmatrix} \tag{5.23}$$

Equation (5.22) or (5.23) clearly shows the coupling between the bending and stretching response of a laminate reflected in the $[B]$ matrix. If $[B] = 0$, then the in-plane response is decoupled from the bending response. As we shall see in a later section, this is indeed the case for laminates which are symmetric about their midplane. We also note that the $\begin{bmatrix} A & B \\ B & D \end{bmatrix}$ matrix of (5.23) is symmetric.

Equation (5.22) can be inverted to give

$$\begin{Bmatrix} \varepsilon^o \\ \kappa \end{Bmatrix} = \begin{bmatrix} A' & B' \\ C' & D' \end{bmatrix} \begin{Bmatrix} N \\ M \end{Bmatrix} \tag{5.24}$$

where the specific forms of the individual matrices can be determined as follows. Recalling (5.15) and (5.20), we have

$$\{N\} = [A]\{\varepsilon^o\} + [B]\{\kappa\} \tag{5.25}$$

$$\{M\} = [B]\{\varepsilon^o\} + [D]\{\kappa\} \tag{5.26}$$

Solving (5.25) for the midplane strains $\{\varepsilon^o\}$ gives

$$\{\varepsilon^o\} = [A]^{-1}\{N\} - [A]^{-1}[B]\{\kappa\} \tag{5.27}$$

Substituting (5.27) into (5.26), we have

$$\{M\} = [B][A]^{-1}\{N\} + ([D] - [B][A]^{-1}[B])\{\kappa\} \tag{5.28}$$

Equations (5.27) and (5.28) can be written in the form

$$\begin{Bmatrix} \varepsilon^o \\ M \end{Bmatrix} = \begin{bmatrix} A^* & B^* \\ C^* & D^* \end{bmatrix} \begin{Bmatrix} N \\ \kappa \end{Bmatrix} \tag{5.29}$$

where

$$\begin{aligned} [A^*] &= [A]^{-1} \\ [B^*] &= -[A]^{-1}[B] \\ [C^*] &= [B][A]^{-1} \\ [D^*] &= [D] - [B][A]^{-1}[B] \end{aligned} \tag{5.30}$$

Writing out the two equations in (5.29), we have

$$\{\varepsilon^o\} = [A^*]\{N\} + [B^*]\{\kappa\} \tag{5.31}$$

$$\{M\} = [C^*]\{N\} + [D^*]\{\kappa\} \tag{5.32}$$

Solving (5.32) for $\{\kappa\}$ gives

$$\{\kappa\} = -[D^*]^{-1}[C^*]\{N\} + [D^*]^{-1}\{M\} \tag{5.33}$$

Substituting (5.33) into (5.31) gives

$$\{\varepsilon^o\} = ([A^*] - [B^*][D^*]^{-1}[C^*])\{N\} + [B^*][D^*]^{-1}\{M\} \tag{5.34}$$

Equations (5.34) and (5.33) can be written in the combined form

$$\begin{Bmatrix} \varepsilon^o \\ \kappa \end{Bmatrix} = \begin{bmatrix} A' & B' \\ C' & D' \end{bmatrix} \begin{Bmatrix} N \\ M \end{Bmatrix} \tag{5.35}$$

where

$$\begin{aligned} [A'] &= [A^*] - [B^*][D^*]^{-1}[C^*] \\ [B'] &= [B^*][D^*]^{-1} \\ [C'] &= -[D^*]^{-1}[C^*] \\ [D'] &= [D^*]^{-1} \end{aligned} \tag{5.36}$$

Combining (5.36) with (5.30) shows the explicit dependence of the coefficients in (5.35) on the $[B]$ matrix.

$$\begin{aligned} [A'] &= [A]^{-1} + [A]^{-1}[B][D^*]^{-1}[B][A]^{-1} \\ [B'] &= -[A]^{-1}[B][D^*]^{-1} \\ [C'] &= -[D^*]^{-1}[B][A]^{-1} \\ [D'] &= ([D] - [B][A]^{-1}[B])^{-1} \end{aligned} \tag{5.37}$$

From the above expressions for $[B']$ and $[C']$ in (5.37), we recognize that

$$[C'] = [B']^T \tag{5.38}$$

Thus, (5.24) (or (5.35)) can be written

$$\begin{Bmatrix} \varepsilon^o \\ \kappa \end{Bmatrix} = \begin{bmatrix} A' & B' \\ B'^T & D' \end{bmatrix} \begin{Bmatrix} N \\ M \end{Bmatrix} \tag{5.39}$$

and in the expanded form as

$$\begin{Bmatrix} \varepsilon_x^o \\ \varepsilon_y^o \\ \gamma_{xy}^o \\ \kappa_x \\ \kappa_y \\ \kappa_{xy} \end{Bmatrix} = \begin{bmatrix} A'_{11} & A'_{12} & A'_{16} & B'_{11} & B'_{12} & B'_{16} \\ A'_{12} & A'_{22} & A'_{26} & B'_{12} & B'_{22} & B'_{26} \\ A'_{16} & A'_{26} & A'_{66} & B'_{16} & B'_{26} & B'_{66} \\ B'_{11} & B'_{12} & B'_{16} & D'_{11} & D'_{12} & D'_{16} \\ B'_{12} & B'_{22} & B'_{26} & D'_{12} & D'_{22} & D'_{26} \\ B'_{16} & B'_{26} & B'_{66} & D'_{16} & D'_{26} & D'_{66} \end{bmatrix} \begin{Bmatrix} N_x \\ N_y \\ N_{xy} \\ M_x \\ M_y \\ M_{xy} \end{Bmatrix} \tag{5.40}$$

It is noted that the full $\begin{bmatrix} A' & B' \\ B'^T & D' \end{bmatrix}$ matrix in (5.40) is symmetric like the $\begin{bmatrix} A & B \\ B & D \end{bmatrix}$ matrix of (5.23). This follows directly because the inverse of a symmetric matrix is symmetric.

Equations (5.22), (5.23), (5.24), and (5.40) are fundamental equations of lamination theory. They describe the response of a laminated plate under the influence of given midplane strains and laminate curvatures or given forces and moments per unit length. If the midplane strains, laminate curvatures, and stacking sequence of a laminate are known, the strains and stresses at any z-location can be determined directly from (5.8) and (5.10). The application of these equations for specific classes of laminates is described in the following sections of this chapter.

5.9 Symmetric Laminates

If the stacking sequence of the layers is symmetric about the midplane of the laminate, the $[B]$ matrix is identically zero. This can be proved by considering the contributions of two identical layers, p and q (same material, fiber orientation, and thickness), located symmetrically about the midplane, as depicted in Fig. 5.6. Since the layers are of the same material and have identical fiber orientations, the individual terms of the reduced stiffness matrices are also identical, i.e.,

$$\bar{Q}_{ij}(p) = \bar{Q}_{ij}(q) \tag{5.41}$$

The symmetric location of the layers with respect to the midplane requires that

$$z_p = -z_{q-1} \tag{5.42}$$

$$z_{p-1} = -z_q \tag{5.43}$$

From the definition of $[B]$, (5.17), the contribution of the two symmetric layers to any coefficient B_{ij} is

$$B_{ij}(p+q) = \frac{1}{2}\bar{Q}_{ij}(p)(z_p^2 - z_{p-1}^2) + \frac{1}{2}\bar{Q}_{ij}(q)(z_q^2 - z_{q-1}^2) \tag{5.44}$$

Using (5.41)–(5.43) in (5.44) gives

$$B_{ij}(p+q) = \frac{1}{2}\bar{Q}_{ij}(p)(z_p^2 - z_{p-1}^2 + z_{p-1}^2 - z_p^2) = 0 \tag{5.45}$$

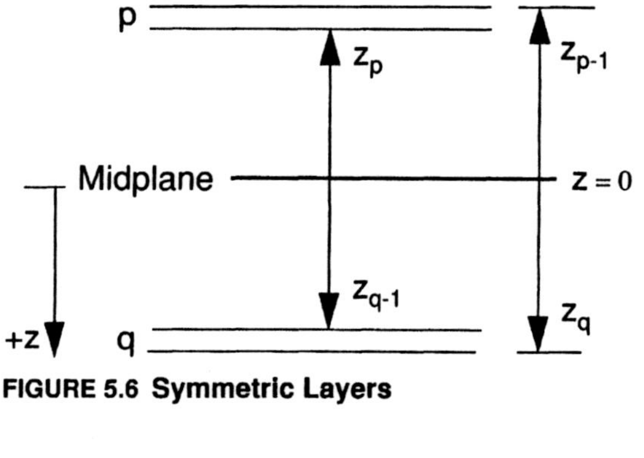

FIGURE 5.6 Symmetric Layers

Thus, the contribution from *any two* symmetric layers is zero for all B_{ij} terms. Hence, *the* $[B]$ *matrix is zero for a symmetric laminate.* In contrast, the contribution to A_{ij} and D_{ij} for the two symmetric layers is nonzero, i.e.,

$$\begin{aligned} A_{ij}(p+q) &= 2\bar{Q}_{ij}(z_p - z_{p-1}) = 2\bar{Q}_{ij}(t_p) \neq 0 \\ D_{ij}(p+q) &= \frac{2}{3}\bar{Q}_{ij}(z_p^3 - z_{p-1}^3) \neq 0 \end{aligned} \tag{5.46}$$

where t_p is the thickness of the pth layer.

Setting $[B] = 0$ in (5.22), the constitutive equations for a symmetric laminate are

$$\{N\} = [A]\{\varepsilon^o\} \tag{5.47}$$

$$\{M\} = [D]\{\kappa\} \tag{5.48}$$

These equations clearly show that $[A]$ represents the in-plane stiffness and $[D]$ represents the bending stiffness of a symmetric laminate.

The constitutive equations for symmetric laminates can be expanded to more clearly show the effects of the individual terms. For in-plane response we have

$$\begin{Bmatrix} N_x \\ N_y \\ N_{xy} \end{Bmatrix} = \begin{bmatrix} A_{11} & A_{12} & A_{16} \\ A_{12} & A_{22} & A_{26} \\ A_{16} & A_{26} & A_{66} \end{bmatrix} \begin{Bmatrix} \varepsilon_x^o \\ \varepsilon_y^o \\ \gamma_{xy}^o \end{Bmatrix} \tag{5.49}$$

or

$$\begin{aligned} N_x &= A_{11}\varepsilon_x^o + A_{12}\varepsilon_y^o + A_{16}\gamma_{xy}^o \\ N_y &= A_{12}\varepsilon_x^o + A_{22}\varepsilon_y^o + A_{26}\gamma_{xy}^o \\ N_{xy} &= A_{16}\varepsilon_x^o + A_{26}\varepsilon_y^o + A_{66}\gamma_{xy}^o \end{aligned} \tag{5.50}$$

As these equations indicate, A_{11} and A_{22} correspond to in-plane stiffness, A_{66} corresponds to shear stiffness, A_{12} is a Poisson-type term, and A_{16} and A_{26} are normal-shear coupling terms. The presence of the normal-shear coupling is a major distinction between the response of fibrous composites and that of isotropic materials. The coupling terms are zero for isotropic materials and for orthotropic materials in the principal material coordinates, but they are nonzero when the principal material directions are off-axis.

Similar expansion of (5.48) for the bending response of a symmetric laminate gives

$$\begin{Bmatrix} M_x \\ M_y \\ M_{xy} \end{Bmatrix} = \begin{bmatrix} D_{11} & D_{12} & D_{16} \\ D_{12} & D_{22} & D_{26} \\ D_{16} & D_{26} & D_{66} \end{bmatrix} \begin{Bmatrix} \kappa_x \\ \kappa_y \\ \kappa_{xy} \end{Bmatrix} \tag{5.51}$$

and

$$\begin{aligned} M_x &= D_{11}\kappa_x + D_{12}\kappa_y + D_{16}\kappa_{xy} \\ M_y &= D_{12}\kappa_x + D_{22}\kappa_y + D_{26}\kappa_{xy} \\ M_{xy} &= D_{16}\kappa_x + D_{26}\kappa_y + D_{66}\kappa_{xy} \end{aligned} \tag{5.52}$$

As for the in-plane response, the D_{11} and D_{22} terms correspond directly to normal bending stiffnesses along the x and y axes, respectively. D_{66} is the twisting (shear) stiffness, D_{12} is a coupling term between the normal moments M_x and M_y and normal curvatures κ_x and κ_y, and D_{16} and D_{26} are coupling terms between normal moments and twist curvatures, κ_{xy}, as well as between twist moments and normal curvatures.

From (5.37) with $[B] = 0$ for a symmetric laminate, we have the results

$$\begin{aligned} [A'] &= [A]^{-1} \\ [B'] &= 0 \\ [C'] &= 0 \\ [D'] &= [D]^{-1} \end{aligned} \tag{5.53}$$

and (5.39) takes the explicit form

$$\begin{Bmatrix} \varepsilon^o \\ \kappa \end{Bmatrix} = \begin{bmatrix} A^{-1} & 0 \\ 0 & D^{-1} \end{bmatrix} \begin{Bmatrix} N \\ M \end{Bmatrix} \tag{5.54}$$

This result is, of course, consistent with the inverted forms of (5.47) and (5.48).

5.10 Special Laminates

Laminates can be classified according to the stacking sequence and fiber orientation of the individual layers. We discussed the very special case of symmetric laminates in the previous section. Here we consider other classes of special laminates. The discussion will be aided by consideration of the explicit equations for the transformed, reduced stiffness coefficients (4.34), which are repeated as (5.55).

$$\begin{aligned} \bar{Q}_{11} &= Q_{11}m^4 + 2(Q_{12} + 2Q_{66})m^2n^2 + Q_{22}n^4 \\ \bar{Q}_{12} &= (Q_{11} + Q_{22} - 4Q_{66})m^2n^2 + Q_{12}(n^4 + m^4) \\ \bar{Q}_{22} &= Q_{11}n^4 + 2(Q_{12} + 2Q_{66})m^2n^2 + Q_{22}m^4 \\ \bar{Q}_{16} &= [(Q_{11} - Q_{12} - 2Q_{66})m^2 + (Q_{12} - Q_{22} + 2Q_{66})n^2]mn \\ \bar{Q}_{26} &= [(Q_{11} - Q_{12} - 2Q_{66})n^2 + (Q_{12} - Q_{22} + 2Q_{66})m^2]mn \\ \bar{Q}_{66} &= (Q_{11} + Q_{22} - 2Q_{12} - 2Q_{66})m^2n^2 + Q_{66}(n^4 + m^4) \end{aligned} \tag{5.55}$$

5.10.1 Specially Orthotropic Laminates

Laminates with $A_{16} = A_{26} = 0$ are called *specially orthotropic* because such laminates do not exhibit coupling between in-plane extensional and shear responses. It is also important to note that D_{16} and D_{26} are not necessarily zero for a specially orthotropic laminate. Thus the term *specially orthotropic* refers to the in-plane response and not the bending response. Several specially orthotropic laminates are discussed in the following sections.

5.10.1.1 Cross-Ply Laminates

Cross-ply laminates consist of layers with fiber orientations of 0° and 90° only. Study of (5.55) indicates that the $\bar{Q}_{16}$ and $\bar{Q}_{26}$ terms involve products of mn (cos θ sin θ). Hence these terms are zero for fiber orientations of 0° or 90°. Thus *laminates consisting of only 0° and 90° layers are specially orthotropic*. In cross-ply laminates the D_{16} and D_{26} terms are also zero since the $\bar{Q}_{16}$ and $\bar{Q}_{26}$ terms are identically zero. It is noted that there is no restriction on the thickness of the layers for cross-ply laminates to be specially orthotropic since the $\bar{Q}_{16}$ and $\bar{Q}_{26}$ terms are identically zero. Examples of cross-ply laminates include (but are not limited to) [0/90], $[0/90]_s$, $[0_3/90]_s$, and $[(0/90)_n]_s$.

It follows from (5.55) that for a cross-ply laminate $\bar{Q}_{11}(0) = Q_{11} = \bar{Q}_{22}(90)$, $\bar{Q}_{22}(0) = Q_{22} = \bar{Q}_{11}(90)$, $\bar{Q}_{12}(0) = Q_{12}(90) = Q_{12}$, and $\bar{Q}_{66}(0) = \bar{Q}_{66}(90) = Q_{66}$. Thus the in-plane stiffness matrix $[A]$ for a symmetric, cross-ply laminate of the form $[0_{n_1}/90_{n_2}]_s$, with equal layer thickness t and total laminate thickness $2H = 2t(n_1 + n_2)$, can be written in the simplified form

$$[A] = 2t\begin{bmatrix} (n_1Q_{11} + n_2Q_{22}) & Q_{12}(n_1 + n_2) & 0 \\ Q_{12}(n_1 + n_2) & (n_2Q_{11} + n_1Q_{22}) & 0 \\ 0 & 0 & Q_{66}(n_1 + n_2) \end{bmatrix} \tag{5.56}$$

For the special case in which there is an equal number of 0° and 90° layers (i.e., $n_1 = n_2$), (5.56) reduces to

$$[A] = H\begin{bmatrix} (Q_{11} + Q_{22}) & 2Q_{12} & 0 \\ 2Q_{12} & (Q_{11} + Q_{22}) & 0 \\ 0 & 0 & 2Q_{66} \end{bmatrix} \tag{5.57}$$

Using (4.16), the stiffness matrix, $[A]$, in (5.56) can be written in terms of engineering constants in the form

$$[A] = \frac{2t}{(1 - \nu_{12}\nu_{21})}\begin{bmatrix} (n_1E_1 + n_2E_2) & \nu_{12}E_2(n_1 + n_2) & 0 \\ \nu_{12}E_2(n_1 + n_2) & (n_2E_1 + n_1E_2) & 0 \\ 0 & 0 & G_{12}(1 - \nu_{12}\nu_{21})(n_1 + n_2) \end{bmatrix} \tag{5.58}$$

And, of course, for $n_1 = n_2$ we have

$$[A] = \frac{H}{(1 - \nu_{12}\nu_{21})}\begin{bmatrix} (E_1 + E_2) & 2\nu_{12}E_2 & 0 \\ 2\nu_{12}E_2 & (E_1 + E_2) & 0 \\ 0 & 0 & 2G_{12}(1 - \nu_{12}\nu_{21}) \end{bmatrix} \tag{5.59}$$

5.10.1.2 Angle-Ply Laminates

Laminates consisting of an equal number of equal-thickness layers at +θ and −θ fiber orientations are called *angle-ply laminates.* Such laminates are specially orthotropic. The fact that A_{16} and A_{26} are zero for angle-ply laminates can be demonstrated through consideration of the definition of A_{ij},

$$A_{ij} = \sum_{k=1}^{N} \bar{Q}_{ij}^{k} t_k \tag{5.60}$$

and the expressions in (5.55) for $\bar{Q}_{16}$ and $\bar{Q}_{26}$. The common factor mn causes layers of opposite sign to have $\bar{Q}_{16}$ and $\bar{Q}_{26}$ values of opposite sign. This is also evident in Fig. 4.5. Thus the sum of +θ and −θ layers of the same thickness is always zero, and the laminate is specially orthotropic. It is emphasized that the +θ and −θ layers must be present in the same thickness for the laminate to be specially orthotropic. Examples of angle-ply laminates include (but are not limited to) [±θ], $[\pm\theta]_s$, and $[(\pm\theta)_n]_s$. Unlike the case for cross-ply laminates, D_{16} and D_{26} are not zero for angle-ply laminates because the +θ and −θ layers are not at the same distance from the midplane.

Following from (5.55), for an angle-ply laminate we have $\bar{Q}_{11}(+\theta) = \bar{Q}_{11}(-\theta)$, $\bar{Q}_{12}(+\theta) = \bar{Q}_{12}(-\theta)$, $\bar{Q}_{22}(+\theta) = \bar{Q}_{22}(-\theta)$, and $\bar{Q}_{66}(+\theta) = \bar{Q}_{66}(-\theta)$. Thus the in-plane stiffness matrix $[A]$ for a symmetric angle-ply laminate of the form $[(\pm\theta)_n]_s$, with equal layer thicknesses t and total laminate thickness $2H = 4nt$, can be written in the simplified form

$$[A] = 2H \begin{bmatrix} \bar{Q}_{11}(\theta) & \bar{Q}_{12}(\theta) & 0 \\ \bar{Q}_{12}(\theta) & \bar{Q}_{22}(\theta) & 0 \\ 0 & 0 & \bar{Q}_{66}(\theta) \end{bmatrix} \tag{5.61}$$

5.10.1.3 Other Specially Orthotropic Laminates

In view of the preceding discussion on the specially orthotropic nature of cross-ply and angle-ply laminates, any combination of cross-ply and angle-ply laminates is specially orthotropic. Other examples of specially orthotropic laminates include

$[0/\pm\theta]_s$ $\quad$ $[0/\pm\theta/90]_s$ $\quad$ $[0/90/\pm\theta]_s$

$[\pm\theta/0/90]_s$ $\quad$ $[\theta/0/-\theta]_s$ $\quad$ $[\theta/90/-\theta]_s$

This list of specially orthotropic laminates is not at all inclusive. It is intended simply to provide examples of some specially orthotropic laminates.

5.10.2 Balanced Laminates

Another term that is often used to describe laminates is *balanced laminate,* defined as any laminate in which all off-axis $+\theta_i$ and $-\theta_i$ layers are present in equal thicknesses t_i. An angle-ply laminate is one type of balanced laminate. However, balanced laminates are not limited to only one ±θ fiber orientation. They can include other off-axis fiber orientations in $\pm\theta_i$ pairs and 0° and 90° layers.

Examples of balanced laminates include

$[0/\pm\theta]_s$	$[0/\pm\theta/90]_s$	$[0/90/\pm\theta]_s$
$[\pm\theta/0/90]_s$	$[\theta/0/-\theta]_s$	$[\theta/90/-\theta]_s$
$[+\theta_1/-\theta_1/+\theta_2/-\theta_2]_s$	$[+\theta_1/-\theta_1/+\theta_2/-\theta_2]$	$[(\pm\theta_1)_4/(\pm\theta_2)_8]$

By their definition, balanced laminates are also specially orthotropic.

5.10.3 Quasi-Isotropic Laminates

A very important and common class of laminates is called *quasi-isotropic* because the in-plane effective elastic response of this class of laminates is isotropic. All symmetric laminates with $2N$ equal-thickness layers ($N \geq 3$) and N equal angles between fiber orientations (Fig. 5.7) are quasi-isotropic. For N equal angles of $\Delta\theta$ between fiber orientations, we can write

$$\Delta\theta = \frac{\pi}{N} \tag{5.62}$$

Examples of quasi-isotropic laminates are those with angles $\Delta\theta$ between fibers of 60° ($\pi/3$), 45° ($\pi/4$), 36° ($\pi/5$), 30° ($\pi/6$), etc.

The proof of the statement on isotropy in the plane follows directly from laminate analysis. Since all layers are of equal thickness, take the layer thickness to be

$$t_k = \frac{2H}{2N} = \frac{H}{N} \tag{5.63}$$

For a symmetric laminate we have, from (5.47),

$$\{N\} = [A]\{\varepsilon^o\} \tag{5.64}$$

where, using (5.63), the A_{ij} are defined as

$$A_{ij} = \sum_{k=1}^{2N} \bar{Q}_{ij}^k t_k = 2\frac{H}{N}\sum_{k=1}^{N} \bar{Q}_{ij}^k \tag{5.65}$$

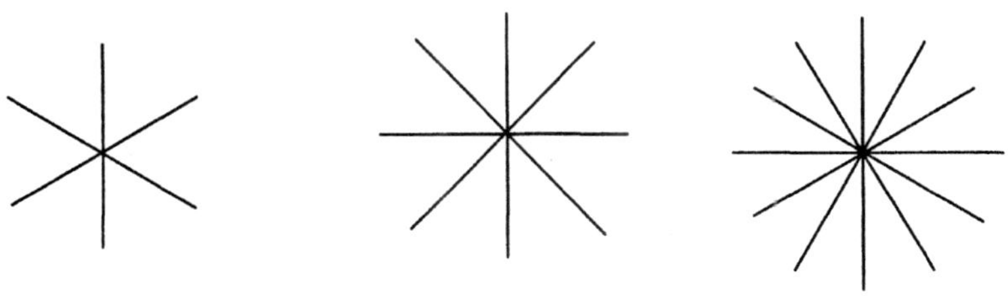

N = 3 (Δθ = 60) N = 4 (Δθ = 45) N = 6 (Δθ = 30)

FIGURE 5.7 Quasi-Isotropic Fiber Orientations

We next use the expressions (4.83) and (4.84) for Q_{ij} in terms of the invariants U_i. As an example, for any layer k,

$$\bar{Q}_{11}^{k} = U_1 + U_2 \cos 2\theta_k + U_3 \cos 4\theta_k \tag{5.66}$$

where θ_k is the orientation of the kth layer measured from the global x-axis (Fig. 5.8).

We can now write the expression for A_{11} from (5.65) using (5.66) as

$$A_{11} = \frac{2H}{N}\left[NU_1 + U_2 \sum_{k=1}^{N} \cos 2\theta_k + U_3 \sum_{k=1}^{N} \cos 4\theta_k\right] \tag{5.67}$$

Now consider the stiffness in any arbitrary direction ϕ from the global x-axis (Fig. 5.8):

$$A_{11}(\phi) = \frac{2H}{N}\left[NU_1 + U_2 \sum_{k=1}^{N} \cos 2(\theta_k - \phi) + U_3 \sum_{k=1}^{N} \cos 4(\theta_k - \phi)\right] \tag{5.68}$$

Using trigonometric identities we can write (5.68) as

$$\begin{aligned} A_{11}(\phi) = \frac{2H}{N} \cdot \Bigg[NU_1 &+ U_2\left(\cos 2\phi \sum_{k=1}^{N} \cos 2\theta_k + \sin 2\phi \sum_{k=1}^{N} \sin 2\theta_k\right) \\ &+ U_3\left(\cos 4\phi \sum_{k=1}^{N} \cos 4\theta_k + \sin 4\phi \sum_{k=1}^{N} \sin 4\theta_k\right)\Bigg] \end{aligned} \tag{5.69}$$

Now, for N equally spaced fiber angles θ_k, we can write

$$\theta_k = \frac{k\pi}{N} \qquad (k = 1, \ldots, N) \tag{5.70}$$

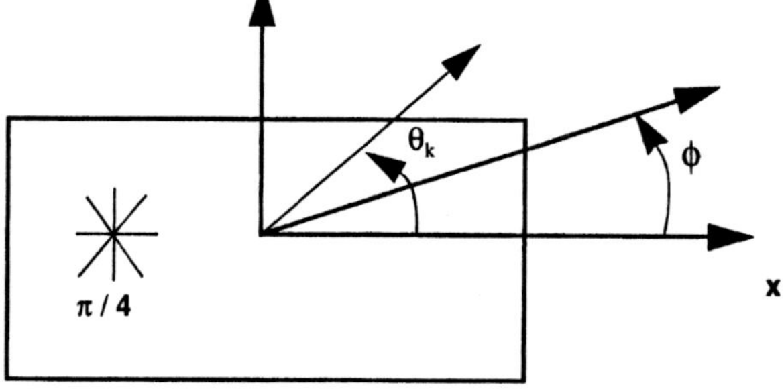

FIGURE 5.8 Arbitrary Direction ϕ in Quasi-Isotropic Laminate

Therefore, the first summation in (5.69) can be written in the form

$$\sum_{k=1}^{N} \cos 2\theta_k = \cos\frac{2\pi}{N} + \cos\frac{4\pi}{N} + \cdots + \cos\frac{2N\pi}{N} \tag{5.71}$$

This series can be written in the form

$$\cos x + \cos 2x + \cdots + \cos Nx = \frac{\sin\left\{\left(N + \frac{1}{2}\right)x\right\}}{2\sin\frac{x}{2}} - \frac{1}{2} \qquad (\text{for } N \geq 3) \tag{5.72}$$

Now for $x = 2\pi/N$, as in the specific case under consideration, the right-hand side (RHS) of (5.72) is

$$RHS = \frac{1}{2}\left[\frac{\sin\left\{2\pi + \frac{\pi}{N}\right\}}{\sin\frac{\pi}{N}} - 1\right] = \frac{1}{2}\left[\frac{\sin\frac{\pi}{N} \cdot 1}{\sin\frac{\pi}{N}} - 1\right] = 0 \tag{5.73}$$

Therefore, for three or more equally spaced fiber angles θ_k,

$$\sum_{k=1}^{N} \cos 2\theta_k = 0 \tag{5.74}$$

In a similar fashion, it can be shown that the three remaining series in (5.69) are also zero, i.e.,

$$\sum_{k=1}^{N} \sin 2\theta_k = \sum_{k=1}^{N} \cos 4\theta_k = \sum_{k=1}^{N} \sin 4\theta_k = 0 \qquad (\text{for } N \geq 3) \tag{5.75}$$

Hence, (5.69) reduces to

$$A_{11}(\phi) = 2HU_1 \tag{5.76}$$

Thus, A_{11} is a constant independent of the direction ϕ in the plane of the laminate.

Likewise, using (5.65) with (5.75) and (4.83), it can be shown that

$$\begin{aligned} A_{12}(\phi) &= 2HU_4 \\ A_{22}(\phi) &= 2HU_1 \\ A_{16}(\phi) &= 0 \\ A_{26}(\phi) &= 0 \\ A_{66}(\phi) &= 2HU_5 \end{aligned} \tag{5.77}$$

Thus, all A_{ij} in a laminate with three or more equally spaced fiber orientations are independent of the direction ϕ and $A_{11}(\phi) = A_{22}(\phi)$; the in-plane response of the laminate is *isotropic*. Such a laminate is called *quasi-isotropic*. Further, since the U_i are only functions of the material type, *all equal-thickness quasi-isotropic laminates of a given material have identical in-plane properties.* Also, since $A_{16} = A_{26} = 0$, all quasi-isotropic laminates are specially orthotropic, and there is no coupling between in-plane normal and shear responses in these laminates.

It is emphasized that a cross-ply laminate ($N = 2$) does not exhibit isotropic response in the plane. The above development for quasi-isotropic laminates refers only to the in-plane elastic response. Indeed, neither the bending response nor the strength of these laminates exhibits isotropic characteristics. That is why they are called *quasi-isotropic* and not *isotropic*.

5.11 Examples for Laminate [*A*], [*B*], and [*D*] Values

In this section we present several examples for the calculation of the $[A]$, $[B]$, and $[D]$ matrices of classical lamination theory. Results are presented for symmetric, unsymmetric, cross-ply, and angle-ply laminates.

Example 5.1 Unsymmetric Cross-Ply Laminate

Express the elements of the $[A]$, $[B]$, and $[D]$ matrices of a $[0/90]_T$ laminate in terms of the elements of the material stiffnesses Q_{ij} if all layers have equal thickness t.

Solution

Consider the laminate as sketched in the figure.

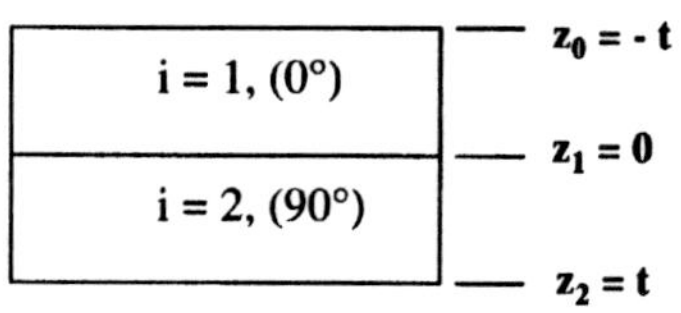

Recalling the definitions of $[A]$, $[B]$, and $[D]$,

$$[A] = \sum_{k=1}^{N} [\bar{Q}]^k (z_k - z_{k-1})$$

$$[B] = \frac{1}{2}\sum_{k=1}^{N} [\bar{Q}]^k (z_k^2 - z_{k-1}^2)$$

$$[D] = \frac{1}{3}\sum_{k=1}^{N} [\bar{Q}]^k (z_k^3 - z_{k-1}^3)$$

For the 0° layer,

$$[\bar{Q}] = \begin{bmatrix} Q_{11} & Q_{12} & 0 \\ Q_{12} & Q_{22} & 0 \\ 0 & 0 & Q_{66} \end{bmatrix}$$

and for the 90° layer,

$$[\bar{Q}] = \begin{bmatrix} Q_{22} & Q_{12} & 0 \\ Q_{12} & Q_{11} & 0 \\ 0 & 0 & Q_{66} \end{bmatrix}$$

Thus, for $[A]$,

$$[A] = \begin{bmatrix} Q_{11} & Q_{12} & 0 \\ Q_{12} & Q_{22} & 0 \\ 0 & 0 & Q_{66} \end{bmatrix}(z_1 - z_0) + \begin{bmatrix} Q_{22} & Q_{12} & 0 \\ Q_{12} & Q_{11} & 0 \\ 0 & 0 & Q_{66} \end{bmatrix}(z_2 - z_1)$$

$$[A] = \begin{bmatrix} Q_{11} & Q_{12} & 0 \\ Q_{12} & Q_{22} & 0 \\ 0 & 0 & Q_{66} \end{bmatrix}(0 - (-t)) + \begin{bmatrix} Q_{22} & Q_{12} & 0 \\ Q_{12} & Q_{11} & 0 \\ 0 & 0 & Q_{66} \end{bmatrix}(t - 0)$$

$$[A] = \begin{bmatrix} Q_{11} + Q_{22} & 2Q_{12} & 0 \\ 2Q_{12} & Q_{11} + Q_{22} & 0 \\ 0 & 0 & 2Q_{66} \end{bmatrix} t$$

For $[B]$,

$$[B] = \frac{1}{2}\left(\begin{bmatrix} Q_{11} & Q_{12} & 0 \\ Q_{12} & Q_{22} & 0 \\ 0 & 0 & Q_{66} \end{bmatrix}(z_1^2 - z_0^2) + \begin{bmatrix} Q_{22} & Q_{12} & 0 \\ Q_{12} & Q_{11} & 0 \\ 0 & 0 & Q_{66} \end{bmatrix}(z_2^2 - z_1^2)\right)$$

$$[B] = \frac{1}{2}\left(\begin{bmatrix} Q_{11} & Q_{12} & 0 \\ Q_{12} & Q_{22} & 0 \\ 0 & 0 & Q_{66} \end{bmatrix}(0 - (-t)^2) + \begin{bmatrix} Q_{22} & Q_{12} & 0 \\ Q_{12} & Q_{11} & 0 \\ 0 & 0 & Q_{66} \end{bmatrix}(t^2 - 0)\right)$$

$$[B] = \frac{1}{2}\begin{bmatrix} Q_{22} - Q_{11} & 0 & 0 \\ 0 & Q_{11} - Q_{22} & 0 \\ 0 & 0 & 0 \end{bmatrix} t^2$$

Clearly, $[B]$ is not identically zero for this unsymmetric laminate unless the material is isotropic with $Q_{11} = Q_{22}$.

For $[D]$,

$$[D] = \frac{1}{3}\left(\begin{bmatrix} Q_{11} & Q_{12} & 0 \\ Q_{12} & Q_{22} & 0 \\ 0 & 0 & Q_{66} \end{bmatrix}(z_1^3 - z_0^3) + \begin{bmatrix} Q_{22} & Q_{12} & 0 \\ Q_{12} & Q_{11} & 0 \\ 0 & 0 & Q_{66} \end{bmatrix}(z_2^3 - z_1^3)\right)$$

$$[D] = \frac{1}{3}\left(\begin{bmatrix} Q_{11} & Q_{12} & 0 \\ Q_{12} & Q_{22} & 0 \\ 0 & 0 & Q_{66} \end{bmatrix}(0 - (-t)^3) + \begin{bmatrix} Q_{22} & Q_{12} & 0 \\ Q_{12} & Q_{11} & 0 \\ 0 & 0 & Q_{66} \end{bmatrix}(t^3 - 0)\right)$$

$$[D] = \frac{1}{3}\begin{bmatrix} Q_{11} + Q_{22} & 2Q_{12} & 0 \\ 2Q_{12} & Q_{11} + Q_{22} & 0 \\ 0 & 0 & 2Q_{66} \end{bmatrix} t^3$$

We see that the 16 and 26 terms are zero for $[A]$, $[B]$, and $[D]$ since the layers are orthotropic. Further, the 11 term equals the 22 term and the 12 term equals the 21 term for all three matrices.

Example 5.2 Unsymmetric Angle-Ply Laminate

Express the elements of the $[A]$, $[B]$, and $[D]$ matrices of a $[\theta/-\theta]_T$ laminate in terms of the elements of the transformed stiffnesses $\bar{Q}_{ij}(\theta)$ if all layers have equal thickness t.

Solution

Consider the laminate as sketched in the figure.

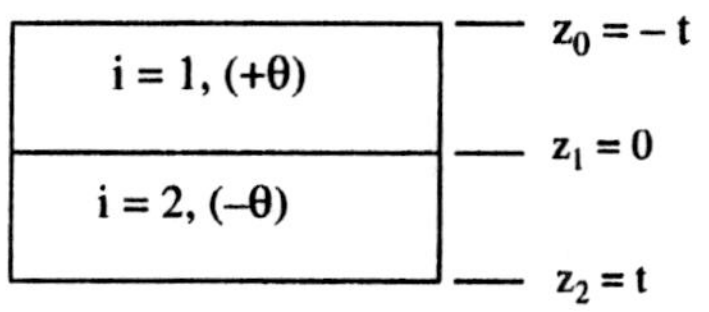

Recalling the definitions of $[A]$, $[B]$, and $[D]$,

$$[A] = \sum_{k=1}^{N} [\bar{Q}]^k (z_k - z_{k-1})$$

$$[B] = \frac{1}{2}\sum_{k=1}^{N} [\bar{Q}]^k (z_k^2 - z_{k-1}^2)$$

$$[D] = \frac{1}{3}\sum_{k=1}^{N} [\bar{Q}]^k (z_k^3 - z_{k-1}^3)$$

For the +θ layer,

$$[\bar{Q}(\theta)] = \begin{bmatrix} \bar{Q}_{11}(\theta) & \bar{Q}_{12}(\theta) & \bar{Q}_{16}(\theta) \\ \bar{Q}_{12}(\theta) & \bar{Q}_{22}(\theta) & \bar{Q}_{26}(\theta) \\ \bar{Q}_{16}(\theta) & \bar{Q}_{26}(\theta) & \bar{Q}_{66}(\theta) \end{bmatrix}$$

and for the −θ layer,

$$[\bar{Q}(-\theta)] = \begin{bmatrix} \bar{Q}_{11}(\theta) & \bar{Q}_{12}(\theta) & -\bar{Q}_{16}(\theta) \\ \bar{Q}_{12}(\theta) & \bar{Q}_{22}(\theta) & -\bar{Q}_{26}(\theta) \\ -\bar{Q}_{16}(\theta) & -\bar{Q}_{26}(\theta) & \bar{Q}_{66}(\theta) \end{bmatrix}$$

Thus, for [A],

$$[A] = \begin{bmatrix} \bar{Q}_{11}(\theta) & \bar{Q}_{12}(\theta) & \bar{Q}_{16}(\theta) \\ \bar{Q}_{12}(\theta) & \bar{Q}_{22}(\theta) & \bar{Q}_{26}(\theta) \\ \bar{Q}_{16}(\theta) & \bar{Q}_{26}(\theta) & \bar{Q}_{66}(\theta) \end{bmatrix}(z_1 - z_0) + \begin{bmatrix} \bar{Q}_{11}(\theta) & \bar{Q}_{12}(\theta) & -\bar{Q}_{16}(\theta) \\ \bar{Q}_{12}(\theta) & \bar{Q}_{22}(\theta) & -\bar{Q}_{26}(\theta) \\ -\bar{Q}_{16}(\theta) & -\bar{Q}_{26}(\theta) & \bar{Q}_{66}(\theta) \end{bmatrix}(z_2 - z_1)$$

$$[A] = \begin{bmatrix} \bar{Q}_{11}(\theta) & \bar{Q}_{12}(\theta) & \bar{Q}_{16}(\theta) \\ \bar{Q}_{12}(\theta) & \bar{Q}_{22}(\theta) & \bar{Q}_{26}(\theta) \\ \bar{Q}_{16}(\theta) & \bar{Q}_{26}(\theta) & \bar{Q}_{66}(\theta) \end{bmatrix}(0 - (-t)) + \begin{bmatrix} \bar{Q}_{11}(\theta) & \bar{Q}_{12}(\theta) & -\bar{Q}_{16}(\theta) \\ \bar{Q}_{12}(\theta) & \bar{Q}_{22}(\theta) & -\bar{Q}_{26}(\theta) \\ -\bar{Q}_{16}(\theta) & -\bar{Q}_{26}(\theta) & \bar{Q}_{66}(\theta) \end{bmatrix}(t - 0)$$

$$[A] = 2\begin{bmatrix} \bar{Q}_{11}(\theta) & \bar{Q}_{12}(\theta) & 0 \\ \bar{Q}_{12}(\theta) & \bar{Q}_{22}(\theta) & 0 \\ 0 & 0 & \bar{Q}_{66}(\theta) \end{bmatrix}t$$

For [B],

$$[B] = \frac{1}{2}\left(\begin{bmatrix} \bar{Q}_{11}(\theta) & \bar{Q}_{12}(\theta) & \bar{Q}_{16}(\theta) \\ \bar{Q}_{12}(\theta) & \bar{Q}_{22}(\theta) & \bar{Q}_{26}(\theta) \\ \bar{Q}_{16}(\theta) & \bar{Q}_{26}(\theta) & \bar{Q}_{66}(\theta) \end{bmatrix}(z_1^2 - z_0^2) + \begin{bmatrix} \bar{Q}_{11}(\theta) & \bar{Q}_{12}(\theta) & -\bar{Q}_{16}(\theta) \\ \bar{Q}_{12}(\theta) & \bar{Q}_{22}(\theta) & -\bar{Q}_{26}(\theta) \\ -\bar{Q}_{16}(\theta) & -\bar{Q}_{26}(\theta) & \bar{Q}_{66}(\theta) \end{bmatrix}(z_2^2 - z_1^2)\right)$$

$$[B] = \frac{1}{2}\left(\begin{bmatrix} \bar{Q}_{11}(\theta) & \bar{Q}_{12}(\theta) & \bar{Q}_{16}(\theta) \\ \bar{Q}_{12}(\theta) & \bar{Q}_{22}(\theta) & \bar{Q}_{26}(\theta) \\ \bar{Q}_{16}(\theta) & \bar{Q}_{26}(\theta) & \bar{Q}_{66}(\theta) \end{bmatrix}(0 - (-t)^2) + \begin{bmatrix} \bar{Q}_{11}(\theta) & \bar{Q}_{12}(\theta) & -\bar{Q}_{16}(\theta) \\ \bar{Q}_{12}(\theta) & \bar{Q}_{22}(\theta) & -\bar{Q}_{26}(\theta) \\ -\bar{Q}_{16}(\theta) & -\bar{Q}_{26}(\theta) & \bar{Q}_{66}(\theta) \end{bmatrix}(t^2 - 0)\right)$$

$$[B] = -\begin{bmatrix} 0 & 0 & \bar{Q}_{16}(\theta) \\ 0 & 0 & \bar{Q}_{26}(\theta) \\ \bar{Q}_{16}(\theta) & \bar{Q}_{26}(\theta) & 0 \end{bmatrix}t^2$$

Clearly, $[B]$ is not identically zero for this unsymmetric laminate unless the material is isotropic with $Q_{16} = Q_{26} = 0$.

For $[D]$,

$$[D] = \frac{1}{3}\left(\begin{bmatrix} \bar{Q}_{11}(\theta) & \bar{Q}_{12}(\theta) & \bar{Q}_{16}(\theta) \\ \bar{Q}_{12}(\theta) & \bar{Q}_{22}(\theta) & \bar{Q}_{26}(\theta) \\ \bar{Q}_{16}(\theta) & \bar{Q}_{26}(\theta) & \bar{Q}_{66}(\theta) \end{bmatrix}(z_1^3 - z_0^3) + \begin{bmatrix} \bar{Q}_{11}(\theta) & \bar{Q}_{12}(\theta) & -\bar{Q}_{16}(\theta) \\ \bar{Q}_{12}(\theta) & \bar{Q}_{22}(\theta) & -\bar{Q}_{26}(\theta) \\ -\bar{Q}_{16}(\theta) & -\bar{Q}_{26}(\theta) & \bar{Q}_{66}(\theta) \end{bmatrix}(z_2^3 - z_1^3)\right)$$

$$[D] = \frac{1}{3}\left(\begin{bmatrix} \bar{Q}_{11}(\theta) & \bar{Q}_{12}(\theta) & \bar{Q}_{16}(\theta) \\ \bar{Q}_{12}(\theta) & \bar{Q}_{22}(\theta) & \bar{Q}_{26}(\theta) \\ \bar{Q}_{16}(\theta) & \bar{Q}_{26}(\theta) & \bar{Q}_{66}(\theta) \end{bmatrix}(0 - (-t)^3) + \begin{bmatrix} \bar{Q}_{11}(\theta) & \bar{Q}_{12}(\theta) & -\bar{Q}_{16}(\theta) \\ \bar{Q}_{12}(\theta) & \bar{Q}_{22}(\theta) & -\bar{Q}_{26}(\theta) \\ -\bar{Q}_{16}(\theta) & -\bar{Q}_{26}(\theta) & \bar{Q}_{66}(\theta) \end{bmatrix}(t^3 - 0)\right)$$

$$[D] = \frac{2}{3}\begin{bmatrix} \bar{Q}_{11}(\theta) & \bar{Q}_{12}(\theta) & 0 \\ \bar{Q}_{12}(\theta) & \bar{Q}_{22}(\theta) & 0 \\ 0 & 0 & \bar{Q}_{66}(\theta) \end{bmatrix} t^3$$

We see that the 16 and 26 terms are zero for $[A]$ and $[D]$, but not for $[B]$.

Example 5.3 Symmetric Cross-Ply Laminate

Express the elements of the $[A]$, $[B]$, and $[D]$ matrices of a $[0/90]_s$ laminate in terms of the elements of the material stiffnesses Q_{ij} if all layers have equal thickness t.

Solution

Consider the laminate as sketched in the figure.

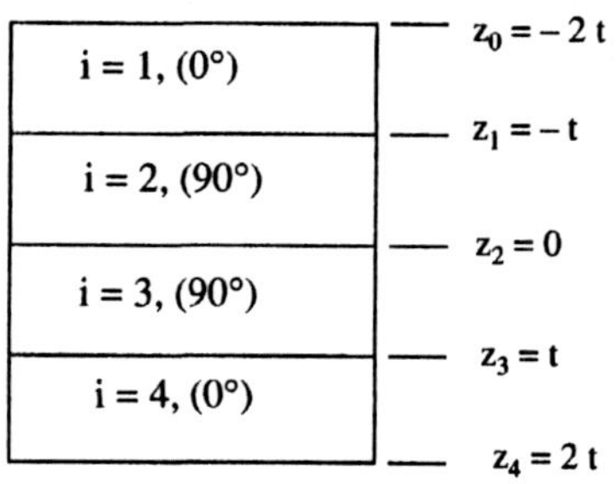

Recalling the definitions of $[A]$, $[B]$, and $[D]$,

$$[A] = \sum_{k=1}^{N} [\bar{Q}]^k (z_k - z_{k-1})$$

$$[B] = \frac{1}{2}\sum_{k=1}^{N} [\bar{Q}]^k (z_k^2 - z_{k-1}^2)$$

$$[D] = \frac{1}{3}\sum_{k=1}^{N}[\bar{Q}]^k(z_k^3 - z_{k-1}^3)$$

For the 0° layer,

$$[Q] = \begin{bmatrix} Q_{11} & Q_{12} & 0 \\ Q_{12} & Q_{22} & 0 \\ 0 & 0 & Q_{66} \end{bmatrix}$$

and for the 90° layer,

$$[Q] = \begin{bmatrix} Q_{22} & Q_{12} & 0 \\ Q_{12} & Q_{11} & 0 \\ 0 & 0 & Q_{66} \end{bmatrix}$$

Thus, for $[A]$,

$$[A] = \begin{bmatrix} Q_{11} & Q_{12} & 0 \\ Q_{12} & Q_{22} & 0 \\ 0 & 0 & Q_{66} \end{bmatrix}[(z_1 - z_0) + (z_4 - z_3)] + \begin{bmatrix} Q_{22} & Q_{12} & 0 \\ Q_{12} & Q_{11} & 0 \\ 0 & 0 & Q_{66} \end{bmatrix}[(z_2 - z_1) + (z_3 - z_2)]$$

$$[A] = \begin{bmatrix} Q_{11} & Q_{12} & 0 \\ Q_{12} & Q_{22} & 0 \\ 0 & 0 & Q_{66} \end{bmatrix}[((-t) - (-2t)) + (2t - t)] + \begin{bmatrix} Q_{22} & Q_{12} & 0 \\ Q_{12} & Q_{11} & 0 \\ 0 & 0 & Q_{66} \end{bmatrix}[(0 - (-t)) + (t - 0)]$$

$$[A] = \begin{bmatrix} Q_{11} & Q_{12} & 0 \\ Q_{12} & Q_{22} & 0 \\ 0 & 0 & Q_{66} \end{bmatrix}(2t) + \begin{bmatrix} Q_{22} & Q_{12} & 0 \\ Q_{12} & Q_{11} & 0 \\ 0 & 0 & Q_{66} \end{bmatrix}(2t)$$

$$[A] = 2\begin{bmatrix} Q_{11} + Q_{22} & 2Q_{12} & 0 \\ 2Q_{12} & Q_{11} + Q_{22} & 0 \\ 0 & 0 & 2Q_{66} \end{bmatrix}t$$

For $[B]$,

$$[B] = \frac{1}{2}\left(\begin{bmatrix} Q_{11} & Q_{12} & 0 \\ Q_{12} & Q_{22} & 0 \\ 0 & 0 & Q_{66} \end{bmatrix}((z_1^2 - z_0^2) + (z_4^2 - z_3^2)) + \begin{bmatrix} Q_{22} & Q_{12} & 0 \\ Q_{12} & Q_{11} & 0 \\ 0 & 0 & Q_{66} \end{bmatrix}((z_2^2 - z_1^2) + (z_3^2 - z_2^2))\right)$$

$$[B] = \frac{1}{2}\left(\begin{bmatrix} Q_{11} & Q_{12} & 0 \\ Q_{12} & Q_{22} & 0 \\ 0 & 0 & Q_{66} \end{bmatrix}[(t^2-4t^2)+(4t^2-t^2)] + \begin{bmatrix} Q_{22} & Q_{12} & 0 \\ Q_{12} & Q_{11} & 0 \\ 0 & 0 & Q_{66} \end{bmatrix}[(0-t^2)+(t^2-0)]\right)$$

$$[B] = \frac{1}{2}\left(\begin{bmatrix} Q_{11} & Q_{12} & 0 \\ Q_{12} & Q_{22} & 0 \\ 0 & 0 & Q_{66} \end{bmatrix}(0) + \begin{bmatrix} Q_{22} & Q_{12} & 0 \\ Q_{12} & Q_{11} & 0 \\ 0 & 0 & Q_{66} \end{bmatrix}(0)\right)$$

$$[B] = [0]$$

Thus $[B] = 0$ for this symmetric laminate.

For $[D]$,

$$[D] = \frac{1}{3}\left(\begin{bmatrix} Q_{11} & Q_{12} & 0 \\ Q_{12} & Q_{22} & 0 \\ 0 & 0 & Q_{66} \end{bmatrix}((z_1^3-z_0^3)+(z_4^3-z_3^3)) + \begin{bmatrix} Q_{22} & Q_{12} & 0 \\ Q_{12} & Q_{11} & 0 \\ 0 & 0 & Q_{66} \end{bmatrix}((z_2^3-z_1^3)+(z_3^3-z_2^3))\right)$$

$$[D] = \frac{1}{3}\left(\begin{bmatrix} Q_{11} & Q_{12} & 0 \\ Q_{12} & Q_{22} & 0 \\ 0 & 0 & Q_{66} \end{bmatrix}((-t^3+8t^3)+(8t^3-t^3)) + \begin{bmatrix} Q_{22} & Q_{12} & 0 \\ Q_{12} & Q_{11} & 0 \\ 0 & 0 & Q_{66} \end{bmatrix}((0+t^3)+(t^3-0))\right)$$

$$[D] = \frac{2}{3}\begin{bmatrix} 7Q_{11}+Q_{22} & 8Q_{12} & 0 \\ 8Q_{12} & Q_{11}+7Q_{22} & 0 \\ 0 & 0 & 8Q_{66} \end{bmatrix} t^3$$

We see that the 16 and 26 terms are zero for $[A]$, $[B]$, and $[D]$ since the layers are orthotropic. Further, the 12 term equals the 21 term for all three matrices.

5.12 Laminate Engineering Constants

Expressions for engineering constants of symmetric laminates can be determined by following a procedure very similar to that used to obtain expressions for the engineering constants of laminae (Section 4.5). Since the laminates of interest are symmetric, the extensional response is uncoupled from the bending response, and hence equation (5.49) governs the in-plane response.

Since it is desired to formulate expressions for engineering constants, we first define the *laminate average stress* $\{\bar{\sigma}\}$ and then use (5.64) to obtain

$$\{\bar{\sigma}\} \equiv \frac{1}{2H}\{N\} = \frac{1}{2H}[A]\{\varepsilon^o\} \tag{5.78}$$

Inverting this equation gives the midplane strains as a function of the average applied stress.

$$\{\varepsilon^o\} = 2H[A]^{-1}\{\bar{\sigma}\} \tag{5.79}$$

For simplicity of exposition, we define the *laminate compliance* $[a^*]$ as

$$[a^*] \equiv 2H[A]^{-1} \tag{5.80}$$

Combining (5.79) and (5.80), the laminate constitutive equation can be written in terms of average stresses, midplane strains, and laminate compliance as

$$\{\varepsilon^o\} = [a^*]\{\bar{\sigma}\} \tag{5.81}$$

In general, $[a^*]$ is fully populated, i.e.,

$$[a^*] = \begin{bmatrix} a_{11}^* & a_{12}^* & a_{16}^* \\ a_{12}^* & a_{22}^* & a_{26}^* \\ a_{16}^* & a_{26}^* & a_{66}^* \end{bmatrix} \tag{5.82}$$

Thus (5.81) has the expanded form

$$\begin{Bmatrix} \varepsilon_x^o \\ \varepsilon_y^o \\ \gamma_{xy}^o \end{Bmatrix} = \begin{bmatrix} a_{11}^* & a_{12}^* & a_{16}^* \\ a_{12}^* & a_{22}^* & a_{26}^* \\ a_{16}^* & a_{26}^* & a_{66}^* \end{bmatrix} \begin{Bmatrix} \bar{\sigma}_x \\ \bar{\sigma}_y \\ \bar{\tau}_{xy} \end{Bmatrix} \tag{5.83}$$

This form is suitable for conducting thought experiments similar to those used for laminae in Section 4.5.

We consider average stresses applied to laminates with one component of stress being nonzero and the two remaining components of stress equal to zero. The results for the laminate engineering constants can be summarized as follows.

For $\bar{\sigma}_x \neq 0$, $\bar{\sigma}_y = 0$, $\bar{\tau}_{xy} = 0$,

- Axial modulus $E_x = \bar{\sigma}_x/\varepsilon_x^o$:

$$E_x = \frac{1}{a_{11}^*} \tag{5.84}$$

- Poisson ratio $\nu_{xy} = -\varepsilon_y^o/\varepsilon_x^o$:

$$\nu_{xy} = -\frac{a_{12}^*}{a_{11}^*} \tag{5.85}$$

- Coefficient of mutual influence $\eta_{xy,x} = \gamma_{xy}^o/\varepsilon_x^o$:

$$\eta_{xy,x} = \frac{a_{16}^*}{a_{11}^*} \tag{5.86}$$

For $\sigma_y \neq 0$, $\sigma_x = 0$, $\tau_{xy} = 0$,

- Transverse modulus $E_y = \bar{\sigma}_y/\varepsilon_y^o$:

$$E_y = \frac{1}{a_{22}^*} \tag{5.87}$$

- Poisson ratio $\nu_{yx} = -\varepsilon_x^o/\varepsilon_y^o$:

$$\nu_{yx} = -\frac{a_{12}^*}{a_{22}^*} \tag{5.88}$$

- Coefficient of mutual influence $\eta_{xy,y} = \gamma_{xy}^o/\varepsilon_y^o$:

$$\eta_{xy,y} = \frac{a_{26}^*}{a_{22}^*} \tag{5.89}$$

For $\bar{\tau}_{xy} \neq 0$, $\bar{\sigma}_x = 0$, $\bar{\sigma}_x = 0$,

- Shear modulus $G_{xy} = \tau_{xy}/\gamma_{xy}^o$:

$$G_{xy} = \frac{1}{a_{66}^*} \tag{5.90}$$

- Coefficient of mutual influence $\eta_{x,xy} = \varepsilon_x^o/\gamma_{xy}^o$:

$$\eta_{x,xy} = \frac{a_{16}^*}{a_{66}^*} \tag{5.91}$$

- Coefficient of mutual influence $\eta_{y,xy} = \varepsilon_y^o/\gamma_{xy}^o$:

$$\eta_{y,xy} = \frac{a_{26}^*}{a_{66}^*} \tag{5.92}$$

These engineering constants are extremely helpful for quantifying the properties of a laminate for design purposes and for comparison with other materials. Actual values of the engineering constants for polymeric and metallic matrix composite laminates are presented in the following sections.

5.12.1 Angle-Ply Laminates

For an angle-ply laminate, (5.61) and (5.80) combine to give

$$[a^*] = \begin{bmatrix} \left(\dfrac{\bar{Q}_{22}}{\bar{Q}_{11}\bar{Q}_{22} - \bar{Q}_{12}\bar{Q}_{12}}\right) & \left(\dfrac{-\bar{Q}_{12}}{\bar{Q}_{11}\bar{Q}_{22} - \bar{Q}_{12}\bar{Q}_{12}}\right) & 0 \\ \left(\dfrac{-\bar{Q}_{12}}{\bar{Q}_{11}\bar{Q}_{22} - \bar{Q}_{12}\bar{Q}_{12}}\right) & \left(\dfrac{\bar{Q}_{11}}{\bar{Q}_{11}\bar{Q}_{22} - \bar{Q}_{12}\bar{Q}_{12}}\right) & 0 \\ 0 & 0 & \dfrac{1}{\bar{Q}_{66}} \end{bmatrix} \tag{5.93}$$

Thus with $a_{16}^{*} = a_{26}^{*} = 0$, the coefficients of mutual influence are identically zero. The remaining engineering properties are described in the following sections.

5.12.1.1 Axial Modulus

From (5.84) and (5.93), the axial modulus for an angle-ply laminate is given by

$$E_x = \frac{1}{a_{11}^{*}} = \frac{\bar{Q}_{11}\bar{Q}_{22} - \bar{Q}_{12}\bar{Q}_{12}}{\bar{Q}_{22}} \tag{5.94}$$

Predictions for the axial modulus of angle-ply, T300/5208 carbon/epoxy laminates are compared with the predictions for unidirectional off-axis laminae in Fig. 5.9. As indicated in the figure, the angle-ply laminate has higher stiffness than the off-axis lamina for fiber angles ranging from 0° to approximately 60°. The experimental results for T300/5208 angle-ply laminates shown in Fig. 5.9 exhibit very good correlation with the theoretical predictions.

The stiffening effect of the angle-ply laminate is demonstrated more clearly in Fig. 5.10, which shows a plot of the ratio $E_x(\pm\theta)/E_x(\theta)$ as a function of the fiber angle. The maximum ratio is in excess of 2.3, at a fiber orientation of approximately 23°. It is noted that this ratio is equivalent to the ratio of the laminate compliance a_{11}^{*} (5.84) for the unidirectional off-axis lamina to that for the angle-ply laminate.

The higher axial stiffness for the angle-ply laminate is a result of the constraining effect (and resulting multi-axial state of stress) that the adjacent layers of the laminate have on each other.

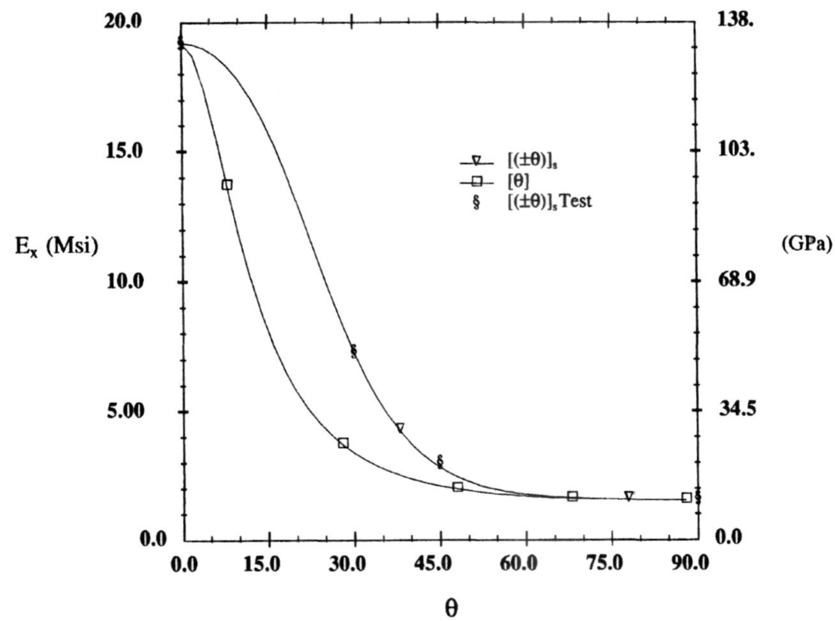

FIGURE 5.9 Axial Modulus Comparisons: T300/5208

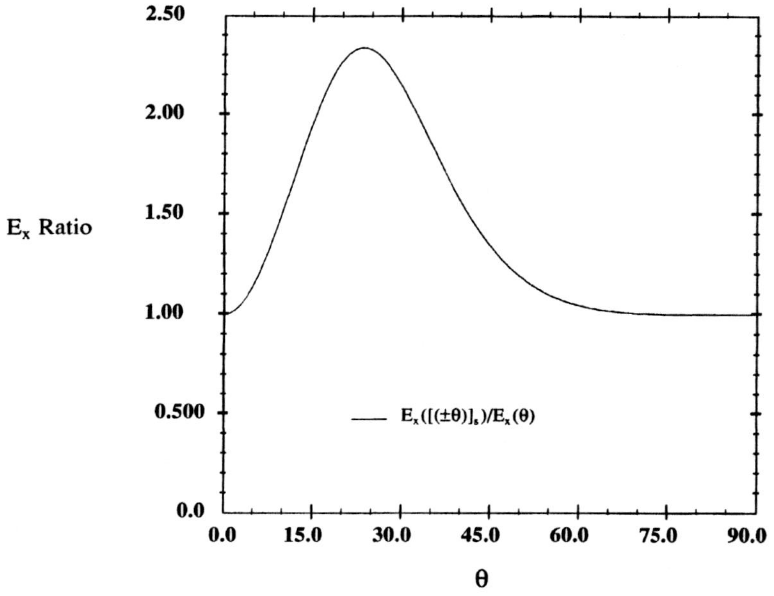

FIGURE 5.10 Ratio $E_x(\pm\theta) / E_x(\theta)$: T300/5208

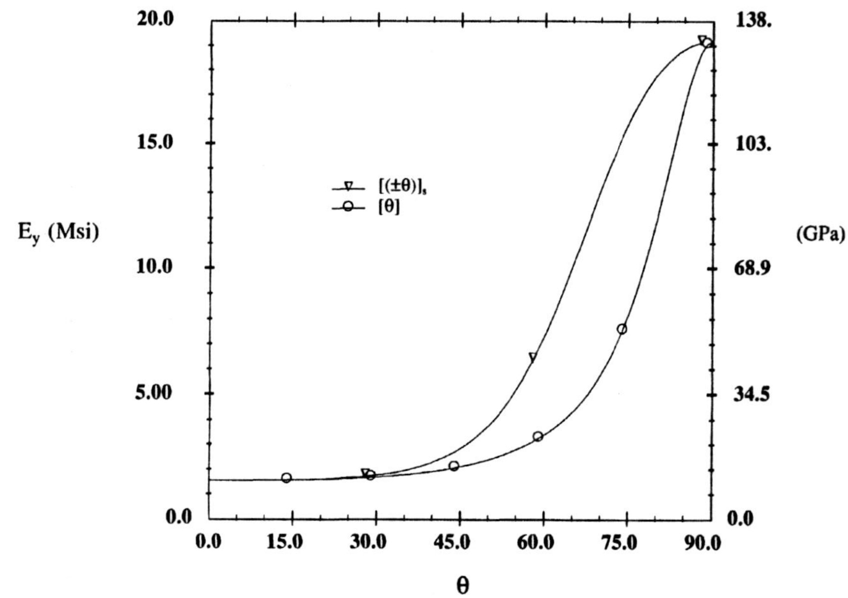

FIGURE 5.11 Transverse Modulus Comparisons: T300/5208

5.12.1.2 Transverse Modulus

From (5.87) and (5.93), the transverse modulus for an angle-ply laminate is

$$E_y = \frac{1}{a_{22}^*} = \frac{\bar{Q}_{11}\bar{Q}_{22} - \bar{Q}_{12}\bar{Q}_{12}}{\bar{Q}_{11}} \tag{5.95}$$

Comparison of the predictions for transverse modulus E_y for angle-ply and unidirectional off-axis laminae is shown in Fig. 5.11. As expected, the curves are the same as those in Fig. 5.9 except for a 90° phase shift.

5.12.1.3 Axial Poisson's Ratio

Combining (5.85) and (5.93) gives the expression for the axial Poisson's ratio for an angle-ply laminate:

$$\nu_{xy} = -\frac{a_{12}^*}{a_{11}^*} = \frac{\bar{Q}_{12}}{\bar{Q}_{22}} \tag{5.96}$$

The comparison of Poisson's ratios shown in Fig. 5.12 is one of the most interesting and surprising results for laminated fibrous composites. As the figure indicates, there is a very large increase in Poisson's ratio when an off-axis ply is laminated with another off-axis ply of opposite sign. This is the case for a wide range of fiber orientations. The angle-ply laminate exhibits a maximum Poisson's ratio in excess of 1.25 (for this particular carbon/epoxy) at an angle of approximately 27°. That the Poisson's ratio can be in excess of 1.0 is very surprising to those accustomed to working with metals, where the maximum Poisson's ratio is 0.5 for plastic flow. The high Poisson's ratio in

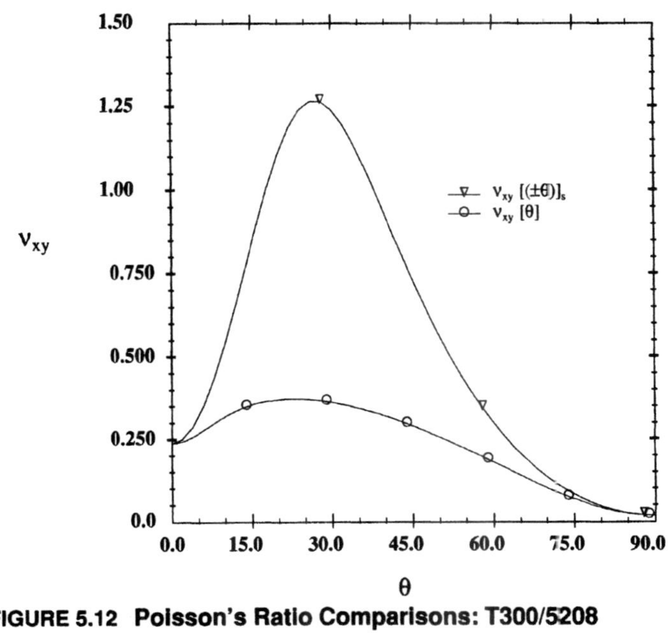

FIGURE 5.12 Poisson's Ratio Comparisons: T300/5208

these laminates is due to the internal state of stress in the individual layers. The laminates behave more like structures with variable internal mechanisms than like homogeneous materials with uniform stress states.

5.12.1.4 Shear Modulus

Combining (5.90) and (5.93) gives the expression for the shear modulus of an angle-ply laminate:

$$G_{xy} = \frac{1}{a^*_{66}} = \bar{Q}_{66} \tag{5.97}$$

Comparisons of the shear modulus $\bar{G}_{xy}$ for angle-ply laminates and off-axis unidirectional laminae are shown in Fig. 5.13. The $[\pm\theta]_s$ laminate is much stiffer than the lamina for essentially all fiber orientations. At $\theta = 45°$, where the shear stiffnesses of both the lamina and the laminate are largest, the stiffness of the laminate is more than 3.5 times that of the lamina for the carbon/epoxy considered. Since pure shear is equivalent to tension and compression along the diagonals (see Fig. 4.13), the ±45° fiber orientation corresponds to tensile and compressive normal stresses in the fiber directions, the directions of maximum stiffness. The results in this figure clearly indicate that ±45° fiber orientations are desired in structures requiring high shear stiffness.

5.12.1.5 Coefficients of Mutual Influence

All four coefficients of mutual influence of angle-ply laminates are zero because they are directly proportional to either a^*_{16} or a^*_{26}. These terms are identically zero for angle-ply laminates, as indicated in (5.93). In general, the coefficients of mutual influence are zero for all specially orthotropic laminates (i.e., those with $A_{16} = A_{26} = 0$).

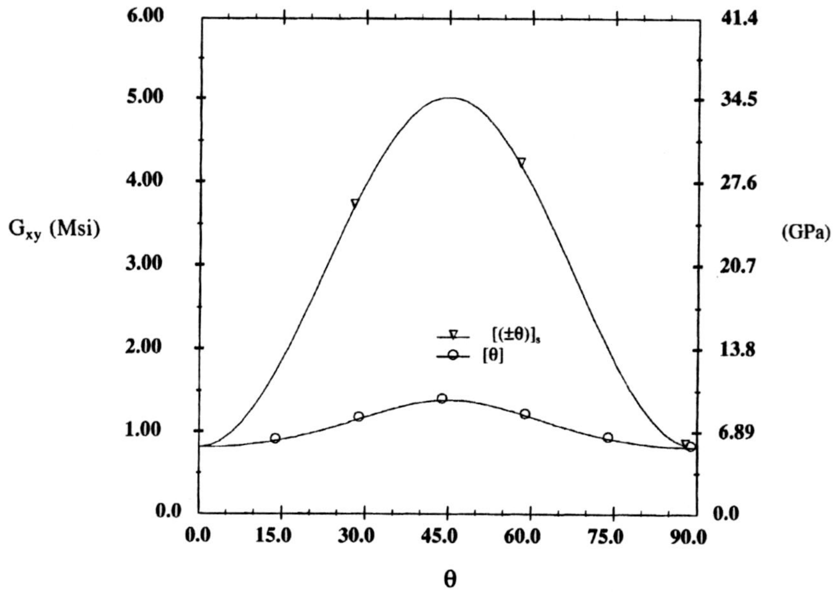

FIGURE 5.13 Shear Modulus Comparisons: T300/5208

5.12.2 General Laminates

Table 5.2 presents lamination theory predictions of engineering constants for $[0/\pm 60]_s$ ($\pi/3$) and $[0/\pm 45/90]_s$ ($\pi/4$) quasi-isotropic laminates, a cross-ply laminate, two angle-ply laminates, and a unidirectional lamina for T300/5208 carbon/epoxy properties. These predictions show that a very large range of properties is possible depending upon the orientation of the individual layers. The $\pi/3$ and the $\pi/4$ laminates have identical properties in all directions, as expected for quasi-isotropic laminates. The modulus of the quasi-isotropic laminates is about 38% of the axial modulus of a unidirectional lamina. The modulus of the cross-ply laminate is 52% of the axial modulus of the lamina, and although it has the same modulus in its principal directions, it is not quasi-isotropic. The in-plane Poisson's ratio ranges from a minimum of 0.02 for the [90] lamina to a maximum of 1.24 for the $[\pm 30]_s$ laminate. The shear modulus ranges from a low of 5.6 GPa (0.82 Msi) for the unidirectional and cross-ply laminates to a high of 43.7 GPa (5.03 Msi) for the ±45 angle-ply laminate. These results clearly demonstrate the possibility of tailoring with fibrous composites.

Laminate	E_x, GPa (Msi)	E_y, GPa (Msi)	ν_{xy}	G_{xy}, GPa (Msi)
$[0/\pm 60]_s$	52.3 (7.59)	52.3 (7.59)	0.30	20.1 (2.92)
$[0/\pm 45/90]_s$	52.3 (7.59)	52.3 (7.59)	0.30	20.1 (2.92)
$[0/90]_s$	71.8 (10.42)	71.8 (10.42)	0.04	5.6 (0.82)
$[\pm 30]_s$	50.5 (7.32)	12.3 (1.78)	1.24	27.4 (3.97)
$[\pm 45]_s$	19.6 (2.85)	19.6 (2.85)	0.74	43.7 (5.03)
[0]	136.5 (19.8)	10.8 (1.56)	0.24	5.6 (0.82)
[90]	10.8 (1.56)	136.5 (19.8)	0.02	5.6 (0.82)

TABLE 5.2 Carbon/Epoxy Laminate Engineering Properties

5.12.3 Comparison of Angle-Ply Engineering Constants

The variation of engineering constants with fiber orientation for polymeric matrix, angle-ply laminates was demonstrated in Figs. 5.9 and 5.11–5.13. These results are very dependent upon the type of material as well as the fiber orientation. In this section we show comparisons for axial modulus E_x, Poisson's ratio ν_{xy}, and shear modulus G_{xy} for a polymeric matrix composite (T300/5208) and a metal matrix composite (SCS-6/Ti-15-3).

Figure 5.14 shows the comparison of the axial moduli of the two materials. The metal matrix composite exhibits a significantly higher modulus for all fiber orientations, ranging from a factor of 1.7 times that of the polymeric matrix value at $\theta = 0°$ to 10.5 times the polymeric value at $\theta = 90°$. It is also interesting to note that the metal matrix composite exhibits a local minimum for the axial modulus at $\theta = 58°$. The differences in these curves are a direct result of the difference in the degree of orthotropy of the two materials. As indicated in Table 1.3, the ratio of E_1/E_2 for T300/5208 is 12.3, whereas the ratio for SCS-6/Ti-15-3 is only 1.5.

The comparison for Poisson's ratio ν_{xy} is shown in Fig. 5.15. Again, the material type has a major influence on the engineering constants. Whereas the polymeric matrix composite exhibits wide variations as a function of fiber orientation, with values of Poisson's ratio ranging from a low of 0.02 at $\theta = 90°$ to a high of 1.27 at $\theta = 27°$, the metal matrix composite exhibits much smaller variations, with all values in a relatively narrow band between 0.396 at $\theta = 38°$ and 0.177 at $\theta = 90°$.

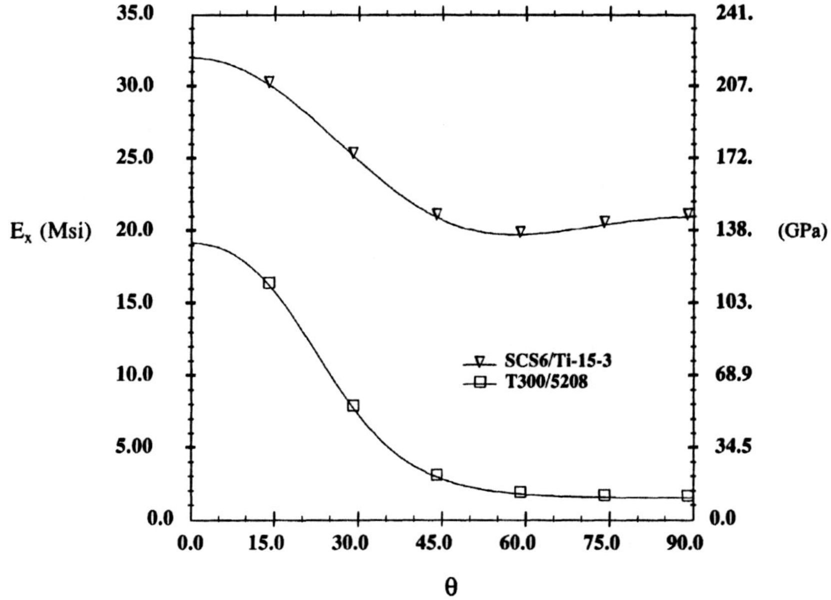

FIGURE 5.14 Angle-Ply Axial Modulus Comparisons

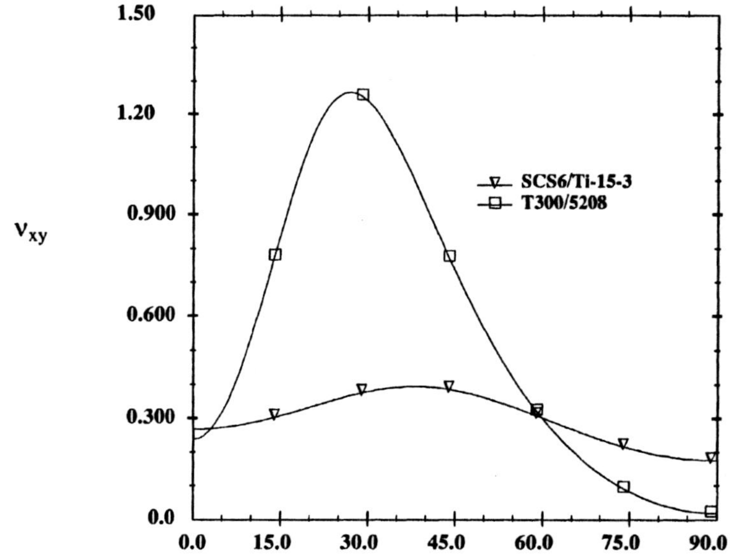

FIGURE 5.15 Angle-Ply Poisson's Ratio Comparisons

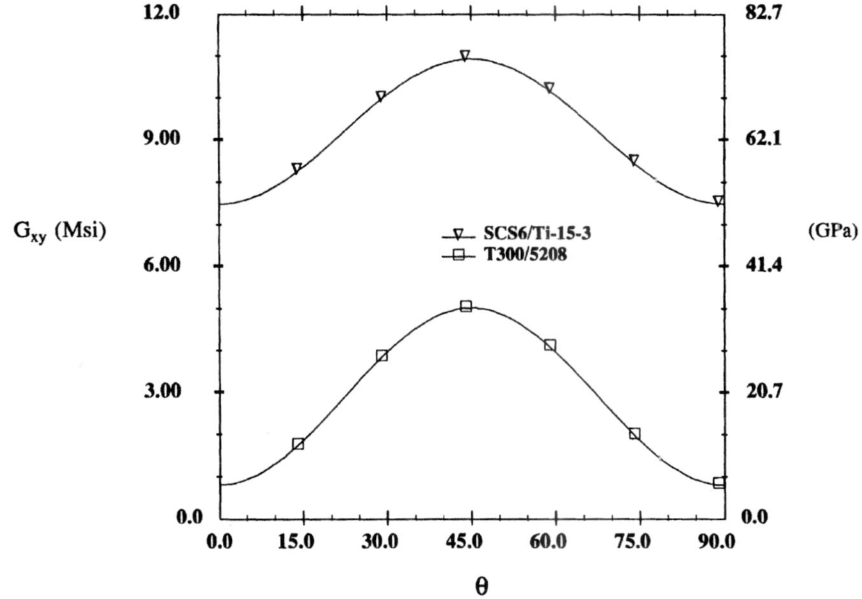

FIGURE 5.16 Angle-Ply Shear Modulus Comparisons

Comparisons for the shear modulus in Fig. 5.16 show that the shear modulus of the metal matrix composite is significantly higher than that of the polymeric matrix material for all fiber orientations. Both materials exhibit maximum values at $\theta = 45°$.

Comparisons between predicted and experimental values for axial modulus and Poisson's ratio for angle-ply laminates of several different materials are presented in Chapter 7. The results indicate good correlation between theory and experiment.

5.13 Stress Distributions

In addition to the overall laminate response as described by the preceding equations, we often want to know the state of stress through the thickness of a laminate. If the midplane strains and curvatures are known, the stress state can be determined directly from the constitutive equation (5.10), which is repeated here:

$$\{\sigma\}^k = [\bar{Q}]^k\{\varepsilon^o\} + [\bar{Q}]^k z\{\kappa\} \tag{5.98}$$

If the forces $\{N\}$ and moments $\{M\}$ are given, the midplane strains and curvatures are determined from (5.24), which is repeated here:

$$\begin{Bmatrix} \varepsilon^o \\ \kappa \end{Bmatrix} = \begin{bmatrix} A' & B' \\ C' & D' \end{bmatrix} \begin{Bmatrix} N \\ M \end{Bmatrix} \tag{5.99}$$

Combining (5.98) and (5.99) gives the most general expression for the stresses at any z-location in a laminate with given forces and moments:

$$\{\sigma\}_x^k = [\bar{Q}]^k[[A']\{N\} + [B']\{M\} + z([C']\{N\} + [D']\{M\})] \tag{5.100}$$

It is apparent from (5.98) and (5.100) that, for the most general laminate and loading conditions, the in-plane components of stress vary linearly through the thickness of individual layers. This is, of course, a direct result of the linear variation in the strains resulting from the Kirchhoff assumptions and the assumption of linear elastic material response in each layer. It is also apparent, however, that the stresses will, in general, be discontinuous from layer to layer because of variation in the stiffness $[\bar{Q}]^k$.

We will demonstrate discontinuities in the stress distributions of laminates by considering symmetric T300/5208 carbon/epoxy laminates subjected first to in-plane loading and then to bending loads. The laminates to be considered include cross-ply, angle-ply, and quasi-isotropic. All laminates are specially orthotropic, and have equal ply thickness t and the same total thickness $2H$.

For a symmetric laminate with $[B] = 0$, the in-plane and bending responses are uncoupled and the midplane strains $\{\varepsilon^o\}$ and curvatures $\{\kappa\}$ are determined for given forces $\{N\}$ and moments $\{M\}$ from the decoupled (5.99). The resulting midplane strains are

$$\begin{Bmatrix} \varepsilon_x^o \\ \varepsilon_y^o \\ \gamma_{xy}^o \end{Bmatrix} = \begin{bmatrix} A_{11} & A_{12} & A_{16} \\ A_{12} & A_{22} & A_{26} \\ A_{16} & A_{26} & A_{66} \end{bmatrix}^{-1} \begin{Bmatrix} N_x \\ N_y \\ N_{xy} \end{Bmatrix} \tag{5.101}$$

and the laminate curvatures are

$$\begin{Bmatrix} \kappa_x \\ \kappa_y \\ \kappa_{xy} \end{Bmatrix} = \begin{bmatrix} D_{11} & D_{12} & D_{16} \\ D_{12} & D_{22} & D_{26} \\ D_{16} & D_{26} & D_{66} \end{bmatrix}^{-1} \begin{Bmatrix} M_x \\ M_y \\ M_{xy} \end{Bmatrix} \tag{5.102}$$

For a symmetric laminate, (5.100) reduces to

$$\{\sigma\}_x^k = [\bar{Q}]^k\{[A']\{N\} + z[D']\{M\}\} \tag{5.103}$$

The individual stress components can be calculated from (5.103) for any z-location, as measured from the midplane of the laminate. As the equation indicates, for nonzero moments $\{M\}$, the stresses vary linearly with z in each layer. The stresses are generally discontinuous from layer to layer because of the distinct values of $[\bar{Q}]^k$ associated with each layer. If the moment $\{M\}$ is zero, the stresses are constant within each layer. In some cases, such as adjacent $\pm\theta$ layers, individual components of stress will be continuous from layer to layer because $[\bar{Q}]^k$ is continuous. These effects will be demonstrated in the examples to follow.

5.13.1 Stresses Due to N_x Loading

For in-plane loading of specially orthotropic laminates ($A'_{16} = A'_{26} = 0$) with only N_x nonzero, the stresses from (5.103) take the form

$$\begin{Bmatrix} \sigma_x \\ \sigma_y \\ \tau_{xy} \end{Bmatrix}^k = \left[\bar{Q}\right]^k \begin{Bmatrix} A'_{11} \\ A'_{12} \\ 0 \end{Bmatrix} N_x \tag{5.104}$$

From (5.101), we see that the shear strain $\gamma^o_{xy} = 0$ for N_x loading of specially orthotropic laminates. Thus, using (5.101) and (5.104), the stresses can be written explicitly in terms of the laminate midplane strains ε^o_x and ε^o_y or, equivalently, in terms of the laminate Poisson's ratio ν_{xy} and the axial strain ε^o_x. The results are

$$\begin{Bmatrix} \sigma_x \\ \sigma_y \\ \tau_{xy} \end{Bmatrix}^k = \left[\bar{Q}\right]^k \begin{Bmatrix} A'_{11} \\ A'_{21} \\ 0 \end{Bmatrix} N_x = \left[\bar{Q}\right]^k \begin{Bmatrix} \varepsilon^o_x \\ \varepsilon^o_y \\ 0 \end{Bmatrix} = \begin{bmatrix} \bar{Q}_{11} & \bar{Q}_{12} & \bar{Q}_{16} \\ \bar{Q}_{12} & \bar{Q}_{22} & \bar{Q}_{26} \\ \bar{Q}_{16} & \bar{Q}_{26} & \bar{Q}_{66} \end{bmatrix}^k \begin{Bmatrix} 1 \\ -\nu_{xy} \\ 0 \end{Bmatrix} \varepsilon^o_x \tag{5.105}$$

Normalized through-thickness distributions of in-plane components of stress in $[0_2/90_2]_s$ cross-ply, $[\pm 45]_s$ angle-ply, and $[0/\pm 45/90]_s$ quasi-isotropic laminates subjected to in-plane loading N_x are shown in Figs. 5.17 and 5.18. As indicated in the figures, all stress components are constant within each individual layer for this in-plane loading of a symmetric laminate. However, the stresses generally are discontinuous between layers. Stresses are continuous between layers only when the adjacent layers have identical properties, such as in the case of adjacent $\pm\theta$ layers. For the adjacent $\pm 45°$ layers of the angle-ply and quasi-isotropic laminates, the normal components of stress are continuous between layers, but the τ_{xy} shear components are discontinuous, being equal in magnitude but of opposite sign, as dictated by the $\bar{Q}_{16}$ and $\bar{Q}_{26}$ stiffness terms.

5.13.1.1 Cross-Ply Laminates

In the case of the $[0_2/90_2]_s$ cross-ply laminate, (5.104) reduces to

$$\begin{Bmatrix} \sigma_x \\ \sigma_y \\ \tau_{xy} \end{Bmatrix}^k = \frac{1}{\Delta} \begin{bmatrix} \bar{Q}_{11} & \bar{Q}_{12} & 0 \\ \bar{Q}_{12} & \bar{Q}_{22} & 0 \\ 0 & 0 & \bar{Q}_{66} \end{bmatrix}^k \begin{Bmatrix} 2Q_{66}(Q_{11}+Q_{22}) \\ -4Q_{12}Q_{66} \\ 0 \end{Bmatrix} N_x \tag{5.106}$$

where Δ is

$$\Delta = 4tQ_{66}[(Q_{11}+Q_{22})^2 - 4Q_{12}^2] \tag{5.107}$$

From (5.106) and the transformation equations (5.55), the stresses in the individual layers of a $[0/90]_s$ laminate subjected to loading N_x can be written in terms of the principal stiffnesses as

$$\sigma_x(0°) = \frac{N_x}{\Delta}[2Q_{11}Q_{66}(Q_{11}+Q_{22}) - 4Q_{12}^2Q_{66}] \tag{5.108}$$

$$\sigma_x(90°) = \frac{N_x}{\Delta}[2Q_{22}Q_{66}(Q_{11}+Q_{22}) - 4Q_{12}^2Q_{66}] \tag{5.109}$$

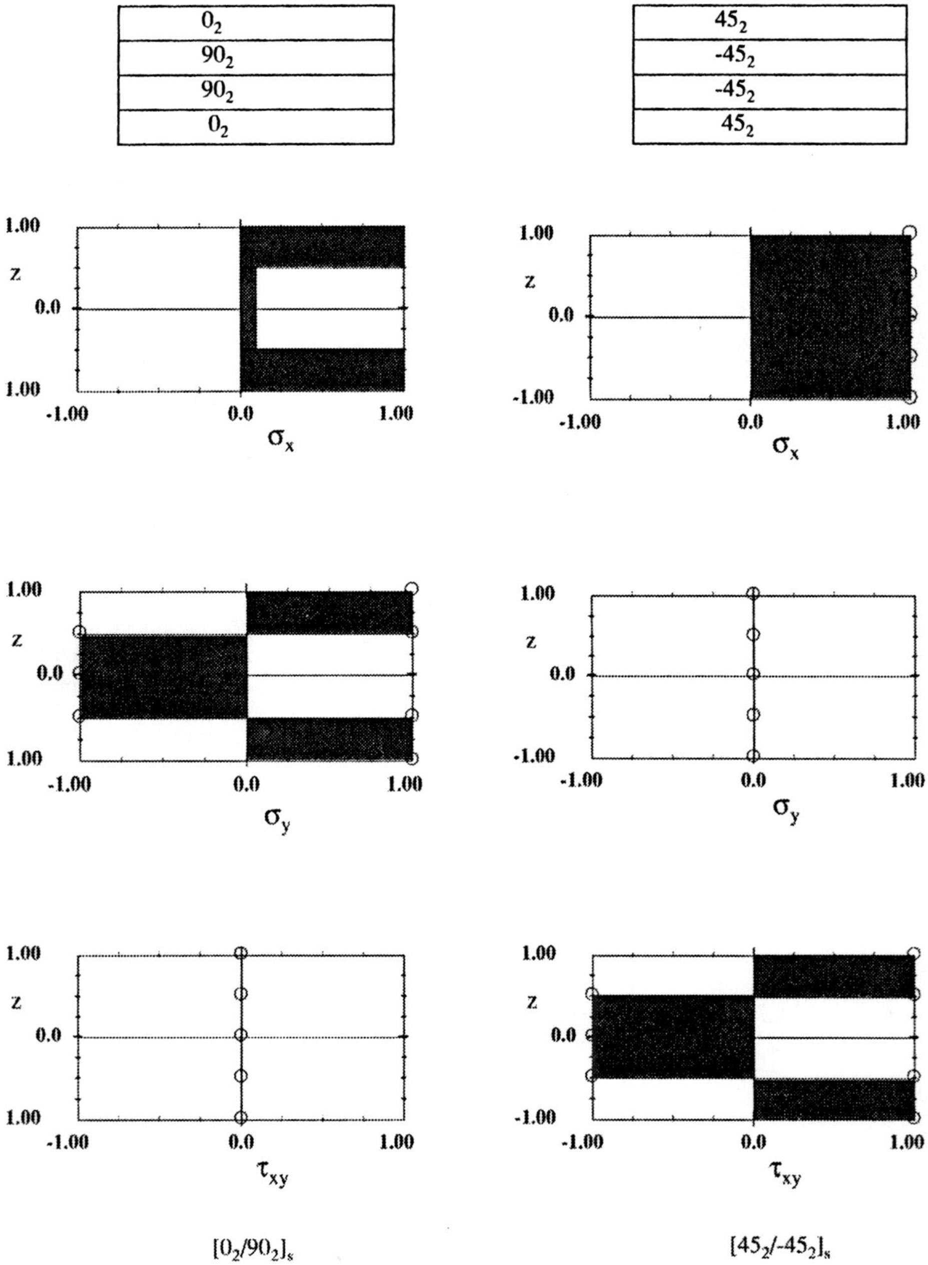

FIGURE 5.17 Stress Distributions for N_x Loading: T300/5208

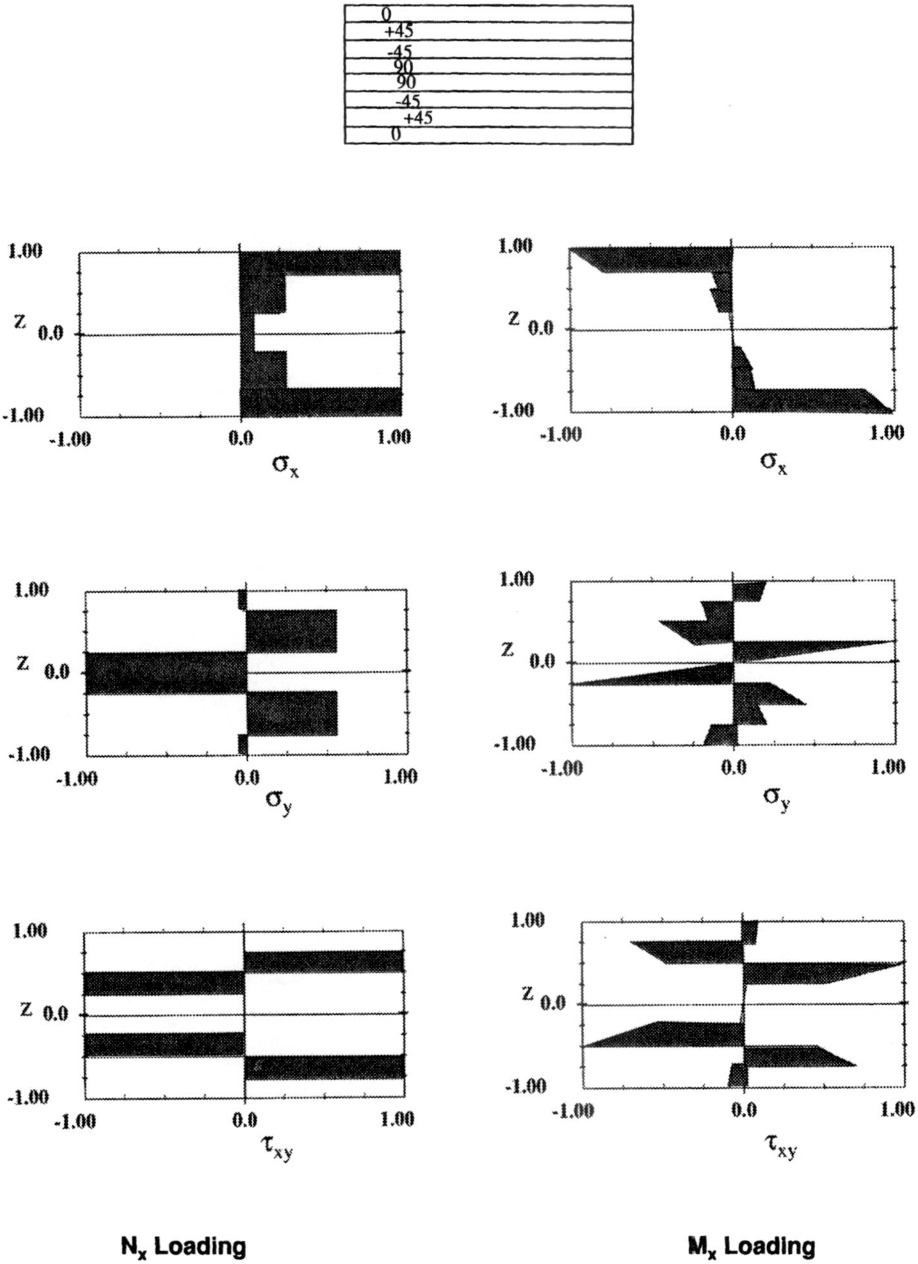

FIGURE 5.18 Stresses in a T300/5208 [0/±45/90]$_s$ Laminate

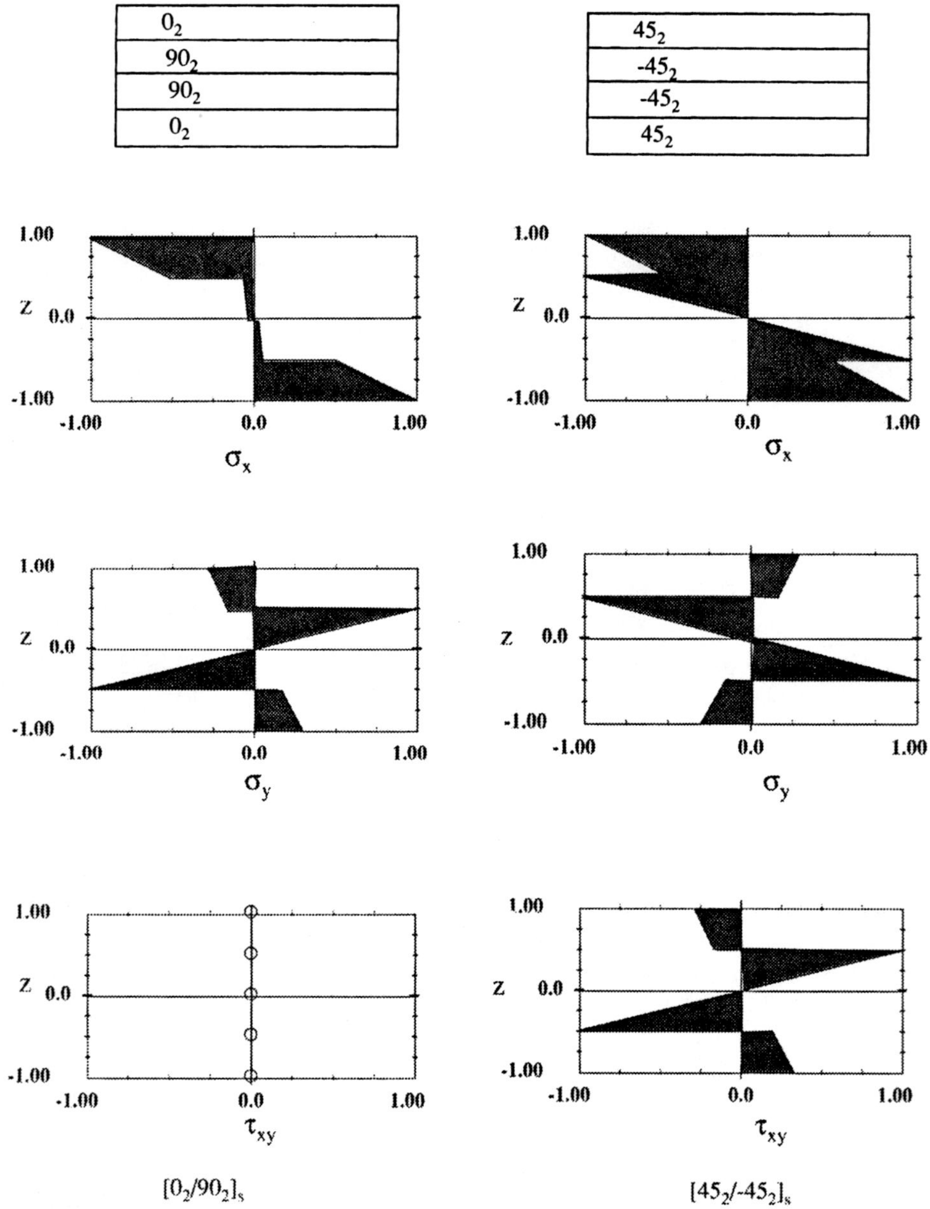

FIGURE 5.19 Stress Distributions for M_x Loading: T300/5208

$$\sigma_y(0°) = \frac{2Q_{12}Q_{66}}{\Delta}(Q_{11} - Q_{22})N_x \tag{5.110}$$

$$\sigma_y(90°) = \frac{-2Q_{12}Q_{66}}{\Delta}(Q_{11} - Q_{22})N_x \tag{5.111}$$

$$\tau_{xy}(0°) = 0 \tag{5.112}$$

$$\tau_{xy}(90°) = 0 \tag{5.113}$$

The transverse σ_y stresses in the 0° and 90° layers are equal in magnitude but opposite in sign, thus satisfying the equilibrium condition $N_y = 0$. The axial stresses σ_x are discontinuous, reflecting the large difference in stiffness (modulus) of the layers. The shear stresses are identically zero throughout the laminate. This is a direct result of the lack of normal-shear coupling ($\bar{Q}_{16} = \bar{Q}_{26} = 0$) for 0° and 90° fiber orientations. These results are reflected in the stress distributions for the cross-ply laminate in Fig. 5.17.

5.13.1.2 Angle-Ply Laminates

Combining the layer constitutive equation (5.9) with (5.81) and (5.93), the stresses in the kth layer of an angle-ply laminate subjected to the axial load $N_x = 2H\bar{\sigma}_x$ are

$$\begin{Bmatrix} \sigma_x \\ \sigma_y \\ \tau_{xy} \end{Bmatrix}^k = \begin{Bmatrix} 1 \\ 0 \\ \dfrac{\bar{Q}_{16}^k\bar{Q}_{22}^k - \bar{Q}_{26}^k\bar{Q}_{12}^k}{\bar{Q}_{11}^k\bar{Q}_{22}^k - \bar{Q}_{12}^k\bar{Q}_{12}^k} \end{Bmatrix}^k \bar{\sigma}_x \tag{5.114}$$

Thus, the angle-ply laminate exhibits uniform axial stress $\sigma_x^k = \bar{\sigma}_x$, zero transverse stress σ_y^k, and equal but opposite shear stresses τ_{xy}^k in the ±θ° layers. The axial σ_x^k stresses are identical because of the identical properties $\bar{Q}_{11}$ and $\bar{Q}_{12}$ in the +θ and −θ layers. From a physical point of view, the transverse σ_y stresses are zero because Poisson's ratio is the same for the +θ and −θ layers; hence there is no constraint to the lateral contraction. The shear stresses τ_{xy}^k are of equal magnitude and opposite sign, satisfying the equilibrium condition $N_{xy} = 0$. It is noted that all angle-ply laminates exhibit the same form of stress distributions, independent of the fiber angle θ; i.e., σ_x^k are uniform, σ_y^k are zero, and τ_{xy}^k are equal in magnitude and opposite in sign.

5.13.1.3 Quasi-Isotropic Laminates

A quasi-isotropic laminate represents a combination of cross-ply and angle-ply laminates. The axial stresses σ_x (Fig. 5.18) are proportional to the axial modulus of the respective layer. However, the transverse stress σ_y is not zero in any of the layers because there is now a mismatch in Poisson's ratio between all layers. The shear stress is zero in the 0° and 90° plies because they are orthotropic and equilibrium is satisfied by the ±45° layers.

5.13.2 Stresses Due to Moment Loading

Distributions of normalized stresses for the three laminates under consideration when subjected to the loading $M_x \neq 0$, $M_y = M_{xy} = 0$, and $\{N\} = 0$ are shown in Figs. 5.18 and 5.19. The linear z dependence within each layer is clearly evident as is the fact that the maximum stresses are not necessarily at the furthest distance from the midplane. This is a result of the variable stiffness of the individual layers. The discontinuity in stiffness between layers results in very complex patterns of stress distribution.

For the $[0_2/90_2]_s$ laminate (Fig. 5.19), the axial stress σ_x shows what might have been anticipated. The stress is maximum at the top and bottom surfaces and varies linearly within each ply. There is a large discontinuity between the 0 and 90° layers, and the slope of the linear variation within the 90° layer is much lower than that in the 0° layer, reflecting the differences in the axial moduli. The transverse σ_y stress varies in a similar but more complex manner with the maximum stresses at the top and bottom of the 90° layer, because the 90° layer has the largest modulus E_y. The shear stress is identically zero as it was for in-plane loading because there is no normal-shear coupling for these fiber orientations.

The distributions in the $[\pm 45]_s$ and $[0/\pm 45/90]_s$ laminates exhibit many of the same features, but in a still more complex manner. The shear stress is nonzero throughout both laminates. The axial stress σ_x displays a most interesting result for the $[\pm 45]_s$ laminate. There is a sharp discontinuity in σ_x at the ±45 interface. This discontinuity is the result of the different sign on $\bar{Q}_{16}$ for the +45° and –45° layers (Fig. 4.5). These latter two laminates clearly show the influence of the coupling terms on the through-thickness distribution of stresses in laminated orthotropic (anisotropic) materials.

5.13.3 Comment on Stress Distributions

The results presented are for very simple loadings of symmetric laminates. It is possible for the stress distributions to be much more complex, and it becomes quite difficult to use intuition without the aid of rigorous analysis. Prediction of failure for these laminates is not simply a matter of determining the maximum stress in a given layer. Appropriate attention must also be given to the anisotropic nature of strength.

The stress distributions of the previous cases generally indicate nonzero transverse stresses σ_y, in direct violation of the boundary condition that the lateral surface of the laminate is free of stress. This fact points to one of the limitations of lamination theory. A basic assumption of lamination theory is that each layer is in a state of plane stress. The individual layers of a finite-width laminate under uniaxial loading cannot be in equilibrium if only plane stress is permitted. The nonzero σ_y stresses in interior regions away from the free edge are acceptable; however, as the edge is approached, the σ_y stresses must go to zero everywhere on the free surface. Equilibrium then requires that nonzero out-of-plane stresses exist near the free edge. These out-of-plane stresses are called *interlaminar stresses*. As will be explained in Chapter 8 on interlaminar stresses, a three-dimensional state of stress exists in a boundary layer along free edges. Thus lamination theory is not valid in these regions. It is noted that the distribution of σ_y stresses does satisfy the conditions $N_y = 0$ and $M_y = 0$ on the free edge.

5.14 Stresses and Strains in Angle-Ply Laminates

5.14.1 Stresses in Principal Material Coordinates

It is instructive to study the dependence of stresses and strains in the principal material coordinates of an equal-layer-thickness $[\pm\theta]_s$ angle-ply laminate, under the uniform axial loading $N_x = 2H\bar{\sigma}_x$ (Fig. 5.20), as a function of the fiber orientation, θ, and degree of material orthotropy represented by the ratio E_1/E_2. We will see that the state of stress, in the principal material coordinates of each layer, varies significantly with both fiber orientation and material orthotropy.

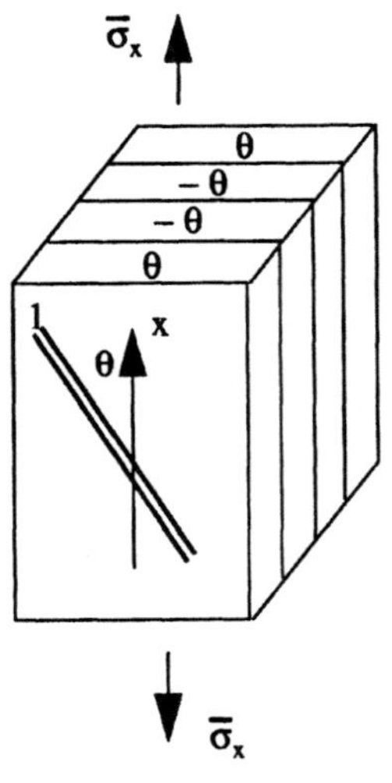

FIGURE 5.20 Angle-Ply Coupon under Axial Load

From (5.114), the stresses in the kth layer of the $[\pm\theta]_s$ tensile coupon are

$$\begin{Bmatrix} \sigma_x \\ \sigma_y \\ \tau_{xy} \end{Bmatrix}^k = \begin{Bmatrix} 1 \\ 0 \\ \dfrac{\bar{Q}_{16}\bar{Q}_{22} - \bar{Q}_{26}\bar{Q}_{12}}{\bar{Q}_{11}\bar{Q}_{22} - \bar{Q}_{12}\bar{Q}_{12}} \end{Bmatrix}^k \bar{\sigma}_x \tag{5.115}$$

We note from (5.115) that an isotropic or orthotropic layer ($\bar{Q}_{16} = \bar{Q}_{26} = 0$) exhibits constant axial stress σ_x and zero transverse and shear stress.

Combining (5.115) with the stress transformation equation (4.23) gives the stresses in the principal material coordinates:

$$\begin{Bmatrix} \sigma_1 \\ \sigma_2 \\ \tau_{12} \end{Bmatrix}^k = \begin{Bmatrix} m^2 + 2mn\dfrac{\bar{Q}_{16}\bar{Q}_{22} - \bar{Q}_{26}\bar{Q}_{12}}{\bar{Q}_{11}\bar{Q}_{22} - \bar{Q}_{12}\bar{Q}_{12}} \\ n^2 - 2mn\dfrac{\bar{Q}_{16}\bar{Q}_{22} - \bar{Q}_{26}\bar{Q}_{12}}{\bar{Q}_{11}\bar{Q}_{22} - \bar{Q}_{12}\bar{Q}_{12}} \\ -mn + (m^2 - n^2)\dfrac{\bar{Q}_{16}\bar{Q}_{22} - \bar{Q}_{26}\bar{Q}_{12}}{\bar{Q}_{11}\bar{Q}_{22} - \bar{Q}_{12}\bar{Q}_{12}} \end{Bmatrix}^k \bar{\sigma}_x \tag{5.116}$$

Taking into consideration the θ dependence of m, n, and $\bar{Q}_{ij}$ in (5.116), we must have

$$\begin{aligned} \sigma_1(\theta) &= \sigma_1(-\theta) \\ \sigma_2(\theta) &= \sigma_2(-\theta) \\ \tau_{12}(\theta) &= -\tau_{12}(-\theta) \end{aligned} \tag{5.117}$$

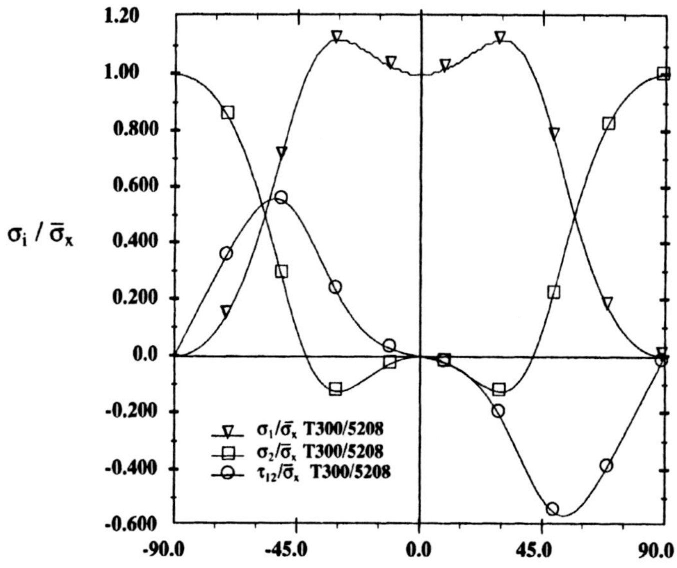

FIGURE 5.21 Stresses in PMC Angle-Ply Laminates under Axial Load

The dependence of stress on fiber orientation and material properties, as predicted by (5.116), for polymer (PMC) and metal (MMC) matrix composites is shown in Figs. 5.21 and 5.22, respectively. The materials selected for study are T300/5208 ($E_1/E_2 = 12.3$) and SCS-6/Ti-15-3 ($E_1/E_2 = 1.5$).

Similarities as well as differences in the θ dependence of the stresses can be noted for the two material systems. The similarities are that the normal stress components, σ_1 and σ_2, are independent of sign for $+\theta$ and $-\theta$ layers whereas the shear components, τ_{12}, exhibit a sign change for positive and negative layers, as required by (5.117). The magnitude of the shear stress τ_{12} is exactly 50% of the applied stress $\bar{\sigma}_x$ at $\theta = \pm 45°$, independent of the material type.

Differences between the stresses in the two material systems are apparent as the fiber angle is increased from 0° to 90°. For the polymer matrix composite (Fig. 5.21), the axial stress, σ_1, increases in magnitude for small off-axis angles, attaining a maximum at ±30° before decreasing rather sharply to zero at 90°. The transverse stress, σ_2, is compressive for angles between 0° and 41° with the largest negative value at ±30°. For fiber angles greater than ±41° the transverse stress increases monotonically to its maximum value at 90°. The shear stress τ_{12} has its maximum magnitudes at ±53°. The variation of normal stresses with fiber orientation in metal matrix composite (Fig. 5.22) differs significantly from that in polymer matrix composite. The axial stress decreases monotonically from its maximum value as the fiber angle is increased from 0°, and the transverse stress is always positive, increasing monotonically from 0° to 90°. The variation of shear stress is similar to that in PMC, with the maximum values at a fiber orientation of ±48°.

Equation (5.116) provides the stresses in material principal coordinates for an angle-ply laminate under axial load in a form convenient for use in a computer program. These same equations can be expressed in terms of the engineering properties of the unidirectional composite and the fiber orientation angle. Representation of the stresses in terms of engineering properties provides insight into the relative importance of the individual properties. Rather than direct substitution for

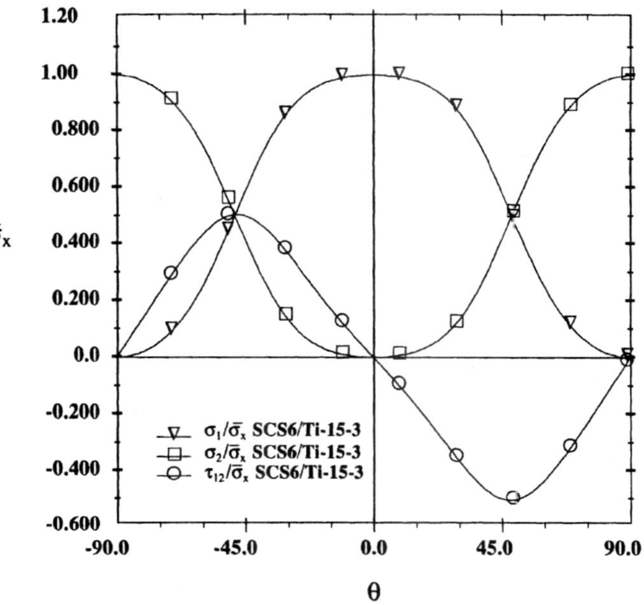

FIGURE 5.22 Stresses in MMC Angle-Ply Laminates under Axial Load

the $\bar{Q}_{ij}$ in terms of engineering properties and fiber orientation angles followed by expansion and simplification of (5.116), we take a different approach.

The stress transformation equations (4.23) give the relationship between stresses in global and principal material coordinates for a layer as

$$\sigma_x = m^2\sigma_1 + n^2\sigma_2 - 2mn\tau_{12} \tag{5.118}$$

$$\sigma_y = n^2\sigma_1 + m^2\sigma_2 + 2mn\tau_{12} \tag{5.119}$$

$$\tau_{xy} = mn\sigma_1 - mn\sigma_2 + (m^2 - n^2)\tau_{12} \tag{5.120}$$

Next, axial force equilibrium of any two layers at $+\theta$ and $-\theta$ and the conditions (5.115) and (5.117) for applied axial stress σ_x require that

$$\bar{\sigma}_x = m^2\sigma_1 + n^2\sigma_2 - 2mn\tau_{12} \tag{5.121}$$

Transverse equilibrium of the same two layers requires that

$$0 = n^2\sigma_1 + m^2\sigma_2 + 2mn\tau_{12} \tag{5.122}$$

Combining (5.121) and (5.122) and using $m^2 + n^2 = 1$ gives the result

$$\bar{\sigma}_x = \sigma_1 + \sigma_2 \tag{5.123}$$

Next, the strain transformation equations (4.24) give the fiber direction strain in the $+\theta$ layer in terms of all three strain components in the $-\theta$ layer, where the transformation angle is 2θ:

$$\varepsilon_1(\theta) = m^2(2\theta)\varepsilon_1(-\theta) + n^2(2\theta)\varepsilon_2(-\theta) + mn(2\theta)\gamma_{12}(-\theta) \tag{5.124}$$

Now the constitutive equations in material principal coordinates are (with the aid of (5.117))

$$\begin{aligned} \varepsilon_1(\theta) &= \frac{\sigma_1}{E_1} - \frac{\nu_{12}}{E_1}\sigma_2 \\ \varepsilon_1(-\theta) &= \frac{\sigma_1}{E_1} - \frac{\nu_{12}}{E_1}\sigma_2 \\ \varepsilon_2(-\theta) &= \frac{\sigma_2}{E_2} - \frac{\nu_{12}}{E_1}\sigma_1 \\ \gamma_{12}(-\theta) &= \frac{\tau_{12}(-\theta)}{G_{12}} \end{aligned} \tag{5.125}$$

Combining (5.122) through (5.125) with trigonometric identities for $\sin(2\theta)$ and $\cos(2\theta)$ gives the layer material principal stresses in terms of lamina engineering properties and $\sin\theta$ (n) and $\cos\theta$ (m).

$$\sigma_1 = \left[\frac{m^2(2m^2-1) + 4m^2n^2\frac{G_{12}}{E_2}\left(\frac{E_2}{E_1}\nu_{12}+1\right)}{4m^2n^2\frac{G_{12}}{E_2}\left(\frac{E_2}{E_1}+2\frac{E_2}{E_1}\nu_{12}+1\right)+(2m^2-1)(m^2-n^2)}\right]\bar{\sigma}_x \tag{5.126}$$

$$\sigma_2 = \left[1 - \frac{m^2(2m^2-1) + 4m^2n^2\frac{G_{12}}{E_2}\left(\frac{E_2}{E_1}\nu_{12}+1\right)}{4m^2n^2\frac{G_{12}}{E_2}\left(\frac{E_2}{E_1}+2\frac{E_2}{E_1}\nu_{12}+1\right)+(2m^2-1)(m^2-n^2)}\right]\bar{\sigma}_x \tag{5.127}$$

$$\tau_{12} = \frac{-1}{2mn}\left[m^2 + \frac{m^2(2m^2-1) + 4m^2n^2\frac{G_{12}}{E_2}\left(\frac{E_2}{E_1}\nu_{12}+1\right)}{4m^2n^2\frac{G_{12}}{E_2}\left(\frac{E_2}{E_1}+2\frac{E_2}{E_1}\nu_{12}+1\right)+(2m^2-1)(m^2-n^2)}(1-2m^2)\right]\bar{\sigma}_x \tag{5.128}$$

It is convenient to define the common term in (5.126) to (5.128) and analyze it for different materials. Thus, defining

$$B = \left[\frac{m^2(2m^2-1) + 4m^2n^2\frac{G_{12}}{E_2}\left(\frac{E_2}{E_1}\nu_{12}+1\right)}{4m^2n^2\frac{G_{12}}{E_2}\left(\frac{E_2}{E_1}+2\frac{E_2}{E_1}\nu_{12}+1\right)+(2m^2-1)(m^2-n^2)}\right] \tag{5.129}$$

the three stress equations (5.126) to (5.128) can be written

$$
\begin{aligned}
\sigma_1 &= B\bar{\sigma}_x \\
\sigma_2 &= (1-B)\bar{\sigma}_x \\
\tau_{12} &= \frac{-1}{2mn}[B(1-2m^2)+m^2]\bar{\sigma}_x
\end{aligned}
\tag{5.130}
$$

Further, for the special case $\theta = 45°$, independent of the material type, the shear stress in (5.128) (or (5.130)) reduces to

$$
\tau_{12} = \frac{-\bar{\sigma}_x}{2} \tag{5.131}
$$

As will be discussed in Chapter 6, the fact that the magnitude of the principal material shear stress is equal to one-half the applied stress in a $[\pm 45]_s$ tensile coupon means that this configuration can be used as a test specimen to measure the shear modulus G_{12} of any fibrous composite.

We note that for the special case $\theta = 45°$, the coefficient B as defined in (5.129) reduces to

$$
B_{45} = \left[\frac{\frac{E_2}{E_1}\nu_{12}+1}{\frac{E_2}{E_1}+2\frac{E_2}{E_1}\nu_{12}+1}\right] \tag{5.132}
$$

For highly orthotropic materials, such as carbon/epoxy, where $E_1 >> E_2$, B for any angle θ may be approximated from (5.129) by B_a, where

$$
B_a = \left[\frac{m^2(2m^2-1)+4m^2n^2\frac{G_{12}}{E_2}}{4m^2n^2\frac{G_{12}}{E_2}+(2m^2-1)(m^2-n^2)}\right] \tag{5.133}
$$

Thus for an angle-ply laminate with $\theta = 45°$ and $E_1 >> E_2$, (5.132) and (5.133) give $B_a = B_{45} = 1.0$; the axial stress σ_1 equals the applied stress, the transverse stress σ_2 is zero, and the shear stress τ_{12} is one-half the applied stress, i.e.,

$$
\begin{aligned}
\sigma_1 &= \bar{\sigma}_x \\
\sigma_2 &= 0 \\
\tau_{12}(\theta) &= \frac{-\bar{\sigma}_x}{2} \\
\tau_{12}(-\theta) &= \frac{\bar{\sigma}_x}{2}
\end{aligned}
\tag{5.134}
$$

The influence of fiber orientation and material properties on the ratio B/B_a is summarized in Fig. 5.23, which shows a comparison of this ratio for a polymer matrix and a metal matrix

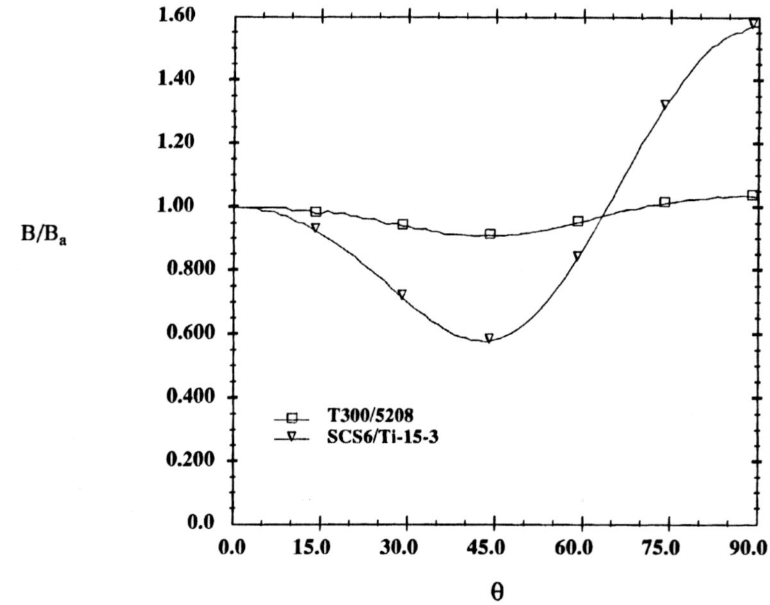

FIGURE 5.23 ***B* and B_a Coefficients for PMC and MMC**

composite. As indicated in the figure, the ratio is nearly constant and equal to 1 for the highly orthotropic polymer matrix composite over most of the range of fiber orientation; the largest difference is approximately 9% at 45°. In contrast, the ratio varies by almost a factor of 3.0 over the 0°–90° range for the less orthotropic metal matrix composite. Thus, the use of the approximate B_a is limited to highly orthotropic materials. It is noted that polymer matrix composites are available with an even higher degree of orthotropy than the T300/5208 used in this example.

5.14.2 Strains in Principal Material Coordinates

Possibly even more interesting than the variations in stresses in an angle-ply laminate under axial loading are the variations of strains. From (5.81) and (5.93), the midplane (and layer) strains are

$$\begin{Bmatrix} \varepsilon_x \\ \varepsilon_y \\ \gamma_{xy} \end{Bmatrix}^k = \begin{bmatrix} \dfrac{\bar{Q}_{22}}{\bar{Q}_{11}\bar{Q}_{22}-\bar{Q}_{12}\bar{Q}_{12}} \\ \dfrac{-\bar{Q}_{12}}{\bar{Q}_{11}\bar{Q}_{22}-\bar{Q}_{12}\bar{Q}_{12}} \\ 0 \end{bmatrix}^k \bar{\sigma}_x \tag{5.135}$$

Combining (5.135) with the strain transformations (4.24) gives the strains in material principal coordinates:

$$\begin{Bmatrix} \varepsilon_1 \\ \varepsilon_2 \\ \gamma_{12} \end{Bmatrix}^k = \begin{bmatrix} m^2 & n^2 & mn \\ n^2 & m^2 & -mn \\ -2mn & 2mn & m^2-n^2 \end{bmatrix}^k \begin{bmatrix} \dfrac{\bar{Q}_{22}}{\bar{Q}_{11}\bar{Q}_{22}-\bar{Q}_{12}\bar{Q}_{12}} \\ \dfrac{-\bar{Q}_{12}}{\bar{Q}_{11}\bar{Q}_{22}-\bar{Q}_{12}\bar{Q}_{12}} \\ 0 \end{bmatrix}^k \bar{\sigma}_x \tag{5.136}$$

This can be written

$$\begin{Bmatrix} \varepsilon_1 \\ \varepsilon_2 \\ \gamma_{12} \end{Bmatrix}^k = \begin{bmatrix} m^2\bar{Q}_{22} - n^2\bar{Q}_{12} \\ n^2\bar{Q}_{22} - m^2\bar{Q}_{12} \\ -2mn\bar{Q}_{22} - 2mn\bar{Q}_{12} \end{bmatrix}^k \frac{\bar{\sigma}_x}{(\bar{Q}_{11}\bar{Q}_{22}-\bar{Q}_{12}\bar{Q}_{12})^k} \tag{5.137}$$

From (5.137) we see that, as expected, the strains in the $+\theta$ and $-\theta$ layers follow the same pattern as the stresses, namely,

$$\begin{aligned} \varepsilon_1(\theta) &= \varepsilon_1(-\theta) \\ \varepsilon_2(\theta) &= \varepsilon_2(-\theta) \\ \gamma_{12}(\theta) &= -\gamma_{12}(-\theta) \end{aligned} \tag{5.138}$$

We note that for the special case of $\theta = 45°$, $\varepsilon_1 = \varepsilon_2$.

The θ dependence of strains in PMC and MMC is shown in Figs. 5.24 and 5.25, respectively. The results in these figures are for tensile loading of angle-ply laminates subjected to the same magnitude of axial stress. Careful study of the figures indicates that both the magnitudes and the ranges of strain are much greater in PMC. In the PMC the maximum magnitude of shear strain γ_{12} at $\theta = \pm 53°$ is more than 10 times the maximum fiber direction strain ε_1, which actually occurs at $\theta = \pm 30°$. In the MMC the maximum magnitude of the shear strain at $\theta = \pm 48°$ is only a little over two times the maximum fiber direction strain, which occurs at $\theta = 0°$. While the same general θ dependence is present for both material systems, the normal strains ε_2 do exhibit a difference with local extrema in the PMC at 90°, ±30°, and 0°; the local extrema in the MMC occur only at 90° and 0°.

5.15 Unsymmetric Laminates

As discussed in detail in Section 5.9, the coupling coefficients B_{ij} of lamination theory are identically zero for laminates that are symmetric about the midplane. When the laminate is unsymmetric, $B_{ij} \neq 0$ and the in-plane (membrane) response is coupled with the bending (flexural) response. The specific type of coupling depends upon which of the B_{ij} are nonzero. The relationship between all coupling terms and the type of coupling present is depicted in Fig. 5.26.

The coupling within the A_{ij} (membrane) and D_{ij} (flexure) terms was discussed in Section 5.9. As indicated in Fig. 5.26, B_{11}, B_{22}, and B_{12} influence the coupling between membrane extension and flexural bending; B_{66} influences the coupling between membrane shear and flexural twist; and the B_{16} and B_{26} terms influence the coupling between membrane extension and flexural twist, as well as the coupling between membrane shear and flexural bending.

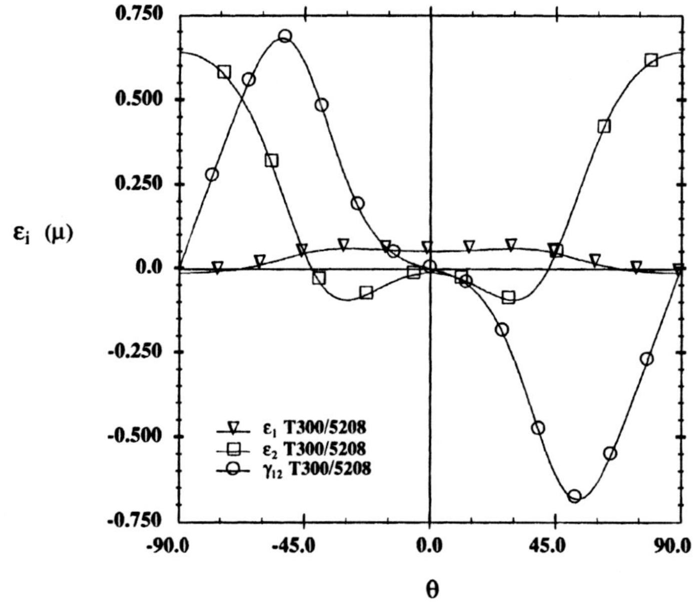

FIGURE 5.24 Strains in PMC Angle-Ply Laminates under Axial Load

The details of the coupling phenomena can be best understood through consideration of the full lamination theory equations (5.40), repeated here:

$$\begin{Bmatrix} \varepsilon_x^o \\ \varepsilon_y^o \\ \gamma_{xy}^0 \\ \kappa_x \\ \kappa_y \\ \kappa_{xy} \end{Bmatrix} = \begin{bmatrix} A'_{11} & A'_{12} & A'_{16} & B'_{11} & B'_{12} & B'_{16} \\ A'_{12} & A'_{22} & A'_{26} & B'_{12} & B'_{22} & B'_{26} \\ A'_{16} & A'_{26} & A'_{66} & B'_{16} & B'_{26} & B'_{66} \\ B'_{11} & B'_{12} & B'_{16} & D'_{11} & D'_{12} & D'_{16} \\ B'_{12} & B'_{22} & B'_{26} & D'_{12} & D'_{22} & D'_{26} \\ B'_{16} & B'_{26} & B'_{66} & D'_{16} & D'_{26} & D'_{66} \end{bmatrix} \begin{Bmatrix} N_x \\ N_y \\ N_{xy} \\ M_x \\ M_y \\ M_{xy} \end{Bmatrix} \tag{5.139}$$

The expanded form (5.139) can be written in the condensed notation as

$$\begin{Bmatrix} \varepsilon^o \\ \kappa \end{Bmatrix} = \begin{bmatrix} A' & B' \\ B'^T & D' \end{bmatrix} \begin{Bmatrix} N \\ M \end{Bmatrix} \tag{5.140}$$

Coupling phenomena are depicted schematically in Fig. 5.27 for the case of uniaxial N_x loading on an unsymmetric laminate. The figure shows the types of curvature that develop when three of the B'_{ij} coefficients are nonzero. The situation is, of course, much more complex for more general loadings.

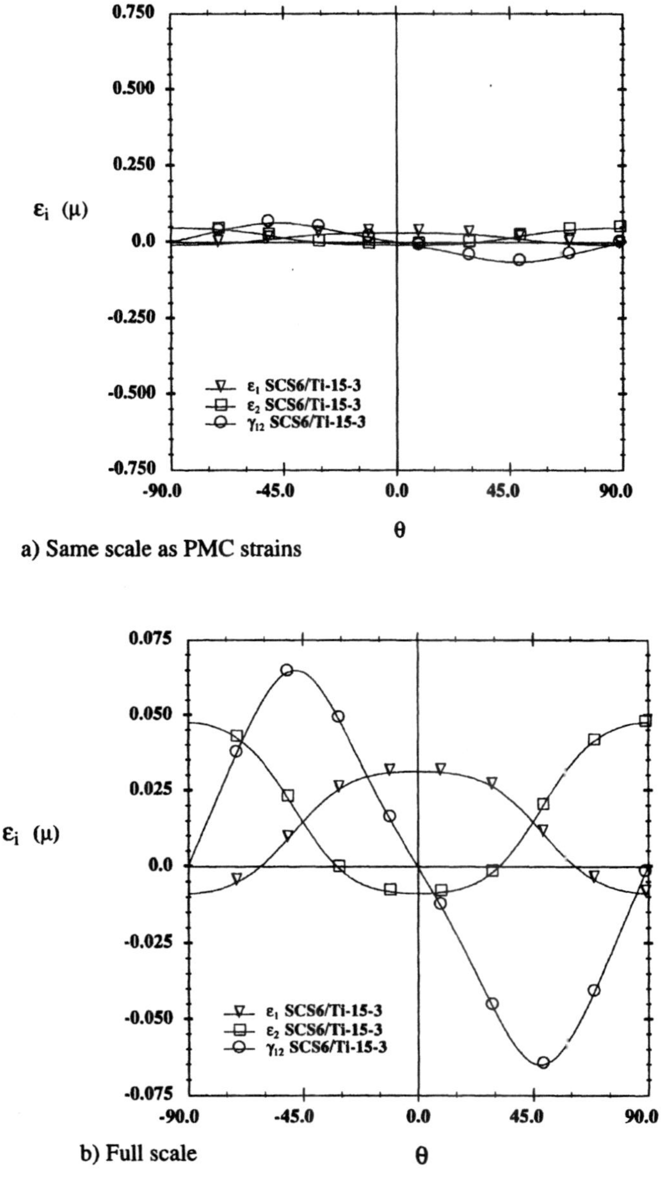

a) Same scale as PMC strains

b) Full scale

FIGURE 5.25 Strains in MMC Angle-Ply Laminates under Axial Load

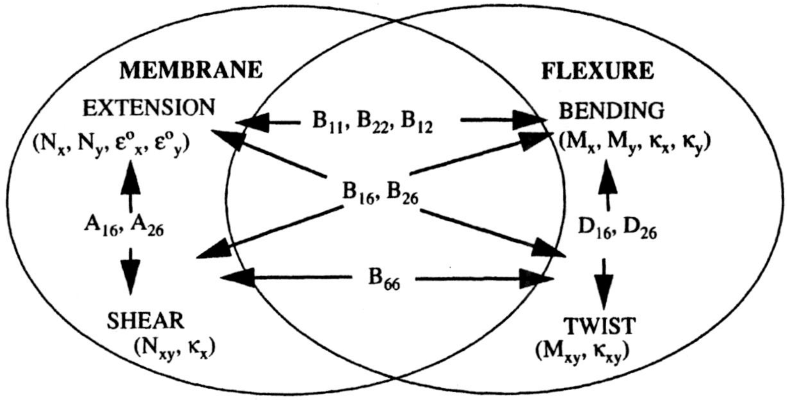

FIGURE 5.26 Laminate Coupling Relationships

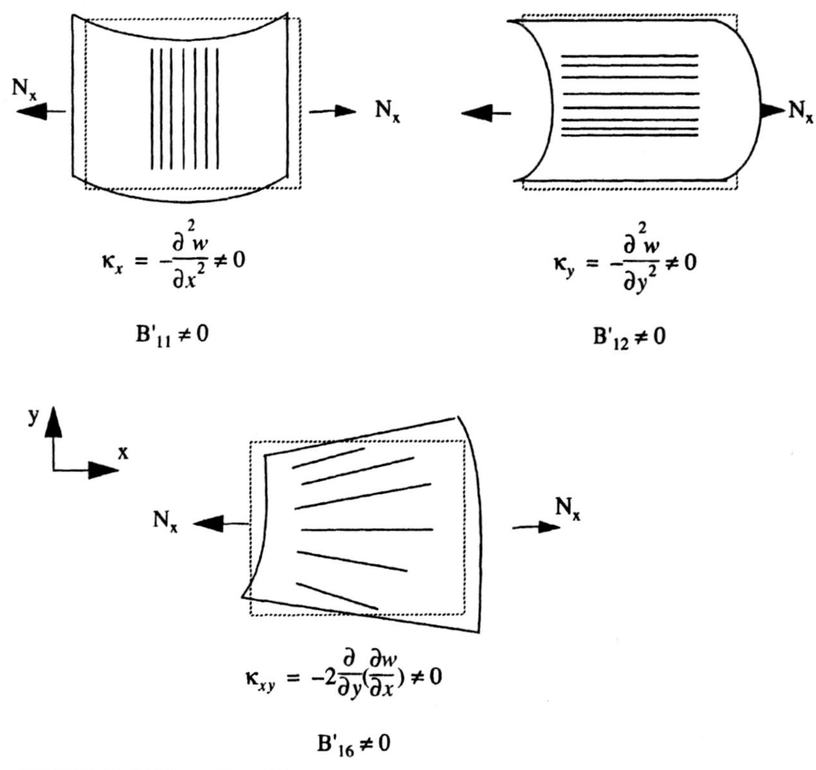

FIGURE 5.27 Coupling Phenomena

The influence of fiber orientation on the coupling terms can be demonstrated by consideration of unsymmetric cross-ply and angle-ply laminates. For an unsymmetric $[0/90]_T$ cross-ply laminate, the nonzero terms of the coefficient matrices are

$$\begin{bmatrix} B_{11} & 0 & 0 \\ 0 & -B_{11} & 0 \\ 0 & 0 & 0 \end{bmatrix} \quad \text{and} \quad \begin{bmatrix} B'_{11} & 0 & 0 \\ 0 & -B'_{11} & 0 \\ 0 & 0 & 0 \end{bmatrix} \tag{5.141}$$

For an unsymmetric $[\theta/-\theta]_T$ angle-ply laminate the nonzero coefficients are

$$\begin{bmatrix} 0 & 0 & B_{16} \\ 0 & 0 & B_{26} \\ B_{16} & B_{26} & 0 \end{bmatrix} \quad \text{and} \quad \begin{bmatrix} 0 & 0 & B'_{16} \\ 0 & 0 & B'_{26} \\ B'_{16} & B'_{26} & 0 \end{bmatrix} \tag{5.142}$$

Individual coefficients can be presented graphically as functions of θ for angle-ply laminates much in the same manner in which all previous coefficients have been presented.

An interesting and practical application of unsymmetric laminates is to assess residual stresses in composite laminates. This application is discussed in the following section. Unsymmetric laminates can also be used to fabricate turbine blades with a predetermined twist.

5.16 Thermo-Elastic Lamination Theory

We can develop the equations of lamination theory for thermal stress analysis following essentially the same procedure as used for pure mechanical loading, but with the use of the thermo-elastic lamina constitutive equation given by (4.98).

We continue to *assume* that the *total strains* $\{\varepsilon\}$ in the laminate follow from the Kirchhoff assumptions on displacements, and that the *total* strains are the superposition of mechanical strains $\{\varepsilon^\sigma\}$ and thermal strains $\{\varepsilon^T\}$. Thus

$$\{\varepsilon\} = \{\varepsilon^o\} + z\{\kappa\} = \{\varepsilon^\sigma\} + \{\varepsilon^T\} \tag{5.143}$$

The lamina constitutive equation (4.98) for the stresses $\{\sigma\}$ in terms of midplane strains $\{\varepsilon^o\}$, curvatures $\{\kappa\}$, and thermal strains $\{\varepsilon^T\}$ is then

$$\{\sigma\} = [\bar{Q}](\{\varepsilon^o\} + z\{\kappa\} - \{\varepsilon^T\}) \tag{5.144}$$

From this point onward the development follows that for mechanical loading, but it is necessary to define new force and moment terms associated with the thermal effects.

5.16.1 Thermal Forces and Moments

From the definition for $\{N\}$ in (5.12) and the constitutive equation (5.144), we have

$$\{N\} = [A]\{\varepsilon^o\} + [B]\{\kappa\} - \int_{-H}^{H} [\bar{Q}]^k \{\varepsilon^T\}^k dz \tag{5.145}$$

where we have indicated explicitly by the superscript k that the integration of the thermal strain term must account for changes in stiffness and thermal strains from layer to layer.

We now define the integral in (5.145),

$$\{N^T\} = \int_{-H}^{H} [\bar{Q}]^k \{\varepsilon^T\}^k dz \tag{5.146}$$

to be the *equivalent thermal force per unit length* and write the expression for in-plane forces in the form

$$\{N\} = [A]\{\varepsilon^o\} + [B]\{\kappa\} - \{N^T\} \tag{5.147}$$

In a very similar fashion, the moment per unit length is

$$\{M\} = [B]\{\varepsilon^o\} + [D]\{\kappa\} - \int_{-H}^{H} [\bar{Q}]^k \{\varepsilon^T\}^k z\, dz \tag{5.148}$$

Defining the *equivalent thermal moment per unit length* $\{M^T\}$,

$$\{M^T\} = \int_{-H}^{H} [\bar{Q}]^k \{\varepsilon^T\}^k z\, dz \tag{5.149}$$

permits us to write

$$\{M\} = [B]\{\varepsilon^o\} + [D]\{\kappa\} - \{M^T\} \tag{5.150}$$

Combining (5.147) and (5.150) and rearranging terms gives us the final form of the equations for thermo-elastic laminate analysis:

$$\begin{Bmatrix} N + N^T \\ M + M^T \end{Bmatrix} = \begin{bmatrix} A & B \\ B & D \end{bmatrix} \begin{Bmatrix} \varepsilon^o \\ \kappa \end{Bmatrix} \tag{5.151}$$

It is noted that (5.151) has the exact same form as the fundamental equation of lamination theory (5.22) if the force and moment are taken to be the sum of the mechanical and thermal loads. Equation (5.151) can, of course, be inverted to give the midplane strains and curvatures in terms of the forces and moments:

$$\begin{Bmatrix} \varepsilon^o \\ \kappa \end{Bmatrix} = \begin{bmatrix} A' & B' \\ B'^T & D' \end{bmatrix} \begin{Bmatrix} N + N^T \\ M + M^T \end{Bmatrix} \tag{5.152}$$

It is now apparent that the only additional consideration for thermo-elastic lamination theory is the calculation of the equivalent thermal force $\{N^T\}$ and the equivalent thermal moment $\{M^T\}$. These quantities are called *equivalent forces* and *equivalent moments* because they have units of force per unit length and moment per unit length, respectively. They, in themselves, are not physically applied forces and moments. Indeed, even in the case of an otherwise free laminate, these quantities exist (are generally nonzero) if the laminate is subjected to a change in temperature.

A physical interpretation of these equivalent quantities can be gained by visualizing a laminate which is completely fixed against midplane strains and laminate curvatures, and subjected to a uniform temperature change ΔT. From (5.151) we then have that the equivalent thermal forces and moments are the negative of the applied mechanical loads that restrict the midplane strains and laminate curvatures, i.e.,

$$\begin{bmatrix} N_x + N_x^T \\ N_y + N_y^T \\ N_{xy} + N_{xy}^T \\ M_x + M_x^T \\ M_y + M_y^T \\ M_{xy} + M_{xy}^T \end{bmatrix} = \begin{bmatrix} 0 \\ 0 \\ 0 \\ 0 \\ 0 \\ 0 \end{bmatrix} \tag{5.153}$$

5.16.2 Laminate Coefficient of Thermal Expansion

For a symmetric laminate, we can define a *laminate coefficient of thermal expansion* (CTE), $\{\alpha\}$, as the laminate midplane strain, $\{\varepsilon^o\}$, per unit (uniform) temperature change, ΔT, i.e.,

$$\{\alpha\} = \frac{\{\varepsilon^o\}}{\Delta T} \tag{5.154}$$

For pure thermal loading of a symmetric laminate, the midplane strain from (5.152) is

$$\{\varepsilon^o\} = [A]^{-1}\{N^T\} \tag{5.155}$$

Combining (5.154) and (5.155) gives

$$\{\alpha\} = \frac{[A]^{-1}\{N^T\}}{\Delta T} \tag{5.156}$$

For a uniform temperature change the equivalent thermal force (5.146) is

$$\{N^T\} = \Delta T\int_{-H}^{H}[\bar{Q}]^k\{\alpha\}^k dz \tag{5.157}$$

Combining (5.156) and (5.157) gives the laminate coefficient of thermal expansion:

$$\{\alpha\} = [A]^{-1}\int_{-H}^{H}[\bar{Q}]^k\{\alpha\}^k dz \tag{5.158}$$

Since both $[Q]^k$ and $\{\alpha\}^k$ are constant in each layer, the integral in (5.158) can be written as a summation much like the stiffness matrix $[A]$, i.e.,

$$\{\alpha\} = [A]^{-1}\sum_{k=1}^{N}[\bar{Q}]^k\{\alpha\}^k t_k \tag{5.159}$$

where, as before, $t_k = z_k - z_{k-1}$ is the thickness of the kth layer. The laminate CTE is an important engineering property which is useful in design and thermal stress analysis. Because of the range of properties of fibrous composites, laminate CTE values also vary widely. Examples of CTE values for unidirectional off-axis and angle-ply T300/5208 carbon/epoxy laminates are shown in Figs. 5.28 and 5.29.

As indicated in Fig. 5.28, the axial, α_x, and transverse, α_y, CTEs for off-axis lamina vary monotonically from a small negative value in the fiber direction to a rather large positive value in the transverse direction. In contrast, for an angle-ply laminate, as the magnitude of the angle θ increases from zero, α_x initially decreases further from the negative value at $\theta = 0$ to a rather large negative (minimum) value at approximately 30°. The CTE then increases with increasing θ, passing through zero at approximately 42°, to the large positive transverse value of the lamina for $\theta = 90°$.

The fact that the CTE can range from positive to negative values depending upon the laminate configurations is a very important design consideration. As indicated in Fig. 5.28, it is possible to choose (design) carbon/epoxy laminates which have zero CTE values in specified directions. This fact has important consequences for structural application in thermal environments.

As indicated in Fig. 5.29, the shear coefficient of thermal expansion, α_{xy}, is identically zero for angle-ply laminates but attains a large negative value at 45° for unidirectional off-axis composites.

Combining (5.157) and (5.158), we see that for a uniform temperature change, the equivalent thermal force can be written in terms of the laminate CTE as

$$\{N^T\} = \Delta T[A]\{\alpha\} \tag{5.160}$$

Thus, if $\{\alpha\}$ is zero, the force required to restrain the laminate when it changes temperature is also zero. Note that this does not imply that the internal stresses are zero throughout the laminate. It only requires that the integral of the stresses through the thickness be zero.

It is also interesting to note that the three laminates $[0/\pm45/90]_s$, $[0/90]_s$, and $[\pm45]_s$ all have identical $\{\alpha\}$ values. The proof of this statement is left as a homework exercise.

5.16.3 Angle-Ply CTE Comparisons

A comparison of the axial coefficient of thermal expansion for angle-ply laminates of polymeric matrix (T300/5208) and metal matrix (SCS-6/Ti-15-3) composites is shown in Fig. 5.30. As indicated in the figure, the wide variations in CTE for polymeric matrix, including the positive to negative variation, are not exhibited by the metal matrix composite. The CTE for the metal matrix composite are always positive, in a very narrow band over the full range of fiber orientations. This figure clearly shows that if a zero or near-zero CTE is desired, a polymeric matrix composite is the material of choice.

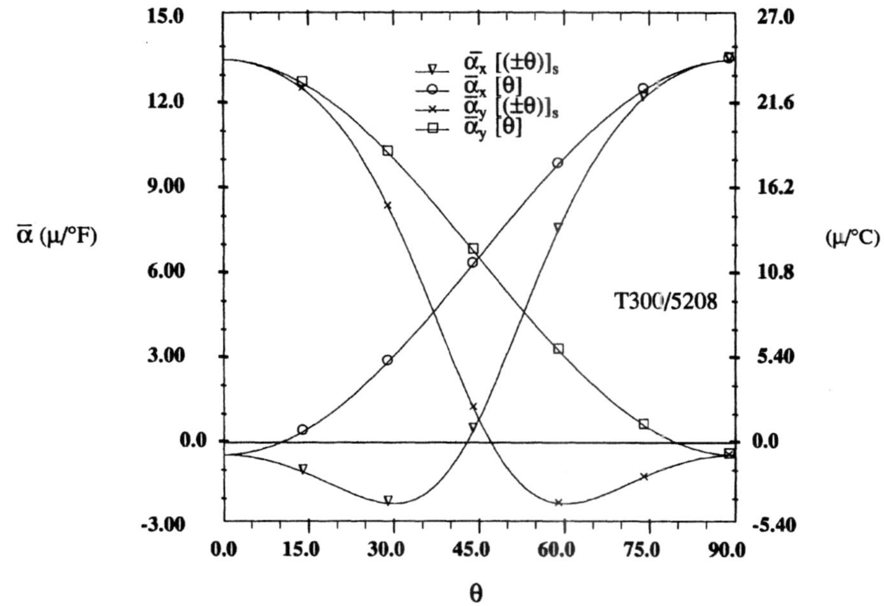

FIGURE 5.28 Lamina and Angle-Ply Laminate CTE Variations

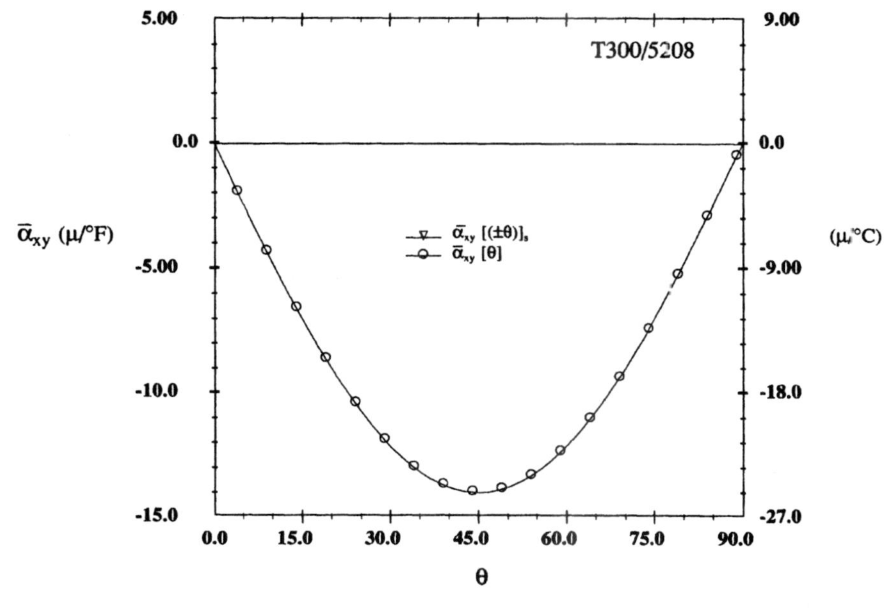

FIGURE 5.29 Shear Thermal Expansion Coefficients

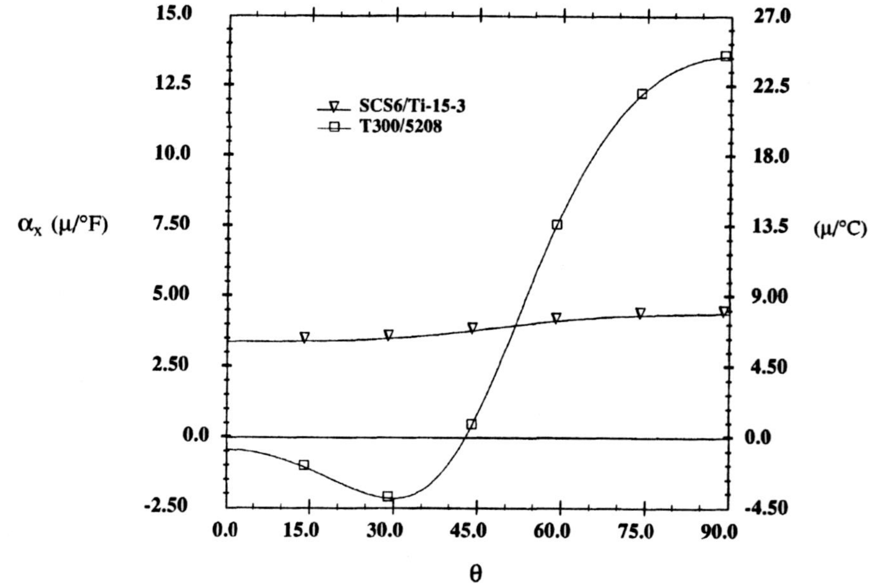

FIGURE 5.30 Angle-Ply CTE Comparisons

5.16.4 Thermal Stresses

The procedure for determining the stress distribution in a laminate subjected to thermomechanical loading follows that for mechanical loading, the only difference being that the forces and moments in (5.100) are the sum of the mechanical and equivalent thermal quantities. Thus

$$\{\sigma\}_x^k = [\bar{Q}]^k(([A'] + z[C'])\{N + N^T\} + ([B'] + z[D'])\{M + M^T\} - \{\varepsilon^T\}_x^k) \tag{5.161}$$

In evaluating the equivalent thermal force and moment, the temperature distribution through the laminate thickness must be included in the evaluation of the integrals (5.146) and (5.149).

5.16.4.1 Uniform Temperature Change

For the special case of uniform temperature distribution throughout the laminate, ΔT is a constant, and the equivalent thermal forces and moments are

$$\{N^T\} = \Delta T\int_{-H}^{H}[\bar{Q}]^k\{\alpha\}^k dz \tag{5.162}$$

$$\{M^T\} = \Delta T\int_{-H}^{H}[\bar{Q}]^k\{\alpha\}^k z\, dz \tag{5.163}$$

For uniform thermal loading of symmetric laminates, we have shown in the previous section that the equivalent thermal force is

$$\{N^T\} = \Delta T[A]\{\alpha\} \tag{5.164}$$

For general laminates, (5.162) can be written

$$\{N^T\} = \Delta T \sum_{k=1}^{N} [\bar{Q}]^k \{\alpha\}^k t_k \tag{5.165}$$

The equivalent thermal moment for uniform temperature change is

$$\{M^T\} = \frac{\Delta T}{2} \sum_{k=1}^{N} [\bar{Q}]^k \{\alpha\}^k (z_k^2 - z_{k-1}^2) \tag{5.166}$$

The summation in (5.166) has the same form as the definition of the $[B]$ matrix (5.16). Therefore, following the same procedure as in Section 5.9, we see that this summation is zero for a symmetric laminate. Hence, the equivalent thermal moment is zero when a symmetric laminate undergoes a uniform temperature change. For this case, the stress equation (5.161) reduces to

$$\{\sigma\}_x^k = [\bar{Q}]^k ([A']\{N^T\} - \{\varepsilon^T\}_x^k) \tag{5.167}$$

Combining equations (4.89), (5.156), and (5.167) gives the simple expression for the thermal stresses:

$$\{\sigma\}_x^k = [\bar{Q}]^k (\{\alpha\} - \{\alpha\}_x^k) \Delta T \tag{5.168}$$

Thus, the stresses in any layer of a symmetric laminate subjected to a uniform temperature change are proportional to the difference between the laminate CTE, $\{\alpha\}$, and the layer CTE, $\{\alpha\}^k$.

5.16.5 Application to Unsymmetric Laminates

For unsymmetric laminates subjected to pure, uniform thermal loading, the applied forces $\{N\}$ and moments $\{M\}$ are zero. The equivalent thermal force $\{N^T\}$ and equivalent thermal moment $\{M^T\}$ are given by (5.165) and (5.166), respectively. Combining these two equations with (5.152) gives the midplane strain and laminate curvature as

$$\{\varepsilon^o\} = [A']\{N^T\} + \lfloor B' \rfloor \{M^T\} \tag{5.169}$$

$$\{\kappa\} = [B'^T]\{N^T\} + \lfloor D' \rfloor \{M^T\} \tag{5.170}$$

or

$$\{\varepsilon^o\} = \Delta T \left\{ [A'] \sum_{k=1}^{N} [\bar{Q}]^k \{\alpha\}^k t_k + \lfloor B' \rfloor \left(\frac{1}{2} \sum_{k=1}^{N} [\bar{Q}]^k \{\alpha\}^k (z_k^2 - z_{k-1}^2) \right) \right\} \tag{5.171}$$

$$\{\kappa\} = \Delta T\left\{[B'^T]\sum_{k=1}^{N}[\bar{Q}]^k\{\alpha\}^k t_k + [D']\left(\frac{1}{2}\sum_{k=1}^{N}[\bar{Q}]^k\{\alpha\}^k(z_k^2 - z_{k-1}^2)\right)\right\} \tag{5.172}$$

Thus, there is a linear relationship between the curvature and the temperature change, with the constant of proportionality being a function of the material properties and the laminate stacking sequence. This relationship can be used to determine the equivalent stress-free temperature of composites fabricated at elevated temperature. Unsymmetric laminates will exhibit curvature at room temperature because of the residual stresses developed during cool-down after fabrication at the elevated temperature. If the material properties and laminate stacking sequence are known and the curvature can be measured, the equivalent stress-free temperature can be calculated from (5.172).

The strain and stress distributions in an unsymmetric laminate subjected to pure thermal loading can be determined using the midplane strains and curvatures from (5.171) and (5.172) in the fundamental strain superposition equation (5.143) and the thermo-elastic constitutive equation (5.144).

Example 5.4 Uniform Thermal Loading of a [0/90] Unsymmetric Laminate

Determine the residual deformations, strains, and stresses that develop during cool-down from the cure temperature in a [0/90] unsymmetric laminate of T300/5208 carbon/epoxy if the temperature change from the elevated temperature at which the layers become bonded to room temperature is −180°F. Consider the temperature distribution to be uniform through the thickness of the laminate.

Solution

The laminate before and after cool-down is depicted in Fig. 5.31. We see that the curvature (in the x-z plane) is concave downward toward the layer with the higher coefficient of thermal expansion in the x-direction. It is noted that the curvature in the orthogonal y-z plane is concave upward, resulting in a "saddle" shape of a plate after cure.

The distribution of total axial strain shown in Fig. 5.32 is linear through the thickness as required by our initial assumption of Kirchhoff displacements. Because the laminate is unsymmetric, the neutral axis ($\varepsilon_x = 0$) has shifted well up into the 0° layer. It should be recalled that the total strain is the sum of the "free" thermal strain that is uniform through the thickness of each layer and the mechanical strain that is associated with stress through the constitutive equation. Obviously, the mechanical strain varies linearly through the thickness.

The distribution of axial stresses, σ_x, is shown in Fig. 5.33. These stresses are the residual thermal stresses after cure. We see that the axial stress in the 0° layer is tensile on the top of the layer but attains a larger-magnitude, compressive stress at the bottom of this layer. The stress in the 90° layer is a small tensile value throughout the layer thickness. *Thus the largest stress magnitude is at the laminate midplane!* The axial stresses are discontinuous across the 0/90 interface, as dictated by the change in material properties across the interface. The distribution of the axial stresses is such that the resulting in-plane force,

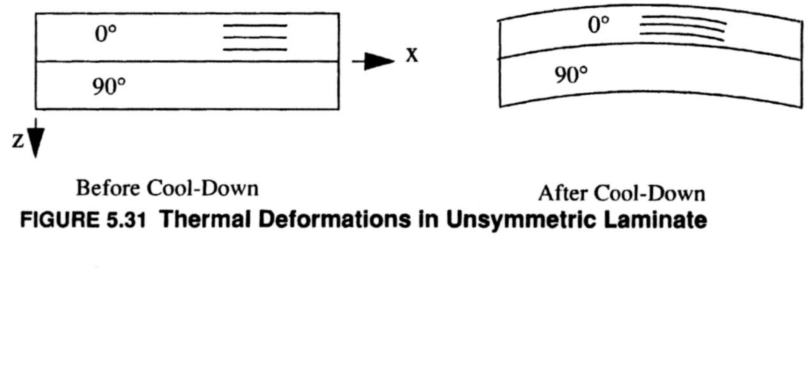

FIGURE 5.31 Thermal Deformations in Unsymmetric Laminate

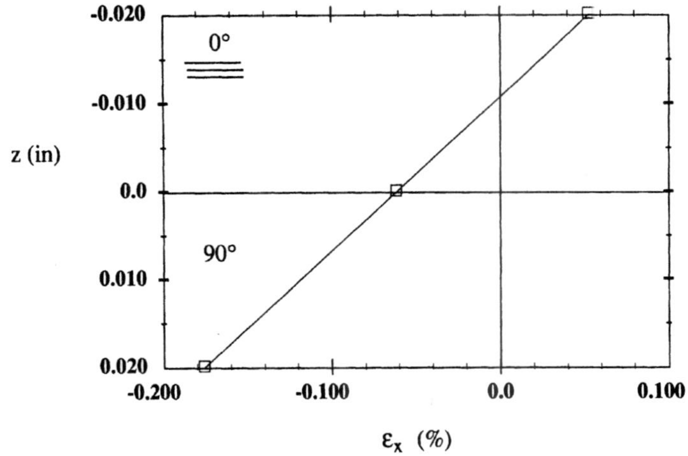

FIGURE 5.32 Total Axial Strains in a [0/90] T300/5208 Laminate: $\Delta T = -180$

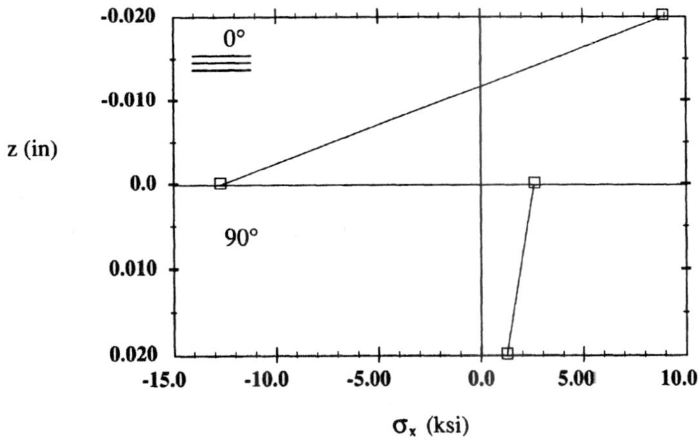

FIGURE 5.33 Axial Stresses in a [0/90] T300/5208 Laminate: $\Delta T = -180$

N_x, and the moment, M_x, are both zero, as required by equilibrium. It is most interesting to note that whereas the total axial strain in the 90° layer is *compressive* throughout the thickness of the layer, the axial component of stress is *tensile!* This is, of course, the result of the fact that stresses are equal to stiffness times the difference between the total strain and the free thermal strain (see Eq. (5.144)).

5.17 Hygrothermal Laminate Analysis

The inclusion of moisture effects in the equations of lamination theory follows directly from the development of the hygrothermal constitutive equation for a lamina, (4.106), and the preceding developments for thermo-elastic lamination theory.

The equivalent hygroscopic forces and moments are defined in the same manner as the thermal forces and moments:

$$\{N^H\} = \int_{-H}^{H} [\bar{Q}]^k \{\varepsilon^H\}^k dz \tag{5.173}$$

$$\{M^H\} = \int_{-H}^{H} [\bar{Q}]^k \{\varepsilon^H\}^k z\, dz \tag{5.174}$$

The hygrothermal lamination theory equations are then

$$\begin{Bmatrix} N + N^T + N^H \\ M + M^T + M^H \end{Bmatrix} = \begin{bmatrix} A & B \\ B & D \end{bmatrix} \begin{Bmatrix} \varepsilon^o \\ \kappa \end{Bmatrix} \tag{5.175}$$

Inverting,

$$\begin{Bmatrix} \varepsilon^o \\ \kappa \end{Bmatrix} = \begin{bmatrix} A' & B' \\ C' & D' \end{bmatrix} \begin{Bmatrix} N + N^T + N^H \\ M + M^T + M^H \end{Bmatrix} \tag{5.176}$$

Likewise, the laminate coefficient of hygroscopic expansion, $\{\beta\}$, for a symmetric laminate with a uniform change in moisture content is

$$\{\beta\} = [A]^{-1} \sum_{k=1}^{N} [\bar{Q}]^k \{\beta\}^k t_k \tag{5.177}$$

Finally, the equation for the stresses in a symmetric laminate under conditions of uniform temperature change ΔT and moisture content ΔM is

$$\{\sigma\}_x^k = [\bar{Q}]^k \{(\{\alpha\} - \{\alpha\}_x^k)\Delta T + (\{\beta\} - \{\beta\}_x^k)\Delta M\} \tag{5.178}$$

The curvature of unsymmetric laminates due to moisture absorption can be determined from (5.172) by replacing ΔT with ΔM and $\{\alpha\}^k$ with $\{\beta\}^k$.

5.18 Laminate Through-Thickness Properties

Lamination theory is concerned with the in-plane response of a laminate which is in a state of plane stress. As shown by Herakovich (1984), we can combine lamination theory with 3-D Hooke's law to obtain an approximate expression for the *effective through-thickness Poisson's ratios* ν_{xz} and ν_{yz} of symmetric laminates. The expressions developed represent a relaxation of the Kirchhoff assumption that $w = w(x, y)$. We are now permitting $w = w(x, y, z)$.

5.18.1 Through-Thickness Poisson's Ratios

We begin with the definition of the through-thickness Poisson's ratio ν_{xz} in terms of the through-thickness strain ε_z and the in-plane strain ε_x in the direction of the applied load, i.e.,

$$\nu_{xz} = -\frac{\varepsilon_z}{\varepsilon_x} \tag{5.179}$$

for the loading $N_x \neq 0$, with all other applied forces and moments being zero. The average (or overall) through-thickness strain ε_z can be written in terms of the total change in thickness, Δw, divided by the laminate thickness, $2H$.

$$\varepsilon_z = \frac{\Delta w}{2H} \tag{5.180}$$

The total thickness change can be written as the integral of the displacements $\varepsilon_z\, dz$ over the thickness of the laminate:

$$\Delta w = \int_{-H}^{H} \varepsilon_z\, dz \tag{5.181}$$

For uniform *plane stress* in each layer of the laminate, the through-thickness strain ε_z^k is constant in any kth layer and can be determined from the 3-D Hooke's law (3.86) as

$$\varepsilon_z^k = \bar{S}_{13}^k \sigma_x^k + \bar{S}_{23}^k \sigma_y^k + \bar{S}_{36}^k \tau_{xy}^k \tag{5.182}$$

The stresses in (5.182) are determined from the plane stress constitutive equations (4.50) for the kth layer:

$$\begin{Bmatrix} \sigma_x \\ \sigma_y \\ \tau_{xy} \end{Bmatrix}^k = \begin{bmatrix} \bar{Q}_{11} & \bar{Q}_{12} & \bar{Q}_{16} \\ \bar{Q}_{12} & \bar{Q}_{22} & \bar{Q}_{26} \\ \bar{Q}_{16} & \bar{Q}_{26} & \bar{Q}_{66} \end{bmatrix}^k \begin{Bmatrix} \varepsilon_x \\ \varepsilon_y \\ \gamma_{xy} \end{Bmatrix}^k \tag{5.183}$$

Now, for a symmetric laminate subjected to the in-plane loading $N_x \neq 0, N_y = N_{xy} = 0$, and $\{M\} = 0$, the strains in the kth layer are

$$\begin{Bmatrix} \varepsilon_x \\ \varepsilon_y \\ \gamma_{xy} \end{Bmatrix}^k = \begin{Bmatrix} \varepsilon_x^o \\ \varepsilon_y^o \\ \gamma_{xy}^o \end{Bmatrix} = [A]^{-1} \begin{Bmatrix} N_x \\ 0 \\ 0 \end{Bmatrix} \tag{5.184}$$

Combining (5.183) and (5.184) and carrying out the matrix multiplication gives the stresses in the kth layer:

$$\begin{Bmatrix} \sigma_x \\ \sigma_y \\ \tau_{xy} \end{Bmatrix}^k = \begin{bmatrix} \bar{Q}_{11} & \bar{Q}_{12} & \bar{Q}_{16} \\ \bar{Q}_{12} & \bar{Q}_{22} & \bar{Q}_{26} \\ \bar{Q}_{16} & \bar{Q}_{26} & \bar{Q}_{66} \end{bmatrix}^k \begin{Bmatrix} A_{11}^{-1} \\ A_{12}^{-1} \\ A_{16}^{-1} \end{Bmatrix} N_x \tag{5.185}$$

Combining (5.182) and (5.185) gives the through-thickness strains in the kth layer:

$$\begin{aligned} \varepsilon_z^k = {} & N_x[A_{11}^{-1}(\bar{S}_{13}^k\bar{Q}_{11}^k + \bar{S}_{23}^k\bar{Q}_{12}^k + \bar{S}_{36}^k\bar{Q}_{16}^k) + A_{12}^{-1}(\bar{S}_{13}^k\bar{Q}_{12}^k + \bar{S}_{23}^k\bar{Q}_{22}^k + \bar{S}_{36}^k\bar{Q}_{26}^k) \\ & + A_{16}^{-1}(\bar{S}_{13}^k\bar{Q}_{16}^k + \bar{S}_{23}^k\bar{Q}_{26}^k + \bar{S}_{36}^k\bar{Q}_{66}^k)] \end{aligned} \tag{5.186}$$

In the preceding equation, all terms are known and constant within any layer of a given symmetric laminate subjected to the prescribed loading. Thus (5.186) can be combined with (5.181) to give the total change in laminate thickness:

$$\Delta w = N_x(A_{11}^{-1}F_1 + A_{12}^{-1}F_2 + A_{16}^{-1}F_6) \tag{5.187}$$

where the F_i are defined as

$$F_i = \sum_{k=1}^{N} [(\bar{S}_{13}^k\bar{Q}_{1i}^k + \bar{S}_{23}^k\bar{Q}_{2i}^k + \bar{S}_{36}^k\bar{Q}_{6i}^k)t_k] \qquad (i = 1, 2, 6) \tag{5.188}$$

Finally, combining (5.179), (5.180), (5.187), and (5.188) with (5.184) for ε_x, we have

$$\nu_{xz} = \frac{-(A_{11}^{-1}F_1 + A_{12}^{-1}F_2 + A_{16}^{-1}F_6)}{2HA_{11}^{-1}} \tag{5.189}$$

or, more simply,

$$\nu_{xz} = \frac{-A_{1i}^{-1}F_i}{2HA_{11}^{-1}} \qquad (i = 1, 2, 6) \tag{5.190}$$

In a similar fashion, it can be shown that the through-thickness Poisson's ratio ν_{yz} for loading in the y-direction is

$$\nu_{yz} = \frac{-A_{2i}^{-1}F_i}{2HA_{22}^{-1}} \tag{5.191}$$

Figure 5.34 shows the through-thickness Poisson's ratios, ν_{xz}, for unidirectional off-axis and angle-ply T300/5208 carbon/epoxy laminates as a function of the fiber orientation. For the unidirectional lamina, the through-thickness Poisson's ratio increases monotonically from the axial

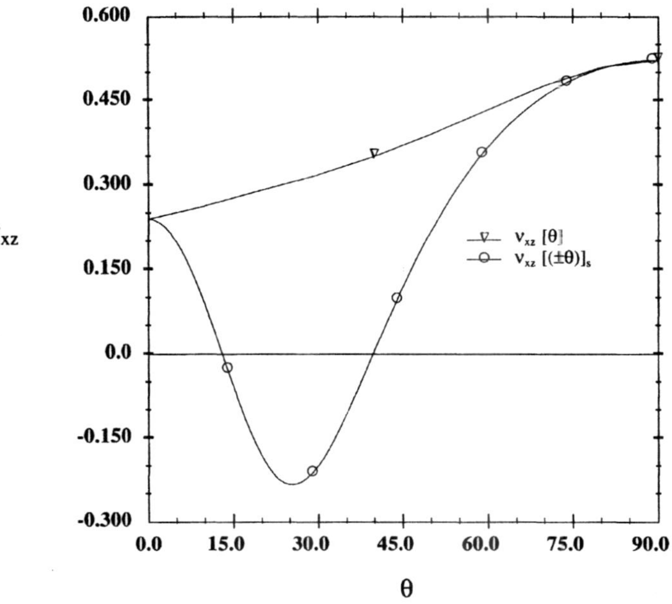

FIGURE 5.34 Through-Thickness Poisson's Ratios ν_{xz}: T300/5208

value to the transverse value as the fiber orientation angle increases. In sharp contrast, the through-thickness Poisson's ratio for angle-ply laminates initially decreases with increasing angle, passes through zero at approximately 13°, attains a minimum negative value of –0.233 at approximately 25°, then passes through zero again at approximately 40° as it increases to the transverse value at 90°. This figure is one example of the great variety of design possibilities with fibrous composites. The minimum value and the angles corresponding to zero through-thickness Poisson's ratios can be altered by appropriate modification of the laminate configuration.

For design purposes, plots can be constructed showing the variation of these values for a family of laminates as functions of layer fiber orientation and thickness. Figure 5.35 shows the variation of ν_{xz} and ν_{yz} with fiber orientation for angle-ply laminates. Clearly, the two values are out of phase by 90° and have identical values at $\theta = 45°$.

As a final indication of the wide range of values possible for Poisson's ratio of fibrous composites, Fig. 5.36 shows the through-thickness ν_{xz} of unidirectional and angle-ply laminates compared with the in-plane ν_{xy} values for these same laminates. It is evident from the figure that for angle-ply laminates the angle corresponding to the maximum in-plane Poisson's ratio ν_{xy} is also the angle corresponding to the minimum through-thickness Poisson's ratio ν_{xz}. These two values range from a maximum of 1.25 to a minimum of –0.23. In contrast, typical structural metals exhibit Poisson's ratios in the range 0.2 to 0.5. Obviously, these results have strong implications for the designer of composite structures, who must be aware of this range of possible values and the implications they may have on the design.

The wide range of Poisson's ratios in a laminated composite is a result of the internal stresses in the individual layers. The variation of stresses in an angle-ply laminate subjected to axial loading was presented in Fig. 5.21. This state of stress is depicted in Fig. 5.37, where for small angles the transverse stress σ_2 is compressive in both the $+\theta$ and $-\theta$ layers. The compressive transverse stresses σ_2 give rise to the *negative* through-thickness Poisson's ratios. Comparison of Figs. 5.21 and 5.36 shows the close correlation between σ_2 and ν_{xz}.

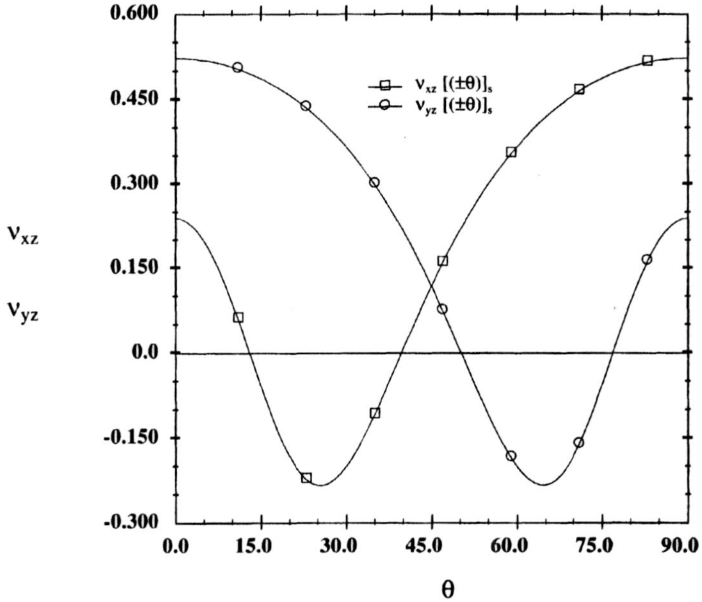

FIGURE 5.35 **Angle-Ply Through-Thickness Poisson's Ratios: T300/5208**

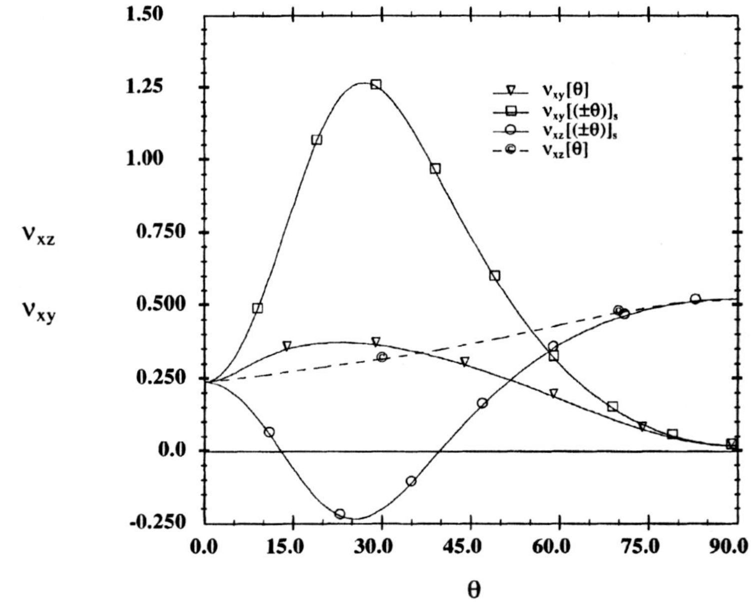

FIGURE 5.36 **Poisson's Ratio Comparisons: T300/5208**

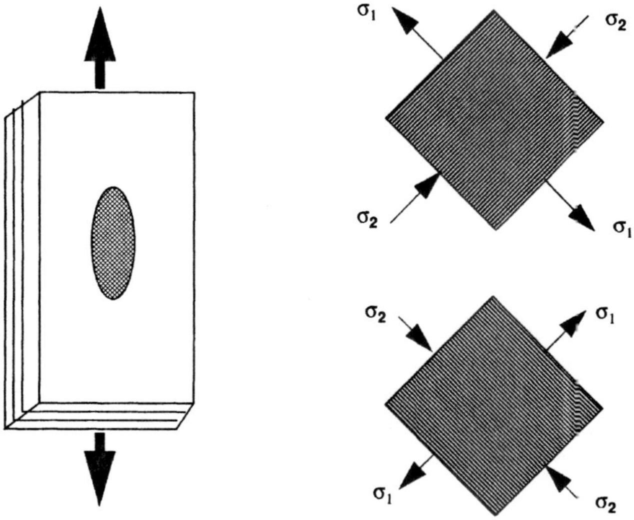

FIGURE 5.37 Stresses in Angle-Ply Laminates

5.18.2 Through-Thickness Coefficient of Thermal Expansion

The development of approximate expressions for the through-thickness coefficient of thermal expansion of a symmetric laminate with uniform temperature throughout the laminate follows the same lines as the development of the through-thickness Poisson's ratio, with the equations modified to include the appropriate thermal terms. The through-thickness coefficient of thermal expansion, α_z, is defined as the total strain (total change in thickness per unit thickness) per unit uniform temperature change ΔT. Thus,

$$\alpha_z = \frac{\Delta w}{2H\Delta T} = \frac{1}{2H\Delta T}\int_{-H}^{H} \varepsilon_z\, dz \tag{5.192}$$

where the strain ε_z is the total strain, i.e., the sum of the mechanical and thermal strains:

$$\varepsilon_z = \varepsilon_z^{\sigma} + \varepsilon_z^{T} \tag{5.193}$$

The three-dimensional thermo-elastic constitutive equation (3.106) for ε_z is

$$\varepsilon_z = \bar{S}_{13}\sigma_x + \bar{S}_{23}\sigma_y + \bar{S}_{33}\sigma_z + \bar{S}_{36}\tau_{xy} + \varepsilon_z^{T} \tag{5.194}$$

and the plane stress condition of laminate analysis requires that $\sigma_z = 0$. Thus, combining (5.192) with the thermo-elastic constitutive equation, the through-thickness coefficient of thermal expansion, α_z, is

$$\alpha_z = \frac{1}{2H\Delta T}\sum_{k=1}^{N} (\bar{S}_{13}^{k}\sigma_x^{k} + \bar{S}_{23}^{k}\sigma_y^{k} + \bar{S}_{36}^{k}\tau_{xy}^{k} + \alpha_z^{k}\Delta T)t_k \tag{5.195}$$

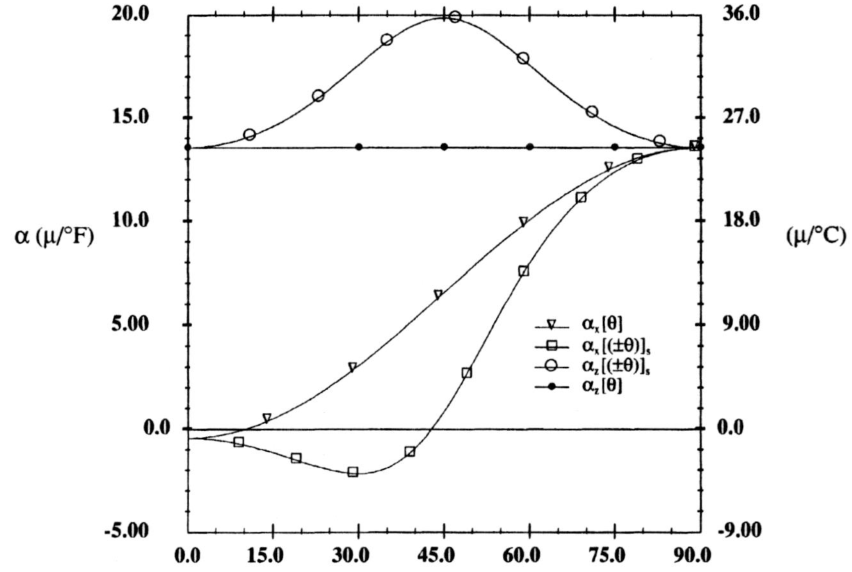

FIGURE 5.38 Coefficients of Thermal Expansion α_z and α_x: T300/5208

where the stresses, from (5.168), are

$$
\begin{aligned}
\sigma_x^k &= [\bar{Q}_{11}^k(\alpha_x - \alpha_x^k) + \bar{Q}_{12}^k(\alpha_y - \alpha_y^k) + \bar{Q}_{16}^k(\alpha_{xy} - \alpha_{xy}^k)]\Delta T \\
\sigma_y^k &= [\bar{Q}_{12}^k(\alpha_x - \alpha_x^k) + \bar{Q}_{22}^k(\alpha_y - \alpha_y^k) + \bar{Q}_{26}^k(\alpha_{xy} - \alpha_{xy}^k)]\Delta T \\
\tau_{xy}^k &= [\bar{Q}_{16}^k(\alpha_x - \alpha_x^k) + \bar{Q}_{26}^k(\alpha_y - \alpha_y^k) + \bar{Q}_{66}^k(\alpha_{xy} - \alpha_{xy}^k)]\Delta T
\end{aligned}
\tag{5.196}
$$

Predictions for through-thickness α_z for unidirectional off-axis and angle-ply T300/5208 laminates are compared with in-plane α_x values in Fig. 5.38. The through-thickness α_z for angle-ply laminates attains a maximum of almost 36 μ/°C (20 μ/°F) at θ = 45°. This is in contrast to the relatively large *negative* value of the in-plane coefficient α_x at θ = 30°. For unidirectional off-axis lamina, α_z is independent of fiber orientation. The difference between α_z for the lamina and laminate is the fact that there are internal stresses in a laminate that is subjected to pure thermal loading, whereas a lamina subjected to pure thermal loading is stress free.

5.18.3 Through-Thickness Coefficient of Hygroscopic Expansion

The development for the through-thickness coefficient of hygroscopic expansion, β_z, follows exactly the development for α_z. The results can be obtained simply by replacing the α's with β's in (5.195) and (5.196). Thus,

$$
\beta_z = \frac{1}{2H\Delta M}\sum_{k=1}^{N}(\bar{S}_{13}^k\sigma_x^k + \bar{S}_{23}^k\sigma_y^k + \bar{S}_{36}^k\tau_{xy}^k + \beta_z^k\Delta M)t_k \tag{5.197}
$$

where the stresses are defined as follows:

$$\begin{aligned}
\sigma_x^k &= [\bar{Q}_{11}^k(\beta_x - \beta_x^k) + \bar{Q}_{12}^k(\beta_y - \beta_y^k) + \bar{Q}_{16}^k(\beta_{xy} - \beta_{xy}^k)]\Delta M \\
\sigma_y^k &= [\bar{Q}_{12}^k(\beta_x - \beta_x^k) + \bar{Q}_{22}^k(\beta_y - \beta_y^k) + \bar{Q}_{26}^k(\beta_{xy} - \beta_{xy}^k)]\Delta M \qquad (5.198)\\
\tau_{xy}^k &= [\bar{Q}_{16}^k(\beta_x - \beta_x^k) + \bar{Q}_{26}^k(\beta_y - \beta_y^k) + \bar{Q}_{66}^k(\beta_{xy} - \beta_{xy}^k)]\Delta M
\end{aligned}$$

5.19 Designing with Laminates

The engineering properties of laminates can vary over a wide range because of the many variables associated with the laminates. In addition to the material properties (including the possibility of hybrid laminates), the fiber orientations, layer thicknesses, and stacking sequence all influence the laminate properties. In order to demonstrate some of the possibilities, we consider symmetric laminates of T300/5208 carbon/epoxy and show the variation of in-plane engineering properties for the class of laminates $[0_{n1}/(\pm\theta)_{n2}/90_{n3}]_s$, where $n1$, $n2$, and $n3$ are the number of layers in each group (or, equivalently, the percentage of each type). Since these laminates are symmetric, stacking sequence does not influence the in-plane response.

The engineering properties E_x, E_y, G_{xy}, ν_{xy}, α_x, and α_y are compared for six laminates in Figs. 5.39–5.44. These figures demonstrate the wide variation in properties for just one type of material and a limited number of laminates. The figures show that a given property can be obtained from a variety of laminates. In particular, zero CTE values are possible for a variety of laminates. The maximum values for shear modulus and Poisson's ratio are obtained from $[\pm\theta]_s$ laminates.

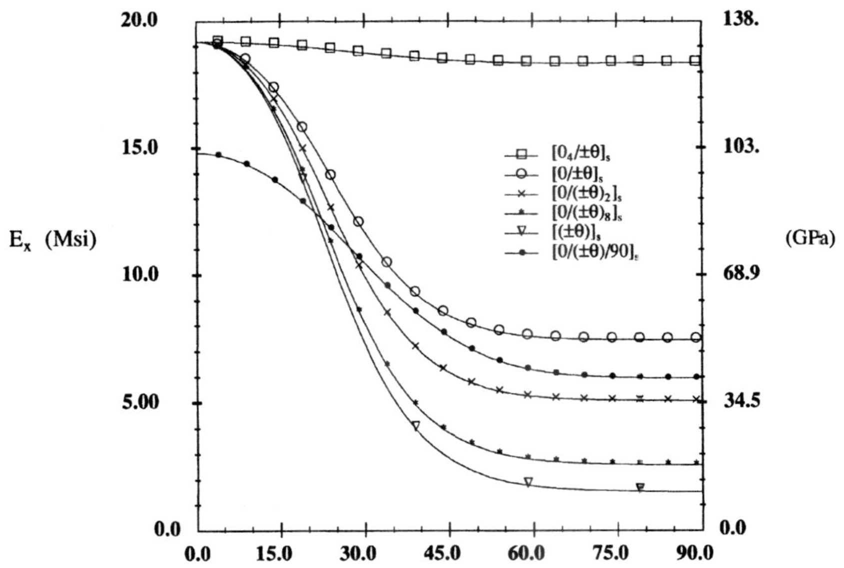

FIGURE 5.39 Axial Modulus of Symmetric T300/5208 Laminates

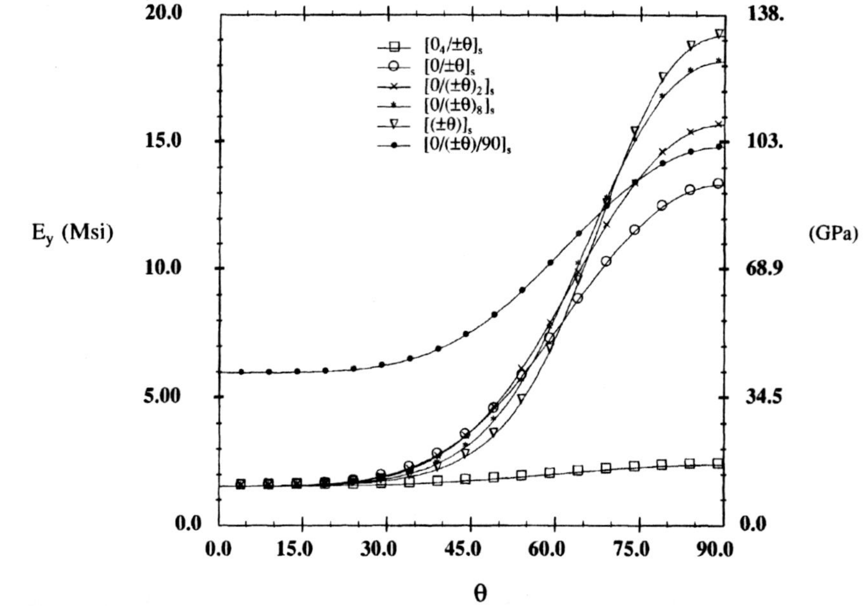

FIGURE 5.40 Transverse Modulus of Symmetric T300/5208 Laminates

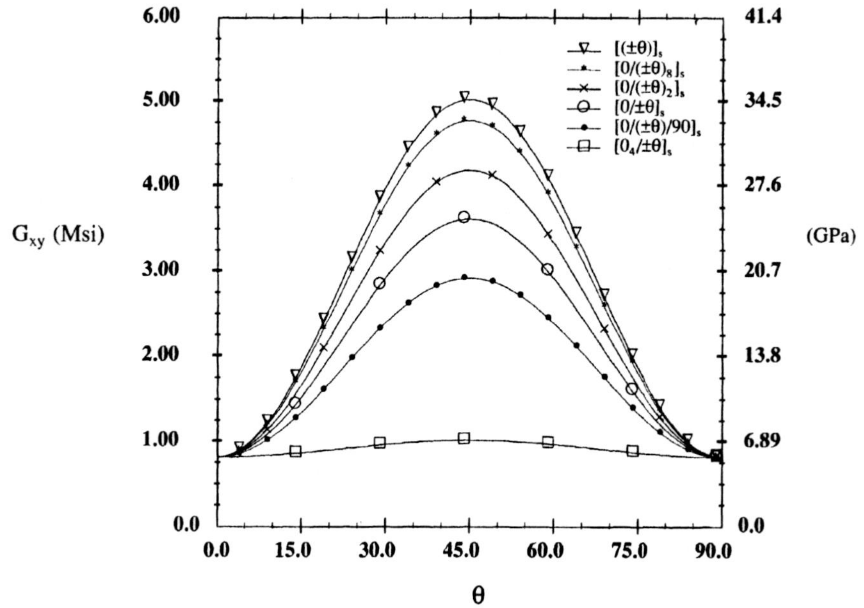

FIGURE 5.41 Shear Modulus of Symmetric T300/5208 Laminates

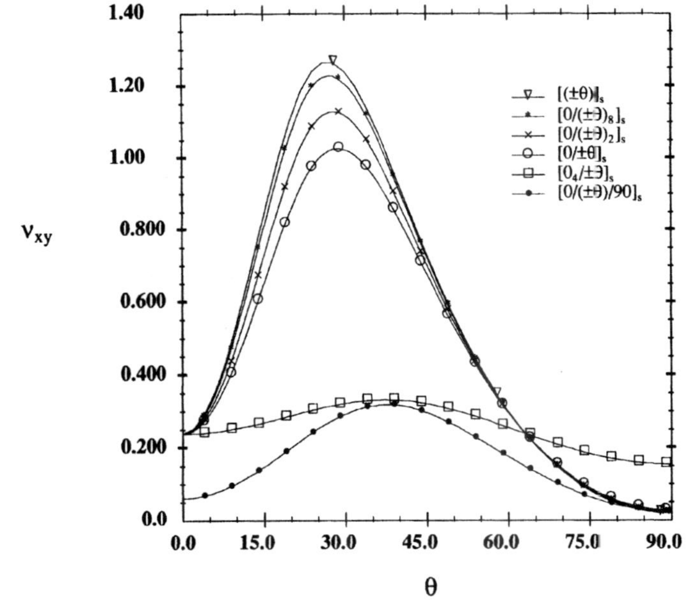

FIGURE 5.42 Poisson's Ratio of Symmetric T300/5208 Laminates

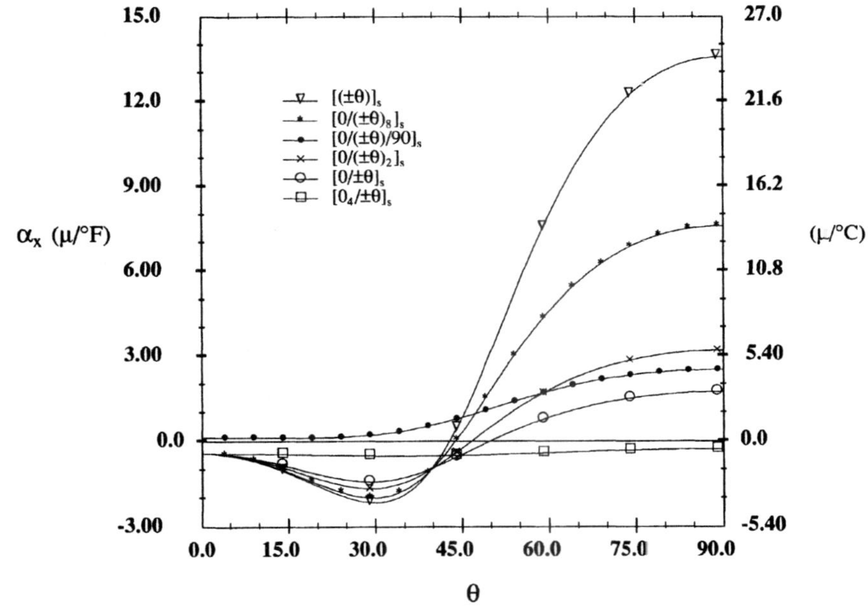

FIGURE 5.43 Axial CTE of Symmetric T300/5208 Laminates

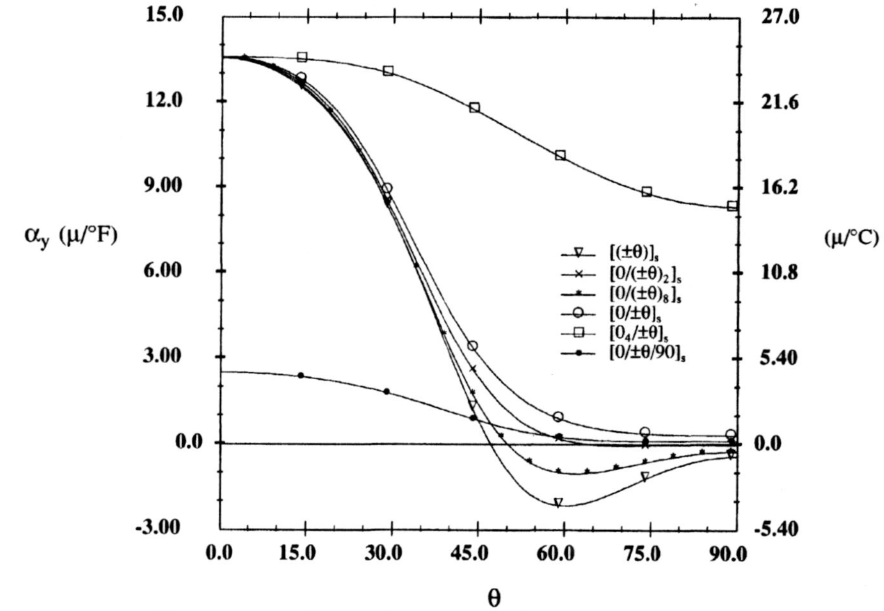

FIGURE 5.44 Transverse CTE of Symmetric T300/5208 Laminates

Similar plots for a specific fiber orientation in the off-axis layers, e.g., $[0_{n1}/(\pm 45)_{n2}/90_{n3}]_s$, when plotted as a function of $n1$, $n2$, and $n3$, are called *carpet plots*. As an example of the influence of percentages, from Fig. 5.39 for $[0_{n1}/(\pm\theta)_{n2}]_s$ laminates, as the percentage of the 0° layer increases from 0% for the $[(\pm\theta)]_s$ laminate to 67% for the $[0_4/\pm\theta]_s$ laminate, the axial modulus increases more than sixfold for fiber orientations θ greater than 40°.

5.20 Summary

A theory has been presented for the elastic response of laminates subjected to in-plane forces, bending moments, and thermal and hygroscopic loading. The theory has been used to predict effective engineering properties, typical stress distributions, and through-thickness Poisson's ratios and coefficients of thermal expansion. Symmetric and unsymmetric laminates as well as quasi-isotropic laminates and design considerations have been discussed.

References

Dong, S. B., Pister, K. S., and Taylor, R. L. (1962), "On the Theory of Laminated Anisotropic Shells and Plates," *J. Aerosp. Sci.*, vol. 29, pp. 969–975.

Herakovich, C. T. (1984), "Composite Laminates with Negative Through-the-Thickness Poisson's Ratios," *J. Compos. Mater.*, vol. 18, pp. 447–455.

Kirchhoff, G. (1850), *J. f. Math.* (Crelle), bd. 40.

Love, A. E. H. (1892), *A Treatise on the Mathematical Theory of Elasticity,* 1st ed., Cambridge University Press (4th ed. 1927); printed by Dover Publications, New York, 1944.

Pister, K. S., and Dong, S. B. (1959), "Elastic Bending of Layered Plates," *J. Eng. Mech. Div.*, EM 4, October, pp. 1–10.

Reissner, E., and Stavsky, Y. (1961), "Bending and Stretching of Certain Types of Heterogeneous Aelotropic Elastic Plates," *J. Appl. Mech.*, vol. 28, pp. 402–408.

5.21 Exercises

For Exercises 5.1 through 5.9, all plies are of the same material and of equal thickness t.

5.1 Express the elements of the $[A]$, $[B]$, and $[D]$ matrices of $[0_2/90]_s$ and $[90/0_2]_s$ laminates in terms of the elements of the material stiffnesses Q_{ij}.

5.2 Express the elements of the $[A]$, $[B]$, and $[D]$ matrices of $[0/90]_s$ and $[90/0]_s$ laminates in terms of the elements of the material stiffnesses Q_{ij}.

5.3 Express the elements of the $[A]$, $[B]$, and $[D]$ matrices of a $[\theta/-\theta]_s$ laminate in terms of the elements of the transformed stiffnesses $\bar{Q}_{ij}(\theta)$.

5.4 Express the elements of the $[A]$, $[B]$, and $[D]$ matrices of a $[-\theta/\theta]_s$ laminate in terms of the elements of the transformed stiffnesses $\bar{Q}_{ij}(\theta)$.

5.5 Express the elements of the $[A]$, $[B]$, and $[D]$ matrices of a $[0/\pm45/90]_s$ laminate in terms of the elements of the stiffnesses Q_{ij}.

5.6 Express the elements of the $[A]$, $[B]$, and $[D]$ matrices of a $[90/\pm45/0]_s$ laminate in terms of the elements of the stiffnesses Q_{ij}.

5.7 Express the elements of the $[A]$, $[B]$, and $[D]$ matrices of a $[45/90/-45/0]_s$ laminate in terms of the elements of the stiffnesses Q_{ij}.

5.8 Express the elements of the $[A]$, $[B]$, and $[D]$ matrices of a $[0/90/\pm45]_s$ laminate in terms of the elements of the stiffnesses Q_{ij}.

5.9 Express the elements of the $[A]$, $[B]$, and $[D]$ matrices of a $[0/90/\pm45]_T$ laminate in terms of the elements of the stiffnesses Q_{ij}.

5.10 Calculate numerical values for the exercises selected by the instructor from Exercises 5.1 to 5.9 if the material is T300/5208 carbon/epoxy.

5.11 Calculate numerical values for the exercises selected by the instructor from Exercises 5.1 to 5.9 if the material is SCS-6/Ti-15-3 silicon-carbide/titanium.

5.12 Write a computer program to compute $[A]$, $[B]$, and $[D]$ for an arbitrary composite laminate. Your computer should have the capability to deal with hybrid laminates of at least two different materials (some plies of one material, other plies of a different material). Compute and print these matrices for the T300/5208 laminates $[0/\pm45/90]_s$ and $[0_3/90_3]_T$.

5.13 Modify your program to compute $\{\varepsilon^o\}$ and $\{\kappa\}$ given $\{M\}$ and $\{N\}$, or to compute $\{M\}$ and $\{N\}$ given $\{\varepsilon^o\}$ and $\{\kappa\}$. Obtain values of $\{\varepsilon^o\}$ and $\{\kappa\}$ for N (lb/in) = $\{10, 5, 2.5\}$ and M (in-lb/in) = $\{25, 25, 0\}$. Use the preceding results to calculate $\{N\}$ and $\{M\}$, which should be your original values. Do this exercise for *both* the symmetric and unsymmetric laminates given in Exercise 5.12.

5.14 Modify your program to compute E_x, E_y, G_{xy}, ν_{xy}, $\eta_{xy,x}$, and $\eta_{xy,y}$ for an arbitrary symmetric laminate. Compute and report results for the laminate $[0/\pm45/90]_s$.

5.15 Modify your program to determine $\{\sigma\}_x^k$ and $\{\sigma\}_1^k$ at the top and bottom of each layer for a generic laminate under arbitrary loading, and plot the results as through-thickness stress plots. Compute, print, and plot these stresses for the $\{N\}$ and $\{M\}$ loads given in Exercise 5.13 for *both* the symmetric and unsymmetric laminates of Exercise 5.12.

5.16 Prove that Eqs. (5.108)–(5.113) give the stresses in cross-ply laminates subjected to N_x loading.

5.17 Prove that Eqs. (5.114) give the layer stresses in a symmetric, angle-ply laminate under axial loading N_x.

5.18 Verify the curves in Figs. 5.21 and 5.22 for the material principal stresses in PMC and MMC angle-ply laminates under tension.

5.19 Verify the curves in Figs. 5.24 and 5.25 for the material principal strains in PMC and MMC angle-ply laminates under tension.

5.20 Show that (5.141) and (5.142) for the bending-stretching coupling coefficients $[B]$ are true by calculating the coefficients for [0/90] and [+30/–30] laminates of SCS-6/Ti-15-3.

5.21 Show that the moment per unit length in a laminate subjected to thermal mechanical loading is given by Eq. (5.148).

5.22 Develop an expression for the thermal moment $\{M^T\}$ in a $[0/90]_T$ laminate subjected to a temperature distribution $\Delta T = \Omega z^2$, where Ω is a known constant. Each layer is of thickness t. Express your answer in terms of Q_{ij} and α_i in the principal material coordinates.

5.23 Develop an expression for the effective laminate coefficient of thermal expansion of a $[0/90]_s$ laminate in terms of the material stiffnesses, Q_{ij}, and coefficients of thermal expansion, α_i.

5.24 Develop an expression for the effective laminate coefficient of thermal expansion of a $[45/-45]_s$ laminate in terms of the material stiffnesses, Q_{ij}, and coefficients of thermal expansion, α_i.

5.25 Prove that the three laminates $[0/\pm45/90]_s$, $[0/90]_s$, and $[\pm45]_s$ all must have identical $\{\alpha\}$ values.

5.26 Modify your program to calculate the laminate coefficient of thermal expansion $\{\alpha\}$ for any symmetric laminate. Verify that your computer program gives the same values for $\{\alpha\}$ for all three laminates in Exercise 5.25.

5.27 Modify your program to calculate the equivalent thermal force $\{N^T\}$ and the equivalent thermal moment $\{M^T\}$ for arbitrary composite laminates subjected to a uniform temperature change.

5.28 Sketch the curvature in the x-z plane of a [90/0] unsymmetric laminate after a cure temperature change of −180°F. Plot the distribution of total strain and stress in the laminate if the material is T300/5208. Relate the results to those of the [0/90] laminate in Example 5.4.

5.29 Use your program to compute the residual curing stresses at 70°F for a $[0/\pm45/90]_s$ T300/5208 laminate, assuming a stress-free temperature of 250°F.

5.30 For the conditions of Exercise 5.29, add an applied force and moment of $\{N\}$ (lb/in) = $\{10, 5, 2.5\}$ and $\{M\}$ (in-lb/in) = $\{25, 25, 0\}$, and compute the midplane strains, laminate curvatures, and both the global x-y and material principal 1-2 stresses at the top, middle, and bottom of each layer. Your output should show the combined stresses as well as the stresses due to the temperature change only and the stresses due to the applied mechanical load only.

5.31 Provide the same results for a $[0_3/90_3]_T$ laminate for the loading of Exercise 5.30.

5.32 Reduce expressions (5.190) and (5.191) for through-thickness Poisson's ratios to their simplest forms for unidirectional off-axis lamina.

5.33 Reduce expression (5.195) for the through-thickness coefficient of thermal expansion to its simplest form for unidirectional off-axis lamina.

CHAPTER 6 TEST METHODS

"Luck is the residue of hard work."
Joyce Davenport, "Hill Street Blues"

6.1 Introduction

In this chapter we present test methods that are commonly used to characterize the thermomechanical response of fibrous composite materials. Because composites are heterogeneous and anisotropic, it has been necessary to develop new specimens and new test methods to measure all the desired material properties. New procedures include determination of fiber volume fraction, void content, moisture content in polymeric matrix composites, off-axis tests, properties and response in orthogonal directions, and nondestructive evaluation (NDE) methods to assess the quality of the fabricated material and damage development during loading. Essentially all test methods for mechanical properties have been modified substantially for use with composites because of the special features exhibited by these materials.

As with the testing of any material, it is desirable to have specimens with a test section of known, uniform stress and strain and without stress concentrations anywhere in the specimen that may lead to early failure. Coming up with such a test section is not always a simple task, especially with anisotropic composites. As we have seen in previous chapters, material anisotropy introduces coupling between axial and shear responses, and unsymmetric laminates exhibit coupling between membrane and bending responses. The (typical) thinness of flat composite laminates dictates that compression and shear specimens be small in scale to minimize the possibility of buckling. However, short-length specimens can have other problems due to end effects. Thus stress analysis of composite test methods is critically important to ensure the accuracy and validity of the results obtained from a test.

The most fundamental engineering properties of interest are the elastic constants in principal material coordinates. If we restrict our attention to plane stress problems, the elastic properties needed are E_1, E_2, ν_{12}, and G_{12}. With these four engineering properties known, all other elastic properties for plane stress analysis can be determined using the theory developed in Chapter 4 for laminae and Chapter 5 for laminates. As will be demonstrated, in principle the required plane stress elastic constants can be determined from two tension tests, one on-axis and the other off-axis. However, three factors make it imperative that additional tests be performed. First, the theory in Chapters 4 and 5 must be verified in the laboratory. Of course, this has to be accomplished only once for each different type of material under consideration (i.e., PMC, MMC, CMC). The second reason that additional tests must be conducted relates to fabrication. We must ensure that the material has been fabricated as desired with correct fiber volume fraction, fiber orientation, and stacking sequence. Third, material response is not limited to linear elastic behavior. Thus, the nonlinear, inelastic behavior must be determined experimentally and correlated with models for the nonlinear response, including the prediction of failure.

In general, the layers in a laminate are off-axis, and thus it is desirable to measure off-axis properties as a function of the fiber orientation angle θ. The primary in-plane elastic properties of interest for an off-axis layer or laminate are E_x, ν_{xy}, G_{xy}, and $\eta_{xy,x}$. These properties can be determined from one tension test and one shear test.

It is also necessary to know the material response as a function of temperature (and possibly other environmental conditions) as well as loading type (i.e., static, dynamic, fatigue). This chapter will focus attention on static, room-temperature testing; however, many of the methods described can be used for dynamic and fatigue testing and at temperatures other than room temperature.

The major emphasis in this chapter is on mechanical property characterization. However, prior to a detailed discussion of the test methods, we briefly address methods for assessing the quality of fabricated materials and measurement of physical properties.

6.2 Quality Assessment

While fabrication methods have improved to the point where experienced personnel can repeatedly fabricate high-quality materials, it is still important to assess the quality of composites prior to laboratory testing. Quality assessment is, of course, even more important for the inexperienced fabricator. The quality assessment can provide information as to the quality of the bond between fiber and matrix, as well as the presence of voids, broken fibers, matrix cracks, and delaminations. Several methods for assessing the quality of materials are discussed in the following sections.

6.2.1 Microscopy

High-magnification visual examination of composite materials can be very informative. It can provide information as to the shape of the fibers and the geometry and uniformity of the fiber spacing, the presence of voids and regions rich or poor in matrix, and fiber alignment. Figure 6.1 shows two micrographs of filament-wound AS4/3501-6 carbon/epoxy at two levels of magnification. At the lower magnification (Fig. 6.1a), the individual layers are evident, with resin-rich regions between the layers. At the higher magnification (Fig. 6.1b), the individual filaments and tows (each consisting of 10,000 or more filaments) are very evident as is the fiber shape (round), packing/distribution, and density, as well as the resin-rich regions between tows. The composite appears to be void free at this level of magnification.

6.2.2 Ultrasonic Inspection

In ultrasonic inspection mechanical waves are propagated at frequencies above the upper limit of human hearing (i.e., above 20 kHz), and the intensity of wave reflection or transmission through the medium under inspection is measured by a receiving transducer. The frequencies used for ultrasonic inspection are typically in the range of 100 kHz to 25 MHz. A portion of ultrasonic wave energy is reflected by defects and interfaces, and the energy transmitted through the material is reduced due to reflection and attenuation. Thus, both reflection and transmission of energy can be used for ultrasonic inspection.

A map (or image) of the intensity provides a measure of the discontinuities in the material. This map is called a C-scan. Transmission of waves through different types of known discontinuities can be calibrated to provide a meaningful assessment of the material. C-scans are now used routinely to assess the integrity of materials and structures. A C-scan showing damage (white bands) extending at ±45° from a centered hole in a $[(\pm 45)_{12}]_s$ AS4/3502 carbon/epoxy laminate subjected to compression is shown in Fig. 6.2.

Discussion of ultrasonic inspection can be found in ASTM (American Society for Testing and Materials) Standards E114-90 (Pulse-Echo), E214-68(91) (Reflection), E317-93 (Pulse-Echo), and E494-92a (Ultrasonic Velocities).

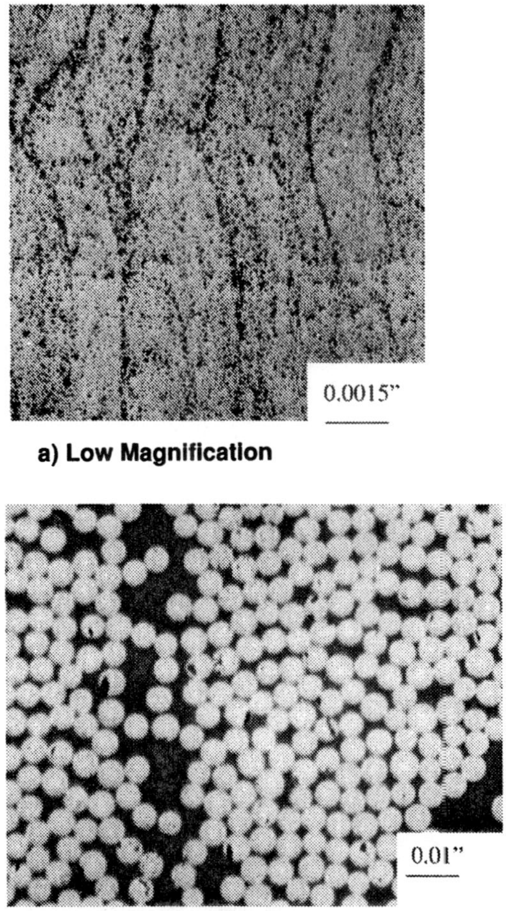

a) Low Magnification

b) High Magnification

FIGURE 6.1 Photomicrographs of Filament-Wound AS4/3501-6 Carbon/Epoxy
(Hirschfeld and Herakovich, 1990)

6.2.3 X-Ray

X-ray is used for inspection of composites much as it is used in other applications. The designation *X-ray* was used by Roentgen in 1895 because of the unknown character of his discovery. X-rays are electromagnetic waves of extremely short wavelength. They are capable of penetrating solid substances and are affected by discontinuities much as other waves. In actual applications, regions of a specimen with lower density, such as voids, defects, and cracks, absorb less radiation, with the result that the intensity of the radiation reaching a photographic film or plate placed on the far side of the sample is higher in these regions. The darker areas of the film indicate the outline of the low-density region.

FIGURE 6.2 Ultrasonic C-Scan of Damage in $[(\pm 45)_{12}]_s$ AS4/3502 Carbon/Epoxy
(Shuart and Williams, 1984)

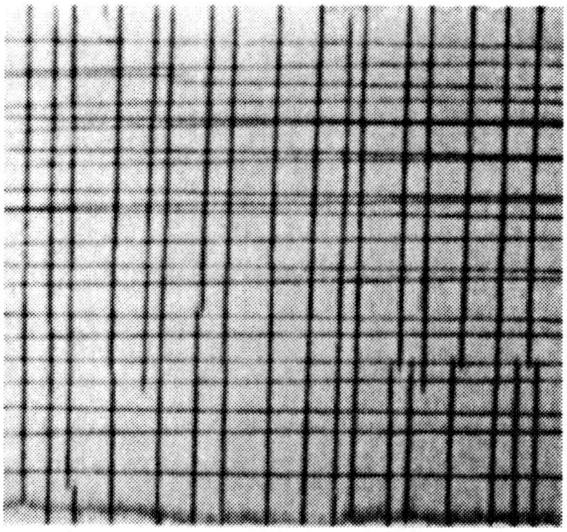

FIGURE 6.3 X-Ray Radiographs of Transverse Cracks in $[0_3/90_3]_s$ Carbon/Polyimide
(Herakovich et al., 1980)

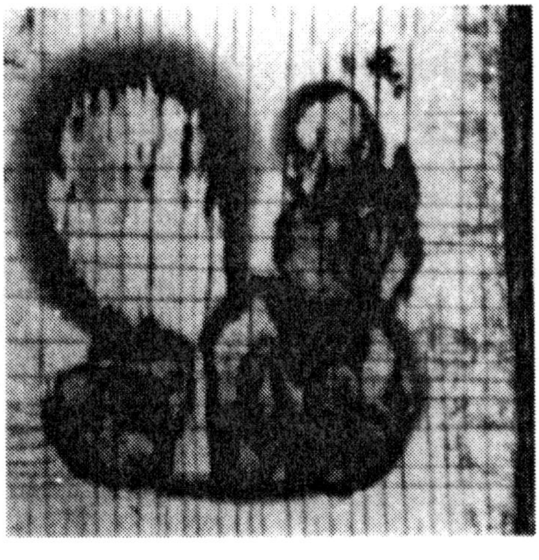

FIGURE 6.4 X-Ray Radiographs of Delaminations in $[(0/90)_3]_s$ Carbon/Polyimide
(Herakovich et al., 1980)

Detection of defects by X-ray can be enhanced through the use of an opaque dye penetrant such as tetrabromoethane (TBE). The penetrant is applied to the surface of the specimen and must be able to penetrate to the damaged area if it is to have the desired enhancement. Thus the damage must be continuous to the surface for this method to be successful.

Radiographs of TBE-enhanced defects in Celion 6000/PMR-15 carbon/polyimide are shown in Figs. 6.3 and 6.4 for specimens that were subjected to a rapid heating to the cure temperature (625°F, 330°C) followed by a liquid nitrogen quench to –320°F (–196°C) (Herakovich et al., 1980). Transverse cracks are evident in both directions of the $[0_3/90_3]_s$ laminate (Fig. 6.3), and delaminations (outlined by the dark rings) as well as transverse cracks are evident in the $[(0/90)_3]_s$ laminate (Fig. 6.4). It is interesting to note that the delaminations do not extend to the free edges of the specimen; the penetrant reaches the delaminations through the transverse cracks. ASTM standards concerned with radiography include E94-93, E142-92, and E1316-94.

6.3 Physical Properties

6.3.1 Density

The density of a material is defined as its mass per unit volume. A test method is described in detail in ASTM Standard D792-91. This method can be used to determine the density of the composite as well as the density of the constituents. The density of a material is determined using its weight in air, W_a, and in water, W_w, and the known densities of air (taken to be negligible) and water, ρ_w (0.9975 g/cm^3 at 23°C). The volume of a specimen is then determined from the difference between the weight of the material in air and the weight in water and from the known density of water. The composite density, ρ_c, is then

$$\rho_c = \frac{W_a}{\left(\dfrac{W_a - W_w}{\rho_w}\right)} = \frac{\rho_w W_a}{W_a - W_w} \tag{6.1}$$

6.3.2 Fiber Volume Fraction

The volume fraction of fiber in a composite is a very important parameter controlling the thermomechanical properties of the lamina. The fiber volume fraction for composites currently in use ranges from 30% to 65% (see Table 1.3). There are several methods for determining fiber volume fraction. One method is to count the number of fibers in several measured representative areas of a polished surface of the composite under high magnification (e.g., Fig. 6.1b), measure the diameter (or cross-sectional area) of one or more fibers, and then calculate the average fiber volume fraction as the percentage of area that is fiber. The advantage of this method is that it is relatively simple and it provides indications of the type and uniformity of fiber spacing as well as indications of the void content. A second method is to digest (dissolve) the matrix by putting a measured volume of composite in an acid bath. Weighing the (dry) fibers remaining after digestion and knowing the density of the fibers then provides the volume of fibers and the fiber volume fraction. The liquid used for matrix digestion must be chosen to ensure that fibers are not digested. The choice of solvent is dependent on the type of matrix and fiber materials. Hot nitric acid is used for carbon/epoxy. Details of the digestion method can be found in ASTM Standards D3171-76 (1990) (polymeric composites) and D3553-76 (1989) (metal matrix composites). A third method for determining fiber volume fraction is to determine the density of the composite and then calculate the fiber volume fraction knowing the density of the fiber and the matrix. This method assumes that the void content is negligible.

6.3.3 Void Content

Polymeric and ceramic matrix composites typically have voids after fabrication. A well-fabricated composite may have a void content of 1% or less, whereas a poorly made composite can have a void content as high as 7%. The void content of a composite can affect its mechanical properties, fatigue resistance, corrosion resistance, thermal properties, and strength. Thus it is important to know the void content of a composite and to develop fabrication methods for maintaining quality control at low void contents.

The void content, V_c, is determined from the experimentally measured composite density, ρ_e, and the theoretical composite density, ρ_t, calculated based upon the known densities of the constituents and the resin content. Several different methods are described in ASTM Standard D2734-91, which requires the use of ASTM Standard D2584-68 (1985) for determination of the resin content. The theoretical density of the composite is the weight, W_c, per unit volume of composite, where the volume of composite is the sum of the volume of resin and the volume of fiber. The volumes of resin and fiber are determined from the known weight of resin, W_r, and of fiber, W_f, and their respective densities, ρ_r and ρ_f. Thus,

$$\rho_t = \frac{W_c}{V_r + V_f} = \frac{W_c}{\dfrac{W_r}{\rho_r} + \dfrac{W_f}{\rho_f}} \tag{6.2}$$

The void content, V_c (in percent), is then

$$V_c = \frac{100(\rho_t - \rho_e)}{\rho_t} \tag{6.3}$$

6.3.4 Moisture Content

Polymeric matrix composites absorb moisture, which results in swelling (expansion) of the material. Moisture may also cause degradation of the matrix or the fiber/matrix interfacial properties. The moisture content of a polymeric matrix composite is specified in terms of the percent of moisture by weight. Thus the procedure for determining moisture content is to weigh a sample at the ambient conditions and then dry the sample to determine the dry weight.

Typical curves showing weight loss as a function of time for T300/5208 carbon/epoxy are shown in Fig. 6.5. The two curves correspond to drying in a vacuum oven and drying in an oven at atmospheric pressure. It is apparent from the figure that the desorption rate is slow in both cases, but it increases significantly under vacuum conditions. Small laboratory specimens of epoxy matrix material can be dried in a matter of a few days in a vacuum oven, whereas, for atmospheric pressure conditions, the drying time extends to two or more weeks. Thus use of a vacuum oven is recommended for drying specimens in the laboratory.

The saturation moisture content of polymeric matrix composites is generally in the range of 1 to 2%. Thus for a typical transverse coefficient of hygroscopic expansion, $\beta_2 = 0.005/\%H_2O$, transverse hygroscopic strains can be as high as 1.0%. Expansion in the fiber direction is limited by the high-modulus fibers. Thus the coefficient of hygroscopic expansion in the fiber direction is typically near zero.

It is good practice to dry polymeric matrix composites prior to testing unless the influence of moisture content is under consideration. Using dry specimens provides a consistent set of data. Because of the slow absorption rate, it is generally not a problem to remove a specimen from the oven after drying, install strain gages (if being used), and then test the specimen. As indicated in Fig. 6.5, the percent weight change over the period of an hour or two is small.

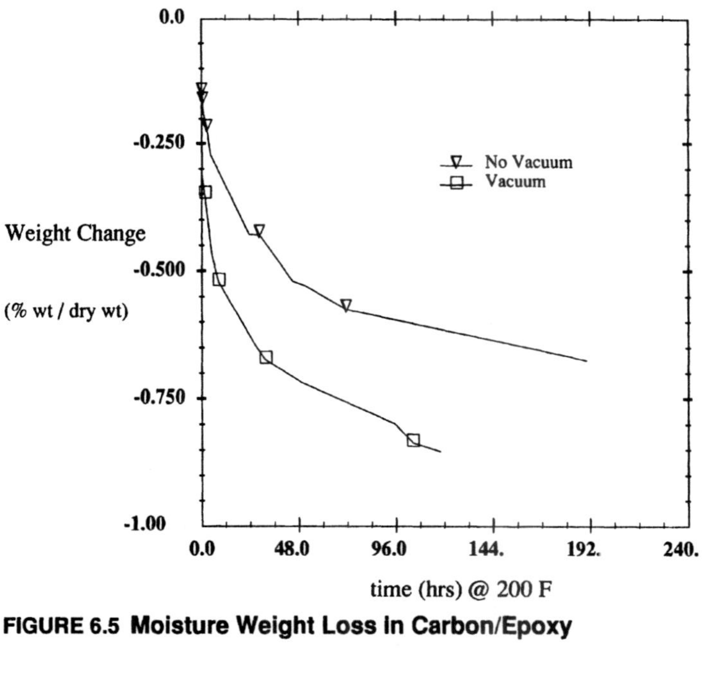

FIGURE 6.5 Moisture Weight Loss in Carbon/Epoxy

The rate of moisture diffusion can be predicted by Fick's law, which can be written

$$\frac{\partial}{\partial t}M = d\frac{\partial^2 M}{\partial z^2} \tag{6.4}$$

where M is the moisture concentration, d is the diffusivity in the z-direction, and t denotes time. The diffusion coefficient, d, must be adjusted for material, flow direction, laminate stacking sequence, temperature, and relative humidity. Experimental results indicate that this model overestimates the moisture absorption (Shen and Springer, 1976).

6.4 Mechanical Property Characterization

6.4.1 General Considerations

The choice of specimen to be used for mechanical testing of any material is dictated by at least two considerations. First, the specimen must provide the desired states of stress and strain. The second consideration is cost, including material, specimen fabrication, and testing. Composites are most often used in the form of thin laminates that can be fabricated using a variety of methods. For the most meaningful property determination, it is desirable to test specimens of the same fiber volume fraction, thickness, and fabrication method as the material being used in the application of interest.

The two most likely candidate material characterization specimens are flat laminates and laminated tubes. The great majority of experimental results for the characterization of composite response have been obtained using flat laminates. While the flat laminate may appear to be the obvious choice for economic reasons, it has severe limitations for some states of stress, including pure shear, compression, and combined stress states.

The tube has many advantages, including the relative ease with which the specimen can be subjected to axial tension or compression, internal pressure, torsion, and multi-axial loading, including all of the aforementioned load types. The disadvantages of the tube are (1) the cost of fabrication and testing and (2) that tube fabrication may result in different microstructure and hence different effective properties compared with flat laminates.

A serious problem with the determination of composite strength is the fact that both flat laminates and tubes are susceptible to early failure resulting from stress concentrations associated with load introduction and material discontinuities at free edges and ply drop-off regions. Stress concentration–induced failures are a major reason for the study of a wide variety of specimen types. Determination of the compression strength of flat laminates has also resulted in a number of different specimens being proposed for compression testing because of the susceptibility to buckling for thin specimens.

6.4.2 Strain Measurement

There are three methods commonly used to measure strains in composite materials: extensometers, strain gages, and optical methods. Extensometers provide a measurement of the average strain over a finite length, typically of the order of 1.0 in (25.4 mm). They are used primarily for measurement of axial strains but can also be used to measure transverse strains.

Strain gages are available as uniaxial gages, bidirectional gages, and rosettes. Uniaxial gages measure the strain in one direction, bidirectional gages measure the strain in two directions, and rosettes measure the strain in three directions. Rosettes thus provide a complete description of the

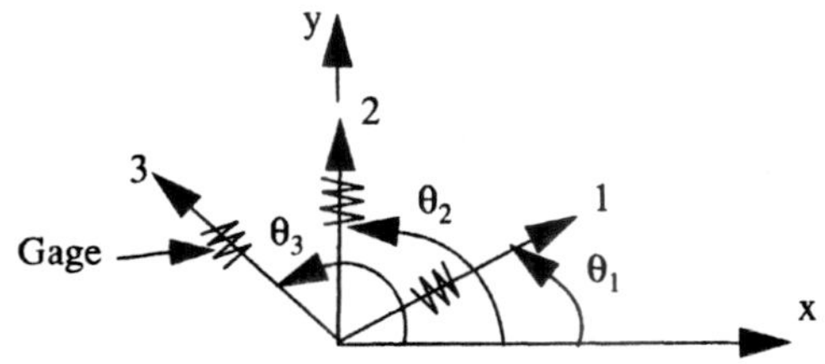

FIGURE 6.6 Strain Gage Rosette

average strain over the region of measurement through the strain transformation equations. Strain gages are available in a range of sizes, with the smallest gage of the order of 1.59 mm (0.0625 in). The bidirectional and rosette gages can be stacked or adjacent. Thus the region over which the strain is measured is dependent upon the choice of gage.

The principle behind a rosette gage with three "legs" such as that depicted in Fig. 6.6 is that the normal strains are measured along any three directions θ_1, θ_2, θ_3, and then the strain transformation equations are employed to determine the global strains ε_x, ε_y, γ_{xy}. From the strain transformation equations (4.24), the normal strains in the directions of the three "legs" of the rosette are

$$\begin{aligned} \varepsilon_1(\theta_1) &= m^2(\theta_1)\varepsilon_x + n^2(\theta_1)\varepsilon_y + m(\theta_1)n(\theta_1)\gamma_{xy}/2 \\ \varepsilon_2(\theta_2) &= m^2(\theta_2)\varepsilon_x + n^2(\theta_2)\varepsilon_y + m(\theta_2)n(\theta_2)\gamma_{xy}/2 \\ \varepsilon_3(\theta_3) &= m^2(\theta_3)\varepsilon_x + n^2(\theta_3)\varepsilon_y + m(\theta_3)n(\theta\ )\gamma_{xy}/2 \end{aligned} \tag{6.5}$$

With the three normal strains ε_1, ε_2, and ε_3 known from measurement, the three equations (6.5) can be solved for the three unknown strains ε_x, ε_y, and γ_{xy}.

Using strain gage rosettes on both sides of flat specimens provides the most complete information as to axial, transverse, and shear strains as well as knowledge of development of specimen curvature. Consideration must, of course, be given to the selection of an appropriate gage for the test temperature. Costs can be reduced by using a rosette on one side of the specimen and a single axial gage on the opposite side to check for bending.

Because of the highly anisotropic nature of composites, measurements can be very sensitive to fiber orientation and gage alignment. Results presented in Chapter 4 (Fig. 4.9) show the sensitivity of axial modulus to fiber orientation. Strain gage correction factors for transverse sensitivity also can be very dependent on alignment. Thus it is very important that gage alignment and fiber orientation be measured accurately and used appropriately in the analysis of experimental results.

Optical methods such as Moiré interferometry (Post et al., 1994) can be used if full-field strain measurements are desired or if it is not possible or desirable to attach strain gages and lead wires to a specimen.

6.4.3 Tensile Testing

The standard specimen for tensile testing of continuous fiber composites is a flat, straight-sided coupon (Fig. 6.7). Tabs are recommended for gripping the specimen. The tabs can be fabricated from a variety of materials, including fiberglass, copper, aluminum, or the material and laminate being tested. The tabs are not mandatory when strength is not of primary concern; emory cloth is often used in place of tabs in such cases. Tensile testing of flat coupons is detailed in ASTM

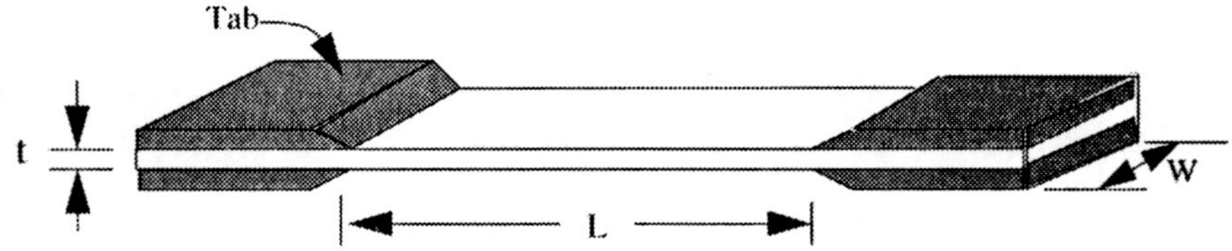

a) Geometry

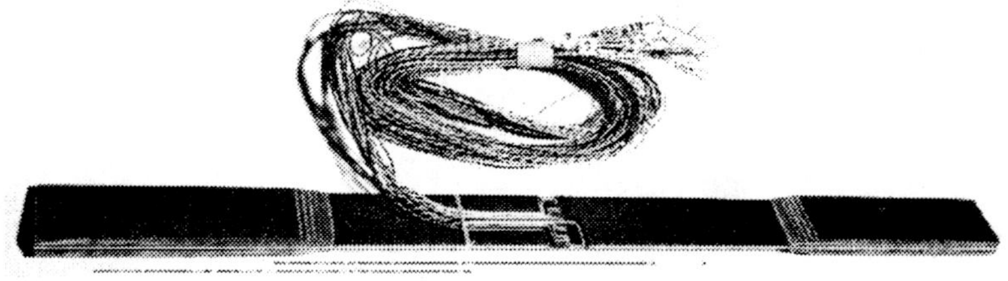

b) Photograph of Specimen with Strain Gages

FIGURE 6.7 Tensile Coupon

Standard D3039/D3039M-93. A photograph of a composite tensile coupon with fiberglass tabs and strain gages mounted across the face of the specimen is shown in Fig. 6.7b.

For orthotropic laminae and laminates, this specimen can provide data on the axial modulus, E_x; the in-plane and through-thickness Poisson's ratios, ν_{xy} and ν_{xz}; any nonlinear, inelastic response; and the tensile ultimate stress, σ_x^{ult}, and the tensile ultimate strain, ε_x^{ult}. For unidirectional off-axis laminae, the specimen also provides data on the coefficient of mutual influence and the in-plane shear response. The width of tensile specimens typically ranges from 0.5 in (12.7 mm) to 1.0 in (25.4 mm), with the choice often dictated by the availability and cost of the material. The coupon specimens can be gripped by flat, wedge-action, or pneumatic grips.

For a straight-sided coupon, failure may occur anywhere along the specimen. For a series of tests on a group of nominally identical specimens, some failures typically occur in the grip region at higher stress levels than some of the failures that occur in the gage section. Failure in the grips can be due to the random distribution of flaws in the specimen, and it can be argued that the combined state of stress in the grip region results in higher stresses at the micromechanics level, which initiate the failures. The straight-sided coupon is used for materials with a high degree of orthotropy rather than the traditional bone-shaped specimen for homogeneous, isotropic materials because cutting fibers to shape the shoulder of the bone can result in significant stress concentration and a corresponding reduction in tensile strength. However, bone-shaped specimens have been used successfully for chopped-fiber, whisker, and some metal matrix composite tensile tests (ASTM D3552-77 (1989)). Tensile testing for the transverse properties of hoop-wound polymer matrix composite cylinders is detailed in ASTM D5450/D5450M-93.

6.4.3.1 Orthotropic Laminae and Laminates

For orthotropic, symmetric laminates (including 0° and 90° laminae), analysis (Section 5.12) shows that the effective axial modulus and Poisson's ratio are given by

$$E_x = \frac{1}{a_{11}^*}$$
$$\nu_{xy} = \frac{-a_{12}^*}{a_{11}^*} \qquad (6.6)$$

These same two properties are measured directly from a tensile test on a specimen of thickness t under axial force per unit length N_x as

$$E_x = \frac{\sigma_x}{\varepsilon_x} = \frac{N_x}{t\varepsilon_x}$$
$$\nu_{xy} = \frac{-\varepsilon_y}{\varepsilon_x} \qquad (6.7)$$

Thus the tensile test not only provides direct experimental determination of elastic properties, it also provides an opportunity for comparison of theory and experiment. And since the number of possible laminates is infinite, an experimentally verified theory must be available in order to optimize the laminate configuration.

The tensile strength is defined as the average stress at failure. Thus, in terms of the maximum applied force per unit length, N_x^{max}, and the thickness, t, the ultimate stress, $\bar{\sigma}_x^{\text{ult}}$, is

$$\bar{\sigma}_x^{\text{ult}} = \frac{N_x^{\text{max}}}{t} \qquad (6.8)$$

It is important to note that failure of laminates is often influenced by interlaminar stresses along the free edges of the coupon (see Chapter 8). Thus great care should be taken in reporting laminate strength data to include the laminate stacking sequence, exact specification of the direction of loading, and the type of failure (i.e., in-plane or delamination).

6.4.3.2 Off-Axis Laminae

The tensile test on a unidirectional off-axis lamina is a rather specialized test because of the presence of axial-shear coupling associated with the nonzero a_{16}^* term of the laminate compliance (5.82) (or, alternatively, coupling associated with the coefficient of mutual influence $\eta_{xy,x}$ (4.69)). In addition to the elastic constants E_x, ν_{xy}, and ν_{xz}, the off-axis tensile test can be used to study the coefficient of mutual influence, $\eta_{xy,x}$, and the nonlinear response and strength of an off-axis lamina, all as a function of the fiber orientation angle, θ. The test can also be used to study the material response in principal material coordinates. The specimen geometry should be such that the fibers do not extend continuously from one grip to the other. (See Horgan, 1982, for a discussion of end effects in anisotropic materials.)

The fundamental question in conducting a tensile test on a unidirectional off-axis coupon is, "what are the boundary conditions on the specimen?" If a pure, uniform state of axial stress $\sigma_x \neq 0$, $\sigma_y = \tau_{xy} = 0$ can be applied to the ends and sides of a specimen, and the specimen is free to assume any desired deformation pattern, the state of stress will be uniform and constant throughout the specimen. Under such circumstances, the deformation pattern will be like that shown in Fig. 6.8a.

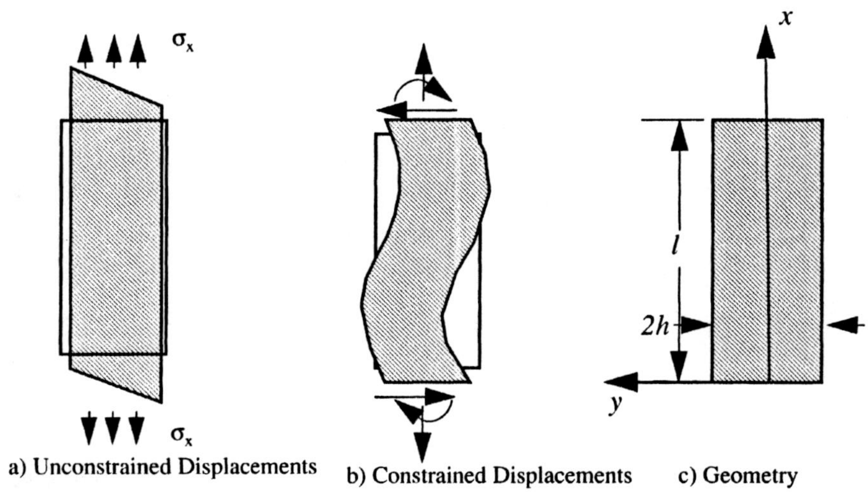

FIGURE 6.8 Off-Axis Tensile Coupon

For uniform, far-field axial stress loading $\sigma_x \neq 0$, the stresses in principal material coordinates are determined from the transformation equations (4.23) to be

$$\begin{aligned} \sigma_1 &= m^2 \sigma_x \\ \sigma_2 &= n^2 \sigma_x \\ \tau_{12} &= -mn\sigma_x \end{aligned} \tag{6.9}$$

Since these stresses are statically determinate, the state of stress remains constant throughout the loading history, including during any nonlinear, inelastic response. If the strains in the specimen are measured with a rosette strain gage, the normal and shear strains can be determined in any desired coordinates using the strain transformation equations (4.24). The global elastic constants associated with axial stress loading are determined directly using the measured stress and strains:

$$\begin{aligned} E_x &= \frac{\sigma_x}{\varepsilon_x} \\ \nu_{xy} &= \frac{-\varepsilon_y}{\varepsilon_x} \\ \eta_{xy,x} &= \frac{\gamma_{xy}}{\varepsilon_x} \end{aligned} \tag{6.10}$$

The plane stress constitutive equations in principal material coordinates are given by (4.17)–(4.19), and the E-ν reciprocal relationship is given by (4.15). Combining these equations with the stress transformation equations (6.9) gives the strains in terms of the applied stress, elastic properties, and fiber orientation angle in the form

$$\varepsilon_1 = \frac{\sigma_1}{E_1} - \frac{\nu_{21}\sigma_2}{E_2} = (m^2 - n^2\nu_{12})\frac{\sigma_x}{E_1}$$

$$\varepsilon_2 = -\frac{\nu_{12}\sigma_1}{E_1} + \frac{\sigma_2}{E_2} = \left(-\frac{\nu_{12}m^2}{E_1} + \frac{n^2}{E_2}\right)\sigma_x \tag{6.11}$$

$$\gamma_{12} = \frac{\tau_{12}}{G_{12}} = \frac{-mn\sigma_x}{G_{12}}$$

Since the applied stress $\sigma_x \neq 0$ and all three components of strain are known from measurements, the shear modulus is determined directly from the third equation in (6.11) for any fiber orientation. The remaining equations then represent two equations in the three unknowns E_1, E_2, and ν_{12}. Therefore only two tests are required to determine the four elastic constants. If a unidirectional 0° specimen is used to determine E_1 and ν_{12}, then a test on any off-axis lamina (other than 90°) can be used to determine both E_2 and G_{12}. As will be discussed later, there is a material-dependent preferred range for the fiber orientation angle of the off-axis test.

Unfortunately, it is not a simple task to apply pure, uniform tensile stress to an off-axis coupon. Tensile specimens are usually gripped in such a manner that the ends of the specimen are constrained as indicated in Fig. 6.8b, and the boundary condition is actually a specification of the axial end displacement (unless very special methods are used to introduce the load). As a result of the end constraints, the specimen assumes a doubly curved displacement field, as shown in Fig. 6.8b (see Nemeth et al., 1983), and the state of stress varies throughout the specimen. Since the strains are measured by strain gages, the problem is to determine the stresses at the point where the strains are measured, in terms of the applied loading.

The analysis of the constrained off-axis coupon presented here follows the original work of Pagano and Halpin (1968). The lamina is assumed to be in a state of plane stress under far-field loading such that the axial strain is ε_x^o. The specimen length, l, width, $2h$, and origin of the coordinate system are defined in Fig. 6.8c.

The sides of the coupon are stress free, and hence we have the boundary conditions

$$\begin{aligned} \sigma_y(x, \pm h) &= 0 \\ \tau_{xy}(x, \pm h) &= 0 \end{aligned} \tag{6.12}$$

Approximate displacement boundary conditions for the axial displacement, u, and transverse displacement, v, are *assumed* to be

$$\begin{aligned} v(0,0) &= \frac{\partial}{\partial y}u(0,0) = 0 \\ v(l,0) &= \frac{\partial}{\partial y}u(l,0) = 0 \\ u(0,0) &= 0 \\ u(l,0) &= \varepsilon_x^o l \end{aligned} \tag{6.13}$$

These boundary conditions do not constrain the displacement along the entire width at ends of the specimen; the specimen is allowed to "pull out" from the grips away from the centerline $y = 0$.

Further, it is *assumed* that the shear stress does not vary along the length of the specimen; it is an unknown function of y only, i.e.,

$$\tau_{xy} = f_1(y) \tag{6.14}$$

Using (6.14) in the plane stress equilibrium equations (2.40) and integrating, we obtain general expressions for the normal components of stress:

$$\sigma_x = -xf_1'(y) + f_2(y) \tag{6.15}$$

$$\sigma_y = g(x) \tag{6.16}$$

where $f_2(y)$ and $g(x)$ are unknown functions of their respective arguments. Substituting the results into the compatibility equation (2.41) gives the third-order partial differential equation

$$-\bar{S}_{11}x f_1'''(y) + \bar{S}_{11} f_2''(y) + 2\bar{S}_{16} f_1''(y) + \bar{S}_{22} g''(x) = 0 \tag{6.17}$$

A solution to this PDE, satisfying the stress boundary conditions (6.12), is

$$\begin{aligned} f_1(y) &= C_0(y^2 - h^2) \\ f_2(y) &= -2\frac{\bar{S}_{16}}{\bar{S}_{11}}C_0 y^2 + C_1 y + C_2 \\ g(x) &= 0 \end{aligned} \tag{6.18}$$

Combining (6.15) and (6.16) with (6.18) gives the stresses:

$$\sigma_x = -2\frac{\bar{S}_{16}}{\bar{S}_{11}}C_0 y^2 + C_1 y + C_2 - 2C_0 xy \tag{6.19}$$

$$\sigma_y = 0 \tag{6.20}$$

$$\tau_{xy} = C_0(y^2 - h^2) \tag{6.21}$$

where C_0, C_1, and C_2 are unknown constants of integration. The constants of integration are determined using the boundary conditions (6.13). Starting from the strain-displacement relationships (2.16) and (2.17) with the strains expressed in terms of the stresses as in (4.52) and the stresses given by (6.19)–(6.21), we proceed as follows. From (2.16) and (2.17), the strain-displacement relationships in global coordinates are

$$\varepsilon_x = \frac{\partial u}{\partial x} \tag{6.22}$$

$$\varepsilon_y = \frac{\partial v}{\partial y} \tag{6.23}$$

$$\gamma_{xy} = \frac{\partial u}{\partial y} + \frac{\partial v}{\partial x} \tag{6.24}$$

From (4.52) the strains in terms of stresses with $\sigma_y = 0$ are

$$\varepsilon_x = \bar{S}_{11}\sigma_x + \bar{S}_{16}\tau_{xy} \tag{6.25}$$

$$\varepsilon_y = \bar{S}_{12}\sigma_x + \bar{S}_{26}\tau_{xy} \tag{6.26}$$

$$\gamma_{xy} = \bar{S}_{16}\sigma_x + \bar{S}_{66}\tau_{xy} \tag{6.27}$$

Integration of (6.22) using (6.19), (6.21), and (6.25) gives the u-displacements in the form

$$u = \bar{S}_{11}C_1xy + \bar{S}_{11}C_2x - \bar{S}_{11}C_0x^2y - \bar{S}_{16}C_0x(y^2 + h^2) + \lambda(y) \tag{6.28}$$

where $\lambda(y)$ is a function to be determined.

Likewise, integration of (6.23) using (6.19), (6.21), and (6.26) gives the v-displacements in the form

$$v = \frac{C_0y^3}{3}\left(\bar{S}_{26} - \frac{2\bar{S}_{12}\bar{S}_{16}}{\bar{S}_{11}}\right) + \frac{\bar{S}_{12}C_1y^2}{2} + (\bar{S}_{12}C_2 - \bar{S}_{26}C_0h^2)y - \bar{S}_{12}C_0xy^2 + \mu(x) \tag{6.29}$$

where $\mu(x)$ is a function to be determined.

Combining (6.19), (6.21), (6.24), and (6.27) gives the functions $\lambda(y)$ and $\mu(x)$:

$$\lambda(y) = \left(-2\frac{\bar{S}_{16}^2}{\bar{S}_{11}} + \bar{S}_{12} + \bar{S}_{66}\right)\frac{C_0y^3}{3} + \frac{\bar{S}_{16}C_1y^2}{2} + (\bar{S}_{16}C_2 - \bar{S}_{66}C_0h^2 - C_3)y + C_5 \tag{6.30}$$

$$\mu(x) = \bar{S}_{11}\frac{C_0x^3}{3} - \bar{S}_{11}\frac{C_1x^2}{2} + C_3x + C_4 \tag{6.31}$$

Combining (6.28) and (6.30) gives the complete expression for the u-displacements:

$$\begin{aligned} u =\ & \bar{S}_{11}C_1xy + \bar{S}_{11}C_2x - \bar{S}_{11}C_0x^2y - \bar{S}_{16}C_0x(y^2 + h^2) \\ & + \left(-2\frac{\bar{S}_{16}^2}{\bar{S}_{11}} + \bar{S}_{12} + \bar{S}_{66}\right)\frac{C_0y^3}{3} + \frac{\bar{S}_{16}C_1y^2}{2} + (\bar{S}_{16}C_2 - \bar{S}_{66}C_0h^2 - C_3)y + C_5 \end{aligned} \tag{6.32}$$

Combining (6.29) and (6.31) gives the complete expression for the v-displacements:

$$\begin{aligned} v =\ & \frac{C_0y^3}{3}\left(\bar{S}_{26} - \frac{2\bar{S}_{12}\bar{S}_{16}}{\bar{S}_{11}}\right) + \frac{\bar{S}_{12}C_1y^2}{2} + (\bar{S}_{12}C_2 - \bar{S}_{26}C_0h^2)y - \bar{S}_{12}C_0xy^2 \\ & + \bar{S}_{11}\frac{C_0x^3}{3} - \bar{S}_{11}\frac{C_1x^2}{2} + C_3x + C_4 \end{aligned} \tag{6.33}$$

Now, using the displacements (6.32) and (6.33) in the boundary conditions (6.13) gives the constants in terms of the applied strain ε_x^o:

$$C_0 = \frac{6\bar{S}_{16}\varepsilon_x^o}{h^2\left[6(\bar{S}_{11}\bar{S}_{66} - \bar{S}_{16}^2) + \bar{S}_{11}^2\left(\frac{h}{l}\right)^2\right]} \tag{6.34}$$

$$C_1 = C_0 l \tag{6.35}$$

$$C_2 = \frac{C_0 h^2}{6\bar{S}_{16}}\left[6\bar{S}_{66} + \bar{S}_{11}\left(\frac{l}{h}\right)^2\right] \tag{6.36}$$

$$C_3 = \frac{\bar{S}_{11} C_0 l^2}{6} \tag{6.37}$$

$$C_4 = C_5 = 0 \tag{6.38}$$

Along the centerline of the specimen, $y = 0$, the displacements are

$$u = (\bar{S}_{11} C_2 - \bar{S}_{16} C_0 h^2)x \tag{6.39}$$

$$v = \frac{\bar{S}_{11} C_0 (2x - l)(x - l)x}{6} \tag{6.40}$$

From (6.40) we see that the centerline lateral displacement is cubic in x with zero values at $x = 0$, $x = l/2$, and $x = l$, resulting in the displacement field shown in Fig. 6.8b. Further, at the quarter point, $x = l/4$, (as an example), and using (6.34) for C_0, the normalized lateral displacement is

$$\frac{v}{h} = \frac{\bar{S}_{11}\bar{S}_{16}\left(\frac{l}{h}\right)^3}{32\left[6(\bar{S}_{11}\bar{S}_{66} - \bar{S}_{16}^2) + \bar{S}_{11}^2\left(\frac{h}{l}\right)^2\right]} \tag{6.41}$$

From (6.41) we see that the normalized lateral displacement increases with increasing specimen aspect ratio l/h.

The axial displacements along the end $x = 0$ are determined from (6.32) to be

$$u(x=0) = \left(-2\frac{\bar{S}_{16}^2}{\bar{S}_{11}} + \bar{S}_{12} + \bar{S}_{66}\right)\frac{C_0 y^3}{3} + \frac{\bar{S}_{16} C_0 l y^2}{2} + (\bar{S}_{16} C_2 - \bar{S}_{66} C_0 h^2 - C_3)y \tag{6.42}$$

Clearly, the axial end displacement is generally nonzero except at $y = 0$. Thus the specimen can "pull out" from the grips in this approximate solution.

The centerline stresses from (6.19) to (6.21) are

$$\begin{aligned} \sigma_x &= C_2 \\ \sigma_y &= 0 \\ \tau_{xy} &= -C_o h^2 \end{aligned} \tag{6.43}$$

where from (6.34), (6.36), and (6.43), the centerline axial stress is

$$\sigma_x = \frac{\left[6\left(\frac{\bar{S}_{66}}{\bar{S}_{11}}\right) + \left(\frac{l}{h}\right)^2\right]\varepsilon_x^o}{\bar{S}_{11}\left\{6\left[\frac{\bar{S}_{66}}{\bar{S}_{11}} - \left(\frac{\bar{S}_{16}}{\bar{S}_{11}}\right)^2\right] + \left(\frac{h}{l}\right)^2\right\}} \tag{6.44}$$

Combining the first and third of (6.43) with (6.34) and (6.36), we can express the centerline shear stress in terms of the centerline axial stress as

$$\tau_{xy} = \beta\sigma_x \tag{6.45}$$

where

$$\beta = \frac{-6\left(\frac{\bar{S}_{16}}{\bar{S}_{11}}\right)}{6\left(\frac{\bar{S}_{66}}{\bar{S}_{11}}\right) + \left(\frac{l}{h}\right)^2} \tag{6.46}$$

Equations (6.19)–(6.21) indicate that the stress distribution in the off-axis coupon as predicted by this approximate solution is, at most, quadratic in y and linear in x, and (6.43) indicates that the stresses are constant along the centerline, $y = 0$. Equations (6.34)–(6.37) show that the constants are functions of the fiber orientation (through the compliance terms) and the specimen aspect ratio l/h. Further, (6.45) indicates that the centerline shear stress is proportional to the centerline axial stress. And (6.46) indicates that the constant of proportionality, β, is a function of the material orthotropy, fiber orientation, and the specimen aspect ratio, l/h. As l/h becomes large, β tends to zero.

The validity and accuracy of this approximate analytical solution was verified by Nemeth et al. (1983) for carbon/polyimide using experimental Moiré interferometry and finite-element stress analysis. Their results showed that as the specimen aspect ratio is increased, the state of stress in an off-axis coupon approaches constant shear stress and zero transverse normal stress. For tensile tests on carbon/polyimide specimens with an aspect ratio of 15, the shear stress at the center of the specimen is approximately 1/100 that of the axial stress, and the transverse normal stress is essentially zero. They recommended a specimen aspect ratio of 15 or more for the determination of elastic properties in polymer matrix composites.

It is important to note that this approximate solution indicates that the state of stress at the center of the specimen, $y = 0$, $x = l/2$, is biaxial with nonzero stresses σ_x and τ_{xy}. Thus the constitutive equations expressed in terms of engineering constants are

$$\begin{aligned}
\varepsilon_x &= \frac{\sigma_x}{E_x} + \frac{\eta_{x,xy}\tau_{xy}}{G_{xy}} \\
\varepsilon_y &= \frac{-\nu_{xy}\sigma_x}{E_x} + \frac{\eta_{y,xy}\tau_{xy}}{G_{xy}} \\
\gamma_{xy} &= \frac{\eta_{xy,x}\sigma_x}{E_x} + \frac{\tau_{xy}}{G_{xy}}
\end{aligned} \tag{6.47}$$

Hence, the strains are the result of biaxial stresses σ_x and τ_{xy}. Thus, if the engineering properties are determined assuming a uniaxial state of stress σ_x, they are in error. The fundamental question is, "how significant is the error?"

When the biaxial stress state (6.43) is substituted into the plane stress constitutive equations (4.53), the resulting expressions for in-plane strains along the centerline are

$$\begin{aligned} \varepsilon_x &= \bar{S}_{11}\sigma_x + \bar{S}_{16}\tau_{xy} = \bar{S}_{11}C_2 - \bar{S}_{16}C_0h^2 \\ \varepsilon_y &= \bar{S}_{12}\sigma_x + \bar{S}_{26}\tau_{xy} = \bar{S}_{12}C_2 - \bar{S}_{26}C_0h^2 \\ \gamma_{xy} &= \bar{S}_{16}\sigma_x + \bar{S}_{66}\tau_{xy} = \bar{S}_{16}C_2 - \bar{S}_{66}C_0h^2 \end{aligned} \tag{6.48}$$

These equations serve as a starting point for determining elastic properties from an off-axis test.

6.4.3.3 Axial Modulus

If we define the "apparent" axial modulus E_x in terms of the axial stress σ_x in (6.43) and the axial strain ε_x of (6.48), the result is

$$E_x^* = \frac{\sigma_x}{\varepsilon_x} = \frac{1}{\bar{S}_{11}[1-\eta]} \tag{6.49}$$

where

$$\eta = \frac{6\left(\dfrac{\bar{S}_{16}}{\bar{S}_{11}}\right)^2}{6\left(\dfrac{\bar{S}_{66}}{\bar{S}_{11}}\right) + \left(\dfrac{l}{h}\right)^2} = -\left(\frac{\bar{S}_{16}}{\bar{S}_{11}}\right)\beta \tag{6.50}$$

Now the actual axial modulus is, by definition, $E_x = 1/\bar{S}_{11}$, and thus (6.49) can be written

$$E_x = E_x^*(1-\eta) \tag{6.51}$$

Hence η is a measure of the error between the apparent axial modulus E_x^* and the actual modulus E_x. From (6.50) we see that the error η is zero if the fiber orientation is 0° or 90° (i.e., $\bar{S}_{16} = 0$). We also see from (6.50) that the error tends to zero as the specimen aspect ratio l/h becomes large. Further, for a given specimen aspect ratio, there is a fiber orientation θ that corresponds to the *maximum* error because of the θ dependence of the transformed compliance coefficients.

The expression (6.49) for the axial modulus was developed based upon the axial stress along the specimen centerline ($y = 0$). In actuality, the stress that is measured by a testing machine is the average axial stress $\bar{\sigma}_x = P/A$, where with the aid of (6.19) and for a specimen thickness t we have

$$\bar{\sigma}_x = \frac{P}{A} = \frac{1}{2ht}\int_{-h}^{h}\sigma_x(l,y)t\,dy = \left[C_2 - \frac{2}{3}\left(\frac{\bar{S}_{16}}{\bar{S}_{11}}\right)h^2C_0\right] \tag{6.52}$$

Using (6.34) and (6.36) for the constants C_2 and C_0 in (6.52) gives a relationship between the overall specimen average stress, $\bar{\sigma}_x$, and the centerline stress, σ_x:

$$\sigma_x = \frac{\bar{\sigma}_x}{1 - \frac{2}{3}\eta} \tag{6.53}$$

Now, if the *experimental modulus*, $\bar{E}_x$, is calculated based upon the average axial stress $\bar{\sigma}_x$ and the measured axial strain ε_x, then with the aid of (6.53) we have

$$\bar{E}_x = \frac{\bar{\sigma}_x}{\varepsilon_x} = \frac{\sigma_x\left(1 - \frac{2}{3}\eta\right)}{\varepsilon_x} \tag{6.54}$$

Now, combining (6.49), (6.51), and (6.54), the ratio of actual E_x to the experimental modulus $\bar{E}_x$ is

$$\frac{E_x}{\bar{E}_x} = \frac{1 - \eta}{1 - \frac{2}{3}\eta} \tag{6.55}$$

We note from (6.50) that the error parameter η is a function of the principal material properties (E_1, E_2, ν_{12}, G_{12}) and specimen fiber orientation θ and aspect ratio l/h. Thus the principal material properties must be known for the error to be determined as a function of fiber orientation and specimen aspect ratio.

As an example of the error introduced by ignoring the grip-induced biaxial state of stress in an off-axis coupon, the ratio of actual modulus E_x to the experimental modulus $\bar{E}_x$ as a function of fiber orientation, specimen aspect ratio $(l/2h)$, and shear modulus for Celion 6000/PMR-15 carbon/polyimide is shown in Fig. 6.9. These results show that the error is largest for fiber orientation in the range 10–15° and decreases with increasing aspect ratio. For the cases considered, the maximum error ranges from 2 to 10%. For fiber orientations greater than 45° the error is negligible for aspect ratio ≥5, and for aspect ratios ≥20 the error is negligible for nearly all fiber orientations.

6.4.3.4 Poisson's Ratio

The centerline strains (6.48) can also be used to compare the measured Poisson's ratio, $\bar{\nu}_{xy} = -\varepsilon_y/\varepsilon_x$, from off-axis tensile tests with the actual Poisson ratio $\nu_{xy} = -\bar{S}_{12}/\bar{S}_{11}$ defined for a uniaxial state of stress $\sigma_x \neq 0$. The result is

$$\frac{\nu_{xy}}{\bar{\nu}_{xy}} = \frac{1 + \left(\frac{\bar{S}_{16}}{\bar{S}_{11}}\right)\beta}{1 + \left(\frac{\bar{S}_{26}}{\bar{S}_{12}}\right)\beta} \tag{6.56}$$

For the special cases of $\theta = 0°$ or 90°, (6.56) indicates that the measured Poisson's ratio is identical to the actual value, as it should be since there is no coupling in these cases. Also, as the specimen aspect ratio becomes large, β (from (6.46)) tends to zero and the exact value is approached. It is also noted that there is no difference between the apparent Poisson's ratio ν^*_{xy} based upon the centerline axial stress, σ_x, and the measured Poisson's ratio $\bar{\nu}_{xy}$ based upon the average axial stress, $\bar{\sigma}_x$, since the stress is a common factor that "drops out" in the calculation of Poisson's ratio.

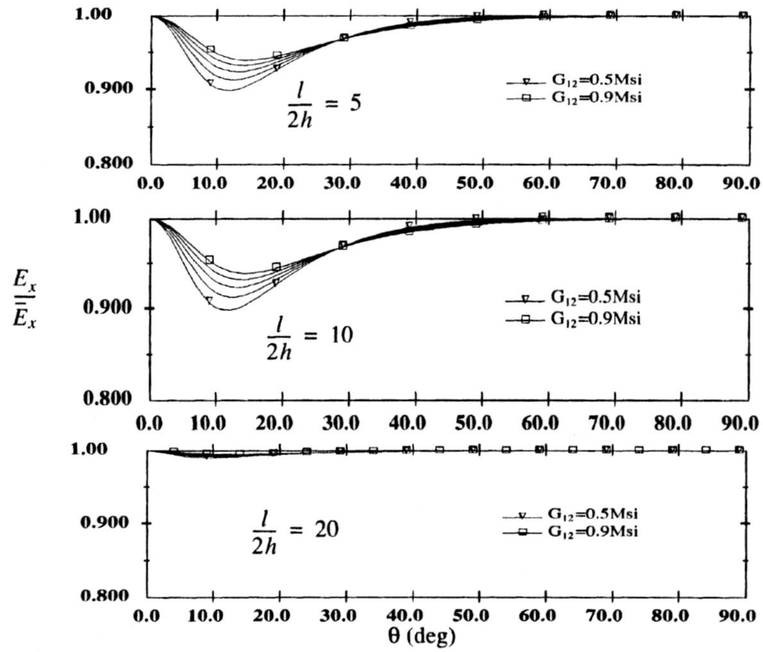

FIGURE 6.9 Error in Off-Axis Axial Modulus for Celion 6000/PMR-15
(Pindera and Herakovich, 1986)

6.4.3.5 Coefficient of Mutual Influence

Substitution of the centerline strains (6.48) into the definition of $\eta_{xy,x}$ gives an expression for the measured coefficient of mutual influence, $\bar{\eta}_{xy,x}$, in terms of the compliance coefficients, the specimen aspect ratio, and the actual value $\eta_{xy,x}$:

$$\frac{\eta_{xy,x}}{\bar{\eta}_{xy,x}} = \frac{1 + \left(\frac{\bar{S}_{16}}{\bar{S}_{11}}\right)\beta}{1 + \left(\frac{\bar{S}_{66}}{\bar{S}_{16}}\right)\beta} \tag{6.57}$$

From (6.57) it is evident that $\bar{\eta}_{xy,x}$ is identical to $\eta_{xy,x}$ for an orthotropic lamina and that the measured value $\bar{\eta}_{xy,x}$ approaches the exact value as the specimen aspect ratio becomes large. As with Poisson's ratio, the coefficient of mutual influence is independent of which axial stress (centerline or average) is used.

6.4.4 Shear Testing

Several different methods have been used to determine the shear properties of fibrous composites, including tension of a $[\pm 45]_s$ laminate, tension of an off-axis lamina, rail shear of unidirectional laminae, torsion of a unidirectional tube, and Iosipescu shear of unidirectional laminae and cross-ply laminates. These methods are described in the following paragraphs.

6.4.4.1 $[\pm 45]_s$ Tensile Test

Tension on a $[\pm 45]_s$ coupon can be used to provide measurement of the lamina in-plane shear modulus G_{12} (Rosen, 1972; ASTM Standard D3518/D3518/M-91). From (5.130), the stresses in principal material coordinates for any angle-ply laminate under axial stress $\bar{\sigma}_x$ are

$$\begin{aligned} \sigma_1 &= B\bar{\sigma}_x \\ \sigma_2 &= (1-B)\bar{\sigma}_x \\ \tau_{12} &= \frac{-1}{2mn}(B(1-2m^2)+m^2)\bar{\sigma}_x \end{aligned} \tag{6.58}$$

where B is defined in terms of the elastic properties and fiber orientation ($m = \cos\theta$ and $n = \sin\theta$) in the form

$$B = \left[\frac{m^2(2m^2-1)+4m^2n^2\frac{G_{12}}{E_2}\left(\frac{E_2}{E_1}\nu_{12}+1\right)}{4m^2n^2\frac{G_{12}}{E_2}\left(\frac{E_2}{E_1}+2\frac{E_2}{E_1}\nu_{12}+1\right)+(2m^2-1)(m^2-n^2)}\right] \tag{6.59}$$

For the special case $\theta = 45°$, (6.58) gives the shear stresses:

$$\tau_{12}(\pm\theta) = \mp\frac{\bar{\sigma}_x}{2} \tag{6.60}$$

This result shows that the shear stress in principal material coordinates of a $[\pm 45]_s$ laminate subjected to axial load is statically determinate (i.e., independent of material properties of the specimen) with a magnitude equal to one-half the applied average axial stress.

In view of the linear elastic constitutive equation for orthotropic materials (Eq. (4.11)), the shear response is uncoupled from the normal response, and hence the shear modulus G_{12} can be determined directly from a tensile test on a $[\pm 45]_s$ laminate. The shear strain γ_{12} in principal material coordinates is given by transformation of the measured axial and transverse strains ε_x and ε_y (γ_{xy} is zero for orthotropic laminates under tension, and γ_{12} is independent of γ_{xy} for $\theta = \pm 45°$). The strain transformation gives $\gamma_{12} = -(\varepsilon_x - \varepsilon_y)$. Thus

$$G_{12} = \frac{\bar{\sigma}_x}{2(\varepsilon_x - \varepsilon_y)} \tag{6.61}$$

Equation (6.61) can also be written in terms of the effective properties of the $[\pm 45]_s$ laminate. Dividing all terms in (6.61) by ε_x and recognizing the definitions of E_x and ν_{xy}, we have

$$G_{12} = \frac{\frac{\bar{\sigma}_x}{\varepsilon_x}}{2\left(1-\frac{\varepsilon_y}{\varepsilon_x}\right)} = \frac{E_x}{2(1+\nu_{xy})} \tag{6.62}$$

6.4.4.2 Shear Response from Off-Axis Lamina

The unidirectional off-axis tensile coupon can be used to determine the shear response of a composite in the principal material coordinates. As shown in Section 6.4.3.2, the state of stress in the principal material coordinates of an off-axis coupon can be determined using the transformation equations (6.9) for uniaxial far-field stress loading or by transforming the centerline stresses (6.43) from the approximate elasticity solution for constrained displacement loading. In the principal material coordinates the shear response is uncoupled from the normal response with the elastic constitutive equation

$$G_{12} = \frac{\tau_{12}}{\gamma_{12}} \tag{6.63}$$

Thus, with the shear stress known in terms of the loading and the shear strain γ_{12} determined after transformation of the measured strains ε_x, ε_y, and γ_{xy}, the shear modulus can be determined. Different results will, of course, be obtained depending upon the shear stress used in (6.63).

If the end-constraint–induced shear stress τ_{xy} is neglected and the stress transformation is based solely upon the average far-field axial stress $\bar{\sigma}_x$, the corresponding shear stress in principal material coordinates, $\bar{\tau}_{12}$, is

$$\bar{\tau}_{12} = -mn\bar{\sigma}_x \tag{6.64}$$

The apparent shear modulus $\bar{G}_{12}$ is then

$$\bar{G}_{12} = \frac{-mn\bar{\sigma}_x}{\gamma_{12}} \tag{6.65}$$

However, for the biaxial, centerline state of stress (6.43) and (6.45) from the approximate elasticity solution, the stress transformation equations (4.23) give the principal material shear stress, τ_{12}, in terms of the nonzero stresses σ_x and τ_{xy} in the form

$$\tau_{12} = -\sigma_x mn + \tau_{xy}(m^2 - n^2) \tag{6.66}$$

When this shear stress is expressed in terms of the average axial stress using (6.45) and (6.53), the result is

$$\tau_{12} = \frac{-\bar{\sigma}_x}{\left(1 - \frac{2}{3}\eta\right)}[mn - \beta(m^2 - n^2)] \tag{6.67}$$

Since the measured shear strain is independent of the analysis method, the ratio of actual to apparent shear modulus is

$$\frac{G_{12}}{\bar{G}_{12}} = \frac{\tau_{12}}{\bar{\tau}_{12}} = \frac{\left[mn + \eta\left(\frac{\bar{S}_{11}}{\bar{S}_{16}}\right)(m^2 - n^2)\right]}{mn\left(1 - \frac{2}{3}\eta\right)} \tag{6.68}$$

It is evident that, in general, both sides of (6.68) are functions of the actual shear modulus G_{12}. This is true because η and $\bar{S}_{ij}$ are functions of G_{12} (see (4.44)–(4.49)). Thus (6.68) cannot be solved explicitly, and implicit methods must be used. It is also evident from (6.68) that for the special case of $\theta = 45°$ the ratio $G_{12}/\bar{G}_{12}$ approaches 1.0 as the specimen aspect ratio becomes large (i.e., η becomes small).

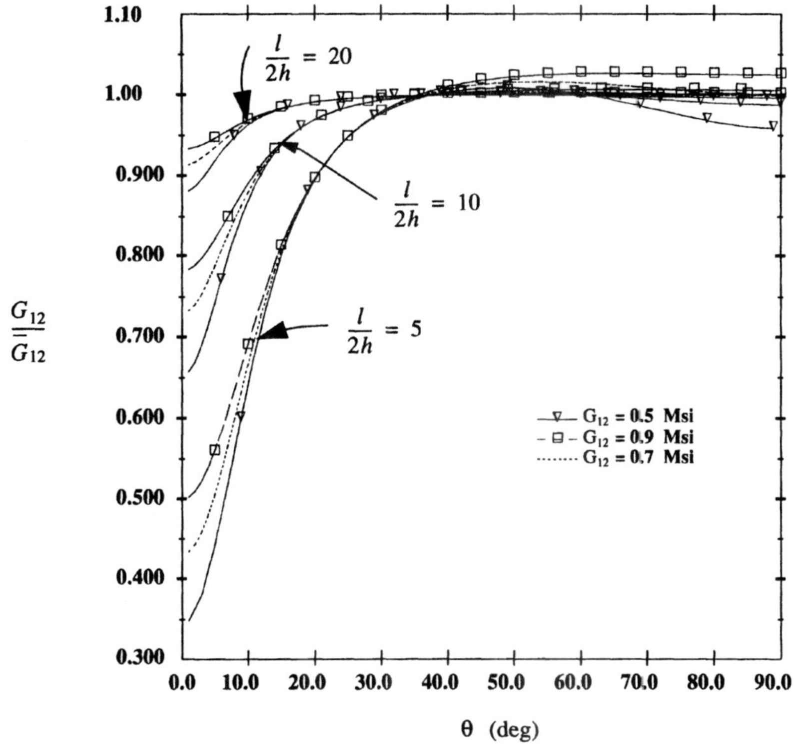

FIGURE 6.10 Shear Modulus G_{12} Dependence on Off-Axis Parameters
(Pindera and Herakovich, 1986)

The curves in Fig. 6.10 show a parametric study of the ratio $G_{12}/\bar{G}_{12}$ as a function of the actual shear modulus and specimen aspect ratio for a carbon/polyimide (Celion 6000/PMR-15) (Pindera and Herakovich, 1986). The curves were plotted for three different values of shear modulus ranging from 5 Msi (34.5 GPa) to 9 Msi (62.0 GPa) and three specimen aspect ratios ranging from 5.0 to 20.0. It is evident that the error decreases with increasing specimen aspect ratio and that the error is zero or near zero for all aspect ratios at $\theta = 45°$. Based upon these results, the off-axis tensile coupon with an aspect ratio of 10 or more and a fiber orientation in the range 25–45° is recommended for determination of the shear modulus G_{12} in polymer matrix composites. It is noted that this recommendation is material dependent.

6.4.4.3 Thin-Walled Tube in Torsion

The thin-walled tube (Fig. 6.11) is the ideal specimen for testing composite laminae in shear because it can provide a uniform state of pure shear stress. Thus the tube can be used to determine accurately the shear response of composites. Unfortunately, failure of tubes, like other specimens, typically initiates in or near the grips, and failure is not the result of a pure shear stress state.

The selection of specimen tube size is dictated by the capacity of the test machine. In particular, the tube diameter must be sized such that the torsional strength is within the machine capacity. This is often the limiting condition in testing tubes.

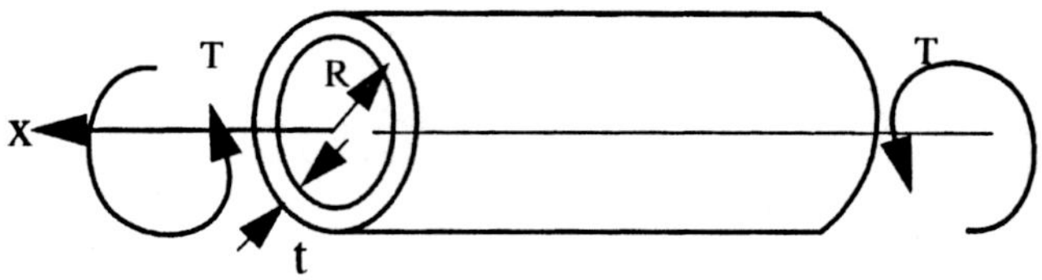

FIGURE 6.11 Tube in Torsion

For a thin-walled unidirectional tube (0° or 90°) under pure torsional loading (Fig. 6.11), the average shear stress $\bar{\tau}_{12}$ (= $\bar{\tau}_{x\theta}$) is related to the torque T by

$$\bar{\tau}_{12} = \frac{T}{2\pi R^2 t} \tag{6.69}$$

where R is the radius and t is the thickness of the tube.The shear modulus is determined directly from the calculated shear stress (6.69) and measured shear strain. More details on this test are presented in ASTM Standard D5448/D5448M-93.

6.4.4.4 Iosipescu Shear Test

The Iosipescu (1967) shear specimen was originally designed as a round specimen with a V-notch groove for shear testing of metals. It was first proposed for use with flat composite laminates by Bergner et al. (1977) and Herakovich and Bergner (1980). The flat specimen has two identical V-notches symmetrically placed about the centerline at midlength (Fig. 6.12). A fixture (Fig. 6.13) for implementation of the specimen was developed by Adams and Walrath (1982) (see ASTM D5379/D5379M-93). When the specimen is loaded in shear, a region of nearly uniform, pure shear stress is present in the test section (between the notches). The actual stress distribution is a function of the material properties and fiber orientation.

Profiles of stress distributions between the notches of [0] and [90] aramid/epoxy and carbon/polyimide Iosipescu specimens are shown in Fig. 6.14 (shear stresses) and Fig. 6.15 (normal stresses) (Pindera et al., 1987). The stresses in these figures (which were determined from a finite-element stress analysis) are normalized with respect to the average shear stress, $\bar{\tau}_{xy}$, between the notches. Careful examination of the results indicates that the shear stress distribution exhibits a nearly uniform region in the center of the specimen. The distribution is more uniform in the center of the [90] lamina, but the [0] lamina exhibits smaller overall stress gradients. For the [0] lamina the test section shear stress is less than the average stress, whereas for the [90] lamina the test sec-

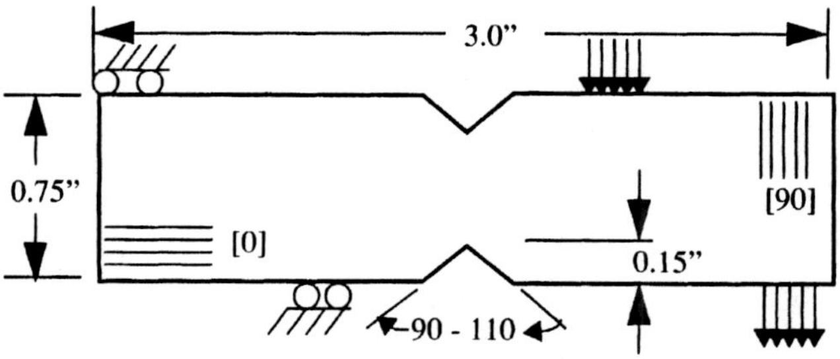

FIGURE 6.12 Iosipescu Shear Specimen

FIGURE 6.13 Iosipescu Specimen Mounted in Fixture

tion shear stress is greater than the average stress. These results indicate that it is appropriate to use correction factors applied to the average stresses.

Table 6.1 (Pindera et al., 1987) presents a comparison of shear modulus and shear strength values for aramid/epoxy and carbon/polyimide as obtained from off-axis tensile and Iosipescu shear specimens. Correction factors and corrected moduli are also presented. The results show excellent correlation between the corrected shear modulus values for aramid/epoxy; the correlation for the carbon/polyimide is good, but not as good as for the lower-modulus aramid/epoxy.

The table also indicates a wide variation in maximum shear stress as obtained from the different test methods, a clear indication that the specimens are failing in different modes due to specimen-induced stress concentrations. For unidirectional laminae, the Iosipescu specimen is susceptible to failure by splitting along the fiber direction. Figure 6.16 shows failed [0] and [90] aramid/epoxy laminae. The maximum shear stress was 6.75 ksi (46.5 MPa) for the [0] lamina, compared with only 3.72 ksi (25.6 MPa) for the [90] lamina. Neither of these maximum stresses corresponds to the shear strength of the material, as the failure is not a pure shear failure in either case. Failure is strongly influenced by the stress concentrations present at the crack initiation site. For the [0] lamina there is a large stress concentration in the shear stress at the notch root, and for the [90] lamina there is a concentration of the transverse normal stress σ_2 (σ_x in Fig. 6.15b).

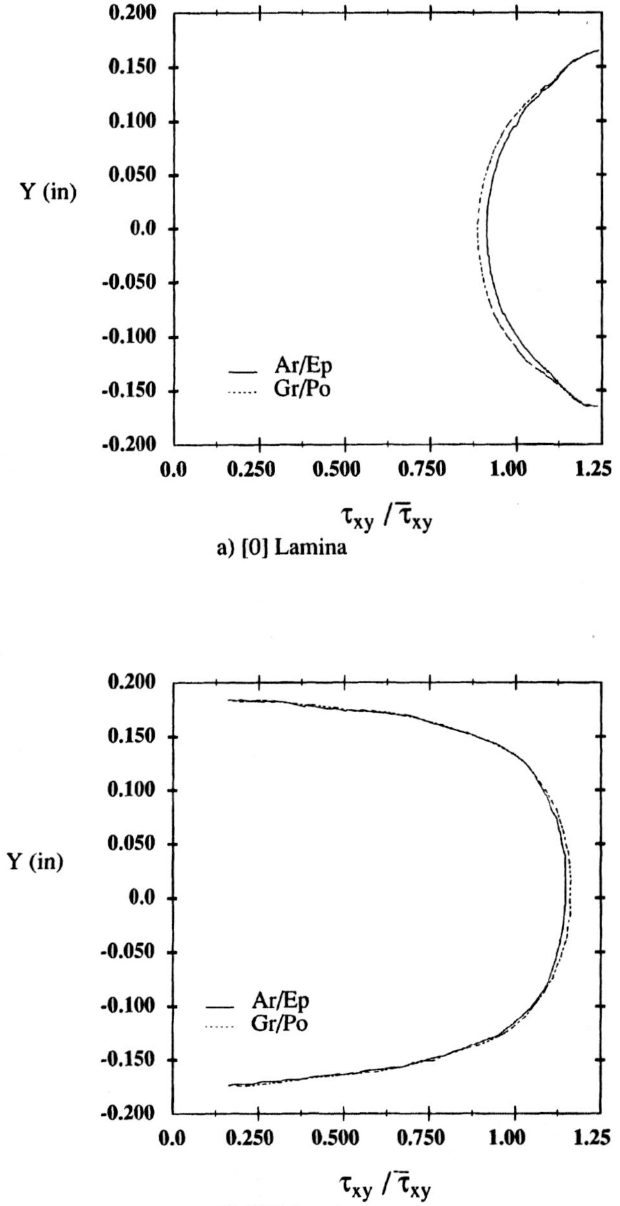

a) [0] Lamina

b) [90] Lamina

FIGURE 6.14 Shear Stress Profiles: Iosipescu Specimen

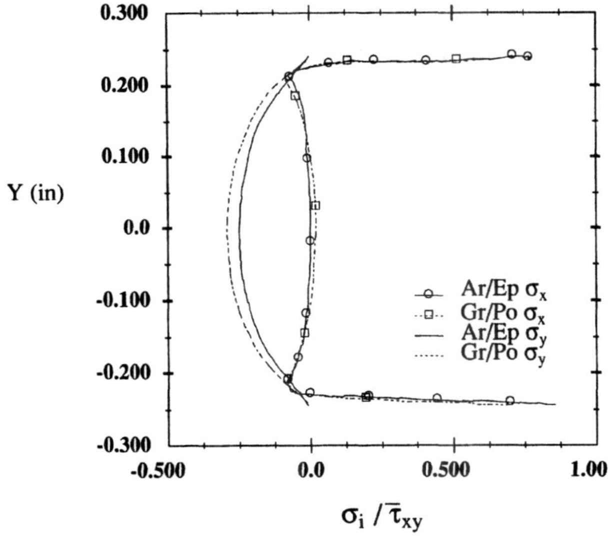

a) [0] Lamina

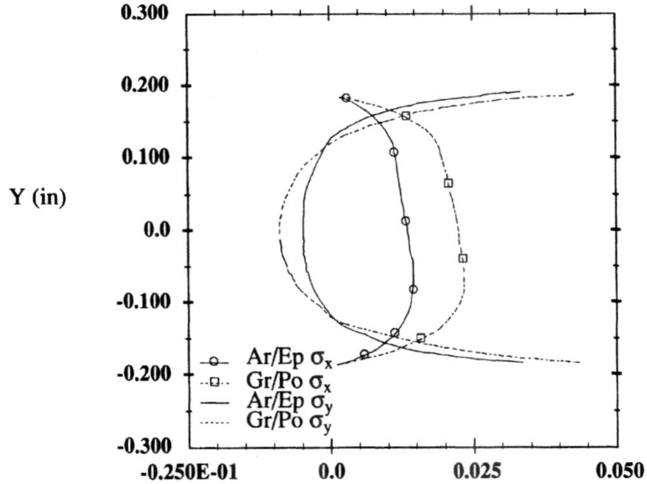

b) [90] Lamina

FIGURE 6.15 Normal Stress Profiles: Iosipescu Specimen

Material	Specimen	Apparent Shear Modulus, Msi (GPa)	Maximum Shear Stress, ksi (MPa)	Correction Factor, $G_{12}/\bar{G}_{12}$	Corrected Shear Modulus, Msi (GPa)
Aramid/ epoxy	10° off-axis	0.238 (1.64)	6.36 (43.8)	0.9435	0.225 (1.55)
	45° off-axis	0.224 (1.54)	3.52 (24.3)	1.0010	0.224 (1.54)
	0° Iosipescu	0.240 (1.65)	6.75 (46.5)	0.8970	0.215 (1.48)
	90° Iosipescu	0.187 (1.29)	3.72 (25.6)	1.1535	0.216 (1.49)
Carbon/ polyimide	10° off-axis	0.925 (6.38)	10.39 (71.6)	0.8812	0.815 (5.62)
	45° off-axis	0.723 (4.98)	1.98 (13.6)	1.0038	0.725 (5.00)
	0° Iosipescu	0.802 (5.53)	9.75 (67.2)	0.8866	0.711 (4.90)

TABLE 6.1 Shear Moduli and Correction Factors

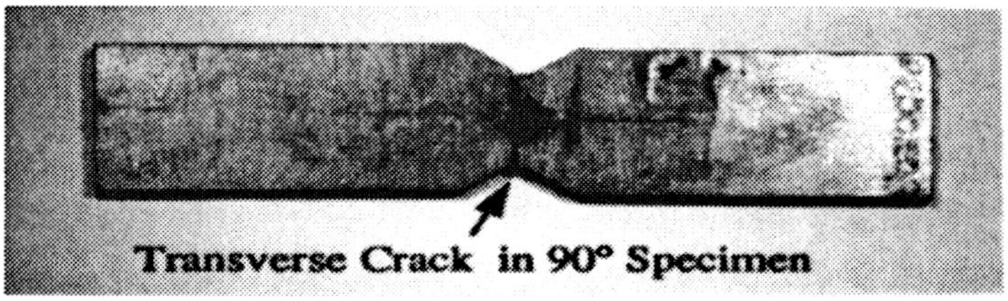

FIGURE 6.16 Failed Iosipescu Shear Specimens: Aramid/Epoxy

6.4.4.5 Rail Shear Test

Two configurations of rail shear tests have been used to study the shear response of composites. They are the two-rail fixture and the three-rail fixture shown in Fig. 6.17. Detailed specifications for the specimens are given in ASTM Standard D4255-83. In both cases, the specimen is a symmetric, specially orthotropic laminate ($A_{16} = A_{26} = 0$) that is bolted and/or bonded to *rigid* steel rails. The thickness of the laminate must be sufficient to eliminate the possibility of buckling associated with the diagonal compression of pure shear. Shear is introduced in the specimens through application of axial loads (typically compressive) as indicated in Fig. 6.17. For a specimen of length L and thickness t, the average shear stress, $\bar{\tau}_{xy}$, for a load P applied to two-rail and three-rail configura-

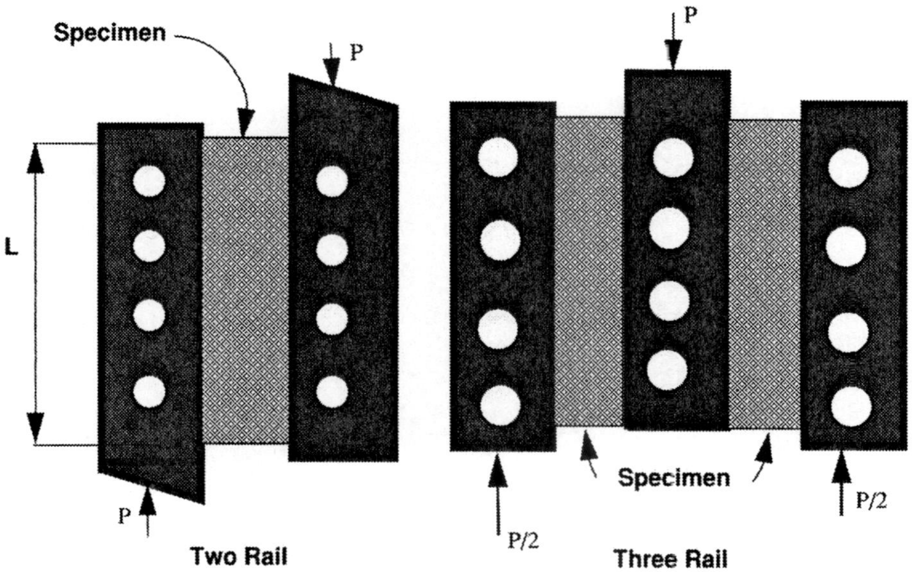

FIGURE 6.17 Rail Shear Test Fixtures

tions is

$$\bar{\tau}_{xy} = \frac{P}{Lt} \quad \text{(two-rail)}$$
$$\bar{\tau}_{xy} = \frac{P}{2Lt} \quad \text{(three-rail)} \qquad (6.70)$$

It is important to note that this specimen is susceptible to early failures due to stress concentrations introduced at bolt locations and free edges.

6.4.5 Compression Testing

Flat straight-sided coupons, tubes, or circular rods can be used for compression tests on fibrous composites, with the rods being used only for unidirectional laminae. Compression testing of fibrous composites can result in buckling both at the individual fiber level (referred to as *fiber kinking*) and at the global laminate level. Compression tests on flat composite laminates are complicated by the fact that the laminates are typically thin and thus more susceptible to global buckling. Rather extreme measures have to be taken to design test fixtures which provide exceptionally good alignment and limit the susceptibility to buckling. Compression specimens also are typically kept short in order to reduce the possibility of buckling failures (ASTM D3410-87).

6.4.5.1 Predicted Buckling Loads

Buckling analysis of flat, unit-width [0] compression specimens which are unsupported along the free edges and subjected to the axial load per unit width N_x leads to the following equations for simple support end conditions (no lateral deflection but freedom to rotate) and clamped support end conditions (no lateral deflection or rotation) (Ashton and Whitney, 1970), where L is the specimen length and t the thickness.

For simple support end conditions,

$$N_x = D_{11}\left(\frac{\pi}{L}\right)^2 \tag{6.71}$$

where D_{11} is the bending stiffness defined in (5.21) or in terms of the critical stress σ_{cr}:

$$\sigma_{cr} = \left(\frac{\pi^2 E_1}{12}\right)\left(\frac{t}{L}\right)^2\left(\frac{1}{1-\nu_{12}\nu_{21}}\right) \tag{6.72}$$

For clamped support end conditions, the stability criterion is determined as the stationary value of the functional:

$$\frac{1}{2}\int_0^a\int_0^b\left[D_{11}\left(\frac{\partial^2 w}{\partial x^2}\right)^2 + 2D_{12}\left(\frac{\partial^2 w}{\partial x^2}\right)\left(\frac{\partial^2 w}{\partial y^2}\right) + D_{22}\left(\frac{\partial^2 w}{\partial y^2}\right)^2 + 4D_{66}\left(\frac{\partial^2 w}{\partial x \partial y}\right)^2 + N_x\left(\frac{\partial w}{\partial x}\right)^2\right]dy\,dx \tag{6.73}$$

where D_{ij} are bending stiffness terms (5.21) and the out-of-plane displacement w is assumed to be of the form

$$w = \sum_{i=1}^{m}\sum_{j=1}^{n} a_{ij}\varphi_i(x)\psi_i(y) \tag{6.74}$$

where φ_i and ψ_i are the characteristic shapes for free vibration of beams satisfying specified boundary conditions. Minimization of the functional (6.73) results in a set of algebraic eigenvalue problems, where the lowest eigenvalue corresponds to the buckling load and the corresponding eigenvector describes the buckling mode.

As an example of buckling predictions, the curves in Fig. 6.18 (Fox et al., 1987) show a comparison of the predicted critical buckling stresses for [0] carbon/epoxy specimens of thickness t and length L, as determined from (6.72) and (6.73) and the Euler column buckling equations for simply supported and clamped boundary conditions (Timoshenko and Gere, 1961). As the figure shows, there is excellent agreement between the theories for this particular case.

Simply supported Euler column buckling stress:

$$\sigma_{cr} = \frac{\pi^2 E}{12}\left(\frac{t}{L}\right)^2 \tag{6.75}$$

Clamped support Euler column buckling stress:

$$\sigma_{cr} = \frac{\pi^2 E}{3}\left(\frac{t}{L}\right)^2 \tag{6.76}$$

6.4.5.2 Compression Test Fixtures

Two compression specimens and fixtures that have been used with success are a fixture designed at NASA Langley Research Center by Clark and Lisagore (1980) (Fig. 6.19) and a specimen designed

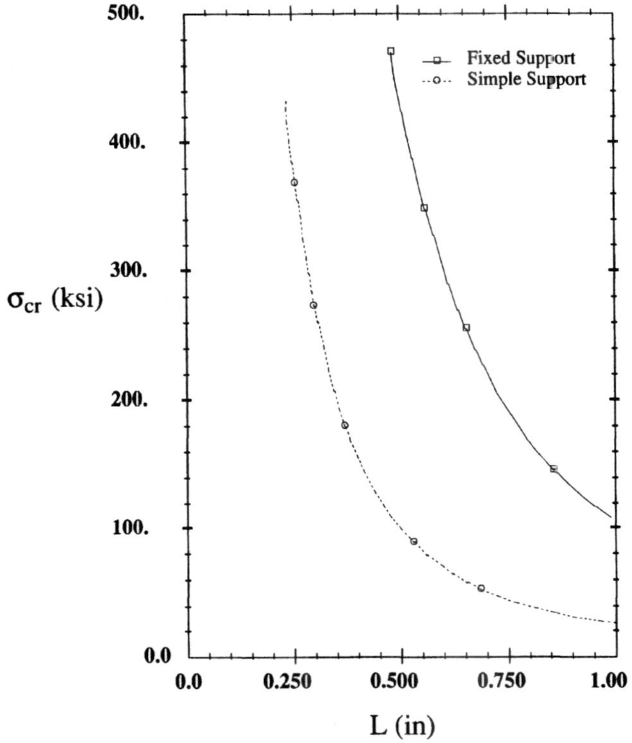

FIGURE 6.18 Critical Buckling Stresses for [0] Carbon/Epoxy

by Lin and Pindera (1988) (Fig. 6.20). The fixture by Lin and Pindera is unique in that it has the capability for tension-compression cyclic loading of flat specimens. Common to these specimens is that they have highly accurate axial alignment and short gage lengths. Even with the extreme measures that have been taken to ensure proper alignment and loading, it is often the case that compression failures take the form of localized fiber buckling (sometimes called fiber kinking) (Fig. 6.21) or brooming of the fibers at the end of the specimen. The short gage lengths used for compression testing can be a limiting factor because of end effects. If the specimen is not long enough, a region of uniform stress and strain will not be attained anywhere along the length of the specimen.

Sandwich, honeycomb core beams in bending (Fig. 6.22) have been used in an attempt to overcome the short-gage-length problem of compression testing. The top flange (or skin) of the sandwich beam in four-point bending acts as a compression specimen (ASTM D5467-93). The disadvantage to this specimen is that it is supported along its length and the compressive bending stress varies through the thickness of the composite specimen. The specimens in Fig. 6.22 are borsic/aluminum laminates bonded to an aluminum honeycomb core (Viswanathan et al., 1975). Only the failure of the [90] laminate is evident in the figure. This specimen failed by transverse cracking and delamination from the honeycomb core along one of the lines of load introduction. The [0] laminate also failed by transverse fracture along a line of load introduction, but no delamination was evident. The $[\pm 30]_s$ laminate failed along a 30° line in the test section. The $[\pm 45]_s$ laminate did not fail as the deformations exceeded the allowable range for the fixture. The transverse

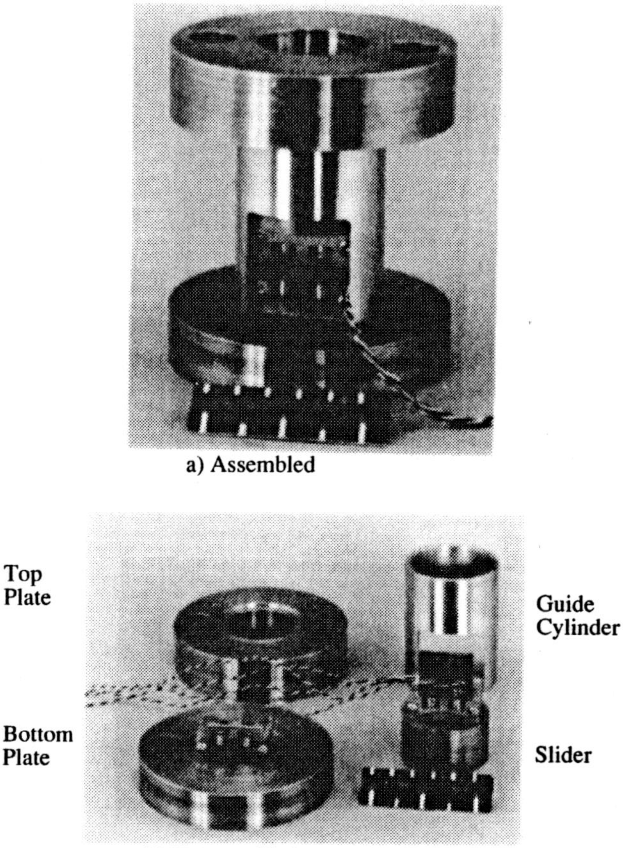

FIGURE 6.19 NASA Compression Fixture
(Clark and Lisagore, 1980)

compressive properties of hoop-wound polymer matrix composite cylinders can be determined according to ASTM D5449/D5449M-93.

6.4.6 Four-Point Bending

Four-point bending (Fig. 6.23) can be used to test laminates in pure bending. The test methods described in ASTM D790-92 cover the determination of flexural properties of polymeric composites in the form of rectangular bars.

The four-point bending test can also be used with honeycomb sandwich beams to test laminates in tension (the bottom flange) or compression (the top flange). Figure 6.22 shows compression specimens of borsic/aluminum laminates (top flange) that were tested in four-point bending (Viswanathan et al., 1975). The honeycomb is aluminum and the bottom flange is titanium (Ti-6Al-4V).

FIGURE 6.20 Tension-Compression Test Fixture
(Lin and Pindera, 1988)

FIGURE 6.21 Fiber Kinking in Unidirectional T300/934 Carbon/Epoxy
(Fox et al., 1987)

6.4.7 Thermal Expansion

Accurate measurement of coefficients of thermal expansion can be obtained using a dilatometer, a high-precision device for measurement of small changes in length. Dilatometers are commercially available for CTE measurements over a range of temperatures from cryogenic (–160°C (–320°F)) to 1600°C (2912°F). Interferometry, such as Moiré, can also be used for high-precision measurement. For CTE values that do not involve the measurement of very small strains, strain gages and extensometers can be used. Strain gages and extensometers generally do not provide sufficient accuracy for the CTE in the fiber direction of unidirectional composites (Bowles et al., 1981).

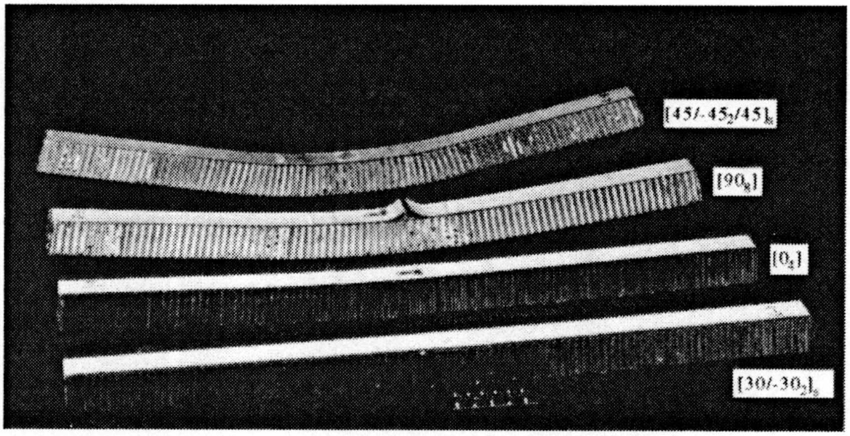

FIGURE 6.22 Honeycomb Core Sandwich Beams after Bending
(Viswanathan et al., 1975)

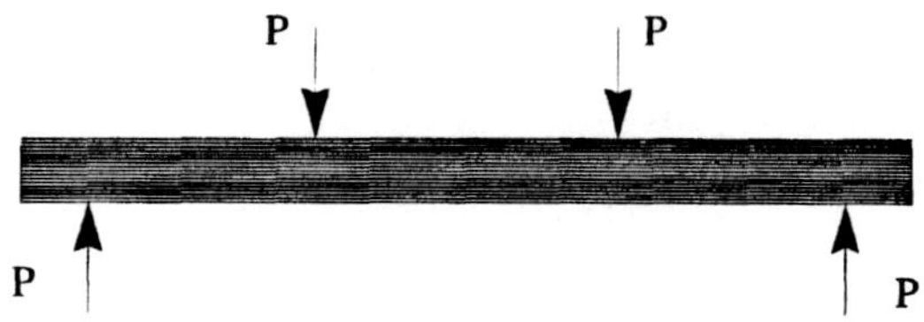

FIGURE 6.23 Four-Point Bending

Details on the determination of linear thermal expansion of reinforced polymer matrix composites are given in ASTM E289-94b and ASTM E831-93.

References

Adams, D. F., and Walrath, D. E. (1982), "Iosipescu Shear Properties of SMC Composite Materials," in *Composite Materials: Testing and Design,* ASTM STP 787 (I. M. Daniel, ed.), American Society for Testing and Materials, Philadelphia, pp. 19–33.

Ashton, J. E., and Whitney, J. M. (1970), *Theory of Laminated Plates,* Technomic Publishing Co., Stamford, CT.

Bergner, H. W., Jr., Davis, J. G., Jr., and Herakovich, C. T. (1977), *Analysis of Shear Test Method for Composite Laminates,* VPI-E-77-14, Virginia Tech, Blacksburg.

Bowles, D. E., Post, D., Herakovich, C. T., and Tenney, D. R. (1981), "Moiré Interferometry for Thermal Expansion of Composites," *Exp. Mech.,* vol. 21, no. 12, pp. 441–447.

Clark, R. K., and Lisagor, W. B. (1980), *Effects of Method of Leading and Specimen Configuration on Compressive Strength of Graphite/Epoxy Composite Materials,* NASA TM-81796, National Aeronautics and Space Administration.

Fox, D. J., Sykes, G. F., Jr., and Herakovich, C. T. (1987), *Space Environmental Effects on Graphite-Epoxy Compressive Properties and Epoxy Tensile Properties,* CCMS-87-11, Center for Composite Materials and Structures, Virginia Tech, Blacksburg.

Herakovich, C. T., and Bergner, H. W., Jr. (1980), "Finite Element Stress Analysis of a Notched Coupon Specimen for In-Plane Shear Behavior of Composites," *Composites*, July, pp. 149–154.

Herakovich, C. T., Davis, J. G., Jr., and Mills, J. S. (1980), "Thermal Microcracking in Celion 6000/PMR-15 Graphite-Polyimide," in *Thermal Stresses in Severe Environments* (D. P. H. Hasselman and R. A. Heller, eds.), Plenum, New York, pp. 649–664.

Hirschfeld, D. A., and Herakovich, C. T. (1990), "Failure Analysis of Notched Graphite-Epoxy Tubes," CCMS-90-04, Center for Composite Materials and Structures, Virginia Tech, Blacksburg.

Horgan, C. O. (1982), "Saint-Venant End Effects in Composites," *J. Compos. Mater.*, vol. 16, pp. 411–422.

Iosipescu, N. (1967), "New Accurate Procedure for Single Shear Testing of Metals," *J. Mater.*, vol. 2, no. 3, p. 537.

Lin, M. W. Y., and Pindera, M. -J. (1988), *Elastoplastic Response of Unidirectional Graphite/Aluminum under Combined Tension-Compression Cyclic Loading,* CCMS-88-01, Center for Composite Materials and Structures, Virginia Tech, Blacksburg.

Nemeth, M. P., Herakovich, C. T., and Post, D. (1983), "On the Off-Axis Tension Test for Unidirectional Composites," *Compos. Technol. Rev.*, vol. 5, no. 2, Summer, pp. 61–68.

Pagano, N. J., and Halpin, J. C. (1968), "Influence of End Constraints in the Testing of Anisotropic Bodies," *J. Compos. Mater.*, vol. 2, no. 1, p. 18.

Pindera, M. -J., Choksi, G., Hidde, J. S., and Herakovich, C. T. (1987), "A Methodology for Accurate Shear Characterization of Unidirectional Composites," *J. Compos. Mater.*, vol. 21, pp. 1164–1184.

Pindera, M. -J., and Herakovich, C. T. (1986), "Shear Characterization of Unidirectional Composites with the Off-Axis Tension Test," *Exper. Mech.*, vol. 26, no. 1, pp. 103–112.

Post, D., Han, B., and Ifju, P. G. (1994), *High Sensitivity Moire: Experimental Analysis for Mechanics and Materials*, Springer-Verlag, New York.

Roentgen, W. C. (1895), "Concerning a New Kind of Ray: Preliminary Report," Minutes of the Physical-Medical Association, Wuerzburg, Germany (Also in *Nature,* January 23, 1896 London).

Rosen, B. W. (1972), "A Simple Procedure for Experimental Determination of the Longitudinal Shear Modulus of Unidirectional Composites," *J. Compos. Mater.*, vol. 6, no. 4, pp. 552–554.

Shen, C. -H., and Springer, G. S. (1976), "Moisture Absorption and Desorption of Composite Materials," *J. Compos. Mater.*, vol. 10, pp. 2–20.

Shuart, M. J., and Williams, J. G. (1984), "Compression Failure Characteristics of ±45 Dominated Laminates with a Circular Hole or Impact Damage," AIAA Paper No. 84-0848, May, Palm Springs, CA.

Timoshenko, S. P., and Gere, J. M. (1961), *Theory of Elastic Stability*, McGraw-Hill Book Company, New York.

Viswanathan, C. N., Davis, J. G., Jr., and Herakovich, C. T. (1975), *Tensile and Compressive Behavior of Borsic/Aluminum Composite Laminates,* VPI-E-75-12, Virginia Tech, Blacksburg.

ASTM Standards

Nondestructive Evaluation

E114-90: Standard Practice for Ultrasonic Pulse-Echo Straight-Beam Examination by Contact Method (vol 03.03).

E214-68 (1991): Standard Practice for Immersed Ultrasonic Examination by the Reflection Method Using Pulsed Longitudinal Waves (vol. 03.03).

E317-93: Standard Practice for Evaluating Performance Characteristics of Ultrasonic Pulse-Echo Testing Systems Without the Use of Electronic Measurement Instruments (vol. 03.03).

E494-92a: Standard Practice for Measuring Ultrasonic Velocity in Materials (vol. 03.03).

E94-93: Standard Guide for Radiographic Testing (vol. 03.03).

E142-92: Standard Method for Controlling Quality of Radiographic Testing (vol. 03.03).

E1316-94: Terminology for Nondestructive Examinations (vol. 03.03).

Physical Properties

D792-91: Standard Test Methods for Density and Specific Gravity (Relative Density) of Plastics by Displacement (vol. 08.01).

D3171-76 (1990): Test Method for Fiber Content of Resin-Matrix Composites by Matrix Digestion (vol. 15.03).

D3553-76 (1989): Test Method for Fiber Content by Digestion of Reinforced Metal Matrix Composites (vol. 15.03).

D2734-91: Test Method for Void Content of Reinforced Plastics (vol. 08.02).

D2584-68 (1985): Test Method for Ignition Loss of Cured Reinforced Resins (vol. 08.02).

Tension

D3039/D3039M-93: Standard Test Method for Tensile Properties of Polymer Matrix Composite Materials (vol. 15.03).

D3552-77 (1989): Standard Test Method for Tensile Properties of Fiber Reinforced Metal Matrix Composites (vol. 15.03).

Shear

D3518/D3518M-91: Standard Practice for In-Plane Shear Stress-Strain Response of Unidirectional Polymer Matrix Composites (vol. 15.03).

D5379/D5379M-93: Standard Test Method for Shear Properties of Composite Materials by the V-Notched Beam-Method (vol. 15.03).

D4255-83: Standard Guide for Testing Inplane Shear Properties of Composite Laminates (vol. 15.03).

Compression

D3410-87: Test Method for Compressive Properties of Unidirectional or Crossply Fiber-Resin Composites (vol. 15.03).

D5467-93: Test Method for Compressive Properties of Unidirectional Fiber-Reinforced Polymer Matrix Composites using a Sandwich Beam (vol. 15.03).

Cylinders

D54481D/5448M-93: Standard Test Method for Inplane Shear Properties of Hoop Wound Polymer Matrix Composite Cylinders (vol. 15.03).

D5450/D5450M-93.: Standard Test Method for Transverse Tensile Properties of Hoop Wound Polymer Matrix Composite Cylinders (vol. 15.03).

D5449/D5449M-93: Standard Test Method for Transverse Compressive Properties of Hoop Wound Polymer Matrix Composite Cylinders (vol. 15.03).

Flexure

D790-92: Standard Test Methods for Flexural Properties of Unreinforced and Reinforced Plastics and Electrical Insulating Materials (vol. 08.01).

Thermal Expansion

E289-94b: Standard Test Method for Linear Thermal Expansion of Rigid Solids with Interferometry (vol. 14.02).

E831-93: Standard Test Method for Linear Thermal Expansion of Solid Materials by Thermomechanical Analysis (vol. 14.02).

Exercises

6.1 Show that Eq. (6.56) is the correct expression for the ratio of actual Poisson's ratio to apparent Poisson's ratio.

6.2 Show that Eq. (6.57) is the correct expression for the ratio of measured to actual coefficient of mutual influence for an off-axis coupon tensile test.

6.3 Show that Eq. (6.61) is the correct expression for the shear modulus G_{12} for a $[\pm 45]_s$ laminate.

6.4 Develop predictions for the ratios $E_x/\bar{E}_x$ and $G_{12}/\bar{G}_{12}$ for off-axis SCS-6/Ti-15-3 metal matrix tensile coupons, and compare the results to those in Figs. 6.9 and 6.10 for carbon/polyimide.

CHAPTER 7 MATERIAL RESPONSE

"I hear, and I forget.
I see, and I remember.
I do, and I understand."
Ancient Chinese proverb

7.1 Introduction

The response of fibrous composite materials can exhibit an exceptionally wide range of behaviors depending on the properties of the fiber and matrix constituents, the fiber orientation of the layers in a laminate, and the loading. The analytical methods developed in the previous chapters have been concerned with the linear elastic response of anisotropic, laminated materials. In this chapter we shall see that composites often exhibit nonlinear, inelastic response prior to ultimate failure. Typical experimental results for the response of polymeric and metallic matrix composites subjected to tensile, compressive, and shear loadings are presented for unidirectional and laminated composites. Both monotonic and cyclic load histories are included.

The primary purpose of this chapter is to demonstrate the wide range of responses that are possible with fibrous composites. We shall see that tensile loading in the fiber direction of a unidirectional composite is generally controlled by the ultimate strain of the fiber, which is typically quite small (i.e., $\varepsilon_f^{\text{ult}}$ = 1–2%). The transverse ultimate tensile strain of unidirectional composite is also typically quite small, as it is usually controlled by the strength of the fiber/matrix interface. Indeed, for all of the experimental results presented in this chapter, the transverse ultimate tensile strain is less (usually considerably less) than the fiber direction ultimate tensile strain.

The axial and transverse responses of unidirectional composites can be linear or nonlinear depending upon the type of fiber and matrix. The axial response is usually dominated by the fiber response. Composites made using fiber tows (including carbon and aramid) typically exhibit a stiffening response in which the axial modulus may increase by as much as 20% prior to failure. This stiffening is associated with straightening of initially curved filaments in the fiber tows. Polymeric matrix composites typically exhibit linear (or nearly linear) response in the transverse direction, whereas metal matrix composites exhibit nonlinear response in the transverse direction. The shear response is typically nonlinear for all types of unidirectional composites.

Compared with the response of unidirectional composites, the response of laminates is more varied, with some laminates exhibiting exceptionally large, permanent strains prior to failure. The most unusual case to be presented in this chapter is the tensile response of a $[\pm 45]_s$ carbon fiber/polyimide matrix (IM7/K3B) laminate, which exhibits tensile strains in excess of 18% prior to failure (for a material with unidirectional tensile ultimate strains in the fiber and transverse directions that are less than 2.0%).

The chapter is subdivided according to material type. The type of results presented varies depending upon the experimental data that were directly available to the author. All experimental results presented were obtained by the author and his co-workers. It is emphasized that the results presented are a function of the fabrication process and the fiber volume fraction and, therefore, should not be interpreted as *the* properties of a given fiber/matrix combination. Also, the ultimate stress and strain values reported are often a function of the test method employed, with the measured ultimates often being a reflection of the test method as well as the actual material strength.

7.2 Polymer Matrix Composites

7.2.1 Unidirectional Aramid/Epoxy

As indicated in Chapter 1, aramid/epoxy consists of aramid fibers in epoxy matrix. The results presented here are taken from Pindera et al. (1987, 1989). The material was supplied by ALCOA and is the aramid/epoxy used in their hybrid composite ARALL. This aramid/epoxy is very similar to Kevlar, manufactured by Dupont. The fiber volume fraction of the material investigated was $V_f = 0.55$.

7.2.1.1 Tensile Response

Figure 7.1 shows the tensile response of unidirectional, off-axis specimens with fiber orientations ranging from 0° to 90°. Elastic properties and strength values from these tests are presented in Table 7.1. Keeping in mind the difficulties associated with off-axis tensile testing described in Chapter 6, we see that the 0° specimen exhibited the largest ultimate stress (165 ksi, 1137 MPa) as well as the largest ultimate strain (1.8%). Except for the smaller off-axis fiber angles (in the range of 5° to 15°), the responses are nearly linear to failure. However, the 0° specimen does exhibit a significant *increase* in modulus prior to failure. The nonlinearity in the response of the 5°, 10°, and

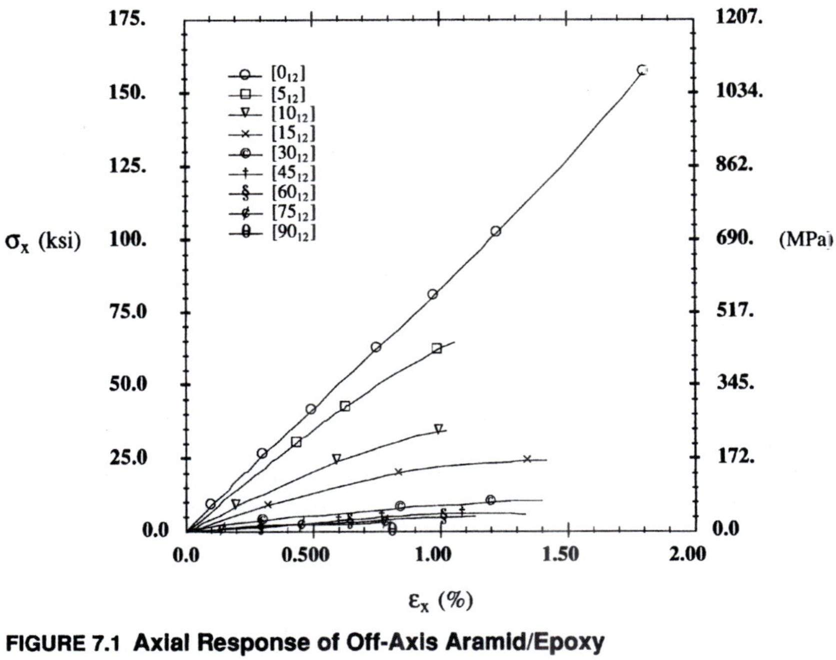

FIGURE 7.1 Axial Response of Off-Axis Aramid/Epoxy

θ, degrees	E_x, GPa (Msi)	ν_{xy}	G_{12}, GPa (Msi)	σ_x^{ult}, MPa (ksi)	ε_x^{ult}, %
0	61.1 (8.8)	0.435	—	1137 (165)	1.8
5	49.4 (7.2)	0.415	1.75 (0.254)	558 (81)	1.5
10	29.4 (4.3)	0.537	1.64 (0.238)	262 (38)	1.0
15	17.9 (2.6)	0.536	1.54 (0.224)	165 (24)	1.5
30	7.0 (1.02)	0.552	1.54 (0.224)	82.7 (12)	1.7
45	4.6 (0.66)	0.495	1.54 (0.224)	48.3 (7)	1.1
60	4.1 (0.59)	0.339	1.55 (0.225)	41.4 (6)	1.3
75	4.3 (0.62)	0.143	1.53 (0.222)	34.5 (5)	0.91
90	4.1 (0.60)	0.056	—	27.6 (4)	0.85
Out-of-Plane Properties					
θ, degrees	E_3, GPa (Msi)	ν_{23}	G_{23}, GPa (Msi)	τ_{23}^{ult}, MPa (ksi)	γ_{23}^{ult}, %
0	4.1 (0.60)	0.37	1.52 (0.22)	27.1 (3.93)	2.6

TABLE 7.1 Aramid/Epoxy Unidirectional Properties

15° specimens is associated with the high shear stress in these specimens (see Chapter 6). The 90° specimen exhibits the smallest ultimate values (4 ksi, 27.6 MPa; 0.85%).

Figure 7.2 shows a comparison of the plane stress predictions (from Eq. (4.58)) and the experimental results for the modulus of unidirectional off-axis aramid/epoxy. As indicated in the figure, the correlation between theory and experiment is excellent for all fiber angles.

Figure 7.3 shows the correlation between theory and experiment for Poisson's ratio ν_{xy} from the same tests. The correlation is very good, but not as good as for the axial modulus. The discrepancies are due mainly to factors associated with experimental accuracy.

The axial response for cyclic loading and unloading of a 10° off-axis aramid/epoxy tensile specimen is shown in Fig. 7.4. As indicated in the figure, nonlinear response and permanent (inelastic) strains are evident. There is some hysteresis and a small decrease in axial modulus over the cyclic load history. The inelastic strains can be associated with inelastic response of the matrix as well as damage development (see Chapter 9).

7.2.1.2 Shear Response

Comparison of the axial shear response, τ_{12} versus γ_{12}, obtained from 0° and 90° Iosipescu specimens as well as 10° and 45° off-axis tensile tests, is shown in Fig. 7.5. The results presented in the figure are typical; ultimate values from replicate tests varied considerably. The ultimate values in Table 7.2 correspond to the maximum values from all tests. It is clear from this figure and the values presented in Table 7.2 that the 0° Iosipescu specimen is the preferred test method for obtaining the fullest extent of the shear response from unidirectional specimens without inducing early failure due to specimen configuration. All of the off-axis tensile specimens and the 90° Iosipescu specimen fail at much lower ultimate stress values due to the presence of high transverse stresses associated with the specimen configuration. The high transverse stresses are present near the grips in the off-axis specimens and at the notch tip for the 90° Iosipescu specimen.

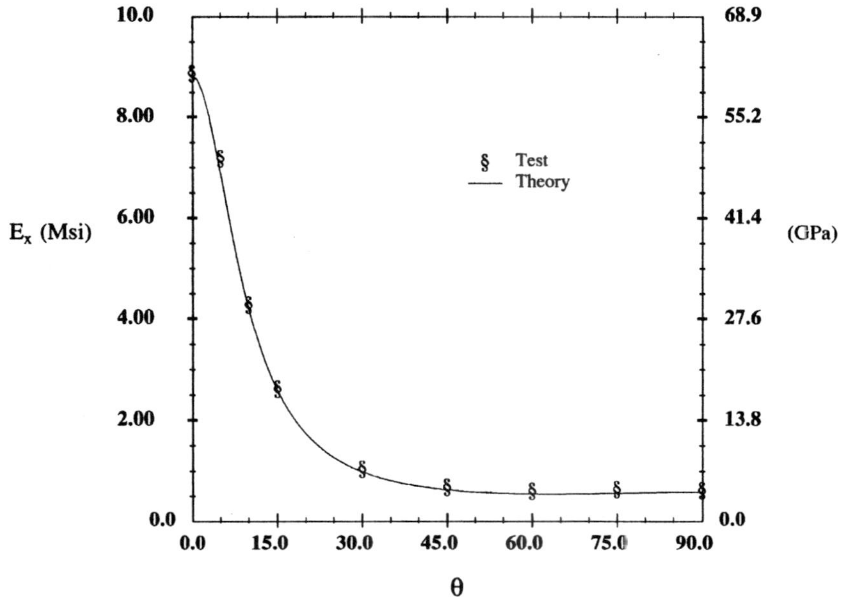

FIGURE 7.2 Aramid/Epoxy Off-Axis Modulus: Theory and Experiment

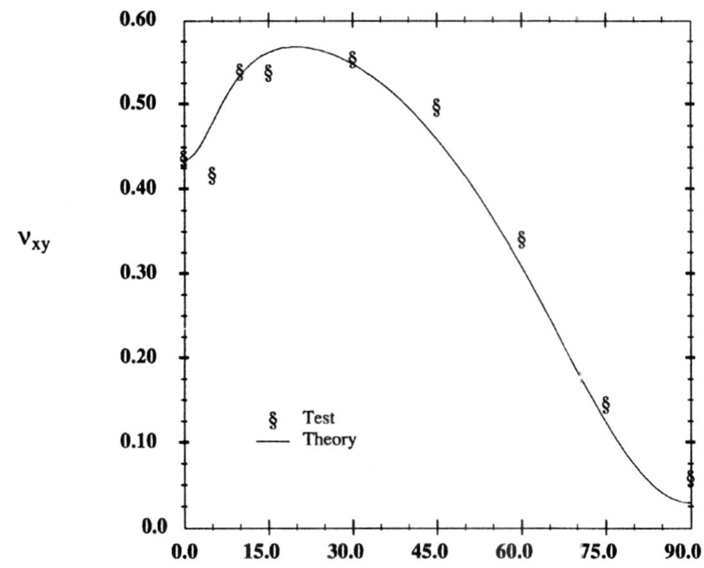

FIGURE 7.3 Aramid/Epoxy Off-Axis Poisson's Ratio Comparisons

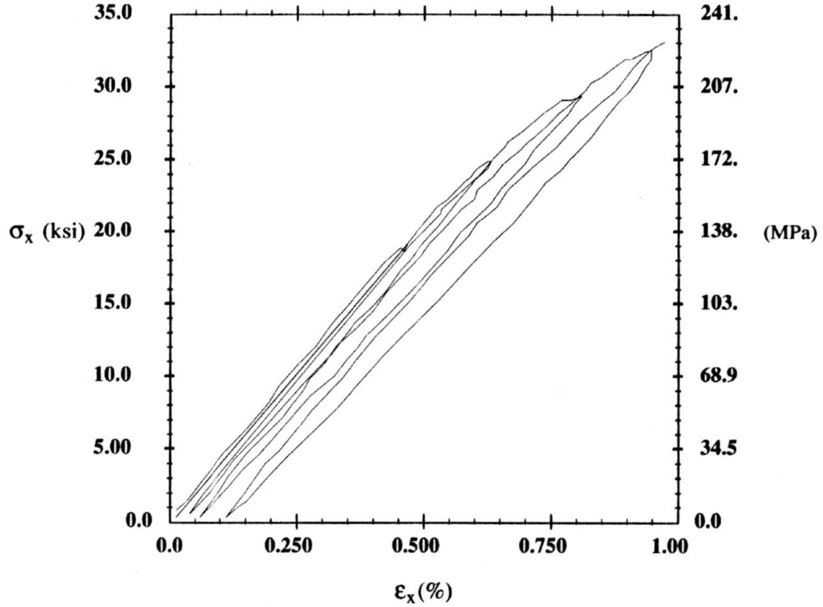

FIGURE 7.4 Aramid-Epoxy 10° Cyclic Axial Response

θ, degrees	Specimen Type	G_{12}, GPa (Msi)	τ_{12}^{ult}, MPa (ksi)	γ_{12}^{ult}, %
Axial Shear				
0	Iosipescu	1.48 (0.215)	46.5 (6.75)	>6.0
90	Iosipescu	1.48 (0.216)	25.6 (3.72)	>2.0
5	Off-Axis	1.54 (0.223)	48.0 (6.96)	>3.0
10	Off-Axis	1.55 (0.225)	43.8 (6.36)	>5.0
15	Off-Axis	1.51 (0.219)	40.2 (5.83)	>5.0
30	Off-Axis	1.54 (0.223)	34.4 (4.99)	3.3
45	Off-Axis	1.54 (0.224)	24.3 (3.52)	1.75
60	Off-Axis	1.55 (0.225)	18.1 (2.63)	1.45
75	Off-Axis	1.53 (0.222)	8.5 (1.23)	0.7
Transverse Shear				
θ, degrees	Specimen Type	G_{23}, GPa (Msi)	τ_{23}^{ult}, MPa (ksi)	γ_{23}^{ult}, %
0	Iosipescu	1.52 (0.22)	27.1 (3.93)	2.6

TABLE 7.2 Aramid/Epoxy Shear Properties

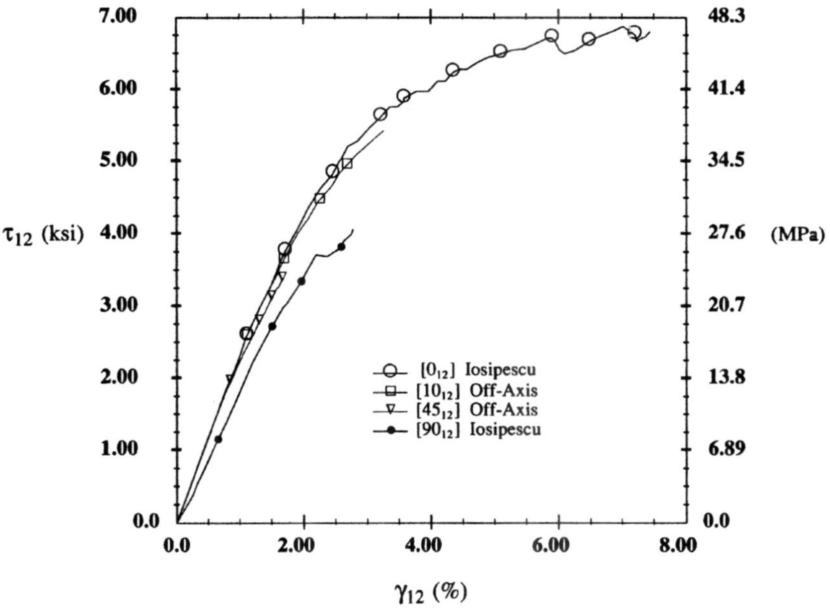

FIGURE 7.5 Aramid/Epoxy Shear Response Comparisons

The shear response is nonlinear, with ultimate shear strains in excess of 6% for the 0° Iosipescu specimen. Damage is also evident in the unstable response at the highest stress levels of the Iosipescu specimens. The off-axis tensile specimens exhibited maximum shear strains in a range from 0.7% to more than 5%.

As shown in Fig. 7.6, cyclic shear response exhibits hysteresis and large permanent strains after unloading. Thus the response is inelastic. There is also some degradation in shear modulus at higher stress levels.

The transverse (out-of-plane) shear response, τ_{23} versus γ_{23}, of aramid/epoxy is nonlinear, with hysteresis and permanent strains after unloading as exhibited in Fig. 7.7. There is degradation in the shear modulus G_{23} of approximately 24% due to damage development prior to complete failure. The results in this figure were obtained using an Iosipescu specimen with the fibers oriented perpendicular to the plane of the specimen. As indicated in Table 7.2, the initial transverse shear modulus G_{23} is nearly the same as the axial shear modulus G_{12}, but the ultimate stress and strain are much lower for transverse shear loading.

7.2.2 Carbon/Epoxy

Experimental results for the tensile response of unidirectional and laminated T300/5208 carbon/epoxy are presented in this section. All tests were conducted on finite-width coupons. Figure 7.8 shows typical results for the axial response of 0°, 90°, 10°, and 30° unidirectional, off-axis specimens. The response is nearly linear for all four fiber orientations, with the largest deviation from linearity being exhibited by the 0° specimen, which exhibits a stiffening with increasing stress. As mentioned previously, this stiffening is associated with fiber straightening. It is evident from the figure and the values given in Table 7.3 from a series of tests that all specimens failed at low strain

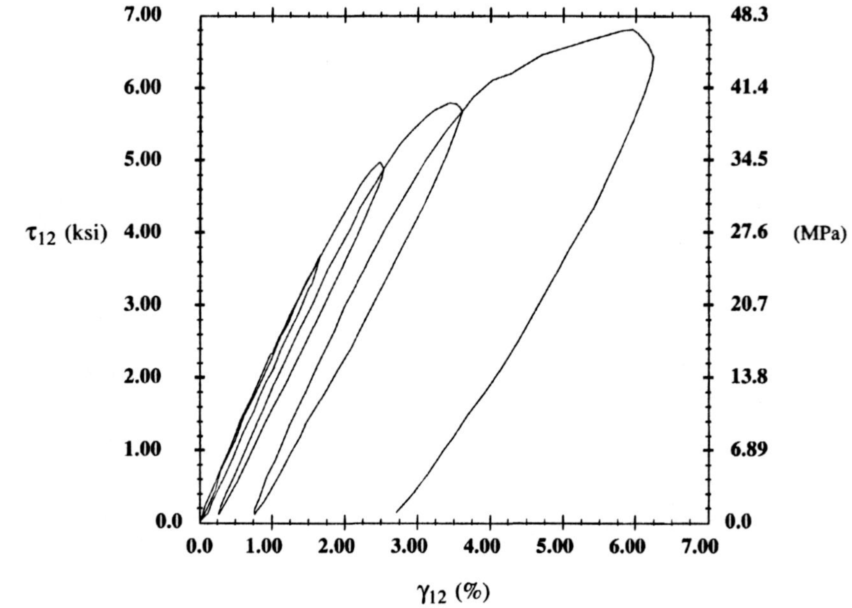

FIGURE 7.6 Cyclic Axial Shear Response of Aramid/Epoxy

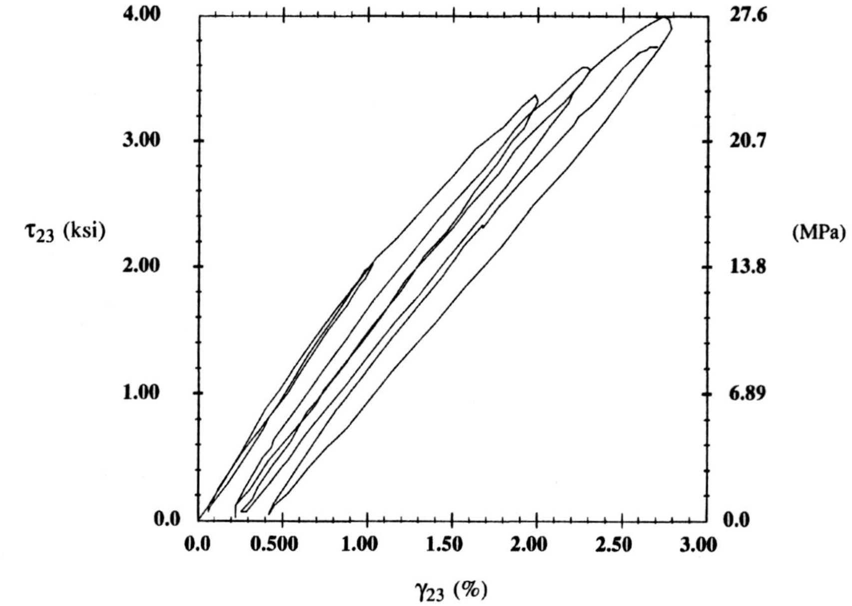

FIGURE 7.7 Aramid/Epoxy Transverse Shear Response

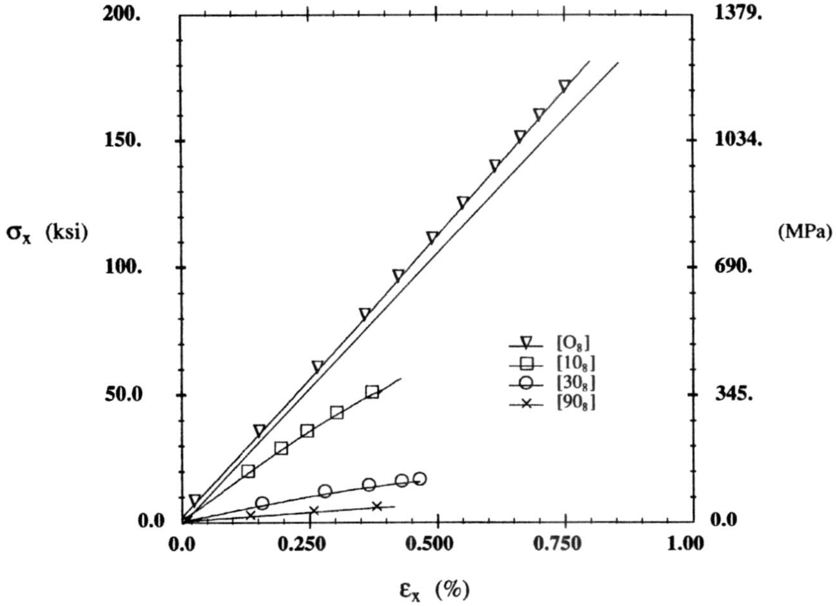

FIGURE 7.8 Unidirectional T300/5208 Tensile Response

θ, degrees	E_x, GPa (Msi)	ν_{xy}	G_{12}, GPa (Msi)	σ_x^{ult}, MPa (ksi)	ε_x^{ult}, %
0	136.5 (19.8)	0.350	NA	1551 (225)	1.05
10	88.9 (12.9)	0.430	6.38 (0.925)	406 (59)	0.57
30	23.2 (3.36)	0.391	5.42 (0.786)	122 (17.7)	0.70
90	9.8 (1.42)	0.047	NA	52 (7.5)	0.50

TABLE 7.3 Properties of Unidirectional T300/5208

levels. The 0° specimen ultimate strain is approximately 1%, with the ultimate strain of other specimens well below 1%. These results clearly show the brittle response of unidirectional carbon/epoxy.

The tensile responses of several T300/5208 laminates are shown in Fig. 7.9. Included in the figure are results for four angle-ply and two quasi-isotropic laminates. The results show that the response varies significantly with fiber orientation. Results for two different stacking sequences with the same fiber orientations are shown for ±10° angle-ply and 0/±45/90 quasi-isotropic laminates. The results in Fig. 7.9 indicate that stacking sequence has little influence on the response until damage begins to have an effect. For the ±10° angle-ply laminates, the response is nearly linear to failure and the moduli are the same, within experimental scatter, for both stacking sequences. However, stacking sequence does influence the ultimate stress by as much as 20%. In the case of the quasi-isotropic laminates, one stacking sequence exhibits essentially linear response to failure,

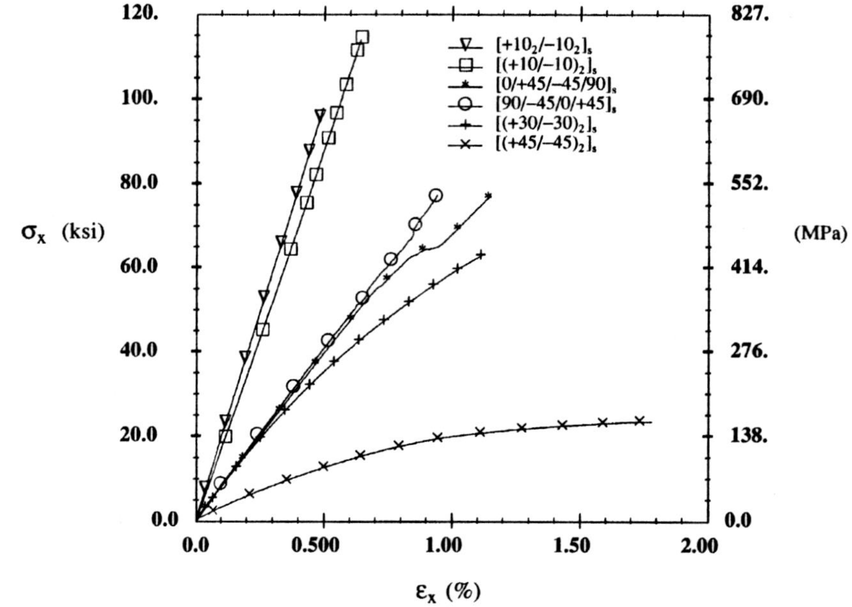

FIGURE 7.9 T300/5208 Laminate Tensile Response

whereas the second stacking sequence exhibits stiffness degradation in advance of final failure. As discussed in Chapter 8, the stacking sequence–dependent differences are the result of the three-dimensional state of stress that is present in a boundary layer along the free edge. It is also evident from this figure that the ±30° and ±45° angle-ply laminates exhibit nonlinear response, with the ±45° laminate exhibiting significantly larger failure strain.

The axial and transverse strains from tensile tests on a $[(45/-45)_2]_s$ and a $[(30/-30)_2]_s$ laminate are shown in Figs. 7.10 and 7.11, respectively. These figures along with the results in Table 7.4 provide experimental verification of the lamination theory predictions (Fig. 5.12) for Poisson's ratio as a function of fiber orientation for angle-ply laminates. For $\theta = 45°$, Poisson's ratio is approximately 0.75, and at $\theta = 30°$ Poisson's ratio is approximately 1.25. Thus, laminates can

Laminate	E_x, GPa (Msi)	ν_{xy}	σ_x^{ult}, MPa (ksi)	ε_x^{ult}, %
$[10_2/-10_2]_s$	127 (18.5)	0.59	655 (95)	0.51
$[(10/-10)_2]_s$	123 (17.8)	0.56	793 (115)	0.66
$[(30/-30)_2]_s$	45 (6.6)	1.12	448 (65)	1.24
$[(45/-45)_2]_s$	16 (2.4)	0.76	145 (21)	1.64
$[0/\pm 45/90]_s$	52 (7.5)	0.27	510 (74)	1.12
$[90/-45/0/45]_s$	51 (7.4)	0.30	517 (75)	0.98

TABLE 7.4 Properties of T300/5208 Laminates

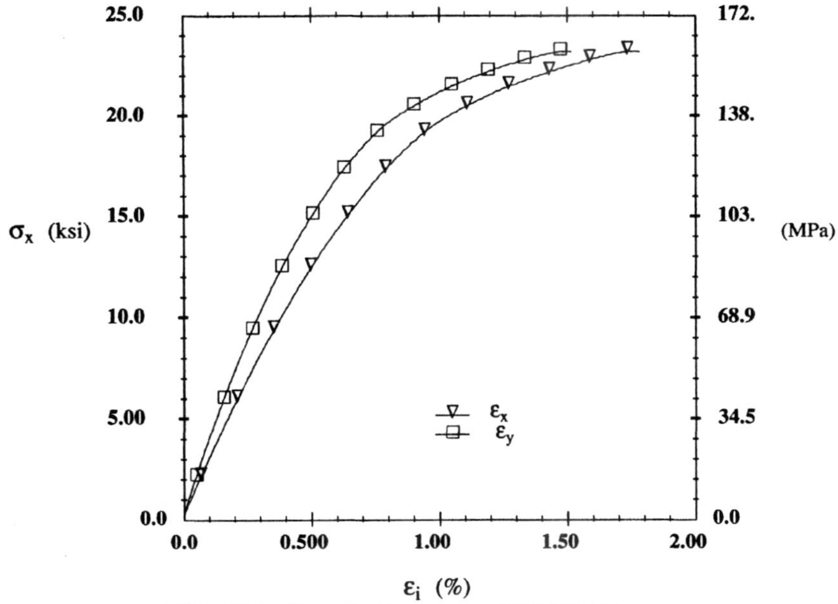

FIGURE 7.10 $[(45/-45)_2]_s$ Tensile Response: T300/5208

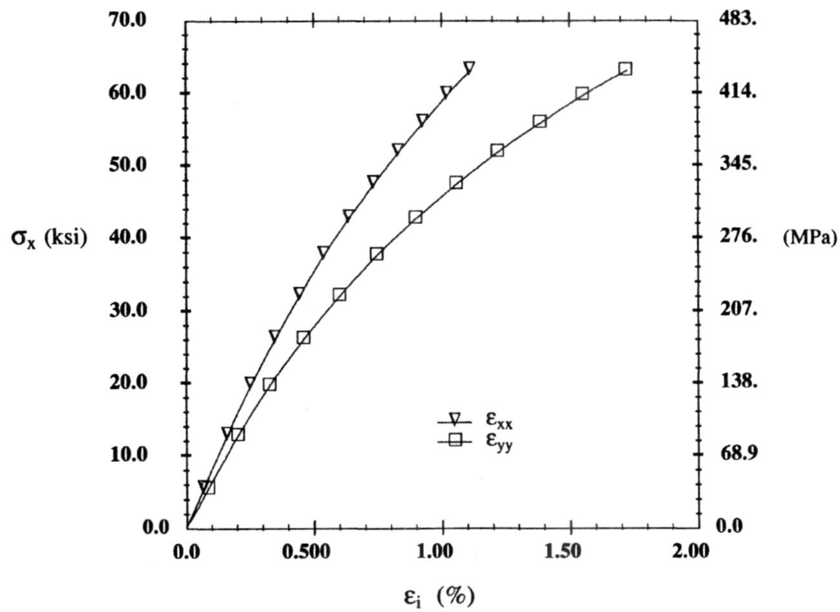

FIGURE 7.11 $[(30/-30)_2]_s$ Tensile Response: T300/5208

exhibit a Poisson's ratio greater than 1.0. As shown in Fig. 7.11, the transverse strain continues to be larger than the axial strain in the nonlinear range for some fiber orientations. This type of Poisson response is quite unlike that observed for isotropic materials.

7.2.3 Carbon/Polyimide

Carbon/polyimide composites are fabricated using carbon fibers and polyimide matrix. The principal reason for selecting this particular resin matrix composite is the higher use temperature (~300°C, 572°F) as compared with epoxy matrix materials (~175°C, 347°F).

7.2.3.1 IM7/K3B Tensile Response

The IM7/K3B composite is fabricated using IM7 carbon fibers and Avimid K3B polyimide matrix. Figures 7.12 and 7.13 show the tensile response of unidirectional and laminated IM7/K3B as obtained by Schroedter and Herakovich (1995) and Wyatt (1997). Figure 7.12 shows the response for strains up to 2%, and Fig. 7.13 shows the complete response for the laminates tested. Table 7.5 provides moduli, Poisson's ratios, and ultimate stress and strain values for the laminates considered. As indicated in the figures and the table, the response of the $[(\pm45)_3]_s$ laminate is significantly different from that of the other laminates in several respects. It has an ultimate strain in excess of 18%, whereas all other laminates fail at strains less than 2%. Also, its response is highly nonlinear, with large permanent strains after unloading (Fig. 7.13). All other laminates of this group exhibited linear or only slightly nonlinear response (Fig. 7.12) and no or small permanent strain after unloading. The large strain to failure of the $[(\pm45)_3]_s$ laminate was accompanied by significant fiber rotation. Just prior to failure the specimen behaved more like a $[(\pm35)_3]_s$ specimen than a $[(\pm45)_3]_s$. The

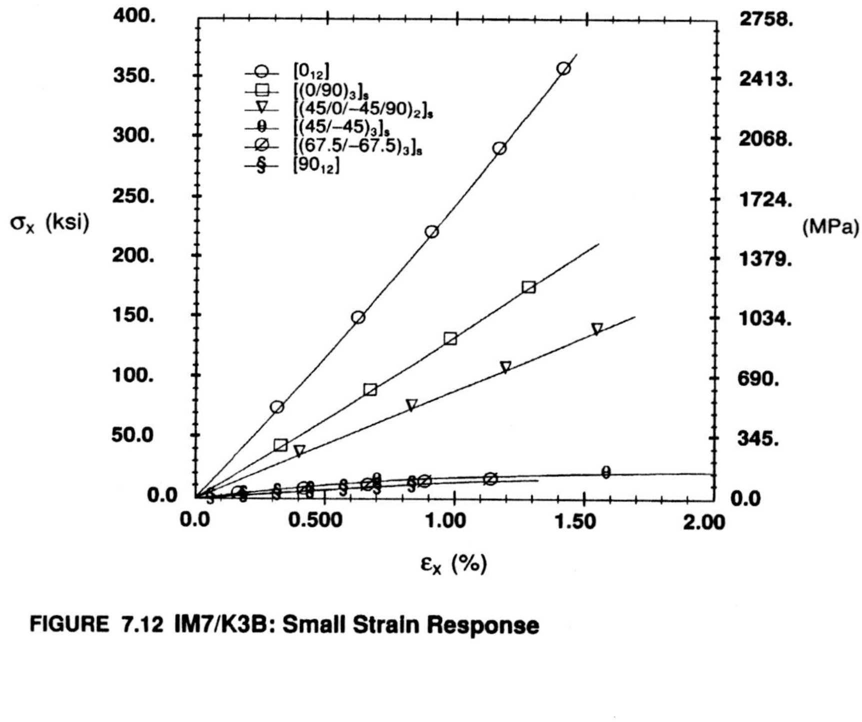

FIGURE 7.12 IM7/K3B: Small Strain Response

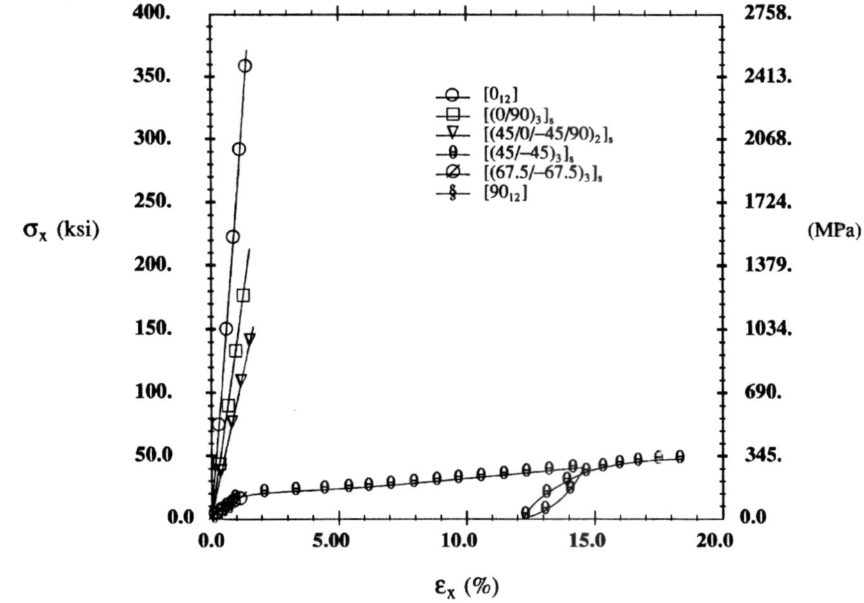

FIGURE 7.13 IM7/K3B: Full Strain Response

Laminate	E_x, GPa (Msi)	ν_{xy}	σ_x^{ult}, MPa (ksi)	ε_x^{ult}, %
$[(\pm 45)_3]_s$	19.8 (2.875)	0.78	333 (48.3)	18.3
$[(0/90)_3]_s$	79.7 (11.56)	0.05	1469 (213.0)	1.60
$[(\pm 67.5)_3]_s$	10.6 (1.56)	0.15	88 (12.7)	0.97
$[(45/0/-45/90)_2]_s$	56.8 (8.24)	0.25	965 (140.0)	1.72
$[0_{12}]$	159.3 (23.1)	0.33	2565 (372.0)	1.44
$[90_{12}]$	8.9 (1.29)	0.014	37 (5.44)	0.42

TABLE 7.5 IM7/K3B Tensile Properties

unloading in Fig. 7.13 is indicative of the permanent strain attainable in this laminate. Note the very high ultimate stress of the [0] specimen for this material, in excess of 2565 MPa (372 ksi).

7.2.3.2 C6000/PMR-15 Tensile Response

Experimentally determined tensile properties of unidirectional off-axis Celion 6000/PMR-15 (Pindera and Herakovich, 1981) are presented in Table 7.6, and experimental laminate properties (Klang, 1981) are presented in Table 7.7. The experimentally determined values for axial modulus and Poisson's ratios are compared with lamination theory in Figs. 7.14 and 7.15. As these curves indicate, there is excellent correlation between the theoretical and experimental for axial modulus and Poisson's ratios. The largest discrepancy between theory and experiment is for Poisson's ratio

θ, degrees	E_x, GPa (Msi)	ν_{xy}	G_{12}, GPa (Msi)	σ_x^{ult}, MPa (ksi)	ε_x^{ult}, %
0	136.5 (19.8)	0.35	—	1551 (225)	1.05
5	118.6 (17.2)	0.41	7.38 (1.07)	703 (102)	0.66
10	88.9 (12.9)	0.43	6.38 (0.925)	407 (59)	0.57
15	58.6 (8.5)	0.45	5.92 (0.858)	283 (41)	0.64
30	23.2 (3.36)	0.39	5.42 (0.786)	122 (17.7)	0.70
45	13.4 (1.95)	0.34	4.98 (0.723)	83 (12.1)	0.55
60	10.5 (1.53)	0.22	4.89 (0.710)	61 (8.8)	0.63
75	10.0 (1.45)	0.10	4.92 (0.713)	62 (9.0)	0.66
90	9.8 (1.42)	0.05	—	52 (7.5)	0.50

TABLE 7.6 Tensile Properties of Unidirectional C6000/PMR15

Laminate	E_x, GPa (Msi)	ν_{xy}	σ_x^{ult}, MPa (ksi)	ε_x^{ult}, %
$[(\pm 10)_2]_s$	125.5 (18.2)	0.63	862 (125)	0.7
$[(\pm 30)_2]_s$	48.3 (7.0)	1.31	496 (72)	1.4
$[(\pm 45)_2]_s$	17.9 (2.6)	0.83	131 (19)	1.9
$[(\pm 60)_2]_s$	12.4 (1.8)	0.30	76 (11)	0.9

TABLE 7.7 Tensile Properties of C6000/PMR15 Laminates

at small off-axis angles. This discrepancy is most likely due to the lack of a uniform axial stress state in low-angle off-axis specimens as discussed in Section 6.4.3.2.

Comparison of the ultimate tensile stress for IM7/K3B and Celion 6000/PMR-15 clearly shows the influence of the fiber strength on the strength of fibrous composites. The strength of the IM7/K3B is 1.65 times that of the Celion 6000/PMR-15.

7.3 Metal Matrix Composites

The responses of two different metal matrix composites are presented in the following sections. One of the materials, boron/aluminum, does not exhibit any substantial damage prior to failure, whereas the other material, silicon-carbide/titanium, exhibits a high degree of damage prior to failure. Both materials exhibit a high degree of nonlinear response for loading other than in the fiber direction. Only a small degree of nonlinearity is observed for fiber-direction loading because the response is dominated by the fibers, which are linear elastic. For boron/aluminum the presentation is limited to unidirectional material under tensile, compressive, and shear loadings. For silicon-carbide/titanium both unidirectional and laminated materials are considered, including biaxial loading of a $[\pm 45]_s$ tube.

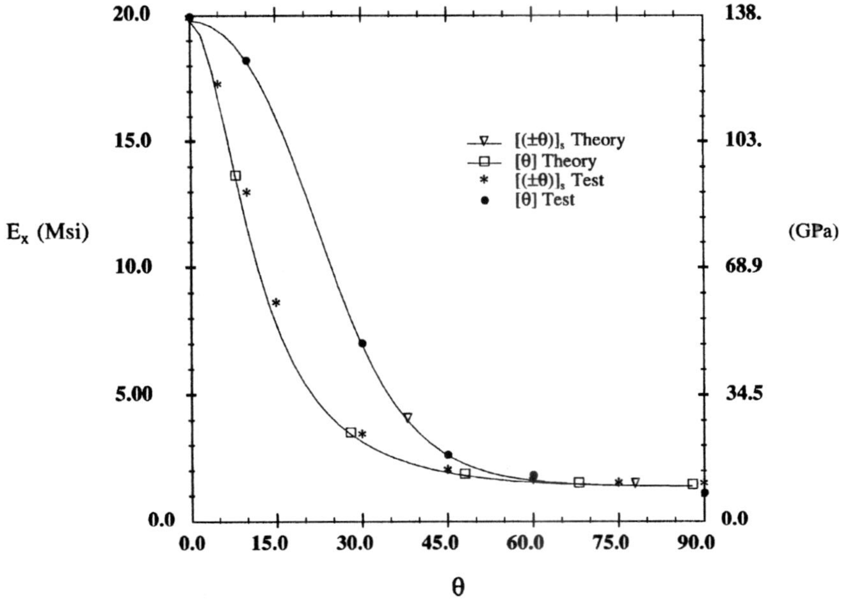

FIGURE 7.14 Axial Modulus Comparisons: C6000/PMR15

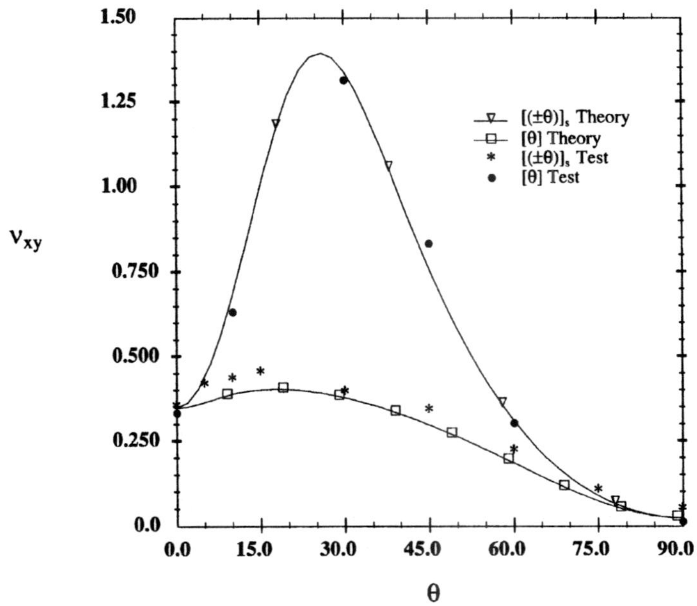

FIGURE 7.15 Poisson's Ratio Comparisons: C6000/PMR15

7.3.1 Unidirectional Boron/Aluminum

The following experimental results for the response of unidirectional boron/aluminum are taken from Becker et al. (1987). The material was fabricated (by DWA Composite Specialty, Inc.) using 0.005 in (142 μm) boron fibers in 6061 0T aluminum alloy. The fiber volume fraction was V_f = 0.46.

7.3.1.1 Tensile Response

Typically, two monotonic tension tests and one cyclic tension test were conducted for each fiber orientation. The small number of tests was dictated by the small amount of material available. Elastic properties and maximum stress and strain values for tension and compression tests are presented in Table 7.8. The elastic properties were very consistent for all tests of a given fiber orientation, but the failure values varied considerably between the two monotonic tests and the cyclic tests. The values presented in the table are the averages for the elastic properties and the maximum values of all tests for the ultimate stresses and strains. The maximum stress and strain attained are presented to demonstrate the capability of the material and because, as described in Chapter 6, tests on composites often exhibit grip failures and debonding of strain gages, which nullifies accurate measurement of the actual material failure. In most cases, the maximum strains from cyclic tests were considerably higher than those from the corresponding monotonic test.

Figure 7.16 shows representative results for the tensile response of B/Al specimens with fiber orientations 0°, 10°, 15°, 30°, 45°, 60°, and 90°. All specimens exhibit some nonlinear response, with that in the 0° specimen being quite small, reflecting the fact that the response of this specimen is fiber dominated. For fiber orientations of 30° and larger, the response is highly nonlinear, approaching elastic–perfectly plastic response as the fiber angle is increased. The 45° specimen exhibits the largest strain to failure. For fiber orientations greater than 45°, the specimens fail at low strains due to the high transverse stresses that develop as the fiber orientation is increased. The highly nonlinear response of specimens with fiber orientations of 30° and larger reflects the matrix dominance of these specimens. The response approaches that of pure aluminum, but with lower strain to failure because of the low fiber/matrix interfacial strength. For intermediate angles the nonlinear response reflects the influence of shear stress as well as transverse stress.

The nonlinearity of this material system is associated with matrix plasticity and not damage, as evidenced by the fact that there is no degradation of stiffness during unloading. Figure 7.17 shows the cyclic, loading, and unloading tensile response of an off-axis 45° specimen. Careful study of the

	Tension				Compression		
θ, degrees	E_x, GPa (Msi)	ν_{xy}	σ_x^{ult}, MPa (ksi)	ε_x^{ult}, %	E_x, GPa (Msi)	σ_x^{ult}, MPa (ksi)	ε_x^{ult}, %
0	227 (32.9)	0.237	1606 (233)	1.07	230 (33.3)	1338 (194)	0.61
10	216 (31.4)	0.257	669 (97)	1.69	221 (32.1)	1269 (184)	0.77
15	211 (30.6)	0.272	462 (67)	1.87	215 (31.2)	889 (129)	0.74
30	183 (26.5)	0.318	193 (28)	0.78	186 (27.0)	545 (79)	2.11
45	152 (22.0)	0.317	172 (25)	1.78	161 (23.4)	386 (56)	7.81
60	137 (19.9)	0.271	124 (18)	0.39	NA	NA	NA
90	139 (20.2)	0.150	145 (21)	0.26	141 (20.4)	310 (45)	2.59

TABLE 7.8 Axial Properties of Unidirectional Boron/Aluminum

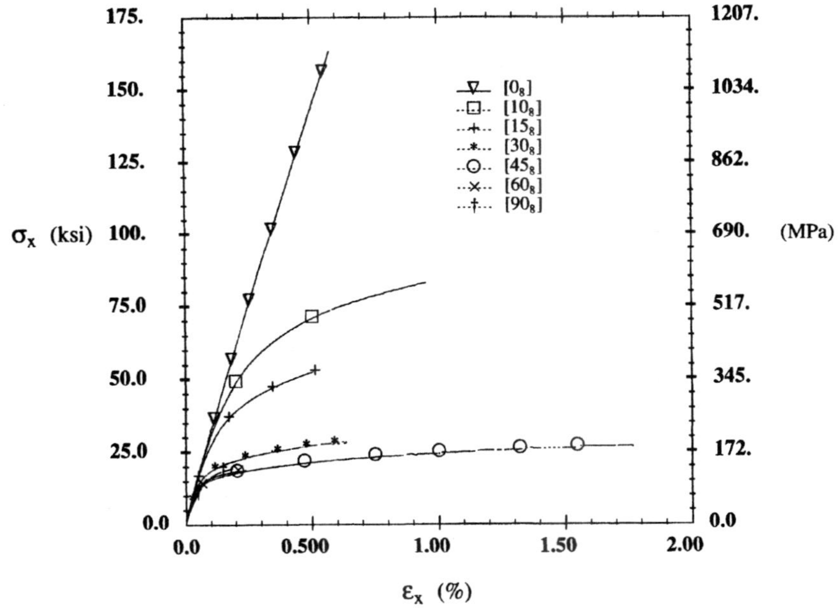

FIGURE 7.16 Tensile Response of Boron/Aluminum

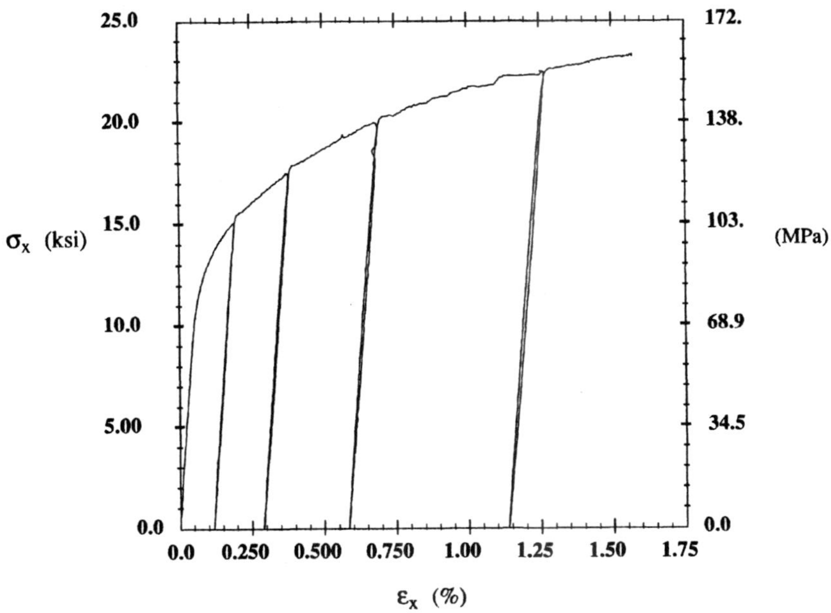

FIGURE 7.17 Cyclic Response of 45° Off-Axis Boron/Aluminum

results indicates that there is no degradation of modulus with increasing stress level. These results indicate that there is no damage to the material; the nonlinearity is due to inelastic response of the matrix. A small amount of hysteresis is noted in the final loading cycle.

The elastic properties determined from these tests exhibit very good correlation with theory. Figures 7.18 and 7.19 show comparisons of theory and experiment for axial modulus and Poisson's ratio, respectively. The θ dependence correlation was very similar to that in Figs. 7.2 and 7.3 for aramid/epoxy (after appropriate adjustment for magnitudes).

7.3.1.2 Shear Response

The shear response of unidirectional boron/aluminum as determined from a 0° Iosipescu shear specimen and several off-axis tension specimens is shown in Fig. 7.20. As was the case for aramid/epoxy (Fig. 7.5), the 0° Iosipescu specimen most consistently provides the largest strains to failure (Table 7.9). The response is highly nonlinear, with the nonlinearity being associated with plastic response of the matrix as indicated by the constant shear modulus of cyclic shear loading of Fig. 7.21. The variable results for the different off-axis specimens under tensile loading is a direct result of the ratio of transverse stress σ_2 to shear stress τ_{12}, as a function of fiber orientation θ. From the stress transformation equations (6.9) for uniform far-field stress σ_x, this ratio is

$$\frac{\sigma_2}{\tau_{12}} = \tan\theta \tag{7.1}$$

Thus, the ratio increases with increasing angle. For angles less than 45° the transverse stress is less than the shear stress, and the opposite is true for angles greater than 45°.

The axial shear modulus G_{12} as determined from the off-axis tension tests varied between 8.2 Msi and 8.6 Msi. The Iosipescu specimen results ranged from 5.7 Msi to 6.7 Msi. This differ-

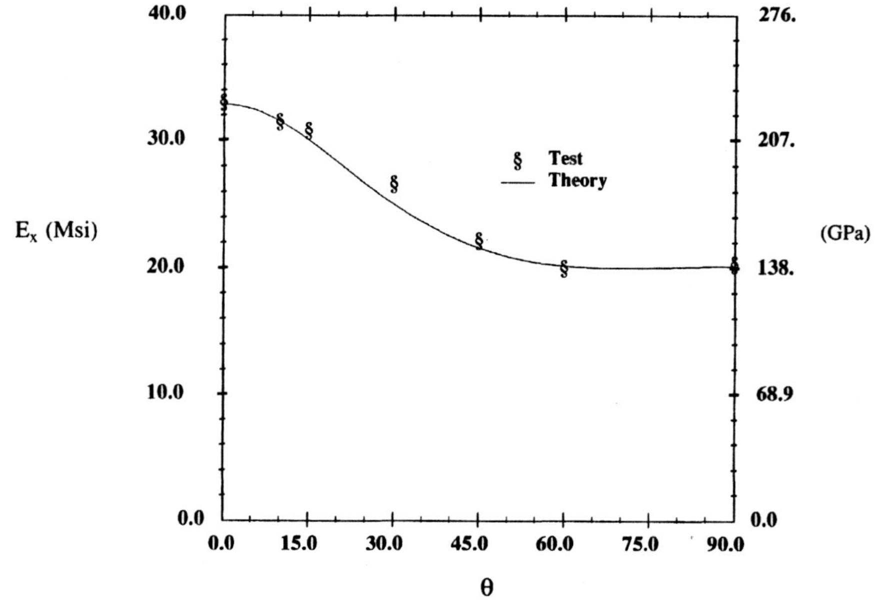

FIGURE 7.18 Boron/Aluminum Off-Axis Modulus: Theory and Experiment

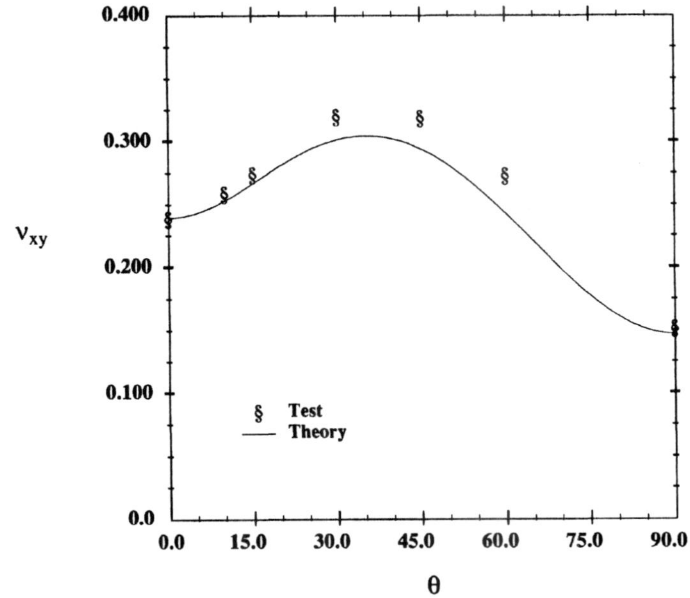

FIGURE 7.19 Boron/Aluminum Off-Axis ν_{xy}: Theory and Experiment

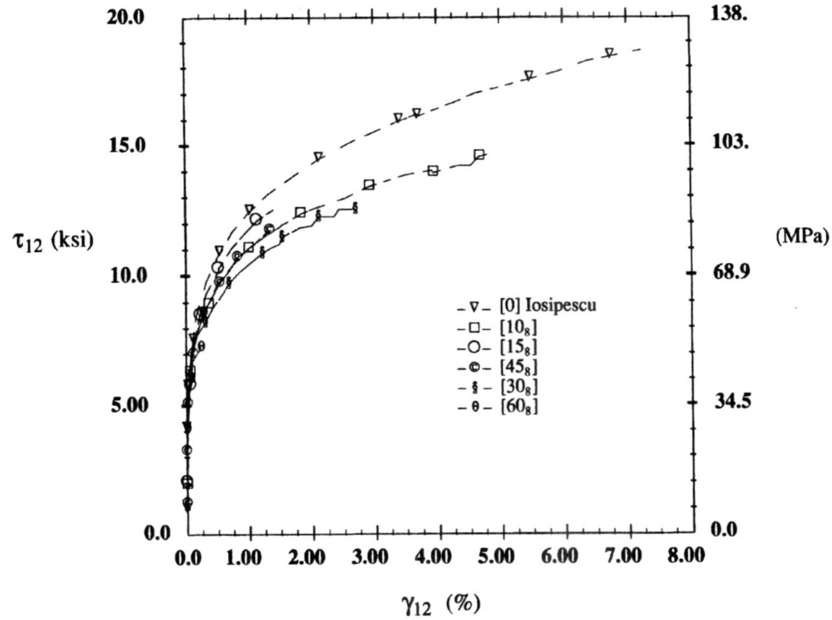

FIGURE 7.20 Boron/Aluminum Axial Shear Response

θ, degrees	G_{12}, GPA (Msi)	τ_{12}^{y}, MPa (ksi)	γ_{12}^{y}, %	τ_{12}^{max}, MPa (ksi)	γ_{12}^{max}, %
Off-Axis Tension					
10	58.6 (8.5)	21.4 (3.1)	0.046	108 (15.7)	5.5
15	57.9 (8.4)	24.8 (3.6)	0.046	111 (16.1)	5.7
30	59.3 (8.6)	22.7 (3.3)	0.039	78.6 (11.4)	1.3
45	57.9 (8.4)	24.1 (3.5)	0.043	86.2 (12.5)	2.8
60	56.5 (8.2)	22.1 (3.2)	0.041	52.4 (7.6)	0.28
Iosipescu					
0	46.2 (6.7)	26.2 (3.8)	0.055	137 (19.9)	6.5
75	44.1 (6.4)	22.7 (3.3)	0.050	99 (14.4)	9.0
80	45.5 (6.6)	16.5 (2.4)	0.035	120 (17.4)	4.0
85	46.2 (6.7)	9.6 (1.4)	0.021	121 (17.5)	3.8
90	39.3 (5.7)	5.5 (0.8)	0.014	105 (15.2)	6.9

TABLE 7.9 Shear Properties of Boron/Aluminum

ence is a function of the specimen and material type, as has been discussed by Pindera (1989). We note that the shear modulus G_{12} of boron/aluminum is more than 30 times the shear modulus of aramid/epoxy.

The results in Fig. 7.21 show that the cyclic, axial shear response is highly nonlinear with some strain hardening. Close examination of this figure indicates that the unloading shear modulus remains constant and equal to the initial elastic shear modulus. This is a clear indication that no significant damage has occurred due to shear loading of this boron/aluminum composite. It is also noteworthy that the response during unloading/reloading is very stable and exhibits no hysteresis.

7.3.1.3 Compressive Response

Curves showing the compressive response of unidirectional boron/aluminum for fiber angles ranging from 0° to 90° are shown in Fig. 7.22. These compressive results were obtained from very short specimens to reduce the potential for buckling. Thus end effects have some influence on the results. Nevertheless, the results are consistent with theory and with the tensile results.

The response is nonlinear for all fiber orientations, with the yield stress decreasing as the fiber angle increases; the exception is the 45° specimen, which exhibited the lowest yield stress and the largest strains to failure. Representative curves comparing the compressive response to the tensile response of unidirectional boron/aluminum are presented in Figs. 7.23 and 7.24. These results show that the tensile and compressive moduli are essentially identical for any given fiber orientation. The response is nonlinear for all specimens, with the proportional limit decreasing as the fiber orientation increases. The tensile ultimates are typically lower than the compressive ultimates due to transverse tensile failures.

7.3.2 Silicon-Carbide/Titanium

Results are presented in this section for two types of silicon-carbide/titanium, $\Sigma/\beta21s$ (sigma fiber in $\beta21s$ titanium matrix) and SCS-6/Ti-15-3 (SCS-6 fiber in Ti-15-3 titanium matrix). The fiber volume fraction of the $\Sigma/\beta21s$ investigated ranged from 29 to 34%, and the fiber volume fraction of the SCS-6/Ti-15-3 was 40%.

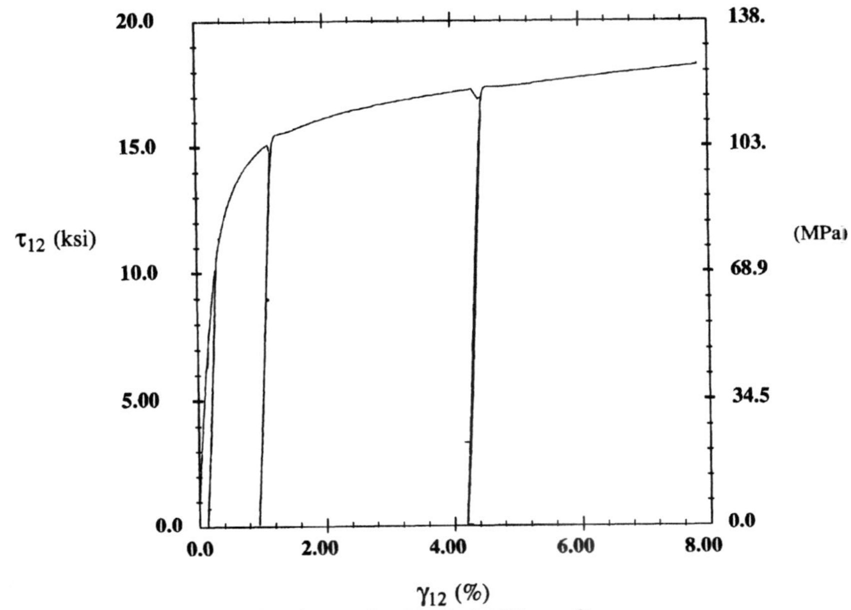

FIGURE 7.21 Boron/Aluminum Cyclic Axial Shear Response

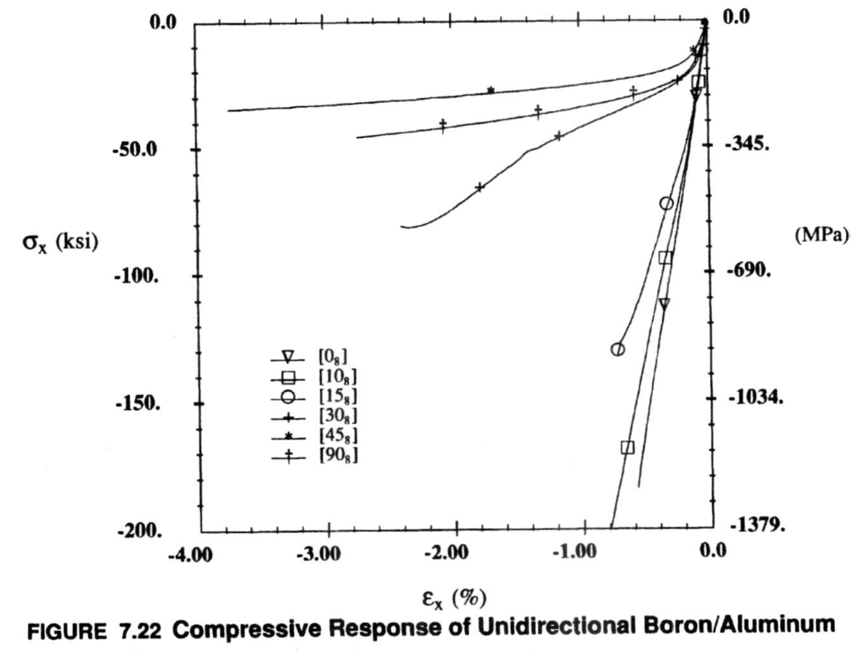

FIGURE 7.22 Compressive Response of Unidirectional Boron/Aluminum

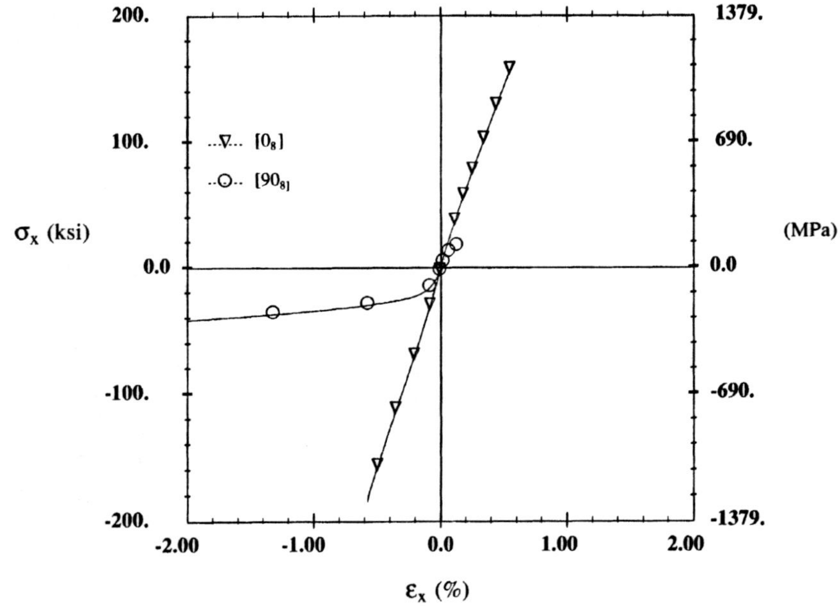

FIGURE 7.23 Tensile and Compressive Response of 0° and 90° Boron/Aluminum

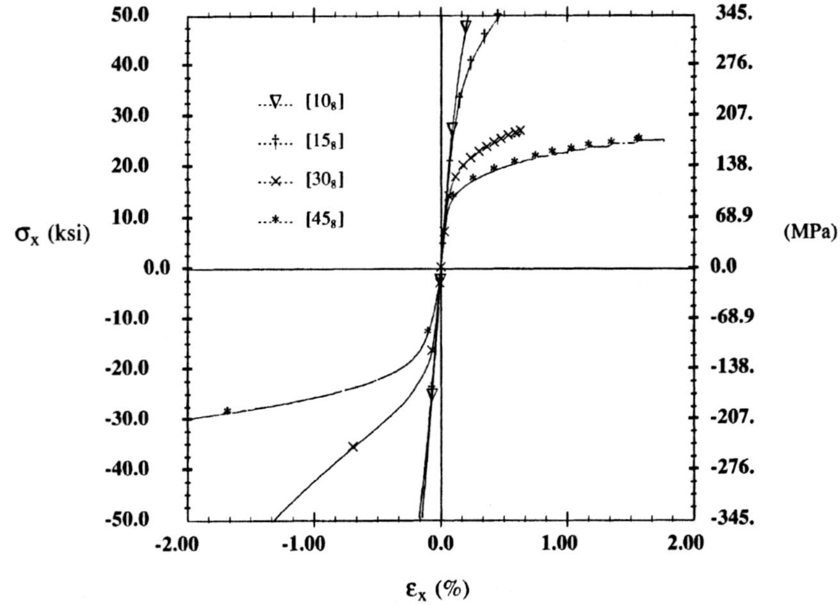

FIGURE 7.24 Tensile and Compressive Response of Off-Axis Boron/Aluminum

The tensile responses for three fiber orientations of unidirectional Σ/β21s silicon-carbide/titanium, [0], [45], and [90], are shown in Fig. 7.25. Property values from these tests are presented in Table 7.10. The figure shows that, as expected, the response varies considerably with fiber orientation, with the 45° off-axis specimen susceptible to early failure due to stress concentrations at the grip. The ultimate strain in the fiber direction for the [0] specimen is less than 1%, but still considerably more than the ultimate strain for the [45] and [90] specimens. The response is nonlinear for all three fiber orientations, with the proportional limit being a strong function of the fiber orientation. The transverse response is essentially bilinear.

The tensile responses of three Σ/β21s laminates, $[0/90]_s$, $[\pm 45]_s$, and $[0/\pm 45/90]_s$, are shown in Fig. 7.26. Again, the response for all three laminates is nonlinear, with the $[\pm 45]_s$ laminate exhibiting the largest degree of nonlinearity and largest strain to failure (>2%). The $[0/90]_s$ and $[0/\pm 45/90]_s$ laminates failed at very low strain levels (see Table 7.10).

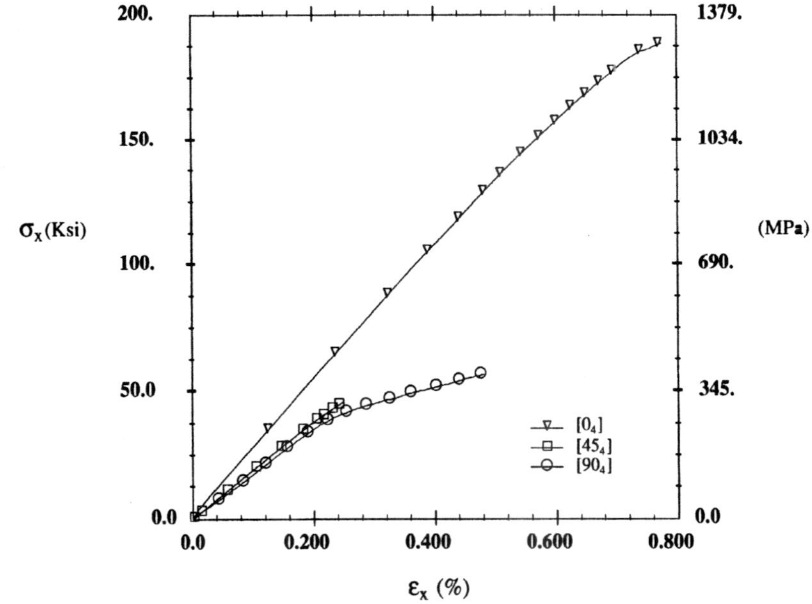

FIGURE 7.25 Tensile Responses of Unidirectional Σ/β21s

Laminate	E_x, GPa (Msi)	ν_{xy}	σ_x^{ult}, MPa (ksi)	ε_x^{ult}, %
[0]	185 (26.8)	0.25	1303 (189)	0.78
[90]	137 (19.9)	0.21	393 (57)	0.46
[18]	165 (24)	0.30	917 (133)	0.72
[45]	136 (19.7)	0.35	317 (46)	0.24
$[\pm 45]_s$	147 (21.4)	0.32	627 (91)	1.44
$[0/\pm 45/90]_s$	147 (21.4)	0.28	462 (67)	0.39
$[0/90]_s$	159 (23)	0.22	689 (100)	0.57

TABLE 7.10 Properties of Σ/β21s from Tension Tests

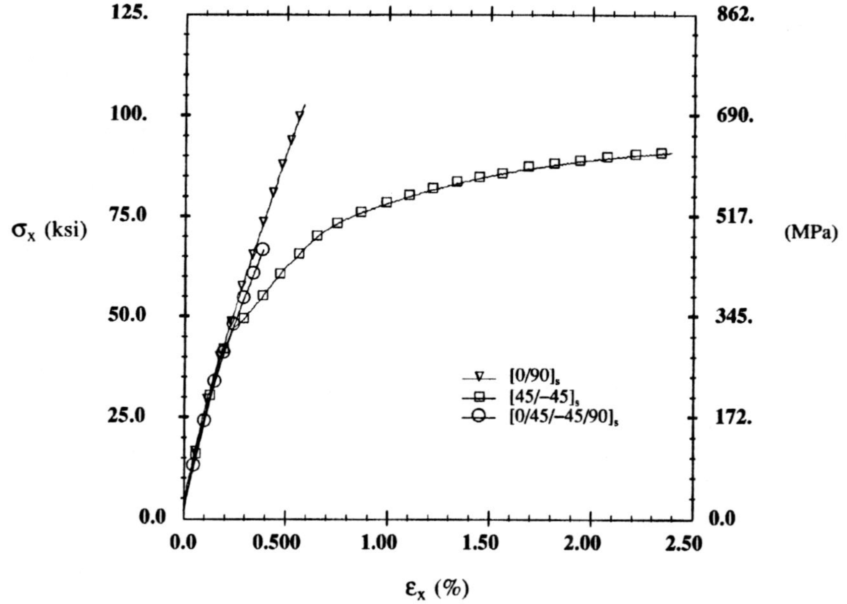

FIGURE 7.26 Tensile Responses of Σ/β21s Laminates

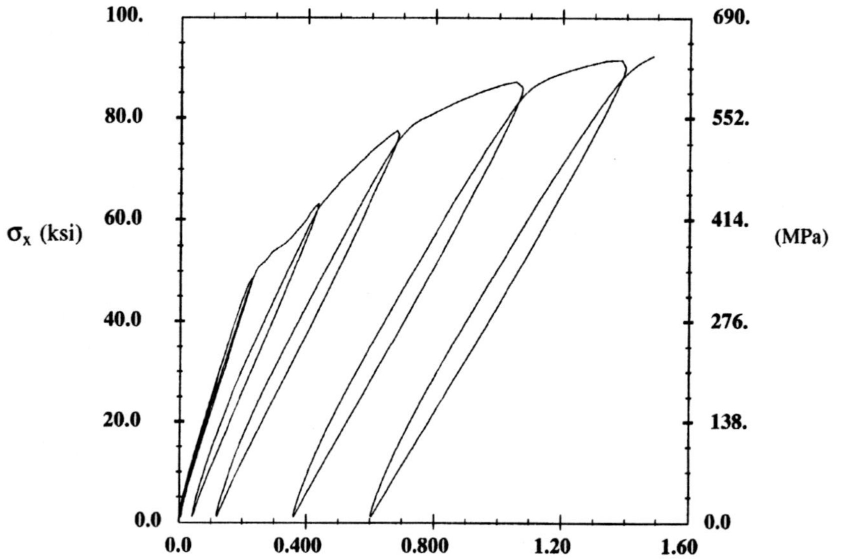

FIGURE 7.27 Cyclic Tensile Response of [±45]$_s$ Σ/β21s

Laminate	E_x, GPa (Msi)	ν_{xy}	G_{xy}, GPa (Msi)	σ_x^{ult}, MPa (ksi)	ε_x^{ult}, %	γ_x^{ult}, %
$[0_4]$	216 (31.3)	0.28	52.9 (7.67)	820 (119)	0.42	1.18
$[90_4]$	149 (21.6)	0.19	52.9 (7.67)	317 (46)	—	—
$[\pm 45]_s$	114 (16.5)	0.37	64.5 (9.35)	469 (68)	3.7	—

TABLE 7.11 Properties of SCS-6/Ti-15-3 Tubes

The cyclic, tensile loading/unloading response of a $[\pm 45]_s$ $\Sigma/\beta 21s$ laminate is shown in Fig. 7.27. This figure shows that the nonlinear response results in permanent strains and significant degradation of the axial modulus with increasing strain. The modulus degradation is a clear indication that damage is present and increases with increasing strain. The evolution of damage in a $[\pm 45]_s$ laminate of $\Sigma/\beta 21s$ titanium matrix composite is contrary to the response of unidirectional boron/aluminum, which did not exhibit any modulus degradation during cyclic tensile or shear loading.

Elastic properties and strength values obtained from SCS-6/Ti-15-3 tubes (Lissenden et al., 1993) are presented in Table 7.11.

7.4 Summary

Experimental results have been presented showing that the mechanical response of fibrous composite materials exhibits a wide range of possibilities. The response can be linear elastic and exhibit brittle failure, or the response can be highly nonlinear with large inelastic strains prior to failure. Constituent properties, fiber volume fraction, fiber orientations, and laminate stacking sequence all have an influence on the response up to failure. It has been shown that lamination theory predicts the elastic properties of composites very accurately.

References

Becker, W., Pindera, M. -J., and Herakovich, C. T. (1987), *Mechanical Response of Unidirectional Boron/Aluminum under Combined Loading,* CCMS-87-06, Center for Composite Materials and Structures, Virginia Tech, Blacksburg.

Klang, E. C. (1981), *Experimental Studies of Edge Cracking of Celion 6000/PMR-15 Graphite-Polyimide Tensile Specimens,* Interim Report 81-1, NASA–Virginia Tech Composites Program, Virginia Tech, Blacksburg.

Lissenden, C. J., Herakovich, C. T., and Pindera, M. -J. (1993), *Inelastic Deformation of Metal Matrix Composites,* Applied Mechanics Report AM-93-03, University of Virginia, Charlottesville.

Pindera, M. -J. (1989), "Shear Testing of Fiber Reinforced Metal Matrix Composites," in *Metal Matrix Composites: Testing, Analysis, and Failure Modes,* ASTM STP 1032 (W. S. Johnson, ed.), American Society for Testing and Materials, Philadelphia, pp. 19–42.

Pindera, M. -J. Gurdal, Z., Hidde, J. S., and Herakovich, C. T. (1987), *Mechanical and Thermal Characterization of Unidirectional Aramid/Epoxy,* CCMS-86-29, Center for Composite Materials and Structures, Virginia Tech, Blacksburg.

Pindera, M. -J., Gurdal, Z., Hidde, J. S., and Herakovich, C. T. (1989), "Mechanical Response of Aramid/Epoxy under Tensile, Compressive and Shear Loading," *J. Reinf. Plast. Compos.*, vol. 8, pp. 410–420.

Pindera, M. -J., and Herakovich, C. T. (1981), *An Endochronic Theory for Transversely Isotropic Fibrous Composites*, VPI-E-81-27, Virginia Tech, Blacksburg.

Pindera, M. -J., and Herakovich, C. T. (1986), "Shear Characterization of Unidirectional Composites with the Off-Axis Tension Test," *Exp. Mech.*, vol. 26, no. 1, pp. 103–112.

Schroedter, R. D., III., and Herakovich, C. T. (1995), *Damage Evolution in IM7/K3B Carbon-Polyimide Composite*, HTC-95-03, Center for High Temperature Composites, University of Virginia, Charlottesville.

Wyatt, D. E. (1997), "Tensile Response of IM7/K3B Composites," Senior thesis, University of Virginia, Charlottesville.

CHAPTER 8 INTERLAMINAR STRESSES

"There is more here than meets the eye."
Timeworn English expression

8.1 Introduction

The presence of free edges in laminated material systems introduces an additional level of complexity. The state of stress in the vicinity of free edges is three-dimensional, with nonzero through-thickness stresses. The through-thickness stresses include the interlaminar normal stress, σ_z, and two interlaminar shear stresses, τ_{yz} and τ_{zx} (Fig. 8.1). These through-thickness stresses are called *interlaminar stresses*. It must be emphasized that while these stresses are called *interlaminar*, indicating that they act between layers, they are not limited to the interface between layers. Stress continuity requires that they vary in a continuous manner through the thickness of all layers of the laminate.

Stress-free boundaries typically present the most severe interlaminar stresses. However, interlaminar stresses are also present in other structural configurations, including internal ply-drop in

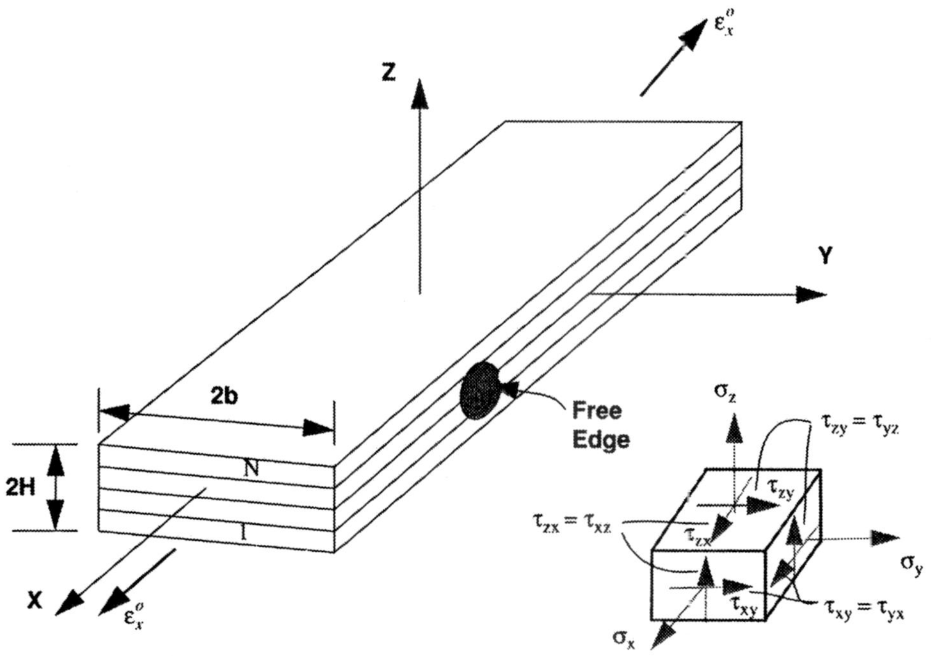

FIGURE 8.1 Finite-Width Laminated Coupon under Axial Load

tapered laminates and geometric discontinuities such as the flange termination region of stiffened panels. We will confine our discussion here to the analysis of interlaminar stresses near free edges of finite-width coupons.

Approximate elasticity solutions show that the interlaminar stresses are generally confined to a boundary layer region adjacent to the free edge, and that the interlaminar components of stress can be singular at the intersection of a free edge with the interface between two distinct layers. Since the state of stress in this boundary layer region is not a state of plane stress, the classical lamination theory of Chapter 5 is not valid in this region. Away from the free edge (outside the boundary layer region) the state of stress reverts to a planar state and lamination theory is again valid.

Interlaminar stresses are not unique to anisotropic materials. They may be present in any nonhomogeneous material with free edges. The interlaminar stresses are caused by the mismatch in material properties of bonded adjacent materials (layers) and gradients in in-plane components of stress. For isotropic layers, the mismatch in properties is usually small and interlaminar stresses are typically ignored. In contrast to isotropic materials, orthotropic composite laminae, made of high-modulus fibers in a less stiff matrix, exhibit a very broad range of properties as a function of fiber orientation (Chapter 4). As will be shown through equilibrium considerations, the mismatch in Poisson's ratios between layers requires the existence of interlaminar shear (τ_{yz}) and interlaminar normal stresses (σ_z). Further, off-axis laminae exhibit axial shear coupling as quantified by the coefficient of mutual influence. The potentially large mismatch in coefficients of mutual influence for polymeric matrix composites can lead to very large interlaminar shear stresses τ_{zx}. Other material property mismatches that can result in interlaminar stresses include the coefficients of thermal and hygroscopic expansion.

In real materials that exhibit inelastic response associated with matrix plasticity and damage (as contrasted with the idealized linear elastic materials under consideration here), the interlaminar stresses are not singular, but they do exhibit very large gradients near free edges. The large interlaminar stresses in the boundary layer are more severe for polymer matrix composites than for metal matrix composites. The polymer matrix composites are more highly anisotropic, resulting in a greater mismatch of layer properties.

The interlaminar stresses in the boundary layer can be critical in structural applications because they can lead to delamination-type failures at loads well below those corresponding to in-plane failure. Interlaminar stresses may be present as a result of mechanical, thermal, or moisture loading; they should be considered whenever laminated composite materials are used in the design of a structure which has free edges, including the free edges around holes and cutouts.

Interlaminar stresses near free edges can be controlled to some extent through the choice of materials, fiber orientations, stacking sequence, layer thicknesses, and the use of functionally graded materials. However, when free edges are present, interlaminar stresses can be eliminated completely only through the use of homogeneous materials.

Edge reinforcement techniques have been used to suppress the deleterious effects of interlaminar stresses, but at additional cost. These techniques do not eliminate the interlaminar stresses; they only provide a restraint against delamination. A thorough understanding of the mechanics of the free edge problem is indispensable to the designer of composite structures. With a basic understanding of the problem, laminates (and structures) can be *designed* to minimize interlaminar stresses while still meeting all other structural design requirements.

8.1.1 Historical Review

The first publication concerned with interlaminar stresses in laminated composites appears to be that of Hayashi (1967), who investigated interlaminar shear stresses in an idealized laminate consisting of orthotropic layers separated by isotropic shear layers. Other important early works include that by Bogy (1968), who investigated the singular behavior of stresses at the intersection of a boundary and bonded dissimilar isotropic materials, and the first three-dimensional (numerical) analysis of interlaminar stresses in laminated composites by Pipes and Pagano (1970).

Pipes and Pagano provided the first complete analysis of the problem of an axially loaded, laminated coupon with free edges (Fig. 8.1). They formulated a reduced system of elasticity equations governing the laminate behavior by assuming independence of the stress and strain states on the axial coordinate and then solved the system of equations using the finite-difference method. Their results showed the existence of all three interlaminar stress components in the boundary layer regions along the free edges of finite-width laminated coupons under in-plane tensile loading. They presented results for a variety of fiber orientations and laminate stacking sequences and showed that the width of the boundary layer is approximately equal to the thickness of the laminate, that the interlaminar normal stress σ_z and the interlaminar shear stress τ_{zx} can exhibit singular behavior as the free edge is approached, and that the sign and magnitude of the interlaminar stresses are functions of the laminate configuration, including material type, fiber orientations, layer thicknesses, and stacking sequence.

The free edge problem has been studied on a continuing basis ever since the original work in the late 1960s. The finite-difference solution of Pipes and Pagano was followed quickly by a three-dimensional finite-element solution by Rybicki (1971). Later, it was recognized that the tensile coupon problem also could be formulated as a two-dimensional finite-element problem because of the independence of the stress and strain states on the axial coordinate. The finite-element formulation for cross-ply laminates as a two-dimensional problem was presented initially by Foye and Baker (1971), and the two-dimensional finite-element formulation for laminates including off-axis layers was presented initially by Herakovich et al. (1976).

Noteworthy approximate analytical solutions include a perturbation solution by Tang (1975), a variational approach by Pagano (1978), a solution employing complex stress potentials and eigenfunction series (Wang and Choi, 1982), and solutions based upon statically admissible stress states (Kassapoglou and Lagace, 1986; Rose and Herakovich, 1993). The free edge problem has also been investigated experimentally; see, e.g., Pipes and Daniel (1971), Oplinger et al. (1974), and Herakovich et al. (1984). The experimental investigations provided physical evidence of a boundary layer with large strain gradients near free edges.

8.2 Finite-Width Coupon

The simplest problem to consider for studying edge effects in composites is the symmetrically laminated, axially loaded, finite-width coupon of Fig. 8.1. The N-ply laminate has total thickness $2H$ and finite width $2b$. For simplicity of exposition, we shall consider only laminates which are symmetric about the midplane and orthotropic ($A_{16} = A_{26} = 0$). Individual layers of the laminate are unidirectional (transversely isotropic) laminae of fibrous composites oriented at an angle θ_i with respect to the global x-axis of the coupon. It is assumed that all layers are perfectly bonded together and behave elastically under the application of load. Tensile load is applied at the ends of a long coupon, resulting in a constant, uniform axial strain ε_x^o along the length of the coupon. Away from the ends, the states of stress and strain are independent of the axial coordinate x. The problem is quasi-three-dimensional in that all six components of stress and strain may be nonzero, but the stress analysis can be restricted to a generic two-dimensional cross section because of the x independence of stress and strain. In interior regions away from the edges, all layers are in a state of plane stress, and the intralaminar stresses σ_x, σ_y, and τ_{xy} can be determined from classical lamination theory. We note that z is taken positive upward in this development.

8.2.1 Equilibrium Considerations

Equilibrium considerations can be used to provide the basic justification for the existence of interlaminar stresses in a laminated composite with free edges. Consider the tensile loaded coupon of Fig. 8.1. If we accept the fact (confirmed by elasticity theory) that a planar state of stress, independent of the axial coordinate, is recovered away from the free edge, then we can show that equilibrium of individual layers (or groups of layers) requires the existence of nonzero interlaminar stresses over some portion of interfacial planes. Integration of these interlaminar stresses over the region of application provides forces and moments that must be in equilibrium with the forces and moments calculated from the planar state of stress away from the edge. And since the state of stress away from the edges (i.e., along $y = 0$) is planar, these stresses can be calculated from lamination theory. These equilibrium considerations are developed in detail in the following paragraphs for a unit length of a tensile coupon. A free-body diagram of a section cut from the unit length is shown in Fig. 8.2. It is noted that the stresses σ_x, τ_{zx}, and τ_{xy} acting over the end surfaces $x = \pm L$ are self-equilibrating by virtue of the x independence of the stress state.

8.2.1.1 Interlaminar F_{yz} Shear Force

Figure 8.3 shows partial free-body diagrams of a group of layers from the tensile coupon in Fig. 8.1; the group includes the top free surface and free edge of the laminate. Shown in Fig. 8.3a are the non-self-equilibrating y-component stresses acting on a generic cross-sectional plane above any surface $z = z^*$. Taking into consideration the x independence of stresses, y-force equilibrium per unit length reduces to

$$\int_0^b \tau_{yz}(z^*)dy = -\int_{z^*}^{z_N} \sigma_y \, dz \tag{8.1}$$

This equilibrium equation shows that the *interlaminar shear stress* τ_{yz} *(z^*) must be nonzero over some portion of the surface* $0 \le y \le b$ at $z = z^*$ if the σ_y above this surface are not self-equilibrating.

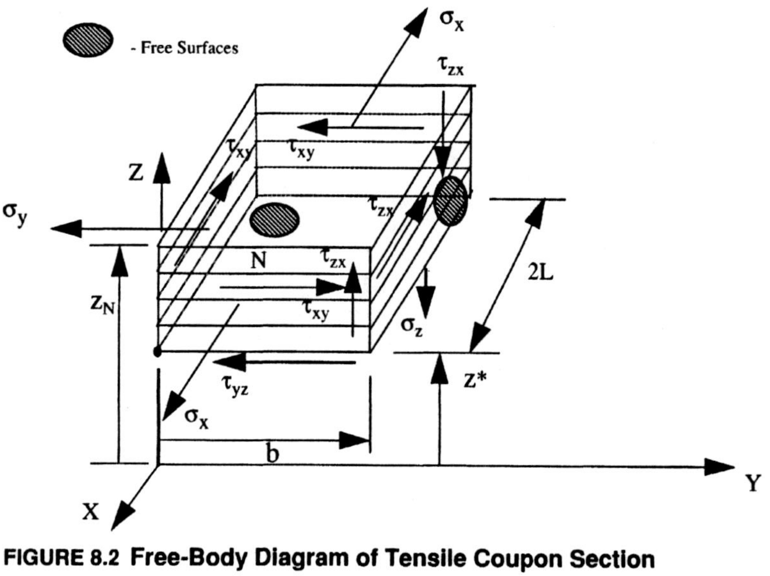

FIGURE 8.2 Free-Body Diagram of Tensile Coupon Section

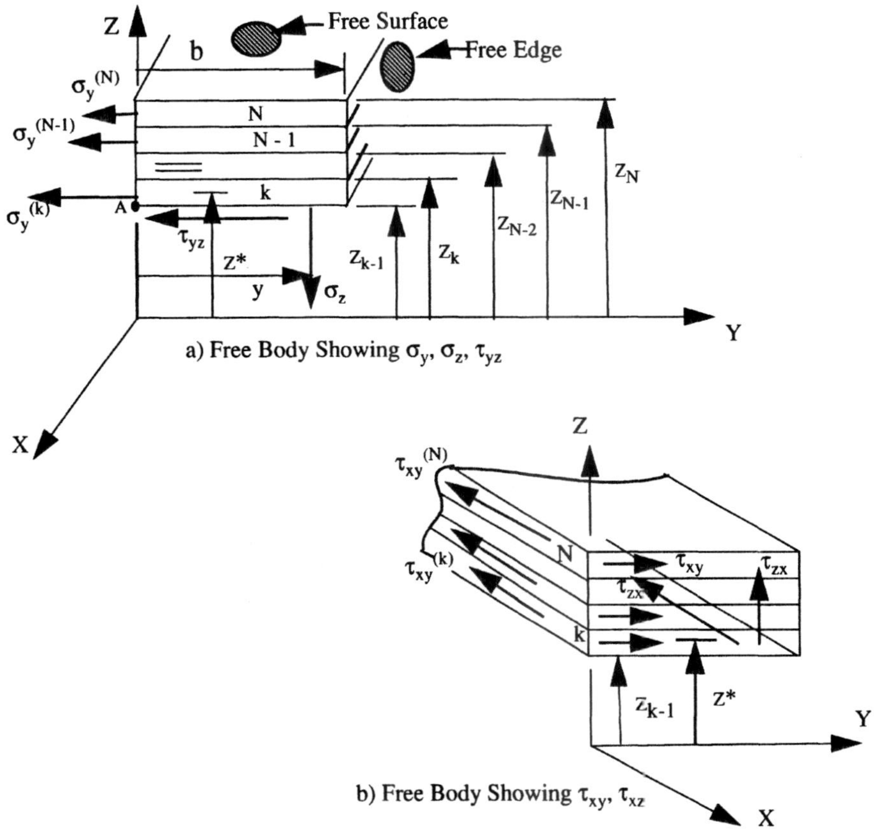

FIGURE 8.3 Partial Free-Body Diagrams

Since the integrals in (8.1) have the units of force per unit length, it is convenient to define the *interlaminar shear force* F_{yz} *at any surface* $z = z^*$ as

$$F_{yz}(z^*) = \int_0^b \tau_{yz}(z^*)dy \tag{8.2}$$

Combining (8.1) and (8.2) gives the definition of the interlaminar shear force F_{yz} at any location $z = z^*$ in terms of the lamination theory stresses σ_y:

$$F_{yz}(z^*) = -\int_{z^*}^{z_N} \sigma_y \, dz \tag{8.3}$$

More specifically, we can express the interlaminar shear force $F_{yz}^{(k)}$ at the kth interface (where $z^* = z_{k-1}$) in terms of in-plane stresses in the layers above the interface and the layer thicknesses t_j. The in-plane stresses above the interface are known (or determined) from lamination theory.

Uniform Strain Loading

For loadings such that strain in the laminate is pure midplane strain (e.g., axial or uniform thermal loading of a symmetric laminate), the lamination theory stresses are constant through the thickness of each layer, and we can write

$$F_{yz}^{(k)} = -\int_{z_{k-1}}^{z_N} \sigma_y \, dz = -\sum_{j=k}^{N} \sigma_y^{(j)} t_j \tag{8.4}$$

Appropriate modification of the thickness term of the kth layer in (8.4) gives the interlaminar shear force at *any* location $z = z^*$ (not only the interface between layers) in terms of stresses determined from lamination theory.

It is clear from (8.4) that, for uniform strain loading, the interlaminar shear force at $z = z^*$ is only a function of the fiber orientations and thicknesses of the layers above z^* and not the stacking sequence of these layers. This equation also clearly shows that the interlaminar shear force will, in general, vary through the thickness of the laminate, i.e., as z^* (or k) varies.

The free edge boundary condition requires that $\tau_{yz} = 0$ for all z at $y/b = 1.0$. Hence, the interlaminar shear stress τ_{yz} is zero at the free edge, may be nonzero near the free edge, but must return to zero in interior regions away from the free edge where a plane state of stress is recovered. Further, if the free-body diagram in Fig. 8.3 is for the full laminate width rather than the half-width, equilibrium requires that

$$\int_{-b}^{b} \tau_{yz}(z^*) dy = 0 \tag{8.5}$$

Hence τ_{yz} must be an odd function of y.

Curvature Loading

For loadings that give rise to nonzero curvatures $\{\kappa_x, \kappa_y, \kappa_{xy}\}$ with the midplane strains being zero, such as moment loading $M_x \neq 0$, the lamination theory stresses are linear in z, and the interlaminar shear force, F_{yz}, of (8.3) at an interface $z^* = z_{k-1}$ takes the form

$$F_{yz}(z^*) = -\sum_{j=k}^{N} \left((\bar{Q}_{12}^{(j)} \kappa_x + \bar{Q}_{22}^{(j)} \kappa_y + \bar{Q}_{26}^{(j)} \kappa_{xy}) \frac{(z_j^2 - z_{j-1}^2)}{2} \right) \tag{8.6}$$

Appropriate modification of the term involving the z-coordinates in (8.6) gives the interlaminar shear force at any location $z = z^*$ in terms of quantities known or determined from lamination theory.

8.2.1.2 Interlaminar M_z Moment

Moment equilibrium about the x-axis through a point such as A for the partial free-body diagram of Fig. 8.3a requires that the interlaminar normal stress σ_z be related to the laminate stresses σ_y through

$$\int_{z^*}^{z_N} \sigma_y \cdot (z - z^*) dz = \int_0^b \sigma_z(z^*) y \, dy \tag{8.7}$$

This moment equilibrium equation suggests a definition for the *interlaminar moment* M_z (the moment associated with the stress σ_z) at any surface $z = z^*$ in the form

$$M_z(z^*) = \int_0^b \sigma_z(z^*)y\,dy \tag{8.8}$$

Using the moment equilibrium equation (8.7) we can write the interlaminar moment M_z at any location z^* in terms of the stresses σ_y determined from lamination theory as

$$M_z(z^*) = \int_{z^*}^{z} \sigma_y \cdot (z - z^*)dz \tag{8.9}$$

Uniform Strain Loading

Similar to the interlaminar shear force F_{yz}, the interlaminar moment $M_z^{(k)}$ at the kth interface (where $z^* = z_{k-1}$) for an axially loaded symmetric laminate can be expressed as a summation of constant layer stresses $\sigma_y^{(j)}$:

$$M_z^{(k)} = \int_{z_{k-1}}^{z_N} \sigma_y \cdot (z - z^*)dz = \sum_{j=k}^{N} \sigma_y^{(j)} \cdot (z_j - z_{j-1})\left[\frac{(z_j + z_{j-1})}{2} - z_{k-1}\right] \tag{8.10}$$

It is clear from (8.8) and (8.10) that the *interlaminar moment is a function of the stacking sequence* of the layers above the plane of interest. Also, the interlaminar moment will, in general, vary through the thickness of the laminate.

We have now established conditions for y-force equilibrium and moment equilibrium about an x-axis of the generic free-body diagram of Fig. 8.3. z-force equilibrium additionally requires that the total force associated with the distribution of σ_z along any plane $z = z^*$ be zero. Thus, the distribution of σ_z must be equivalent to a couple. The sign of this couple is a function of the through-thickness distribution of σ_y at $y = 0$. Hence, the interlaminar normal stresses vary through the thickness of the laminate and may be positive, negative, or zero at any point (y,z). Further, moment equilibrium of a free-body diagram of the full width shows that σ_z must be an even function of y.

Curvature Loading

For loadings that give rise to nonzero curvatures $\{\kappa_x, \kappa_y, \kappa_{xy}\}$ with the midplane strains being zero, the linear z dependence of the stresses results in the following form of (8.9) at an interface $z^* = z_{k-1}$:

$$M_z(z^*) = \sum_{j=k}^{N}\left[(\bar{Q}_{12}^{(j)}\kappa_x + \bar{Q}_{22}^{(j)}\kappa_y + \bar{Q}_{26}^{(j)}\kappa_{xy})\left[\left(\frac{z_j^3 - z_{j-1}^3}{3}\right) - z^*\left(\frac{z_j^2 - z_{j-1}^2}{2}\right)\right]\right] \tag{8.11}$$

8.2.1.3 Interlaminar F_{zx} Shear Force

Summation of forces in the x-direction of the free-body diagram in Fig. 8.3b requires that unbalanced stresses τ_{xy} along the laminate centerline ($y = 0$) be equilibrated by nonzero stresses τ_{zx} along the interface $z = z^*$. This equilibrium equation can be expressed as

$$\int_0^b \tau_{zx}(z^*)dy = -\int_{z^*}^{z_N} \tau_{xy}\, dz \tag{8.12}$$

This equation leads to the definition of the *interlaminar shear force* F_{zx} at any surface $z = z^*$:

$$F_{zx}(z^*) = \int_0^b \tau_{zx}(z^*)dy \tag{8.13}$$

Combining (8.12) and (8.13) gives the interlaminar force F_{zx} in terms of the shear stresses τ_{xy} determined from lamination theory:

$$F_{zx}(z^*) = -\int_{z^*}^{z_N} \tau_{xy}\, dz \tag{8.14}$$

Uniform Strain Loading

As for the previously defined interlaminar force and moment, $F_{zx}^{(k)}$ at the kth interface for uniform strain loading can be written in terms of the known (or determined) lamination theory stresses as

$$F_{zx}^{(k)} = -\int_{z_{k-1}}^{z_N} \tau_{xy}\, dz = -\sum_{j=k}^{N} \tau_{xy}^{(j)} t_j \tag{8.15}$$

where t_j is the thickness of the jth layer and we have again used the fact that the lamination theory stresses are constant through the thickness of each layer for in-plane loading of a symmetric laminate. Appropriate modification of the thickness term in (8.15) gives the interlaminar shear force F_{zx} at any location $z = z^*$.

From (8.12), nonzero interlaminar shear stresses τ_{zx} must exist over some region of the interface $z = z^*$ if the in-plane shear stresses τ_{xy} are not self-equilibrating above this surface. If the full width of the laminate is taken as the free-body diagram, we see that τ_{zx} must be an odd function of y as was τ_{yz}. However, unlike the case for τ_{yz}, there is no boundary condition on τ_{xz} at $y = b$. Hence τ_{zx} is zero away from the edge where lamination theory is valid, is generally nonzero as the edge is approached, and *is not limited in any manner at the free edge*. As a final comment we note that for a group of layers, with given fiber orientations above $z = z^*$, F_{zx} at $z = z^*$ is independent of the stacking sequence of the layers within the group above $z = z^*$.

Curvature Loading

For curvature loading, the interlaminar shear force F_{zx} at $z^* = z_{k-1}$ takes the form

$$F_{zx}(z^*) = -\sum_{j=k}^{N} \left((\overline{Q}_{16}^{(j)} \kappa_x + \overline{Q}_{26}^{(j)} \kappa_y + \overline{Q}_{66}^{(j)} \kappa_{xy}) \frac{(z_j^2 - z_{j-1}^2)}{2} \right) \tag{8.16}$$

8.2.1.4 Interlaminar Forces and Moments for Strain Loading

The preceding equilibrium considerations demonstrated conditions under which interlaminar shear and normal stresses must be nonzero over some portion of any plane $z = z^*$. It was shown that the existence of these interlaminar stresses often can be determined using lamination theory. There are some cases where interlaminar stresses are nonzero but self-equilibrating, in which case lamination theory cannot prove existence (Rose and Herakovich, 1992).

Elasticity theory must be used to determine the actual distribution of the interlaminar stresses. However, insight into the influence of fiber orientation and stacking sequence on interlaminar stresses can be obtained through a study of through-thickness distributions of the interlaminar forces F_{yz} and F_{zx} and the interlaminar moment M_z. In the following discussion all laminates have the same number of layers and total thickness and were subjected to the same magnitude of axial load $N_x = 1000$ lb/in (175 N/mm).

Cross-Ply Laminates

Through-thickness distributions of the interlaminar forces F_{yz} and F_{zx} and the interlaminar moment M_z in three symmetric, cross-ply carbon/epoxy laminates with an equal number of 90° and 0° plies are shown in Figs. 8.4–8.6. The only difference in the laminates is the stacking sequence. The interlaminar shear force F_{zx} is identically zero throughout all three laminates because $\eta_{xy,x}$ is identically zero for both 90° and 0° ply orientations. Thus, there is no $\eta_{xy,x}$ mismatch and no interlaminar shear force F_{zx}.

The $[0_2/90_2]_s$ (Fig. 8.4) and $[90_2/0_2]_s$ (Fig. 8.5) laminates exhibit a quadratic distribution of M_z with the same maximum magnitude (but opposite sign) at the midplane. The sign of the interlaminar moment is constant throughout each laminate. The interlaminar shear force F_{yz} exhibits a linear

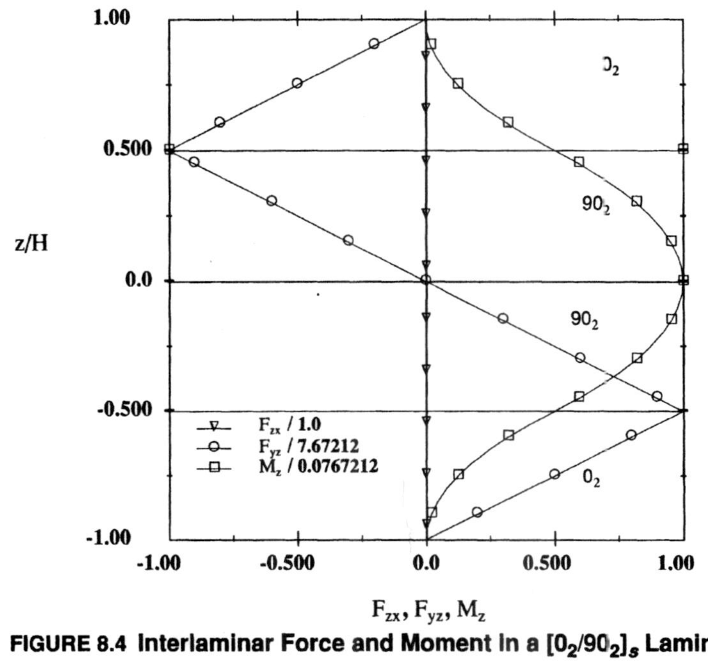

FIGURE 8.4 Interlaminar Force and Moment in a $[0_2/90_2]_s$ Laminate

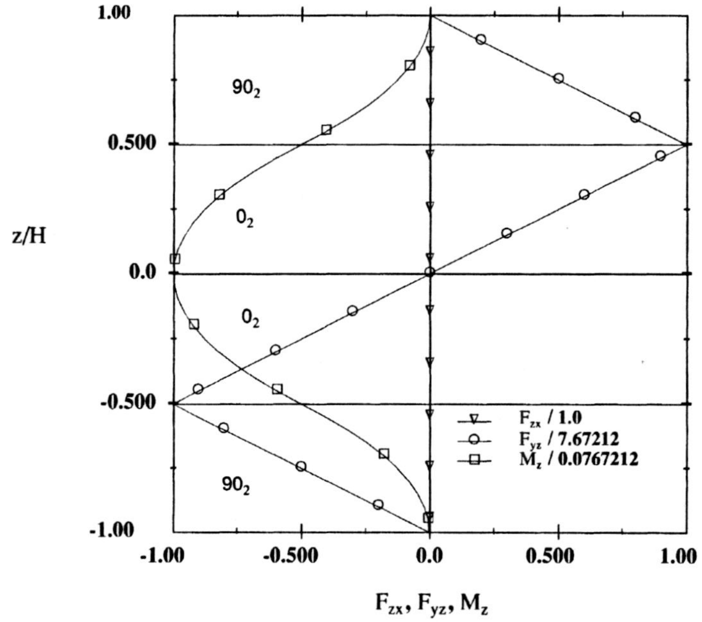

FIGURE 8.5 Interlaminar Force and Moment in a $[90_2/0_2]_s$ Laminate

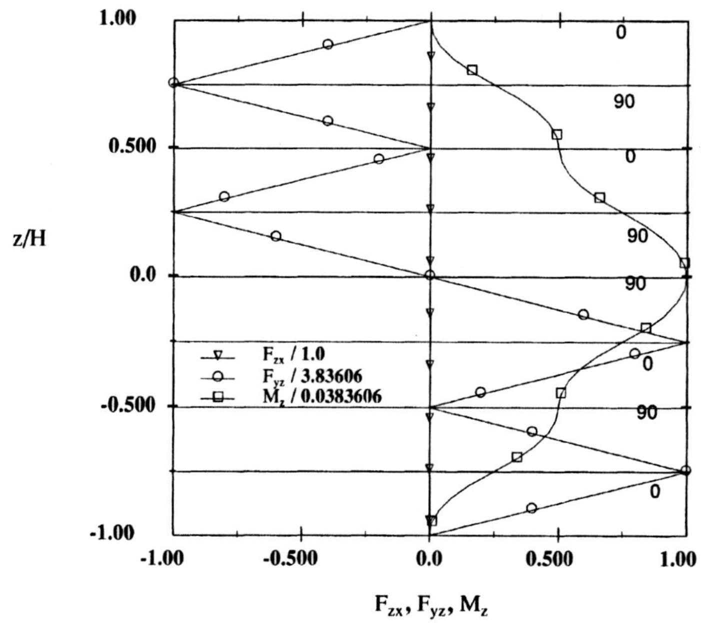

FIGURE 8.6 Interlaminar Force and Moment in a $[(0/90)_2]_s$ Laminate

distribution through each layer with the same maximum magnitude at the 0/90 interfaces. The shear force is zero on the midplane and of opposite sign above and below the midplane. Reversing the stacking sequence has the effect of reversing the sign on all quantities. The quadratic moment distribution and linear shear force distribution are direct consequences of the constant (lamination theory) stresses in each of the layers for in-plane loading of a symmetric laminate. Reversals in the slope of the interlaminar force distributions correspond to inflection points in the moment distributions, and zero force values correspond to extreme values of the moment.

Changing the layer thicknesses and stacking sequence to a $[(0/90)_2]_s$ laminate has a pronounced effect on the distributions (Fig. 8.6). In each layer the moment distribution remains quadratic and the force distribution remains linear. The alternating stacking sequence of the laminate is very evident in the distributions in the form of alternating sign of slope in the F_{yz} plot and reversal of curvature in the M_z plot. The shear force exhibits identical maximum magnitudes at each of the four 0/90 interfaces. The moment maintains the same sign throughout, with inflection points at each 0/90 interface. The maximum moment occurs at the laminate midplane.

The major difference between the results in Figs. 8.4–8.6 is the fact that the magnitudes of the maximum interlaminar force F_{yz} and the interlaminar moment M_z in the alternating-layer $[(0/90)_2]_s$ laminate are half those in the adjacent-layer $[0_2/90_2]_s$ and $[90_2/0_2]_s$ laminates. This is a direct consequence of the one-half reduction in individual layer thicknesses.

Angle-Ply Laminates

Distributions of interlaminar forces and moment for adjacent-layer $[30_2/-30_2]_s$ and alternating-layer $[(30/-30)_2]_s$ angle-ply laminates are shown in Figs. 8.7 and 8.8. The only nonzero force or moment is the interlaminar force F_{zx}. It varies linearly through each layer and exhibits identical maximum magnitudes at each +/–30 interface. These results are typical of the distributions in any

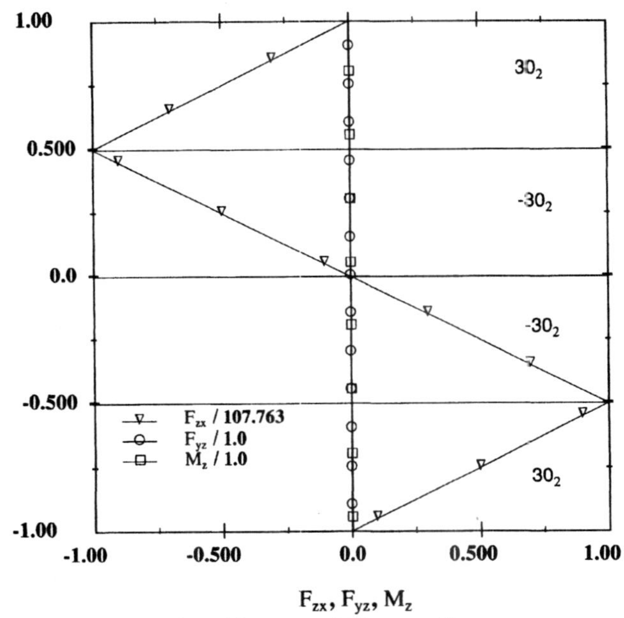

FIGURE 8.7 Interlaminar Force and Moment in a $[30_2/-30_2]_s$ Laminate

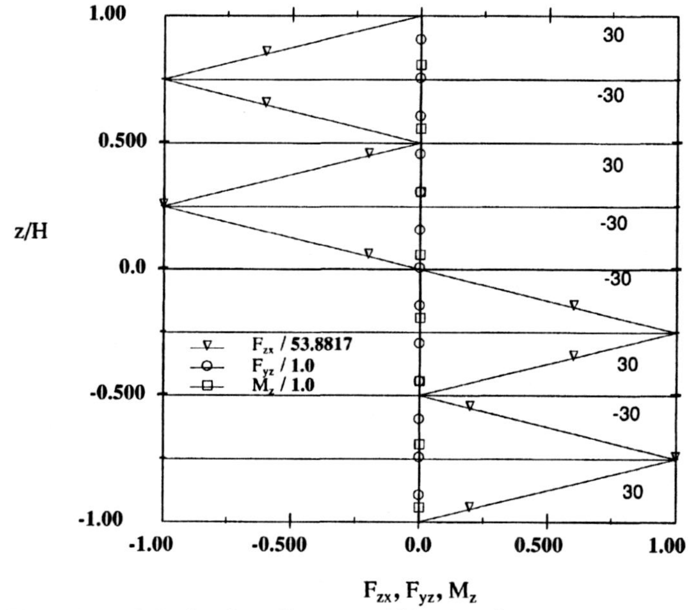

FIGURE 8.8 Interlaminar Force and Moment in a $[(30/-30)_2]_s$ Laminate

angle-ply laminate and follow directly from the fact that while there is no mismatch in Poisson's ratio in an angle-ply laminate, the mismatch in the coefficient of mutual influence gives rise to the nonzero interlaminar shear force F_{zx}. As in the cross-ply laminates, the maximum magnitude of the interlaminar force in angle-ply laminates is halved when the layer thickness is halved.

Quasi-Isotropic Laminates

The strong influence of stacking sequence on the distribution of interlaminar forces and moments can be demonstrated through consideration of two different stacking sequences of quasi-isotropic laminates, $[\pm 45/0/90]_s$ (Fig. 8.9) and $[0/-45/90/45]_s$ (Fig. 8.10). Unlike the cross-ply and angle-ply laminates, all three interlaminar quantities are nonzero in the quasi-isotropic laminates. As expected, both laminates exhibit linear shear force variations though the layers. This is a direct result of the equilibrium considerations discussed previously. The maximum interlaminar shear force always occurs at an interface between layers, but the maximum interlaminar moment does not always occur at an interface. In the $[0/-45/90/45]_s$ laminate the maximum moment occurs in the middle of the 90° layer, and the moment at the midplane ($z = 0$) is a small negative value. In contrast, the maximum moment for the $[\pm 45/0/90]_s$ laminate is at the midplane.

The maximum values of the interlaminar forces and moments generally vary with the stacking sequence. For the two quasi-isotropic laminates under consideration, the maximum interlaminar shear force F_{zx} has the same value, but the interlaminar shear force F_{yz} and the interlaminar moment M_z vary considerably between laminates. These results clearly show the strong influence of stacking sequence on interlaminar effects.

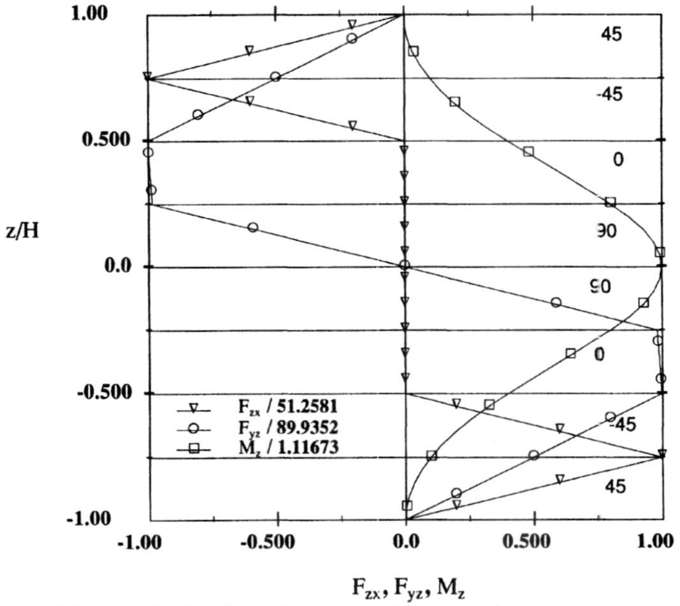

FIGURE 8.9 Interlaminar Force and Moment in a $[\pm 45/0/90]_s$ Laminate

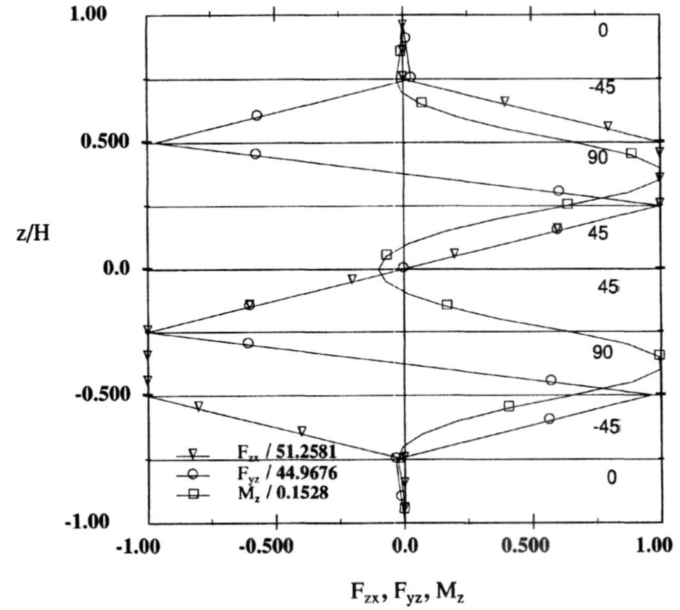

FIGURE 8.10 Interlaminar Force and Moment in a $[0/-45/90/45]_s$ Laminate

8.2.1.5 Interlaminar Forces and Moments for Curvature Loading

As one example of the interlaminar forces and moments that can develop in a finite-width coupon under pure loading $M_x \neq 0$, $M_y = M_{xy} = \{N\} = 0$, the results in Fig. 8.11 show the through-thickness distributions of F_{yz}, F_{zx}, and M_z for $M_x = 100$ in-lb/in on a $[0/90]_s$ T300/5208 carbon/epoxy laminate. For a cross-ply laminate under this loading the shear stresses τ_{xy} are zero throughout, and thus from (8.14) the shear force F_{zx} is zero throughout the thickness. The distributions of F_{yz} and M_z are quadratic and cubic, respectively, with F_{yz} exhibiting discontinuous slopes at the 0/90 interfaces. The maximum shear force F_{yz} is at the midplane $z = 0$, and the distribution of F_{yz} is symmetric about the midplane. This distribution can be compared with that in Fig. 8.9 for uniform axial strain loading of a cross-ply laminate, where the maximum force occurs at the 0/90 interfaces and the force is zero at the midplane.

The distribution of interlaminar moments M_z also differs significantly for the two types of loadings, as demonstrated by Figs. 8.4 and 8.11. The moment distribution for curvature loading is asymmetric about the midplane, with maximum values within the 90° layers at $z/H = \pm 0.4$. In contrast, for the uniform strain loading of Fig. 8.4, the distribution is symmetric about the midplane, with the maximum interlaminar moment occurring at the midplane $z = 0$.

8.2.2 Mismatch Considerations

The presence of interlaminar stresses in a finite-width tensile coupon is the result of the mismatch in material properties between layers. (The mismatch in properties at the micromechanics level is

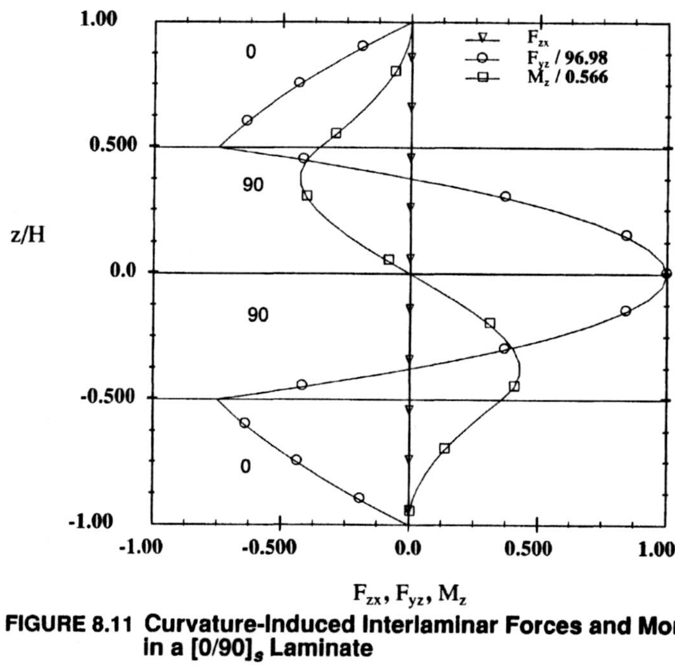

FIGURE 8.11 Curvature-Induced Interlaminar Forces and Moment in a $[0/90]_s$ Laminate

ignored in this discussion.) Coupon specimens of homogeneous materials do not exhibit interlaminar stresses under the application of uniaxial loading because there is no material property mismatch. The two properties of primary interest for laminated composites are Poisson's ratio ν_{xy} and the coefficient of mutual influence $\eta_{xy,x}$. Recall that Poisson's ratio ν_{xy} is defined as the negative ratio of the induced lateral strain ε_y to the axial strain ε_x under the application of axial stress σ_x.

$$\nu_{xy} = -\varepsilon_y/\varepsilon_x \tag{8.17}$$

The coefficient of mutual influence $\eta_{xy,x}$ is defined to be the ratio of the induced in-plane shear strain γ_{xy} to the axial strain ε_x for applied axial stress σ_x.

$$\eta_{xy,x} = \gamma_{xy}/\varepsilon_x \tag{8.18}$$

The variation of these material properties with the fiber orientation, θ, for a unidirectional off-axis carbon/epoxy lamina is shown in Fig. 8.12. It is evident from the figure that Poisson's ratio is an even function of θ and, for this material, ranges from a maximum of 0.37 at $\theta = \pm 23°$ to a minimum of 0.02 at $\theta = \pm 90°$. The coefficient of mutual influence is an odd function of θ, ranging from a minimum of -2.17 at $\theta = +11.5°$ to a maximum of $+2.17$ at $\theta = -11.5°$ (for T300/5208 carbon/epoxy).

A more dramatic representation of the mismatch characteristics is presented in Fig. 8.13. This figure shows the mismatch in Poisson's ratios, $\delta\nu_{xy}$, and the mismatch in coefficients of mutual influence, $\delta\eta_{xy,x}$, for adjacent $+\theta$ and $-\theta$ layers. The mismatch in Poisson's ratios is identically zero for all angles because, as previously mentioned, Poisson's ratio is an even function of the fiber

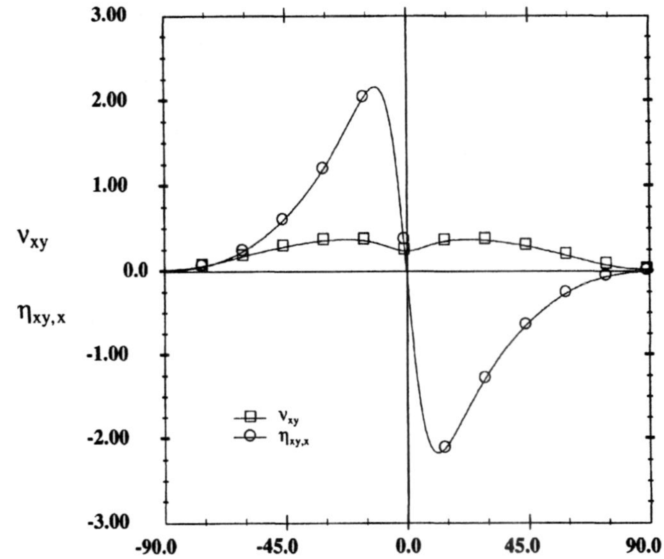

FIGURE 8.12 Off-Axis Poisson's Ratio and Coefficient of Mutual Influence

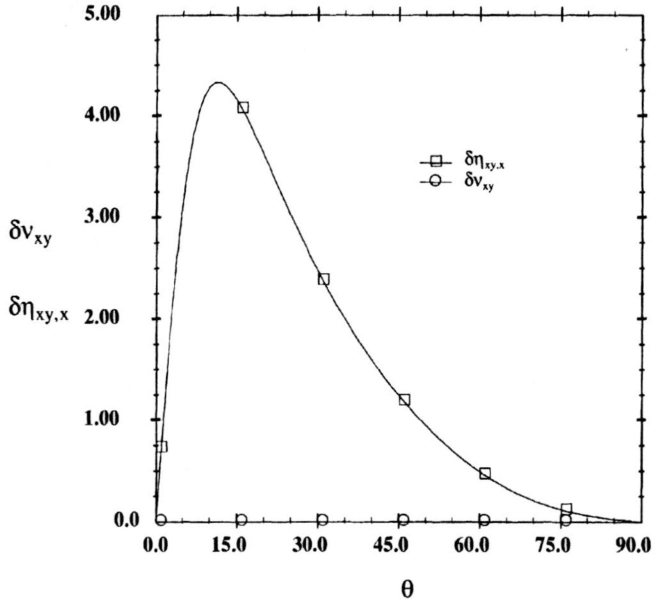

FIGURE 8.13 Mismatch in Properties for +θ/–θ Angle-Ply Laminates

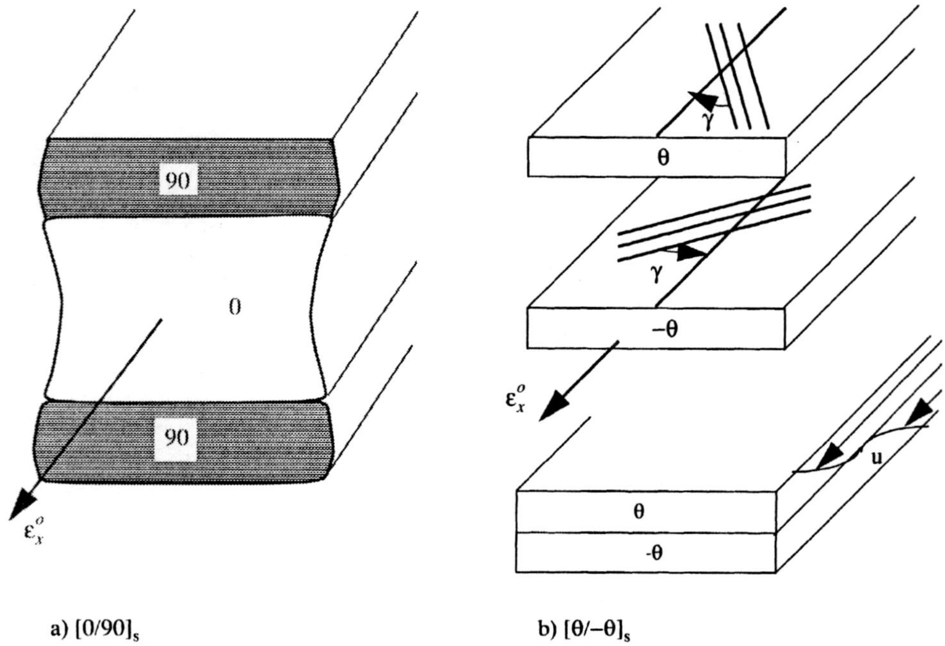

FIGURE 8.14 Displacement Continuity

orientation. It can be shown that the mismatch in Poisson's ratio for θ/90 and θ/0 adjacent-layer combinations has a maximum value of 0.34 for a 22/90 stacking sequence for the carbon/epoxy considered.

As depicted in Fig. 8.14, individual layers of a cross-ply laminate subjected to axial loading exhibit unequal lateral contractions due to differing Poisson's ratios, except at the layer interface, where they are perfectly bonded. The perfect bonding and displacement continuity at the interfaces result in nonzero interlaminar shear stresses τ_{yz}.

Figure 8.14b also depicts the differing shear strains when +θ and –θ angle-ply laminates are subjected to axial loading. The induced shear strains are of opposite sign, and thus the layers have opposing shear displacements. This results in a gradient of axial displacements, u, on the free edge, as indicated in the figure.

The curve for the mismatch in coefficients of mutual influence for +θ/–θ layers (Fig. 8.13) exhibits much higher values (more than 10 times greater) than the maximum Poisson mismatch for the laminates considered. The mismatch in coefficients of mutual influence shows a strong maximum at 11.5° for the +θ/–θ combination. The mismatch in $\eta_{xy,x}$ for θ/90 and θ/0 interfaces is exactly half that of the +θ/–θ configuration because $\eta_{xy,x}$ for the 0° and 90° layers is identically zero and $\eta_{xy,x}$ is an odd function of θ.

8.2.3 Displacement Fields

Side-by-side Moiré fringe patterns showing the axial displacements along the top surface and free edge of three angle-ply carbon/epoxy tensile coupons (including magnifications of the edge patterns) are shown in Fig. 8.15 (Czarnek et al., 1983). On both the top surface and the edge of the coupons, the fringe patterns are independent of the axial coordinate. The displacements are uniform on the top surface of the specimens with the exception of a boundary layer region along the edges. It is evident from the Moiré fringe patterns on the top surface that the nonuniform axial displacements in the boundary layer region are antisymmetric about the coupon centerline for these angle-ply laminates. The experimental results clearly verify the assumption that the strains (gradients of the Moiré fringes) are independent of the axial coordinate. These experimental results also show that the boundary layer width is approximately equal to the laminate thickness for all three laminates.

The shear strains γ_{zx} on the specimen edges are proportional to the gradient of the fringe lines with respect to the through-thickness z-coordinate. On the specimen edges, the fringe patterns show large gradients, indicating very high shear strains (Figs. 8.15 and 8.16). The maximum slope of the fringes, which is finite, occurs at the +θ/–θ interface. The influence of fiber orientation is very evident, as the fringe pattern changes significantly with fiber orientation. The maximum shear strain decreases monotonically from the largest value at θ = 10° to the smallest value at θ = 45°. These experimental results are consistent with the mismatch in coefficients of mutual influence in Fig. 8.13.

8.2.4 Design Considerations

Several important design considerations can be drawn from the preceding discussion of mismatch effects and the distribution of interlaminar forces and moment for finite-width symmetric laminates subjected to in-plane axial loading:

- For angle-ply laminates, the Poisson's ratio mismatch is zero and hence the interlaminar shear force F_{yz} and moment M_z are zero for all z.

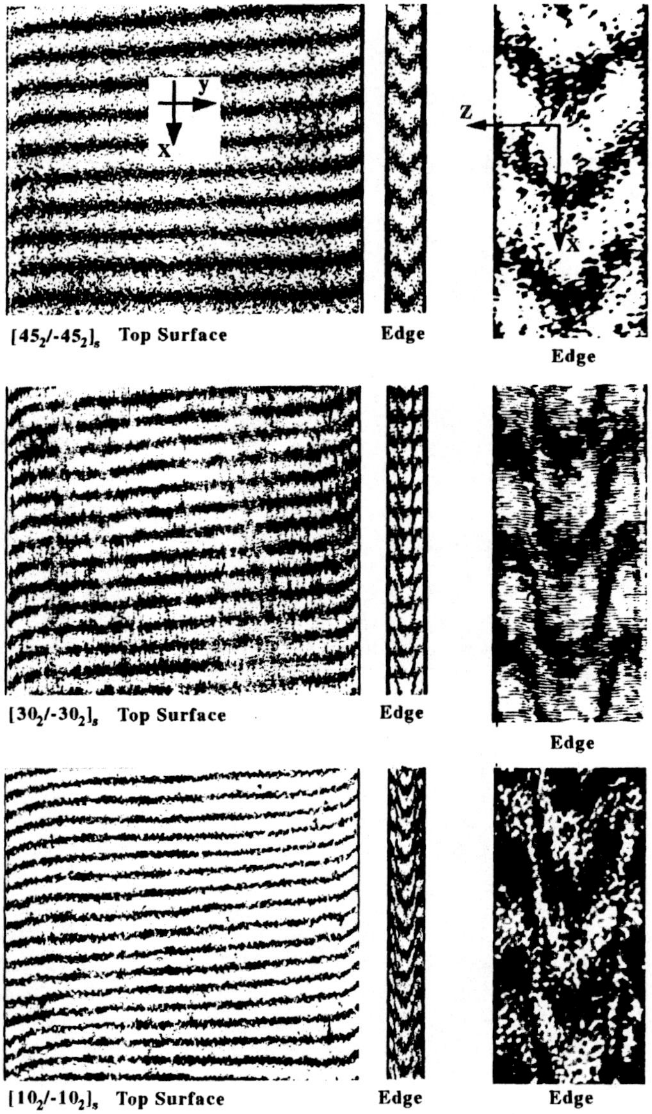

FIGURE 8.15 Moiré Fringe Patterns for $[\theta_2/-\theta_2]_s$ Laminate

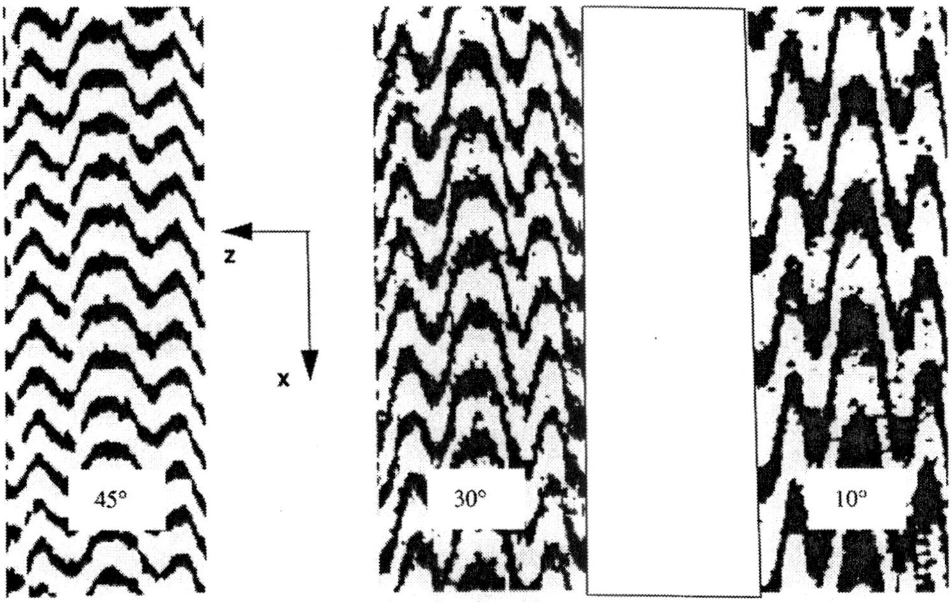

FIGURE 8.16 Free Edge Moiré Fringe Patterns for $[(\pm\theta)_2]_s$ Laminates

- The interlaminar moment at any location $z = z^*$ is a function of the stacking sequence of the layers above the plane of interest. The interlaminar moment varies quadratically through individual layers, and the maximum interlaminar moment may occur within a layer.
- Layers of like orientation should be interspersed between layers of other orientations rather than grouped together. This reduces the maximum magnitude of both the interlaminar shear forces and the interlaminar moment.
- There is no mismatch in the coefficient of mutual influence in cross-ply laminates, and hence the interlaminar shear force F_{zx} is zero for all z in these laminates.
- Interlaminar shear forces at any location $z = z^*$ are a function of the fiber orientations and layer thickness above $z = z^*$ but independent of the stacking sequence of these layers. The interlaminar shear force varies linearly through individual layers, with the maximum values always occurring at layer interfaces.

Design considerations for curvature loading require a more extensive study than has been presented here.

8.3 Finite-Element Analysis

8.3.1 Displacement Formulation

Approximate elasticity solutions of the free edge problem for linear elastic response of balanced, symmetric, finite-width tensile coupons can be developed starting from the condition of x independence of stresses and strains. Integration of the strain-displacement equations (2.14) with all strains independent of x yields the displacement fields

$$\begin{aligned} u(x, y, z) &= x(C_1 y + C_2 z + C_3) + U(y, z) \\ v(x, y, z) &= x(C_4 z + C_5) - C_1\frac{x^2}{2} + V(y, z) \\ w(x, y, z) &= x(-C_4 y + C_6) - C_2\frac{x^2}{2} + W(y, z) \end{aligned} \tag{8.19}$$

Invoking symmetry of displacements about the x-y plane (i.e., $u(x, y, z) = u(x, y, -z)$, $v(x, y, z) = v(x, y, -z)$, and $w(x, y, z) = -w(x, y, -z)$ for all x and y) requires that $C_2 = C_4 = C_6 = 0$, $U(y, z) = U(y,-z)$, $V(y, z) = V(y, -z)$, and $W(y, z) = -W(y, -z)$. Thus the most general displacement field for a midplane symmetric laminate under axial strain is

$$\begin{aligned} u(x, y, z) &= x(C_1 y + C_3) + U(y, z) \\ v(x, y, z) &= xC_5 - C_1\frac{x^2}{2} + V(y, z) \\ w(x, y, z) &= W(y, z) \end{aligned} \tag{8.20}$$

Several investigators (Dana, 1974; Griffin et al., 1981) have shown that for thin, balanced, symmetric laminates under uniform strain loading, the analysis may be limited further to a quarter section of the laminate by assuming symmetry about the x-z plane (i.e., $u(x, y, z) = u(x, -y, z)$, $v(x, y, z) = -v(x, -y, z)$, and $w(x, y, z) = w(x, -y, z)$) without substantial sacrifice in accuracy. Invoking symmetry about the x-z plane requires that $C_1 = C_5 = 0$, $U(y, z) = U(-y, z)$, $V(y, z) = -V(-y, z)$, and $W(y, z) = W(-y, z)$. The most general displacement field then reduces to

$$\begin{aligned} u(x, y, z) &= C_3 x + U(y, z) \\ v(x, y, z) &= V(y, z) \\ w(x, y, z) &= W(y, z) \end{aligned} \tag{8.21}$$

From (8.19)–(8.21) and the fundamental strain-displacement relationships (2.16), we see that C_3 is the uniform axial strain, ε_x^o. The unknown functions $U(y, z)$, $V(y, z)$, and $W(y, z)$ represent the warp displacement over a generic quarter cross-section of the laminate. Thus, for a given uniform axial strain, the unknown warp displacements U, V, and W at any axial location x are functions of y and z only; hence the analysis may be limited to a generic quarter section, as depicted in Fig. 8.17.

For applied far-field axial strain, the boundary conditions on the generic quarter section are as follows:

- Free surface tractions:

$$\begin{aligned} &\sigma_y = \tau_{xy} = \tau_{yz} = 0 \qquad \text{along the free edge } y = b \\ &\sigma_z = \tau_{yz} = \tau_{zx} = 0 \qquad \text{along the top surface } z = H \end{aligned} \tag{8.22}$$

- Displacements:

$$\begin{aligned} &W(y, 0) = 0 \qquad \text{along the midplane } z = 0 \\ &U(0, z) = V(0, z) = 0 \qquad \text{along the centerline } y = 0 \end{aligned} \tag{8.23}$$

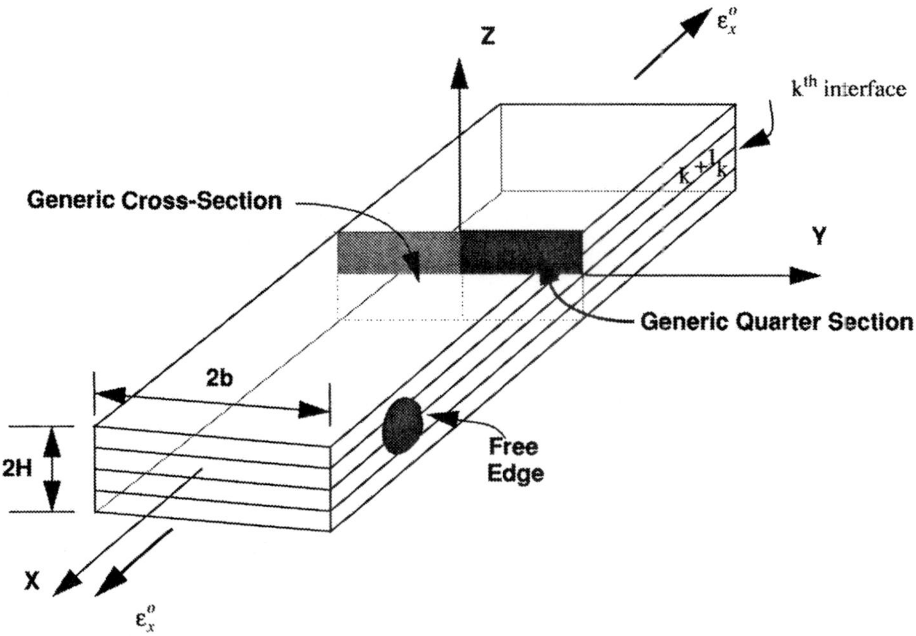

FIGURE 8.17 Generic Cross-Sectional Plane of Tensile Coupon

Since we are dealing with a laminate whose global elastic properties vary from layer to layer, the elasticity solution must be obtained for each layer and the layerwise solutions must be made continuous through satisfaction of interfacial continuity requirements on displacements and tractions. The interfacial continuity requirements at the kth interface (between layers k and $k + 1$, Fig. 8.17) are as follows:

- Traction continuity at $z = z_k$:

$$\sigma_z^{(k)} = \sigma_z^{(k+1)} \qquad \tau_{yz}^{(k)} = \tau_{yz}^{(k+1)} \qquad \tau_{zx}^{(k)} = \tau_{zx}^{(k+1)} \tag{8.24}$$

- Displacement continuity at $z = z_k$:

$$U^{(k)} = U^{(k+1)} \qquad V^{(k)} = V^{(k+1)} \qquad W^{(k)} = W^{(k+1)} \tag{8.25}$$

The continuity requirements (8.24) and (8.25) serve as boundary conditions on the layerwise elasticity solution.

The complete elasticity solution to the free edge problem is the set of functions U, V, and W which satisfy the displacement form of the equilibrium equations (obtained by substitution of the strain-displacement equations (2.16) and constitutive equations (2.37) into the equilibrium equations (2.24)), the compatibility conditions (2.36), the boundary conditions (8.22) and (8.23), and the interfacial traction and displacement continuity conditions (8.24) and (8.25).

The displacement equilibrium equations are second-order, constant-coefficient, linear partial differential equations in terms of the unknown functions U, V, and W. If C_{ij} are the material stiffness coefficients for a given layer, the displacement equilibrium equations for that layer are

$$\begin{aligned} C_{66}U_{,yy} + C_{55}U_{,zz} + C_{26}V_{,yy} + C_{45}V_{,zz} + (C_{36} + C_{45})W_{,yz} &= 0 \\ C_{26}U_{,yy} + C_{45}U_{,zz} + C_{22}V_{,yy} + C_{44}V_{,zz} + (C_{23} + C_{44})W_{,yz} &= 0 \\ (C_{45} + C_{36})U_{,yz} + (C_{44} + C_{23})V_{,yz} + C_{44}W_{,yy} + C_{33}W_{,zz} &= 0 \end{aligned} \tag{8.26}$$

where commas denote partial differentiation with respect to the indicated variables, i.e., $U_{,yy} = \partial^2 U/\partial y^2$.

As mentioned previously, approximate solutions for the free edge problem have been obtained using finite differences, finite elements, and approximate analytical solutions. Here we will present displacement-based finite-element results first and then follow with an approximate solution based upon statically admissible stress states and minimization of the complementary energy. The presentation begins with finite-element results for deformations in the boundary layer, followed by the formulation of the complementary energy approach, and we close with a comparison of stress distributions from the two methods.

8.3.2 Free Edge Deformations

The influence of fiber orientation and stacking sequence on the deformations in the boundary layer near the free edge will be demonstrated through presentation of deformed finite-element grids. Results for cross-ply, angle-ply, and quasi-isotropic laminates of T300/5208 carbon/epoxy are presented in order to demonstrate various facets of the free edge problem. These finite-element results were obtained using a four-node isoparametric element in a computer code written expressly for this problem (Buczek et al., 1983). The approach follows the original work of Renieri and Herakovich (1976).

The same finite-element mesh (Fig. 8.18) was used for all laminates. It consisted of 356 elements, 353 nodes, and 1028 degrees of freedom. A high density of smaller elements was used in the boundary layer region. There are 32 elements through the thickness at the edge. The number of elements per layer ranged from 8 to 24, depending upon the laminate configuration being studied. The triangular transition elements were obtained by double-numbering one of the nodes. All results are for an applied axial strain $\varepsilon_x^o = 0.1\%$.

8.3.2.1 Cross-Ply Laminates

Cross-ply laminates are the least complicated to consider because there are no off-axis layers. Hence the coefficients of mutual influence are identically zero for all layers, and there is no coupling between axial strains ε_x and shear strains γ_{xy} or between the shear strains γ_{yz} and the shear stresses τ_{zx}. Thus there are no interlaminar shear stresses τ_{zx}. This class of laminates also exhibits exact symmetry about the x-z plane, and hence the analysis of one-quarter of the cross section introduces no approximation in the solution, even for thick laminates.

Six configurations have been chosen to demonstrate the influence of edge effects in cross-ply laminates: $[0_2/90_2]_s$, $[90_2/0_2]_s$, $[0/90_3]_s$, $[90/0_3]_s$, $[(0/90)_2]_s$, and $[(90/0)_2]_s$. Deformed finite-element meshes for the six cross-ply laminates are shown in Fig. 8.19. All plots have been scaled such that the maximum V displacement equals the maximum W displacement. This was done to enhance the deformations so that local effects could be distinguished easily. Only the boundary layer portion of each laminate is shown.

The results in Fig. 8.19 clearly show the influence of edge effects in laminates with equal numbers, but different stacking sequences, of 0° and 90° layers. The mismatch in Poisson's ratios from layer to layer is evident in these laminates in the form of distinct variations in lateral contractions (v displacements) through the thickness. The influence of the percentage of 0° and 90° plies also is clearly demonstrated, as is the influence of stacking sequence.

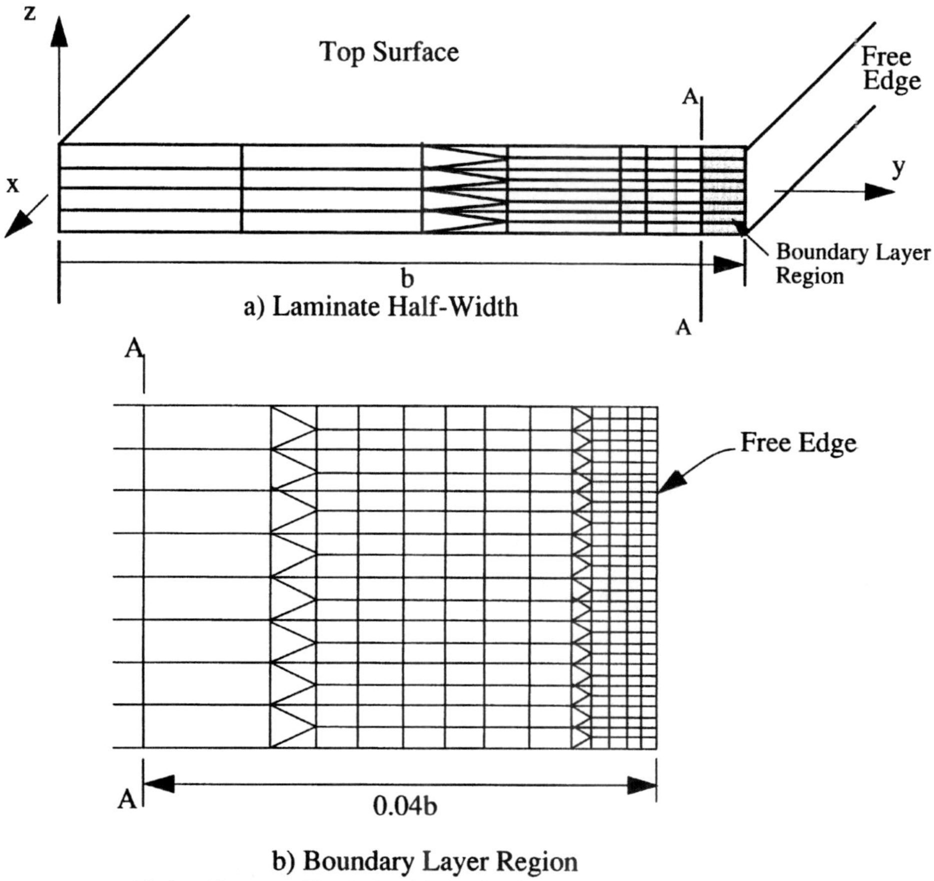

FIGURE 8.18 Finite-Element Mesh for Finite-Width Coupon

Figures such as these are very helpful for understanding edge effects. Close study of the deformed meshes provides insight into the distribution of interlaminar shear strains γ_{yz} and the interlaminar normal strains ε_z. These figures should not be used to assess the width of the boundary layer because different scale factors have been used in the *y*- and *z*-directions.

8.3.2.2 Angle-Ply Laminates

Balanced, symmetric laminates with fibers oriented at $+\theta$ and $-\theta$ to the loading direction provide fundamental insight into the influence of fiber orientation on interlaminar stresses. The interlaminar force and moment considerations discussed previously indicate that the interlaminar normal stress should be small or zero for angle-ply laminates, whereas the interlaminar shear stress τ_{zx} is expected to be quite large. Further, based upon the mismatch considerations of Fig. 8.13 and the Moiré fringe patterns of Fig. 8.16, it is anticipated that the interlaminar shear stress τ_{zx} will be maximum in the 10–15° fiber range. Four fiber orientations have been chosen to demonstrate interlaminar stresses in angle-ply laminates: ±10°, ±30°, ±45°, and ±60°. Two stacking sequences were investigated: $[\theta_2/-\theta_2]_s$ and $[(\theta/-\theta)_2]_s$. The scaling of the deformed meshes causes them to be independent of the stacking sequence for a given fiber orientation; hence, deformed meshes are presented only for the $[\theta_2/-\theta_2]_s$ stacking sequence.

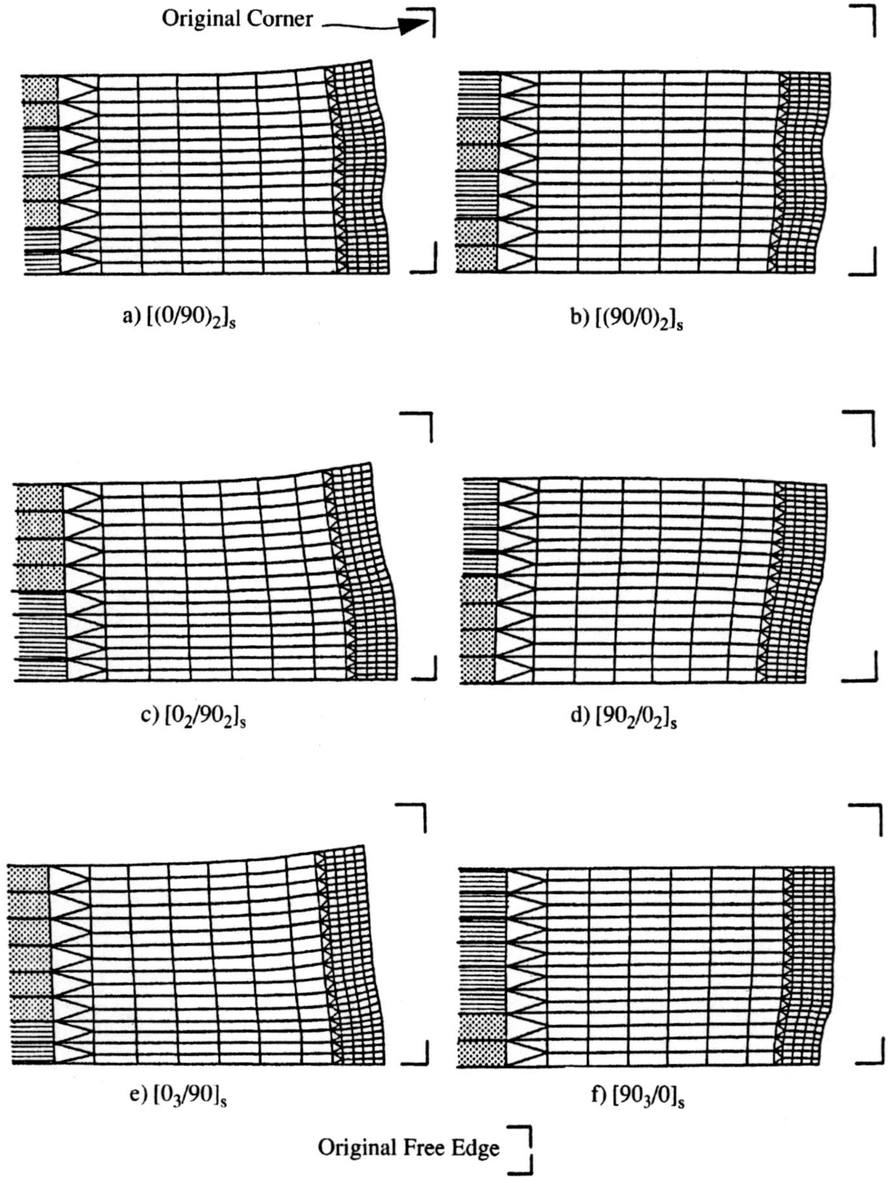

FIGURE 8.19 Deformed Finite-Element Meshes of Cross-Ply Laminates

The most striking feature of the deformed meshes for angle-ply laminates in Fig. 8.20 is that the edges are essentially straight. This is in sharp contrast to the deformed shapes for the cross-ply laminates and is a direct result of the equality of Poisson's ratio in $+\theta$ and $-\theta$ layers.

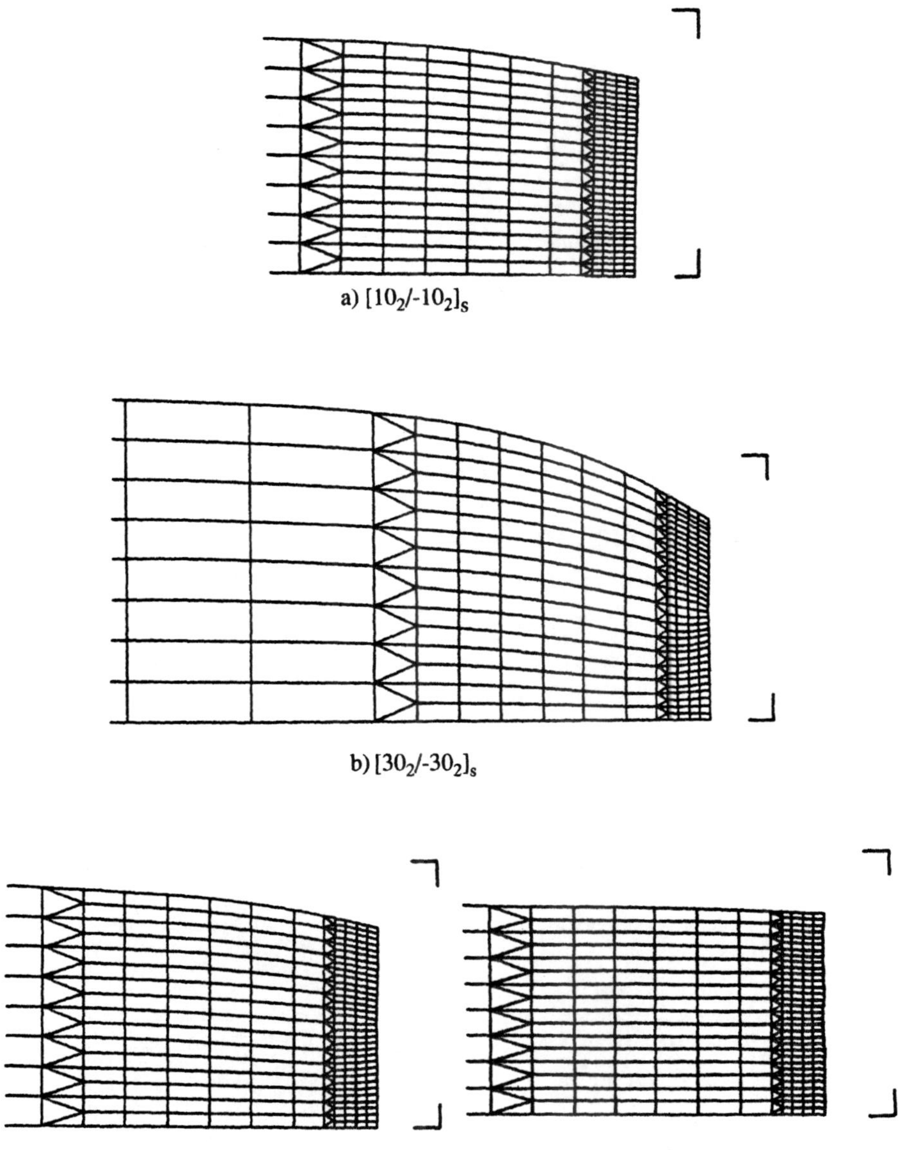

a) $[10_2/\text{-}10_2]_s$

b) $[30_2/\text{-}30_2]_s$

c) $[45_2/\text{-}45_2]_s$

d) $[60_2/\text{-}60_2]_s$

FIGURE 8.20 Deformed Shapes of Angle-Ply Laminates

Another most interesting feature of the deformed shapes for angle-ply laminates is that the thickness of the central region (away from the edges) of the ±30° laminate increases under the application of tensile load. This corresponds to a negative through-thickness effective laminate Poisson's ratio ν_{xz} and is consistent with the approximate expression for through-thickness Poisson's ratio developed in Chapter 5 (see Fig. 5.35). As discussed in Chapter 5, this surprising phenomenon is the result of the laminated material behaving as a structure with compressive transverse stresses σ_2 for some fiber orientations of carbon/epoxy when the laminate is under tensile load (Fig. 5.21). The compressive transverse stresses result in expansion of the individual layers in the thickness direction. As indicated in Figs. 5.35 and 8.20, the through-thickness Poisson's ratio is negative only for a limited range of fiber orientations.

The phenomenon of negative Poisson's ratio is not limited to angle-ply laminates. It can also occur in other laminates involving combinations of angle-ply layers with other fiber orientations, e.g., $[0/\pm\theta]_s$.

The two-dimensional nature of the deformed meshes in Fig. 8.20 does not permit direct assessment of the large through-thickness gradients of axial displacements on the free edge that are evident in the Moiré fringe patterns of Fig. 8.16. However, the influence of these shear deformations on the overall *y* and *z* displacements is clearly evident in the figure.

8.3.2.3 Quasi-Isotropic Laminates

Quasi-isotropic laminates with an equal number of +45°, –45°, 0°, and 90° layers are used in practice because, as shown in Chapter 5, the in-plane elastic properties are isotropic. Assuming interchangeability of +45° and –45° layers, quasi-isotropic laminates of this type can be divided into two groups: those with adjacent ±45° layers and those with interspersed ±45° layers. The 12 unique configurations are presented in Table 8.1.

The deformed shapes of these 12 laminates are shown in Figs. 8.21 and 8.22. It is clear from these figures that stacking sequence can have a dramatic effect on the deformed shape. These figures show that edge effects result mainly in contraction of the free edge for some laminates (e.g., $[90/45/0/-45]_s$). In other laminates the edge effects manifest themselves primarily as expansion (e.g., $[\pm45/0/90]_s$). And in yet other laminates the edge effects cause a combination of expansion and contraction along the edge (e.g., $[0/\pm45/90]_s$). The effect of the large differences in Poisson's ratio from layer to layer is apparent as variable curvature along the edge of the coupon. This is in sharp contrast to the angle-ply laminates, which are free of Poisson mismatch and exhibit essentially straight edges.

Interspersed ±45° Laminates	Adjacent ±45° Laminates
$[90/45/0/-45]_s$	$[90/0/\pm45]_s$
$[0/-45/90/45]_s$	$[90/\pm45/0]_s$
$[45/90/0/-45]_s$	$[0/90/\pm45]_s$
$[45/90/-45/0]_s$	$[0/\pm45/90]_s$
$[45/0/90/-45]_s$	$[\pm45/90/0]_s$
$[45/0/-45/90]_s$	$[\pm45/0/90]_s$

TABLE 8.1 Quasi-Isotropic Laminates

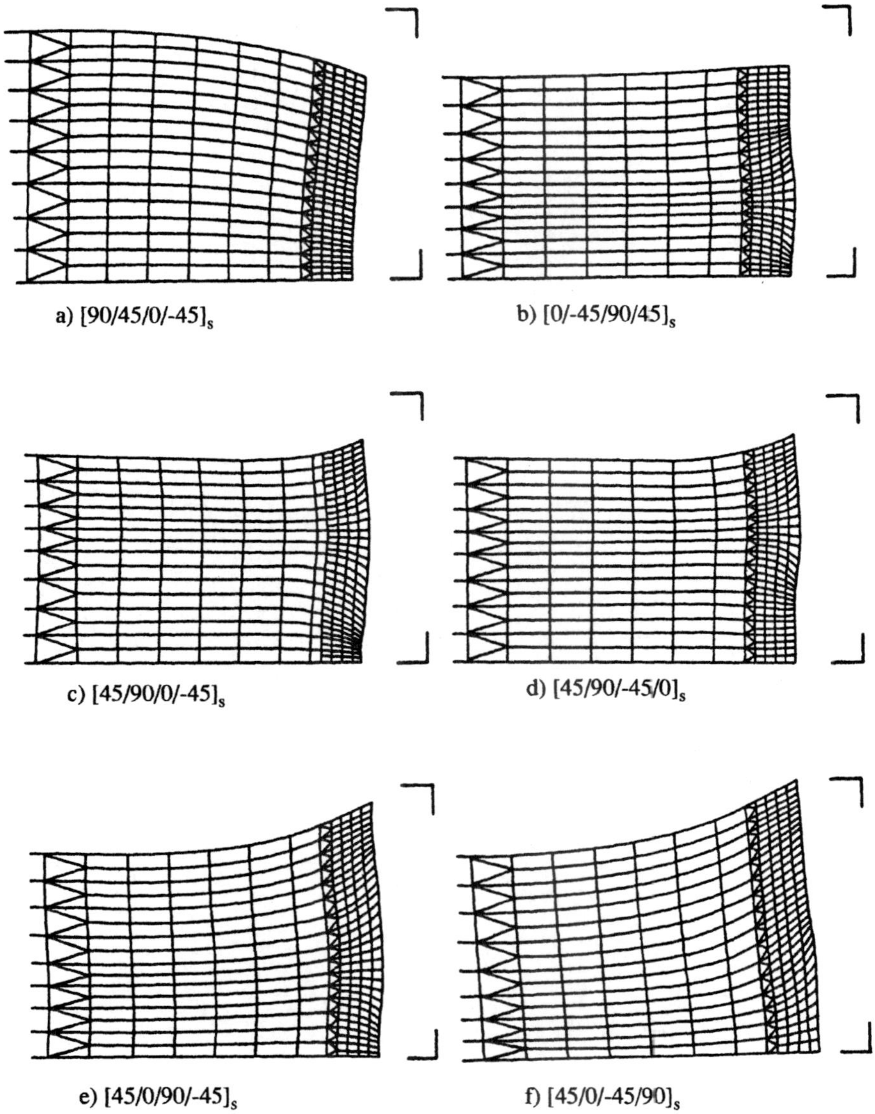

FIGURE 8.21 Deformed Quasi-Isotropic Laminates: Interspersed ±45° Layers

It should be noted that all of these quasi-isotropic laminates have identical through-thickness contraction in the central region away from the edge. Differences in the figures are the result of the normalization scheme used for this presentation.

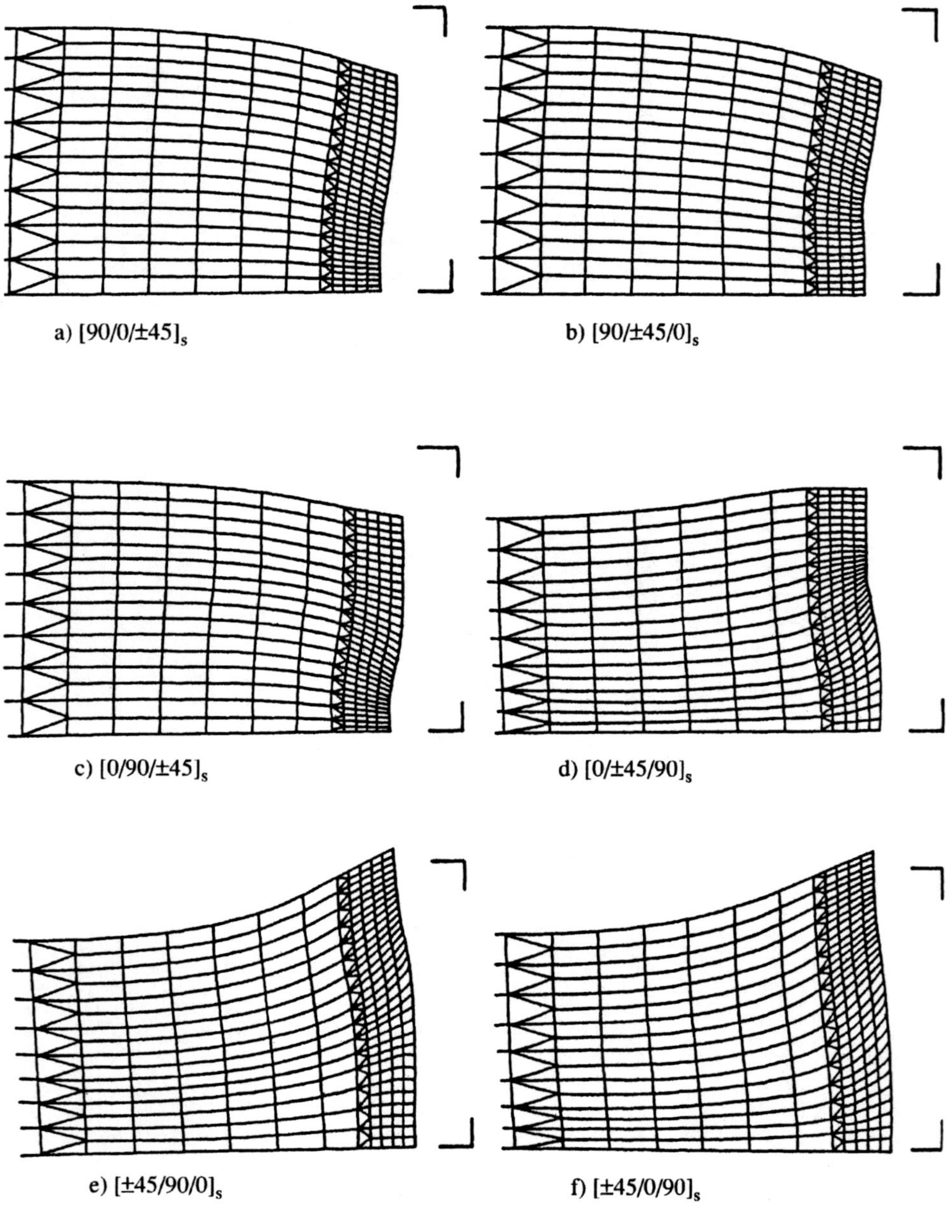

a) $[90/0/\pm45]_s$ b) $[90/\pm45/0]_s$

c) $[0/90/\pm45]_s$ d) $[0/\pm45/90]_s$

e) $[\pm45/90/0]_s$ f) $[\pm45/0/90]_s$

FIGURE 8.22 Deformed Quasi-Isotropic Laminates: Adjacent ±45° Layers

8.4 Complementary Energy Formulation

An approximate analytical solution for the free edge problem based upon complementary energy was presented by Kassapoglou and Lagace (1986). This solution, designated the KL solution herein, assumes admissible stress states that satisfy global and local equilibrium, stress-free boundary conditions, and traction continuity conditions at interfaces between plies. Rose and Herakovich (1991, 1992, 1993) refined the KL solution by adding additional terms to the admissible stress state to include local mismatch effects. Their solution, designated the RH solution herein, also includes provision for curvature associated with bending loads or unsymmetric laminates.

The geometry and coordinate systems used in the solution are shown in Fig. 8.23 for a laminate of half-width b. Local $(x, y, z^{(k)})$ coordinates are defined in each layer, with x parallel to x_1, $y = b - x_2$ measured from the free edge, and $z^{(k)}$ (positive up) having its origin at the bottom of the kth layer. The layers are numbered from 1 at the top to N at the bottom of the laminate. The solution is for a long, finite-width coupon under uniform axial strain loading ε_{11}^{o}. Away from the ends, the states of stress and strain are independent of the axial coordinate x_1. Individual layers are assumed to be homogeneous, monoclinic materials, and the layers are assumed to be perfectly bonded to one another. The half-width b is assumed to be sufficiently large that classical lamination theory stresses are recovered away from the edge. We note that full tensor notation is used throughout this development.

In the RH solution, statically admissible stress states are assumed (for each ply) to be the superposition of three effects: (a) global equilibrium, (b) coefficient of mutual influence mismatch, and (c) Poisson mismatch. It is further assumed that the y and z dependence in the assumed stress states are functionally independent. The complete form for the assumed statically admissible stress state in the RH solution can be written

$$\sigma_{ij}^{(k)} = [\sigma_{ij}^{(k)}]_E + [\sigma_{ij}^{(k)}]_\eta + [\sigma_{ij}^{(k)}]_\nu \tag{8.27}$$

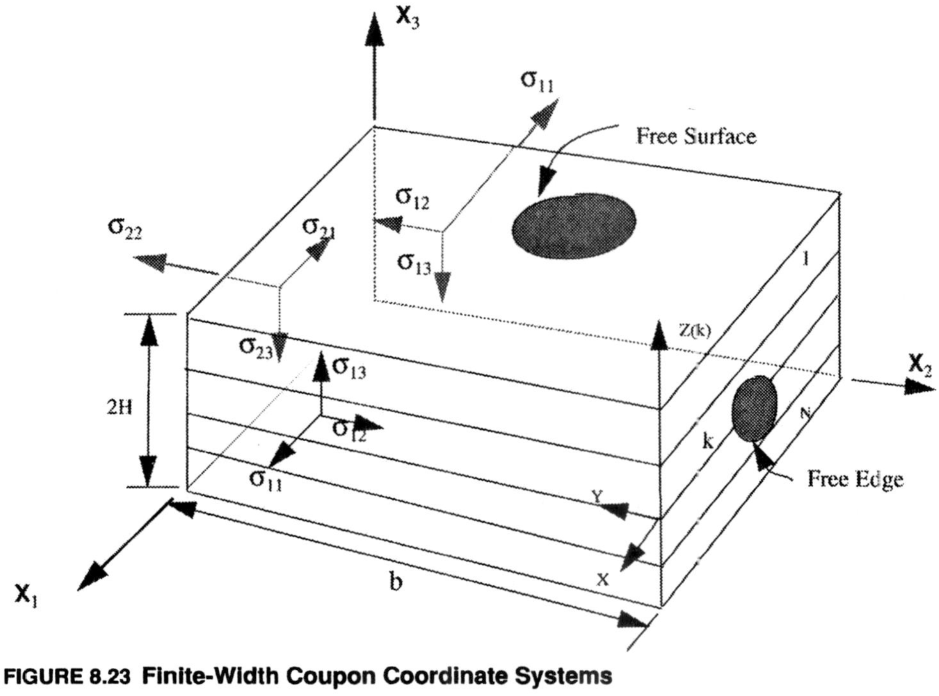

FIGURE 8.23 Finite-Width Coupon Coordinate Systems

where the individual contributions are

- Global equilibrium:

$$[\sigma_{ij}^{(k)}]_E = f_{ij}^{(k)}(y)g_{ij}^{(k)}(z) \qquad (i = 1, 2, 3), (j = 2, 3) \tag{8.28}$$

- Coefficient of mutual influence mismatch:

$$[\sigma_{ij}^{(k)}]_\eta = h_{ij}^{(k)}(y)l_{ij}^{(k)}(z) \qquad (i = 1), (j = 2, 3) \tag{8.29}$$

- Poisson's ratio mismatch:

$$[\sigma_{ij}^{(k)}]_\nu = m_{ij}^{(k)}(y)n_{ij}^{(k)}(z) + m_{ij}^{(k)}(y)p_{ij}^{(k)}(z) \qquad (i = 2, 3), (j = 2, 3) \tag{8.30}$$

The axial stresses in each layer, $\sigma_{11}^{(k)}$, are determined from Hooke's law and the strain displacement equations. In the following, the less mathematically involved KL solution for extension only is developed in order to highlight the procedure without employing the more involved mathematics of the full RH solution. The complete RH solution for extension and bending is then simply presented.

8.4.1 Global Equilibrium Formulation (KL Solution)

The KL global equilibrium solution for extensional loading is based upon assumed admissible stress states in the form given in (8.28) and repeated as (8.31):

$$\sigma_{ij}^{(k)} = f_{ij}^{(k)}(y)g_{ij}^{(k)}(z) \qquad (i = 1, 2, 3), (j = 2, 3) \tag{8.31}$$

Using the assumed forms (8.31) in the pointwise equilibrium equations (2.24), with all stress components independent of x, results in the following system of six ordinary differential equations for the 10 unknown functions $f_{ij}(y)$ and $g_{ij}(z)$ in each layer.

$$\frac{d}{dy}f_{12}^{(k)}(y) = f_{13}^{(k)}(y) \tag{8.32}$$

$$\frac{d}{dy}f_{22}^{(k)}(y) = f_{23}^{(k)}(y) \qquad \frac{d}{dy}f_{23}^{(k)}(y) = f_{33}^{(k)}(y) \tag{8.33}$$

$$\frac{d}{dz}g_{13}^{(k)}(z) = g_{12}^{(k)}(z) \tag{8.34}$$

$$\frac{d}{dz}g_{23}^{(k)}(z) = g_{22}^{(k)}(z) \qquad \frac{d}{dz}g_{33}^{(k)}(z) = g_{23}^{(k)}(z) \tag{8.35}$$

From the preceding it is evident that if four functions are assumed for each layer, the remaining functions for that layer can be determined from the six equations (8.32)–(8.35). It is mathematically convenient to assume $g_{12}^{(k)}$, $g_{22}^{(k)}$, $f_{22}^{(k)}$, and $f_{12}^{(k)}$.

Far from the free edge, classical lamination theory is recovered and (for in-plane loading of a symmetric laminate) the σ_{12} and σ_{22} stresses are constant within each layer. Thus, the $g_{ij}(z)$ functions associated with these stresses must be constants. Without loss of generality we can choose these constants to have a value of 1.0. Defining these constants in the kth layer as $B_1^{(k)}$ and $B_3^{(k)}$ and solving for the remaining g_{ij} functions from (8.32) and (8.33) in terms of constants of integration $B_i^{(k)}$ gives

$$
\begin{aligned}
g_{12}^{(k)}(z) &= B_1^{(k)} = 1 \\
g_{22}^{(k)}(z) &= B_3^{(k)} = 1 \\
g_{13}^{(k)}(z) &= z + B_2^{(k)} \\
g_{23}^{(k)}(z) &= z + B_4^{(k)} \\
g_{33}^{(k)}(z) &= \frac{z^2}{2} + B_4^{(k)} z + B_5^{(k)}
\end{aligned}
\tag{8.36}
$$

Constants $B_2^{(k)}$, $B_4^{(k)}$, and $B_5^{(k)}$ are functions of the laminate stacking sequence.

The remaining functions $f_{ij}^{(k)}(y)$ are determined as follows. Assume $f_{22}^{(k)}(y)$ to have the form

$$
f_{22}^{(k)}(y) = A_1^{(k)} e^{-\phi y} + A_2^{(k)} e^{-\lambda\phi y} + A_3^{(k)}
\tag{8.37}
$$

where $A_i^{(k)}$, ϕ, and λ are unknown constants. From (8.33) we have

$$
f_{23}^{(k)}(y) = \frac{df_{22}^{(k)}}{dy} = -A_1^{(k)} \phi e^{-\phi y} - A_2^{(k)} \lambda \phi e^{-\lambda\phi y}
\tag{8.38}
$$

and

$$
f_{33}^{(k)}(y) = \frac{df_{23}^{(k)}}{dy} = A_1^{(k)} \phi^2 e^{-\phi y} + A_2^{(k)} \lambda^2 \phi^2 e^{-\lambda\phi y}
\tag{8.39}
$$

Likewise, assuming $f_{12}^{(k)}(y)$ to have the form

$$
f_{12}^{(k)}(y) = A_4^{(k)} + A_5^{(k)} e^{-\phi y}
\tag{8.40}
$$

and using (8.32) gives

$$
f_{13}^{(k)}(y) = \frac{df_{12}^{(k)}}{dy} = -A_5^{(k)} \phi e^{-\phi y}
\tag{8.41}
$$

where a total of five unknown constants $A_i^{(k)}$ have been introduced.

All unknown constants $B_i^{(k)}$ and $A_i^{(k)}$ are determined by imposition of the stress-free edge boundary conditions, recovery of the CLT stresses far from the edge, and satisfaction of the traction continuity between layers and stress-free conditions on the top and bottom surfaces. The result from each imposed condition can be summarized as follows.

From (8.31), (8.36), (8.37), and (8.40), recovery of CLT stresses $\tilde{\sigma}_{22}^{(k)}$ and $\tilde{\sigma}_{12}^{(k)}$ for $y >> 1$ requires that

$$
A_3^{(k)} = \tilde{\sigma}_{22}^{(k)}
\tag{8.42}
$$

and

$$A_4^{(k)} = \tilde{\sigma}_{12}^{(k)} \tag{8.43}$$

The free edge condition $\sigma_{12}^{(k)} = 0$ at $y = 0$ with use of (8.40) and (8.43) gives

$$A_5^{(k)} = -\tilde{\sigma}_{12}^{(k)} \tag{8.44}$$

The conditions $\sigma_{22}^{(k)} = 0$ and $\sigma_{23}^{(k)} = 0$ at $y = 0$ in (8.37) and (8.38) with the use of (8.42) give

$$A_1^{(k)} = \frac{-\lambda\tilde{\sigma}_{22}^{(k)}}{(\lambda - 1)} \tag{8.45}$$

$$A_2^{(k)} = \frac{\tilde{\sigma}_{22}^{(k)}}{(\lambda - 1)} \tag{8.46}$$

The remaining $B_i^{(k)}$ constants are determined from the traction continuity requirements between layers and the stress-free conditions on the top and bottom surfaces. For $\sigma_{13}^{(k)}(t^{(k)}) = \sigma_{13}^{(k-1)}(0)$ at the interface between layers k and $k-1$, (8.31), (8.36), (8.41), and (8.44) give

$$B_2^{(k)} = \frac{\sum_{j=k+1}^{N} \tilde{\sigma}_{12}^{(j)} t^{(j)}}{\tilde{\sigma}_{12}^{(k)}} \tag{8.47}$$

For $\sigma_{23}^{(k)}(t^{(k)}) = \sigma_{23}^{(k-1)}(0)$ at the interface between layers k and $k-1$, (8.31), (8.36), (8.38), (8.45), and (8.46) give

$$B_4^{(k)} = \frac{\sum_{j=k+1}^{N} \tilde{\sigma}_{22}^{(j)} t^{(j)}}{\tilde{\sigma}_{22}^{(k)}} \tag{8.48}$$

For $\sigma_{33}^{(k)}(t^{(k)}) = \sigma_{33}^{(k-1)}(0)$ at the interface between layers k and $k-1$, (8.31), (8.36), (8.39), (8.45), (8.46), and (8.48) give

$$B_5^{(k)} = \frac{\sum_{j=k+1}^{N} \tilde{\sigma}_{22}^{(j)} t^{(j)} \left(\frac{t^{(j)}}{2} + \sum_{m=k+1}^{j-1} t^{(m)} \right)}{\tilde{\sigma}_{22}^{(k)}} \tag{8.49}$$

Combining these results gives the final form for the assumed stresses in the kth layer (due to equilibrium considerations only).

$$\sigma_{22}^{(k)} = \tilde{\sigma}_{22}^{(k)}\left[1 - \frac{\lambda}{\lambda - 1}\left(e^{-\phi y} - \frac{e^{-\lambda\phi y}}{\lambda}\right)\right] \tag{8.50}$$

$$\sigma_{33}^{(k)} = \tilde{\sigma}_{22}^{(k)}\phi^2\frac{\lambda}{\lambda - 1}\left(\lambda e^{-\lambda\phi y} - e^{-\phi y}\right)\left(\frac{z^2}{2} + B_4^{(k)}z + B_5^{(k)}\right) \tag{8.51}$$

$$\sigma_{23}^{(k)} = \tilde{\sigma}_{22}^{(k)}\phi\frac{\lambda}{\lambda - 1}\left(e^{-\phi y} - e^{-\lambda\phi y}\right)\left(z + B_4^{(k)}\right) \tag{8.52}$$

$$\sigma_{13}^{(k)} = \tilde{\sigma}_{12}^{(k)}\phi e^{-\phi y}(z + B_2^{(k)}) \tag{8.53}$$

$$\sigma_{12}^{(k)} = \tilde{\sigma}_{12}^{(k)}(1 - e^{-\phi y}) \tag{8.54}$$

In the above expressions the same ϕ and λ were used in the functions assumed for $f_{12}(y)$ and $f_{22}(y)$. Clearly, such an assumption results in a statically admissible stress state. Also, ϕ and λ were taken to be the same constants for all layers. This is required by the traction continuity conditions at layer interfaces. (It is noted that a slightly different approach was taken by Rose and Herakovich, 1992.)

The axial stress is determined by using the x independence of stress and strain in conjunction with the strain-displacement equations (2.16) and constitutive equations (3.86). The final result is

$$\sigma_{11}^{(k)} = \frac{1}{S_{11}^{(k)}}\left[S_{11}^{(k)}\tilde{\sigma}_{11}^{(k)} + S_{12}^{(k)}\tilde{\sigma}_{22}^{(k)} + S_{16}^{(k)}\tilde{\sigma}_{12}^{(k)} - (S_{12}^{(k)}\sigma_{22}^{(k)} + S_{13}^{(k)}\sigma_{33}^{(k)} + S_{16}^{(k)}\sigma_{12}^{(k)})\right] \tag{8.55}$$

Equations (8.50)–(8.55) represent a complete set of statically admissible stresses expressed in terms of two unknown constants ϕ and λ. These constants are determined by minimizing the complementary energy, as explained in the following section. (Note that CLT stresses and therefore also $B_i^{(k)}$ are assumed known.)

8.4.2 Complementary Energy Minimization

The complementary energy, as defined in (2.57), includes the integral of the complementary strain energy density over the volume V of interest minus the integral of the work done over the portion of the boundary, S_D, where either each component of displacement is prescribed or the corresponding component of traction is zero. For a laminated composite the integrals are written as summations of integrals over the layers of the laminate. For axial strain loading ε_{11}^{o} of a unit-length coupon, the boundary integral is nonzero only over the cross section $x_1 = 1.0$ where the axial displacement is prescribed and the shear stresses are taken to be zero. The complementary energy can be written

$$\Pi_c = \sum_{k=1}^{N}\Pi_c^{(k)} = \sum_{k=1}^{N}\left(\frac{1}{2}\left(\int_{A_k}\{\sigma\}^T[\bar{S}]\{\sigma\}dA\right) - \int_{A_k}\sigma_{11}\bar{u}_1\, dA\right) \tag{8.56}$$

or

$$\sum_{k=1}^{N}\Pi_c^{(k)} = \sum_{k=1}^{N}\int_{A_k}\Big((1/2)(R_{22}\sigma_{22}^{(k)2} + R_{33}\sigma_{33}^{(k)2} + R_{66}\sigma_{12}^{(k)2} + R_{44}\sigma_{23}^{(k)2} + R_{55}\sigma_{13}^{(k)2})$$

$$+ (R_{23}\sigma_{33}^{(k)}\sigma_{22}^{(k)} + R_{45}\sigma_{13}^{(k)}\sigma_{23}^{(k)} + R_{26}\sigma_{22}^{(k)}\sigma_{12}^{(k)} + R_{36}\sigma_{12}^{(k)}\sigma_{33}^{(k)}) \tag{8.57}$$

$$+ \frac{1}{\bar{S}_{11}}(\bar{S}_{12}\sigma_{22}^{(k)} + \bar{S}_{13}\sigma_{33}^{(k)} + \bar{S}_{16}\sigma_{12}^{(k)})(\bar{u}_1)\Big)dA$$

where $\bar{u}_1$ is the prescribed displacement at $x_1 = 1.0$, A_k is the cross-sectional area of the kth ply, and $\bar{S}_{ij}$ are compliance coefficients in the reduced notation defined in Chapter 3 ($i, j = 1, 2, 3, \ldots, 6$). The reduced compliances R_{ij} (defined in reduced notation) are

$$R_{ij} = \bar{S}_{ij} - \frac{\bar{S}_{1i}\bar{S}_{1j}}{\bar{S}_{11}} \tag{8.58}$$

The unknowns ϕ and λ are determined from the conditions for stationary values of Π_c, i.e.,

$$\frac{\partial \Pi_c}{\partial \lambda} = 0 \tag{8.59}$$

$$\frac{\partial \Pi_c}{\partial \phi} = 0 \tag{8.60}$$

Written out explicitly, the two simultaneous equations for ϕ and λ are

$$\frac{\partial \Pi_c}{\partial \lambda} = \lambda^4\phi^4 f_2 + 2\lambda^3\phi^4 f_2 + \lambda^2\{2(f_6 + f_9 + f_1) + \phi^2(2f_{11} + f_3 - 2f_{10} - 2f_8)\}$$
$$+ \lambda(4f_6 + 8f_9 + 6f_1) + 2f_6 + 4f_9 + 3f_1 = 0 \tag{8.61}$$

$$\frac{\partial \Pi_c}{\partial \phi} = 3\lambda^3\phi^4 f_2 + \phi^2(\lambda\{\lambda(f_4 + 2f_{11} + f_3 - 2f_{10} - 2f_8) + f_4\})$$
$$+ \lambda^2(2f_6 + f_5 + 2f_7 + 6f_9 + 3f_1)$$
$$+ \lambda(4f_6 + f_5 + 2f_7 + 8f_9 + 5f_1) + 2f_6 + 4f_9 + 3f_1 = 0 \tag{8.62}$$

where the f_i are involved functions of the layer thicknesses $t^{(k)}$, compliances $S_{ij}^{(k)}$, CLT stresses $\tilde{\sigma}_{ij}^{(k)}$, and constants $B_i^{(k)}$ (see Kassapoglou and Lagace, 1986).

Equations (8.61) and (8.62) are nonlinear, algebraic equations. There are 16 pairs of ϕ and λ values that minimize the complementary energy. Only the pairs with real and positive roots are admissible and, among these, the pair that provides the lowest Π_c is the "best" solution.

8.4.3 Extension/Bending Formulation with Mismatch Terms

The overall procedure when mismatch terms are included in the assumed statically admissible stress functions is very similar to that used for the global equilibrium formulation. The main difference is that statically admissible stress functions with additional terms associated with Poisson mismatch and coefficient of mutual influence mismatch are assumed. These additional terms are developed by defining the interlaminar shear stresses and the interlaminar normal stress at each layer interface and deriving the remaining stress components from these definitions and the differential equations of equilibrium. This approach has the advantage that the traction continuity conditions between plies are satisfied by the forms of the assumed stresses. Inclusion of bending causes the lamination theory strains to vary across the thickness of individual layers. This is incorporated in the solution when curvature is present.

The interlaminar stresses arising from material property mismatch at an interface are assumed to be proportional to the mismatch in coefficient of mutual influence or the mismatch in Poisson's ratio between the plies above and below the interface of interest. The mismatch in the coefficient of mutual influence is assumed to affect only the σ_{12} and σ_{13} components of stress, while the Poisson's ratio mismatch is assumed to affect only the σ_{22}, σ_{23}, and σ_{33} components of stress.

For layers numbered as indicated in Fig. 8.24, the mismatches in coefficient of mutual influence at the two interfaces bounding the kth layer are defined

$$\begin{aligned} \delta\eta_{12,1}(k,1) &= \eta_{12,1}(k-1)-\eta_{12,1}(k) \\ \delta\eta_{12,1}(k,2) &= \eta_{12,1}(k)-\eta_{12,1}(k+1) \end{aligned} \tag{8.63}$$

with the surface terms defined to be zero, i.e., $\delta\eta_{12,1}(1,1) = \delta\eta_{12,1}(N,2) = 0$.

The mismatches in Poisson's ratio are defined in similar fashion:

$$\begin{aligned} \delta\nu_{12}(k,1) &= \nu_{12}(k-1)-\nu_{12}(k) \\ \delta\nu_{12}(k,2) &= \nu_{12}(k)-\nu_{12}(k+1) \end{aligned} \tag{8.64}$$

with $\delta\nu_{12}(1,1) = \delta\nu_{12}(N,2) = 0$.

The in-plane functions $h_{ij}^{(k)}(y)$ and $m_{ij}^{(k)}(y)$ of (8.29) and (8.30) are assumed to be explicit combinations of exponential functions, with unknown decay rates, chosen in such a way that the interlaminar stress components arising from local mismatch *integrate* to zero over the coupon half-width. For the $m_{ij}^{(k)}(y)$ functions, the additional restriction that $m_{33}^{(k)}(y)$ has a zero moment about the longitudinal x_1 axis is also imposed. Thus, the mismatch effects permit additional nonzero stress contributions, but these contributions do not alter the global force and moment equilibrium established through the $f_{ij}^{(k)}(y)$ and $g_{ij}^{(k)}(z)$ functions.

The through-thickness functions $l_{ij}^{(k)}(z)$, $n_{ij}^{(k)}(z)$, and $p_{ij}^{(k)}(z)$ are polynomial functions chosen so that the stresses resulting from mismatch decay with distance from an interface. In order to keep the solution to a manageable number of unknowns, the through-thickness decay lengths of the mismatch effects are established a priori and set equal to the thickness t_k of the individual layers in the laminate. Details of the derivations are given in Rose and Herakovich (1992).

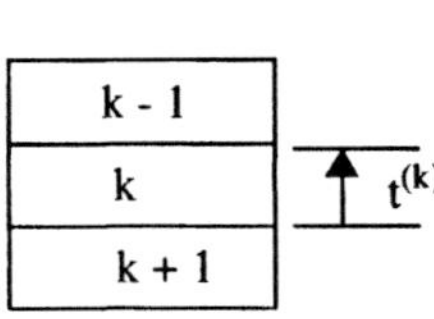

FIGURE 8.24 Layer Identification Sequence

The final forms for the assumed stresses in the RH solution include contributions for global equilibrium, Poisson and coefficient of mutual influence mismatch, and provision for bending defined by the strains $\varepsilon_t^{(k)}$ at the top of the kth layer and $\varepsilon_b^{(k)}$ at the bottom of the kth layer (for uniform in-plane loading $\varepsilon_x = \varepsilon_t^{(k)} = \varepsilon_b^{(k)}$). The assumed stresses are

$$\begin{aligned}\sigma_{12}^{(k)} &= (1-e^{-\phi_1 y})\cdot(B_1^{(k)}z+B_2^{(k)})-\frac{A_1}{\phi_3}\cdot(e^{-\lambda_2\phi_3 y}-e^{-\phi_3 y})\\ &\quad\cdot\left(\frac{2z[\delta\eta kt+\delta\eta kb]}{[t^{(k)}]^2}-\frac{2\delta\eta kb}{t^{(k)}}\right)\end{aligned} \tag{8.65}$$

$$\begin{aligned}\sigma_{13}^{(k)} &= \phi_1 e^{-\phi_1 y}\cdot\left(B_1^{(k)}\frac{z^2}{2}+zB_2^{(k)}+B_5^{(k)}\right)+A_1\cdot(\lambda_2 e^{-\lambda_2\phi_3 y}-e^{-\phi_3 y})\\ &\quad\cdot\left(\frac{z^2\delta\eta kt}{[t^{(k)}]^2}-\delta\eta kb\left(1-\frac{z}{[t^{(k)}]^2}\right)\right)\end{aligned} \tag{8.66}$$

$$\begin{aligned}\sigma_{22}^{(k)} &= \left[1-\frac{\lambda_1}{\lambda_1-1}\left(e^{-\phi_2 y}-\frac{1}{\lambda_1}e^{-\lambda_1\phi_2 y}\right)\right]\cdot(B_3^{(k)}z+B_4^{(k)})-\frac{A_2}{\phi_4}\cdot\delta\nu kt\\ &\quad\cdot\left(\frac{6z}{[z^{(k)}]^2}-\frac{2}{t^{(k)}}\right)\left(e^{-\phi_4 y}+\left(\frac{1-\lambda_4}{\lambda_4-\lambda_3}\right)e^{-\lambda_3\phi_4 y}+\left(\frac{\lambda_3-1}{\lambda_4-\lambda_3}\right)e^{-\lambda_4\phi_4 y}\right)\\ &\quad+(\delta\nu kb)\left(\frac{6z}{[t^{(k)}]^2}-\frac{4}{t^{(k)}}\right)-(V\delta\nu kt-V\delta\nu kb)\\ &\quad\cdot\frac{A_3}{\phi_5^2}\left(\frac{6}{[t^{(k)}]^2}-\frac{12z}{[t^{(k)}]^3}\right)\left(e^{-\phi_5 y}+\left(\frac{1-\lambda_6}{\lambda_6-\lambda_5}\right)e^{-\lambda_5\phi_5 y}+\left(\frac{\lambda_5-1}{\lambda_6-\lambda_5}\right)e^{-\phi_5\lambda_6 y}\right)\end{aligned} \tag{8.67}$$

$$\begin{aligned}\sigma_{23}^{(k)} &= \left[\phi_2\left(\frac{\lambda_1}{\lambda_1-1}\right)(e^{-\phi_2 y}-e^{-\lambda_1\phi_2 y})\right]\cdot\left(B_3^{(k)}\frac{z^2}{2}+B_4^{(k)}z+B_6^{(k)}\right)\\ &\quad+A_2\cdot\left(e^{-\phi_4 y}+\lambda_3\left(\frac{1-\lambda_4}{\lambda_4-\lambda_3}\right)e^{-\lambda_3\phi_4 y}+\lambda_4\left(\frac{\lambda_3-1}{\lambda_4-\lambda_3}\right)e^{-\lambda_4\phi_4 y}\right)\\ &\quad\cdot(\delta\nu kt)\left(\frac{3z^2}{[t^{(k)}]^2}-\frac{2z}{t^{(k)}}\right)+(\delta\nu kb)\left(\frac{3z^2}{[t^{(k)}]^2}-\frac{4z}{t^{(k)}}+1\right)+(V\delta\nu kt-V\delta\nu kb)\\ &\quad\cdot\left(\frac{6A_3}{\phi_5^2}\left(\frac{z}{[t^{(k)}]^2}-\frac{z^2}{[t^{(k)}]^3}\right)\left(e^{-\phi_5 y}+\lambda_5\left(\frac{1-\lambda_6}{\lambda_6-\lambda_5}\right)e^{-\lambda_5\phi_5 y}+\lambda_6\left(\frac{\lambda_5-1}{\lambda_6-\lambda_5}\right)e^{-\phi_5\lambda_6 y}\right)\right)\end{aligned} \tag{8.68}$$

$$\sigma_{33}^{(k)} = \left[\phi_2^2\left(\frac{\lambda_1}{\lambda_1 - 1}\right)\left(-e^{-\phi_2 y} + \lambda_1 e^{-\lambda_1 \phi_2 y}\right)\right] \cdot \left(B_3^{(k)}\frac{z^3}{6} + B_4^{(k)}\frac{z^2}{2} + zB_6^{(k)} + B_7^{(k)}\right)$$

$$- A_2\phi_4 \cdot \left(e^{-\phi_4 y} + \lambda_3^2\left(\frac{1-\lambda_4}{\lambda_4 - \lambda_3}\right)e^{-\lambda_3\phi_4 y} + \lambda_4^2\left(\frac{\lambda_3 - 1}{\lambda_4 - \lambda_3}\right)e^{-\lambda_4\phi_4 y}\right)$$

$$\cdot \left((\delta\nu kt)\left(\frac{z^3}{[z^{(k)}]^2} - \frac{z^2}{t^{(k)}}\right) + (\delta\nu kb)\left(\frac{z^3}{t^{(k^2)}} - \frac{2z^2}{t^{(k)}} + z\right)\right) \tag{8.69}$$

$$- A_3 \cdot \left(\left(e^{-\phi_5 y} + \lambda_5^2\left(\frac{1-\lambda_6}{\lambda_6 - \lambda_5}\right)e^{-\lambda_5\phi_5 y} + \lambda_6^2\left(\frac{\lambda_5 - 1}{\lambda_6 - \lambda_5}\right)e^{-\phi_5\lambda_6 y}\right)\right)$$

$$\cdot \left(V\delta\nu kt\left(\frac{3z^2}{[t^{(k)}]^2} - \frac{2z^3}{[t^{(k)}]^3}\right) - V\delta\nu kb\left(\frac{3z^2}{[t^{(k)}]^2} - \frac{2z^3}{t^{(k^3)}} - 1\right)\right)$$

where we have defined the following quantities:

$$\delta\eta kt = \delta\eta_{12,1}(k, 1)\varepsilon_t^{(k)} \tag{8.70}$$

$$\delta\eta kb = \delta\eta_{12,1}(k, 2)\varepsilon_b^{(k)} \tag{8.71}$$

$$\delta\nu kt = \delta\nu_{12}(k, 1)\varepsilon_t^{(k)} \tag{8.72}$$

$$\delta\nu kb = \delta\nu_{12}(k, 2)\varepsilon_b^{(k)} \tag{8.73}$$

$$V\delta\nu kt = \delta\nu_{12}(k, 2)\varepsilon_t^{(k)} \qquad (\pm) \tag{8.74}$$

$$V\delta\nu kb = \delta\nu_{12}(k, 2)\varepsilon_b^{(k)} \qquad (\pm) \tag{8.75}$$

The $\pm$ in (8.74) and (8.75) indicates that the positive value of the quantity is to be used for layers above the midplane and the negative value is to be used for layers below the midplane.

The constants $B_i^{(k)}$ $(i = 1, 4)$ for the solution including bending are defined in terms of the lamination theory stresses (denoted by ~) away from the free edge. Subscripts t and b denote the top and bottom, respectively, of a ply whose thickness is $t^{(k)}$.

$$B_1^{(k)} = \frac{\tilde{\sigma}_{12t}^{(k)} - \tilde{\sigma}_{12b}^{(k)}}{t^{(k)}} \tag{8.76}$$

$$B_2^{(k)} = \tilde{\sigma}_{12b}^{(k)} \tag{8.77}$$

$$B_3^{(k)} = \frac{\tilde{\sigma}_{22t}^{(k)} - \tilde{\sigma}_{22b}^{(k)}}{t^{(k)}} \tag{8.78}$$

$$B_4^{(k)} = \tilde{\sigma}_{22b}^{(k)} \tag{8.79}$$

The constants $B_i^{(k)}$ (i = 5, 6, and 7) are determined from the interfacial traction continuity conditions. Starting from the bottom free surface and working up, they have the forms

$$B_5^{(k)} = \sum_{j=k+1}^{N} \left(\frac{B_1^{(j)}(t^{(j)})^2}{2} + B_2^{(j)} t^{(j)} \right) \qquad (k = 1, N-1) \tag{8.80}$$

$$B_6^{(k)} = \sum_{j=k+1}^{N} \left(\frac{B_3^{(j)}(t^{(j)})^2}{2} + B_4^{(j)} t^{(j)} \right) \qquad (k = 1, N-1) \tag{8.81}$$

$$B_7^{(k)} = \sum_{j=k+1}^{N} \left(\left(\frac{B_3^{(j)}(t^{(j)})^3}{6} + \frac{B_4^{(j)}(t^{(j)})^2}{2} + \left(\frac{B_3^{(j)}(t^{(j)})^2}{2} + B_4^{(j)} t^{(j)} \right) \right) \left(\sum_{m=k+1}^{j-1} t^{(m)} \right) \right) \tag{8.82}$$

$$(k = 1, N-1)$$

with

$$B_5^{(N)} = B_6^{(N)} = B_7^{(N)} = 0 \tag{8.83}$$

When the complementary energy is minimized with these assumed stress states, the result is a system of 14 nonlinear algebraic equations in the unknowns ϕ_i ($i = 1, 5$), λ_j ($j = 1, 6$), and A_k ($k = 1, 3$). The nonlinear equations must be solved numerically. The "best" solution is the positive pair of ϕ_i and λ_j (and associated A_k) which give the lowest complementary energy. Negative values of ϕ_i and λ_j can be discarded because they correspond to stresses that grow in magnitude with distance from the free edge.

8.5 Interlaminar Stress Distributions

In the following sections results from finite-element analyses are compared with the KL and RH solutions. All results were obtained for a typical carbon/epoxy (properties in Table 8.2) subjected to axial strain loading $\varepsilon_x^o = \varepsilon_{11}^o = 0.1\%$. The finite-element results were obtained using either CLFE2D (Buczek et al., 1983) or a program by Norwood et al. (1990). Results are presented in the form of through-thickness distributions very near the free edge and interfacial stress distributions between layers over the last 20% of the width near the free edge for cross-ply, angle-ply, and quasi-isotropic laminates. All interlaminar stresses are normalized with respect to the average far-field applied axial stress. The normalized stresses are designated with an overbar, i.e., $\bar{\sigma}_{ij}$. As all laminates considered are symmetric, through-thickness distributions are presented for only the upper half of the laminate. It is noted that the magnitudes of the finite-element predictions near singular points are a function of the element type and size, and hence undue reliance should not be given to the values at these points. Away from these points the finite-element values are very reliable.

E_1, GPa (Msi)	$E_2 = E_3$, GPa (Msi)	$G_{12} = G_{13}$, GPa (Msi)	G_{23}, GPa (Msi)	$\nu_{12} = \nu_{13}$	ν_{23}
139 (20.0)	14.5 (2.1)	5.86 (0.85)	5.86 (0.85)	0.21	0.21

TABLE 8.2 Typical Carbon/Epoxy Properties

8.5.1 Cross-Ply Laminates

Since the coefficient of mutual influence, $\eta_{xy,x}$, is identically zero for 0° and 90° layers, cross-ply laminates isolate the influence of Poisson mismatch. The number of unknown parameters in the RH solution for this case reduces from 14 to 10. As has been discussed, the Poisson's ratio mismatch of cross-ply laminates gives rise to nonzero interlaminar shear stress σ_{23} and interlaminar normal stress σ_{33}. The distributions of the interlaminar normal stress along the 0/90 interface of $[0_4/90_4]_s$ and $[90_4/0_4]_s$ laminates are shown in Fig. 8.25, and the through-thickness distributions for the same two laminates are presented in Fig. 8.26.

It is evident from Fig. 8.25 that the interlaminar normal stress exhibits singular behavior near the free edge, has positive and negative segments, and thus is capable of being equivalent to a couple. Also, the distribution exhibits a different form when the stacking sequence is reversed. Reversing the stacking sequence does not result in a simple mirror image of the previous stress distribution. The finite-element and RH solutions exhibit a second reversal of the gradient near the edge that is a result of the local mismatch effects. The width of the boundary layer is approximately 10% of the laminate width.

The near-edge through-thickness distributions of the interlaminar normal stress for the two laminates under consideration (Fig. 8.26) show that the RH and finite-element solutions predict similar, asymmetric distributions (between laminates) with maximum tensile values at locations *within* the 90° layer for both stacking sequences. Neither the asymmetry nor the sign reversal, which are the results of local effects, are captured by the solution based solely upon equilibrium considerations. For both laminates, the KL solution predicts that the sign of the interlaminar normal stress is constant through the laminate thickness and that the maximum magnitude always occurs at the laminate midplane.

The interlaminar distributions of the shear stress σ_{23} for these same two cross-ply laminates are shown in Fig. 8.27. The stress is generally nonzero without sign reversal in the boundary layer. The stress-free boundary condition is recovered at the free edge. The maximum magnitude of the interlaminar shear stress is only about one-third the maximum magnitude of the interlaminar normal stress. All three solutions (KL, RH, and finite-element) predict similar distributions.

The through-thickness distributions of the interlaminar shear stress σ_{23} are shown in Fig. 8.28. The figures show that all three methods predict that the maximum interlaminar shear stress occurs at the 0/90 interface for both laminates. The finite-element and RH predictions are nonlinear with sign reversal within each layer. The KL solution predicts linear distributions of constant sign and significantly lower maximum values.

8.5.2 Angle-Ply Laminates

For a laminate composed of only $+\theta$ and $-\theta$ layers, the lack of Poisson mismatch reduces the number of necessary parameters in the RH solution to four: ϕ_1, ϕ_3, λ_2, and A_1. Further, analysis indicates that $\phi_1 = \phi_3$. The only interlaminar stress of significant interest in angle-ply laminates is the interlaminar shear stress σ_{13}. Typical interlaminar distributions for σ_{13} are presented for a $[(\pm10)_2]_s$ laminate (Fig. 8.29), and through-thickness distributions are presented for $[(\pm10)_2]_s$ and $[10_2/-10_2]_s$ laminates (Fig. 8.30). Generalizations are also presented for the $[(\pm\theta)_2]_s$ class of laminates (Fig. 8.31).

The distributions of σ_{13} along two interfaces of the $[(\pm10)_2]_s$ laminate are presented in Fig. 8.29. The figure shows that this component of stress is nonzero in the boundary layer and exhibits singular behavior near the free edge. The interfacial distribution is a function of the chosen interface, with the first (top) 10/–10 interface exhibiting a negative shear stress and the next interface (–10/10) exhibiting a negative/positive variation along the interface. These differences are a direct result of the combined mismatch and equilibrium considerations. All three predictions at the first interface are nearly identical. However, the equilibrium solution (KL) does not capture

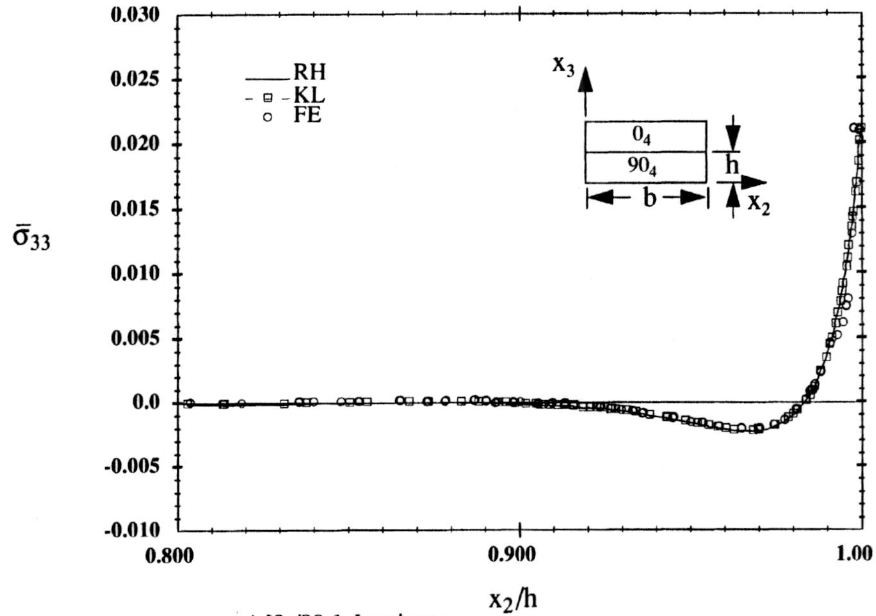

a) $[0_4/90_4]_s$ Laminate

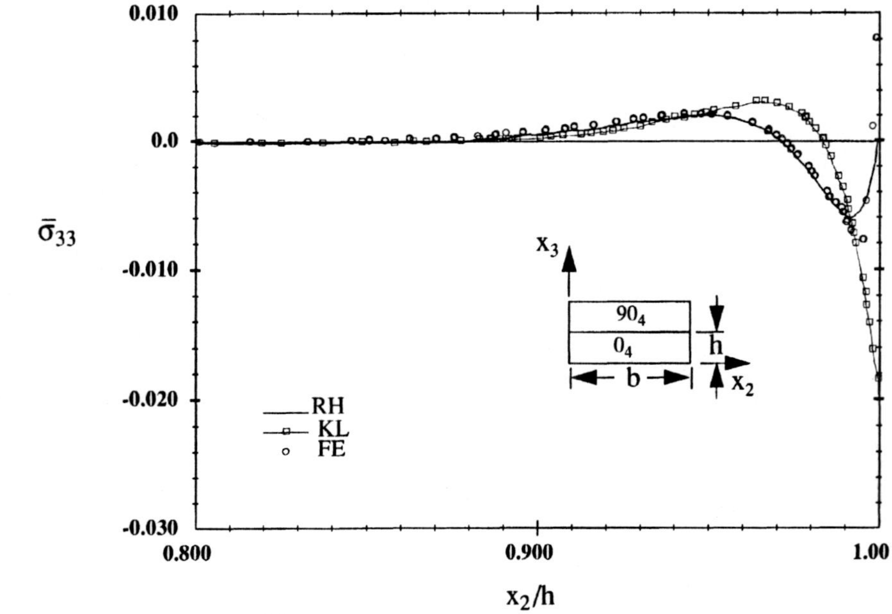

b) $[90_4/0_4]_s$ Laminate

FIGURE 8.25 Normal Stress σ_{33} along Interface of Cross-Ply Laminates

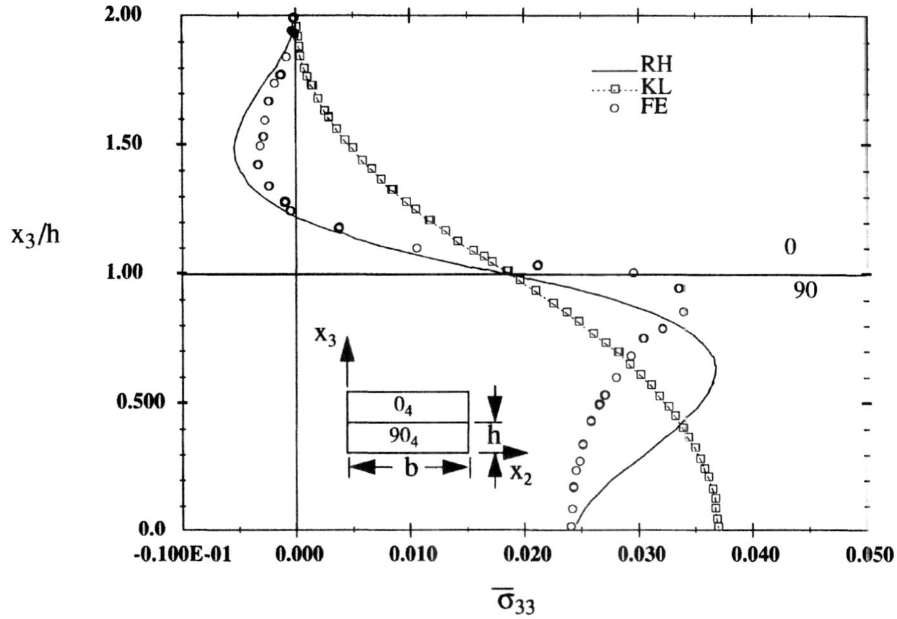

a) $[0_4/90_4]_s$ Laminate

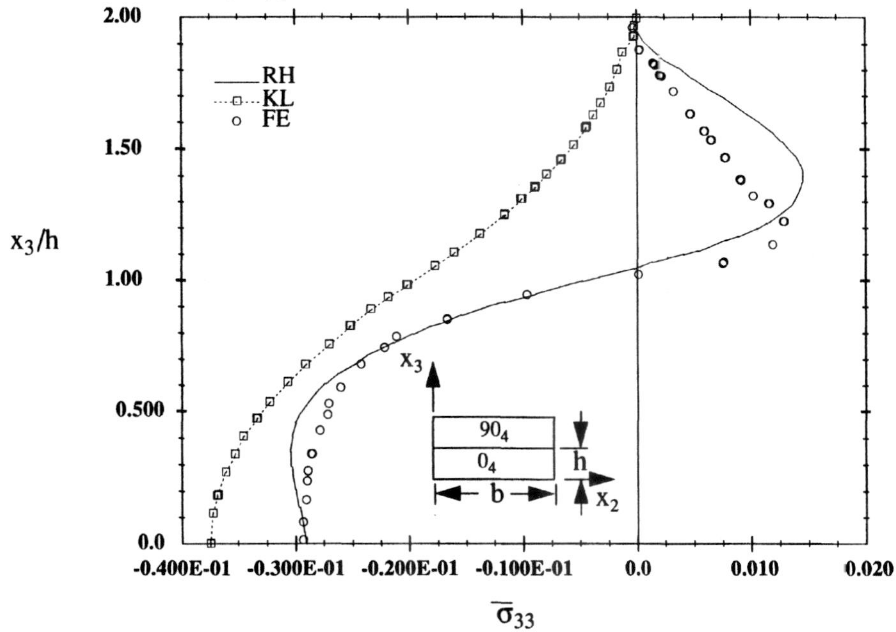

b) $[90_4/0_4]_s$ Laminate

FIGURE 8.26 Through-Thickness Normal Stress $\overline{\sigma}_{33}$ in Cross-Ply Laminates

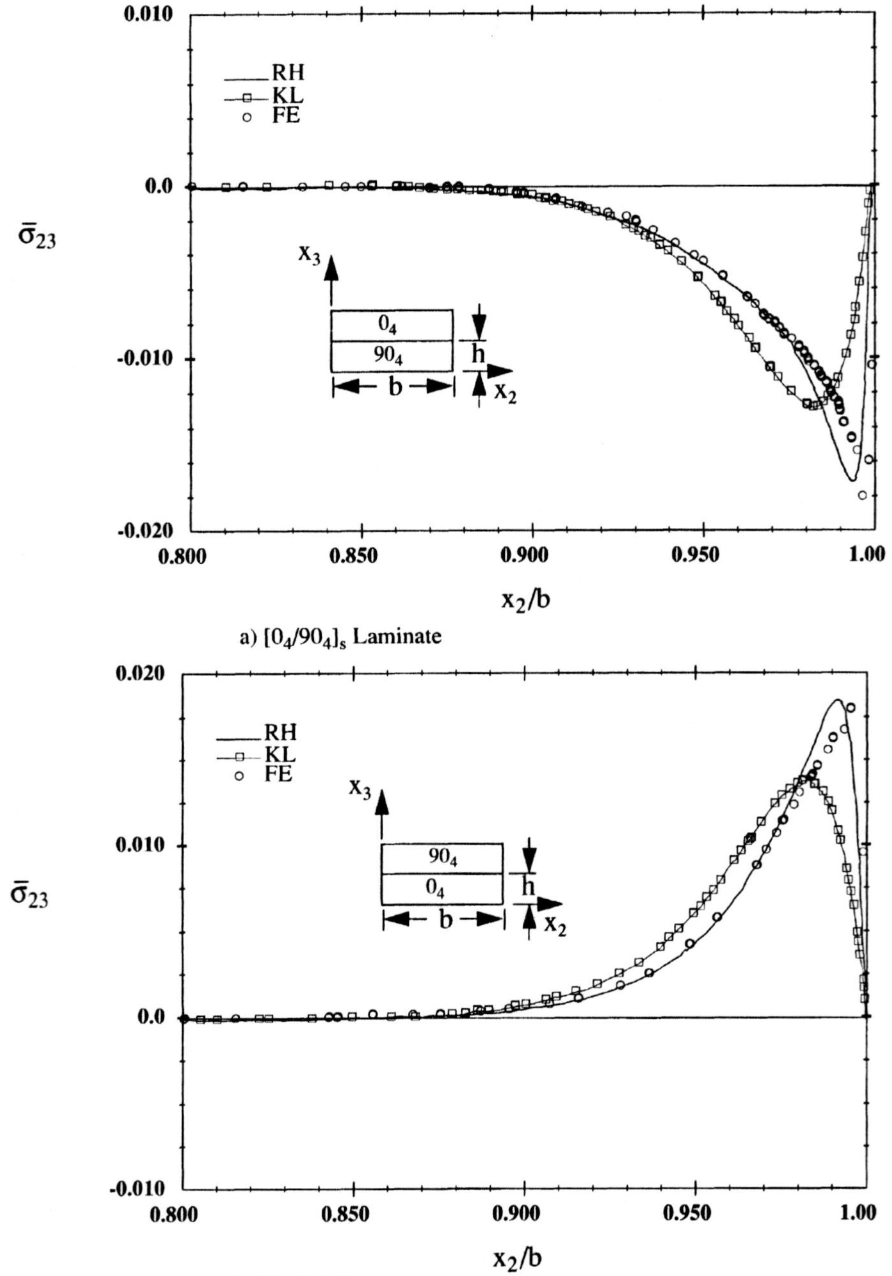

a) $[0_4/90_4]_s$ Laminate

b) $[90_4/0_4]_s$ Laminate

FIGURE 8.27 Shear Stress $\bar{\sigma}_{23}$ along Interface of Cross-Ply Laminat

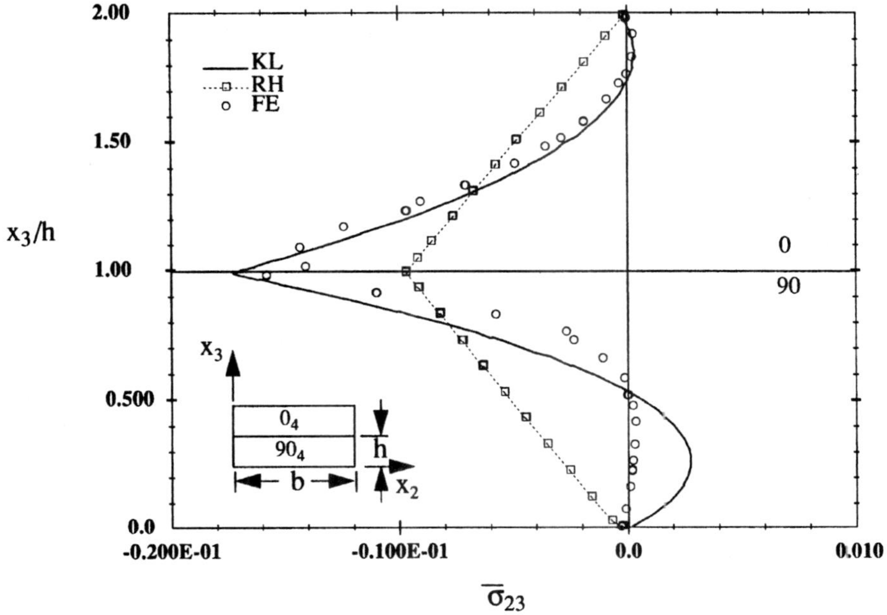

a) $[0_4/90_4]_s$ Laminate

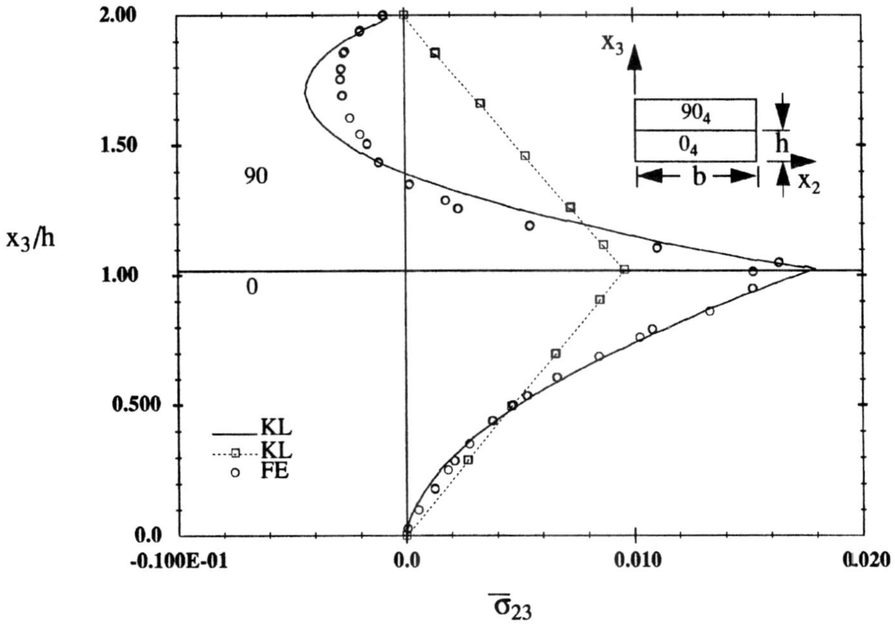

b) $[90_4/0_4]_s$ Laminate

FIGURE 8.28 Through-Thickness Shear Stress $\bar{\sigma}_{23}$ in Cross-Ply Laminates

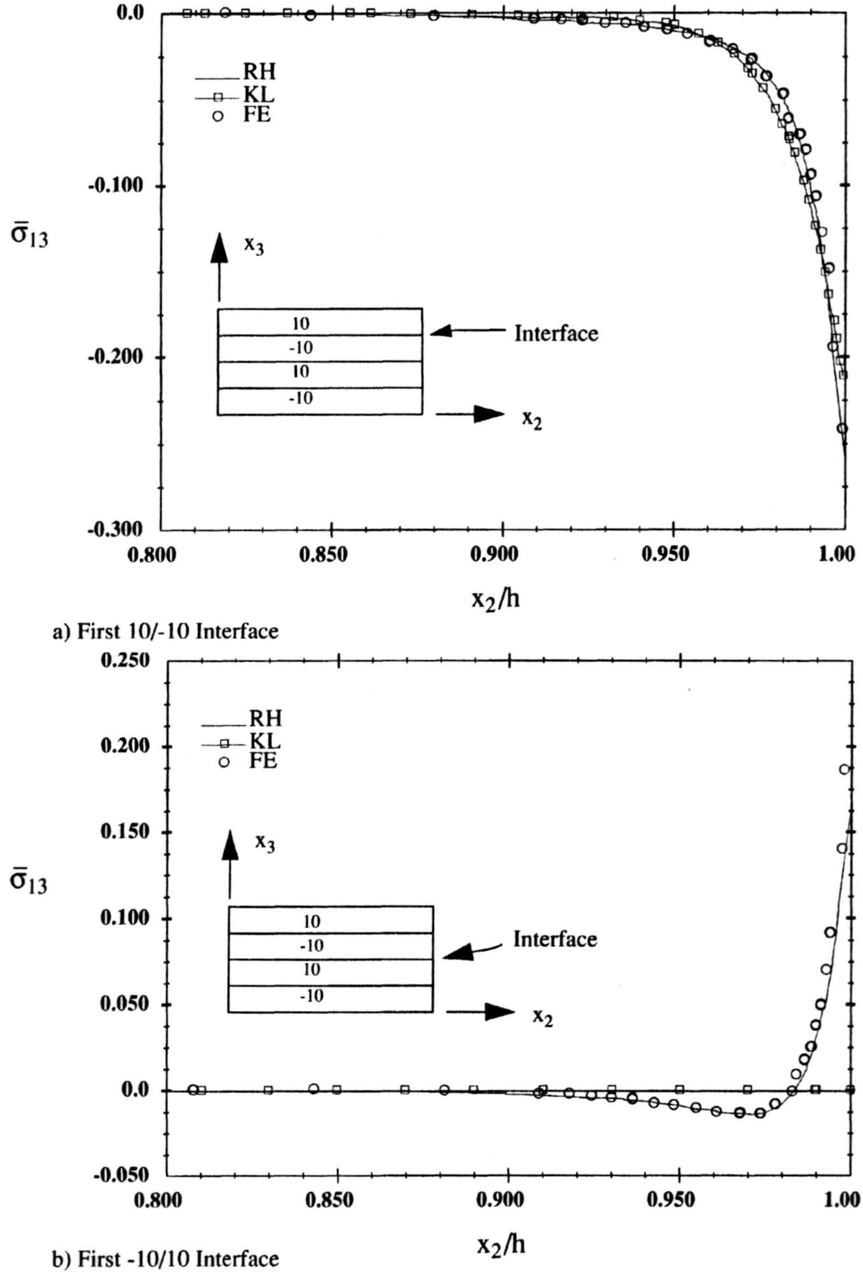

a) First 10/-10 Interface

b) First -10/10 Interface

FIGURE 8.29 Interlaminar σ_{13} in $[(\pm 10)_2]_s$ Laminate

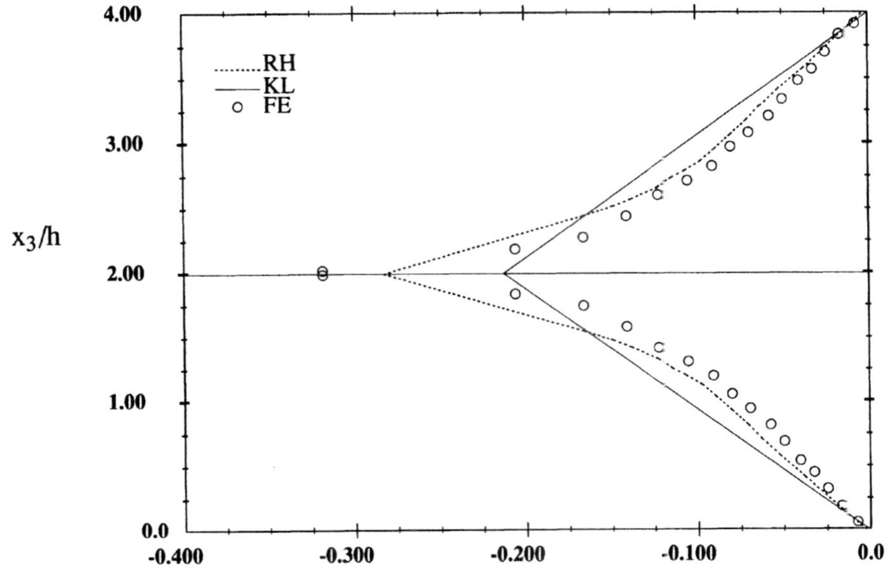

a) Clustered $[10_2/-10_2]_s$ $\overline{\sigma}_{13}$

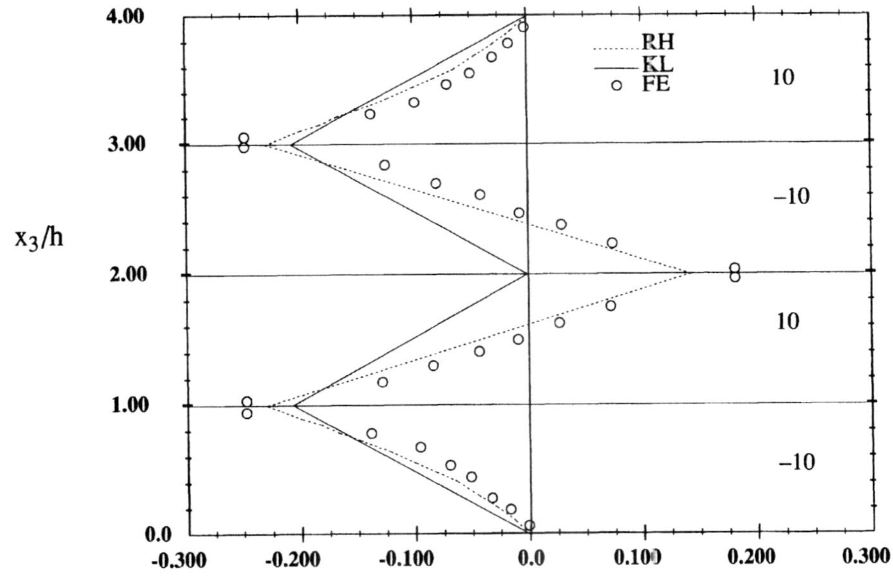

a) Alternating $[\pm 10_2]_s$ $\overline{\sigma}_{13}$

FIGURE 8.30 Through-Thickness σ_{13} in $[(\pm 10)_2]_s$ and $[10_2/-10_2]_s$ Laminates

the interlaminar stress at the –10/+10 (second) interface. Indeed, the solution based solely on equilibrium predicts the interlaminar shear stress to be identically zero along this interface. This prediction is a result of the satisfaction of global equilibrium by the CLT stresses away from the edge. However, interlaminar shear stress must be present because of the mismatch in $\eta_{xy,x}$. And, further, since global equilibrium is satisfied by CLT stresses, the distribution of σ_{13} must not disturb the equilibrium—thus the positive/negative character of the distribution. It is also noteworthy that the maximum stresses at the two interfaces considered are of the same order of magnitude.

The through-thickness distributions for $[10_2/-10_2]_s$ and $[(\pm 10)_2]_s$ laminates in Fig. 8.30 exhibit a distinctly different character as a function of stacking sequence. However, close examination of the curves shows that the gradient of the distribution changes sign at each +/– interface. The largest magnitude occurs at the interface of the $[10_2/-10_2]_s$ laminate, but all local maxima are of the same order of magnitude. As mentioned, the solution based solely on global equilibrium does not recover the nonzero interlaminar shear stress at the –10/+10 interface.

The maximum magnitudes of the interlaminar shear stresses in $[(\pm\theta)_2]_s$ and $[\theta_2/-\theta_2]_s$ angle-ply, carbon/epoxy laminates, as a function of fiber orientation in 5° increments, are shown in Fig. 8.31. The results indicate that the absolute maximum is at $\theta = 15°$ for both stacking sequences, with the maximum shear stress in the $[\theta_2/-\theta_2]_s$ laminate being approximately 25% higher than that in the $[(\pm\theta)_2]_s$ laminate. These predictions demonstrate a very close correlation in trend with the mismatch curve of Fig. 8.13.

It is instructive to study the θ dependence of the parameters $\phi_1 = \phi_3$, λ_2, and A_1 with fiber orientation for angle-ply laminates. The results indicate that λ_2 is independent of the fiber orientation but is a function of the stacking sequence. Values of $\phi_1 = \phi_3$, λ_2, and A_1 for clustered $[\theta_2/-\theta_2]_s$ and alternating $[(\pm\theta)_2]_s$ laminates of T300/5208 carbon/epoxy are given in Figs. 8.32 and 8.33. The parameter ϕ_1 attains a minimum at $\theta = 30°$ for both stacking sequences, and A_1 attains a maximum at $\theta = 25°$ for both stacking sequences. Values of λ_2 for the two stacking sequences were $\lambda_2 = 4.135$ for $[(\pm\theta)_2]_s$ and $\lambda_2 = 3.727$ for $[\theta_2/-\theta_2]_s$.

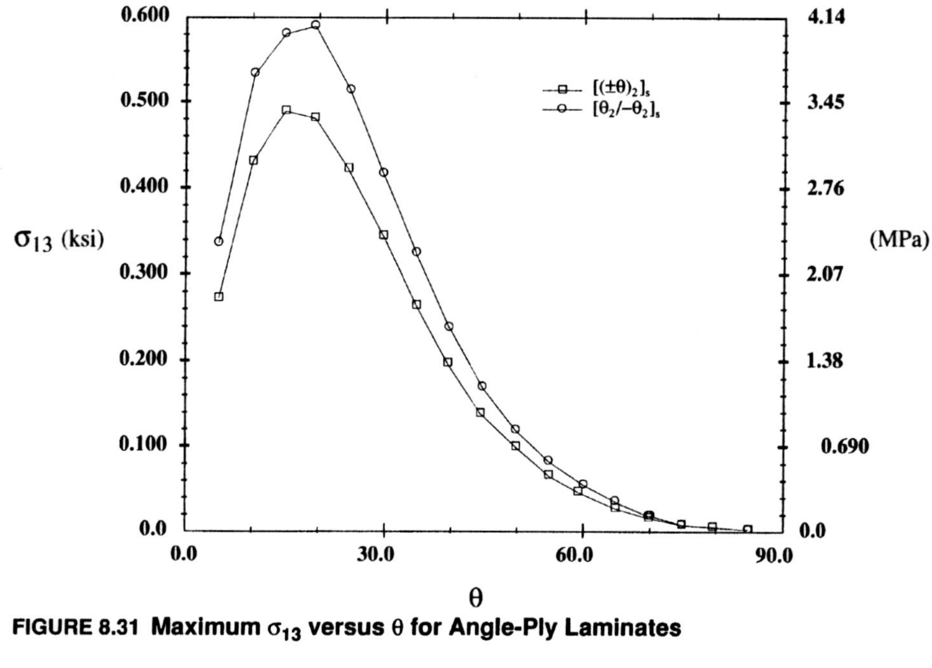

FIGURE 8.31 Maximum σ_{13} versus θ for Angle-Ply Laminates

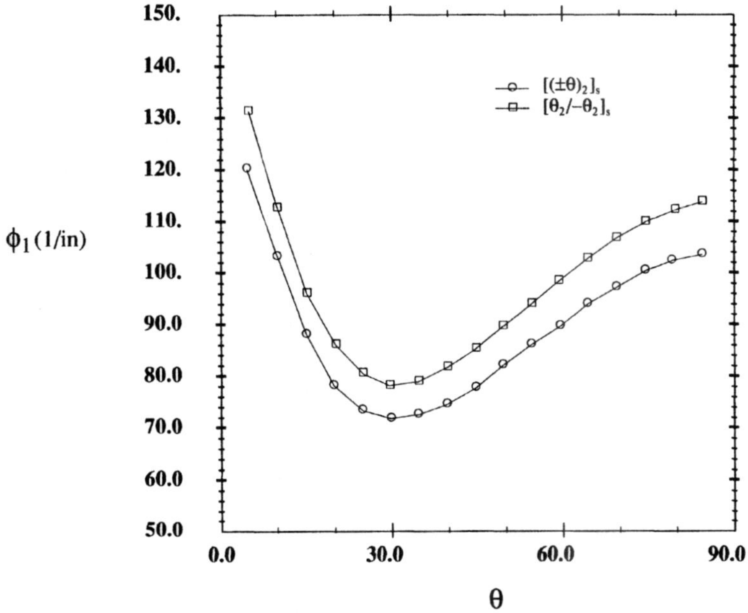

FIGURE 8.32 ϕ_1 versus θ for Angle-Ply Carbon/Epoxy

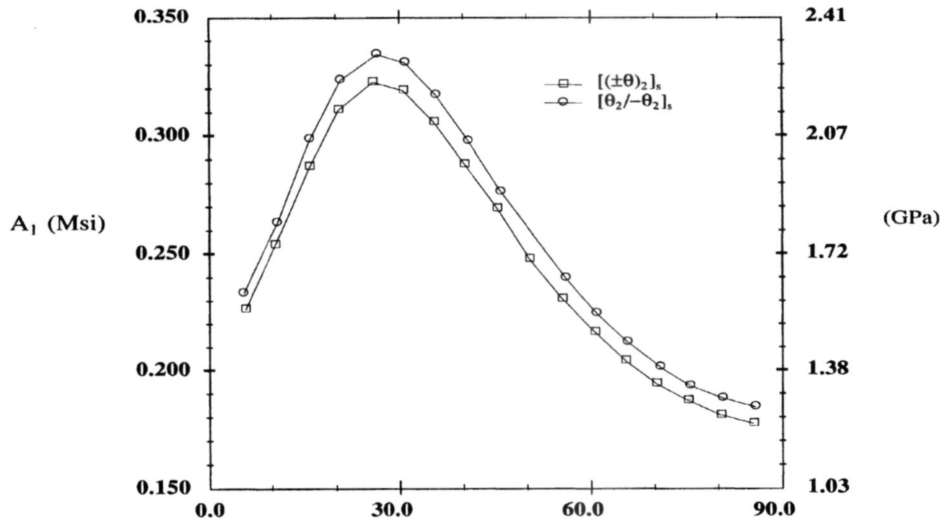

FIGURE 8.33 A_1 versus θ for Angle-Ply Carbon/Epoxy

8.5.3 Quasi-Isotropic Laminates

The $\pi/4$ quasi-isotropic laminates consist of an equal number of symmetric layers at 45° increments, resulting in fiber orientations of 0°, 90°, +45°, and –45°. As mentioned previously, if the +45° and –45° orientations are considered interchangeable, there are 12 unique symmetric stacking sequences. Values of the 14 parameters for the RH solution for 10 of the 12 quasi-isotropic laminates, made with a typical carbon/epoxy, are presented in Tables 8.3 and 8.4.

The distributions of interlaminar stresses as a function of x_2 are very similar to the curves shown in the figures for cross-ply and angle-ply laminates. All interlaminar stresses are generally nonzero in the boundary layer. The interlaminar normal stress σ_{33} and the interlaminar shear stress σ_{13} exhibit singular behavior at the intersection of the free edge and layer interfaces; the interlaminar shear stress σ_{23} is nonzero in the boundary layer but satisfies the stress-free boundary condition at $y = 0$. In view of the similarities with the previous results, no additional plots for the distribution of stresses along layer interfaces are presented.

Near-edge through-thickness stress distributions for three quasi-isotropic laminates ($[45/90/0/-45]_s$, $[90/45/0/-45]_s$, and $[0/\pm45//90]_s$, Figs. 8.34–8.40) will be discussed to show the influence of stacking sequence on the interlaminar stresses. These laminates are representative of the $\pi/4$ quasi-isotropic family. Included are two laminates with interspersed ±45° layers and one with adjacent ±45° layers. The stresses have been normalized with respect to the average far-field axial stress $\bar{\sigma}_{11}$.

The predictions shown in the three figures can be summarized as follows:

- All three components of interlaminar stress are generally nonzero throughout the boundary layer.
- The interlaminar normal stress σ_{33} and the interlaminar shear stress σ_{13} are significant, with maximum values in the range of 15–30% of the far-field applied axial stress.

Parameter	$[0/45/90/-45]_s$	$[90/45/0/-45]_s$	$[45/90/0/-45]_s$	$[45/90/-45/0]_s$	$[45/0/90/-45]_s$	$[45/0/-45/90]_s$
ϕ_1 (1/in)	66.93	50.02	54.86	66.66	53.10	53.32
ϕ_2 (1/in)	81.68	61.75	59.83	82.13	61.39	47.25
ϕ_3 (1/in)	57.73	99.81	47.06	57.31	41.75	67.31
ϕ_4 (1/in)	128.4	56.23	70.08	130.4	77.69	43.01
ϕ_5 (1/in)	59.99	174.5	95.28	72.33	128.1	41.81
λ_1	5.698	1.386	6.725	5.048	3.609	3.227
λ_2	4.223	3.189	7.801	4.178	7.492	4.040
λ_3	3.394	3.661	12.32	3.320	14.18	25.14
λ_4	8.207	21.11	11.90	7.882	7.074	6.388
λ_5	6.748	5.047	1.490	7.223	1.856	8.910
λ_6	18.68	0.352	1.572	12.59	3.578	5.618
A_1 (μ psi)	1.242	3.703	6.204	0.1234	0.0504	0.1564
A_2 (μ psi)	–0.4318	–0.3801	–0.1057	–0.4843	–0.1834	–0.1935
A_3 (μ psi)	–0.0009	–0.1002	–0.0386	0.0178	0.0978	–0.0025

TABLE 8.3 RH Parameters for Quasi-Isotropic Laminates with Interspersed ±45 Layers

Parameter	$[90/\pm45/0]_s$	$[0/90/\pm45]_s$	$[\pm45/90/0]_s$	$[0/\pm45/90]_s$
ϕ_1 (1/in)	54.75	56.90	56.94	61.75
ϕ_2 (1/in)	69.58	64.27	61.75	58.81
ϕ_3 (1/in)	158.6	154.6	99.81	97.46
ϕ_4 (1/in)	65.94	68.45	56.23	56.04
ϕ_5 (1/in)	81.06	71.07	90.95	62.39
λ_1	1.473	2.134	1.996	3.530
λ_2	2.469	2.191	4.203	4.302
λ_3	5.048	3.545	5.488	8.821
λ_4	11.73	17.73	15.96	19.86
λ_5	3.293	4.382	1.722	5.189
λ_6	5.739	6.984	6.155	5.082
A_1 (μ psi)	0.5838	0.6984	0.1682	0.1565
A_2 (μ psi)	–0.2084	–0.4332	0.03639	–0.1505
A_3 (μ psi)	–0.4212	0.0432	–0.1352	–0.1068

TABLE 8.4 RH Parameters for Quasi-Isotropic Laminates with Adjacent ±45 Layers

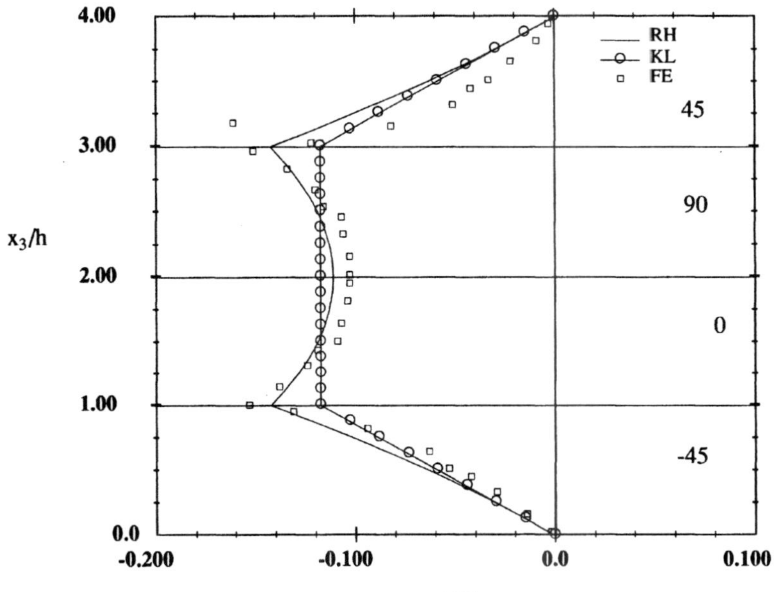

FIGURE 8.34 Near-Edge Interlaminar Stresses σ_{13} in a $[45/90/0/-45]_s$ Laminate

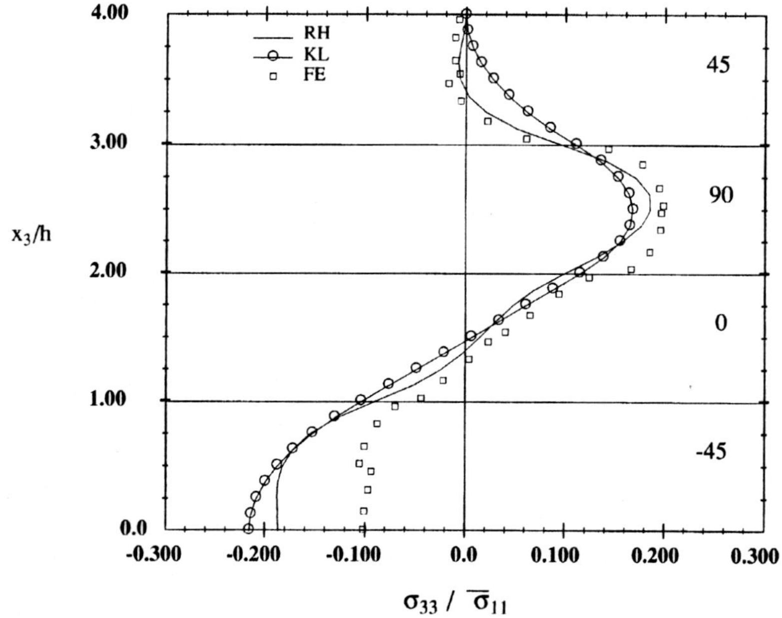

FIGURE 8.35 Near-Edge Interlaminar Stresses σ_{33} in a [45/90/0/–45]$_s$ Laminate

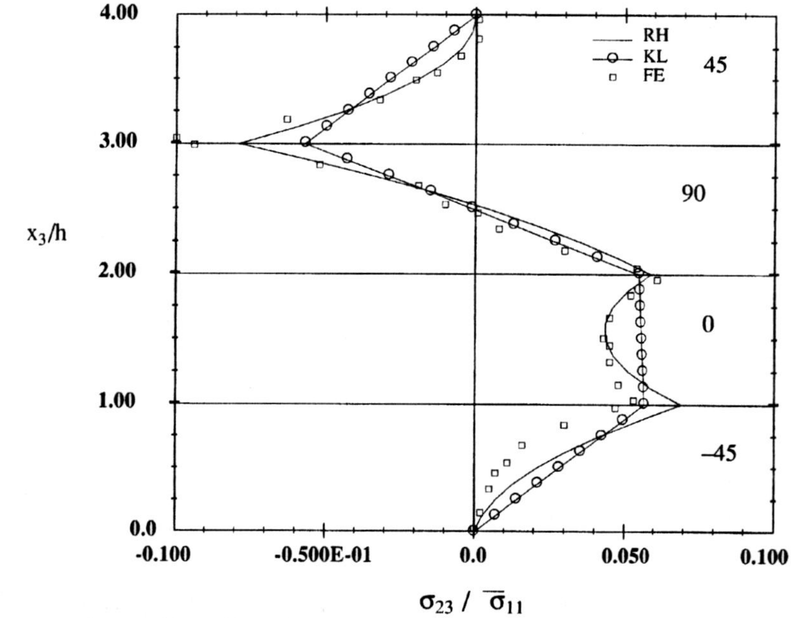

FIGURE 8.36 Near-Edge Interlaminar Shear Stress σ_{23} in a [45/90/0/–45]$_s$ Laminate

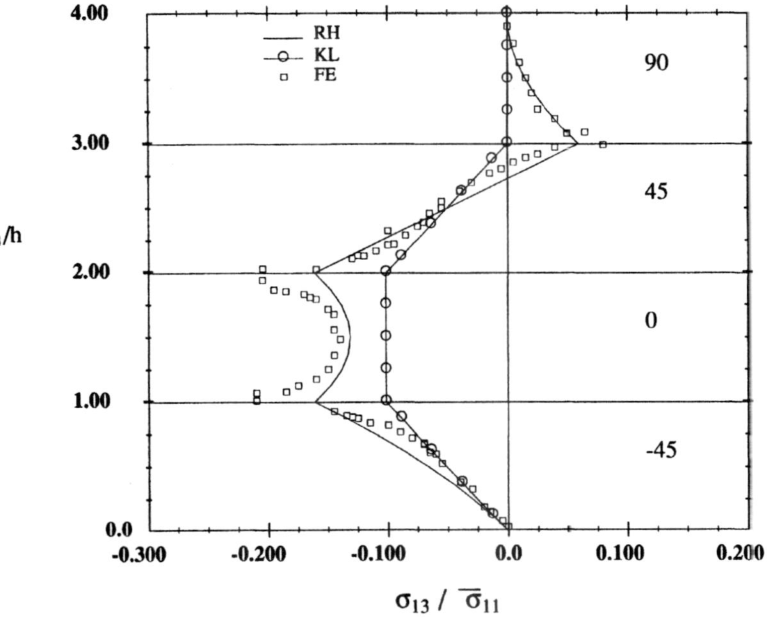

FIGURE 8.37 Interlaminar Stresses σ_{13} in a $[90/45/0/-45]_s$ Laminate

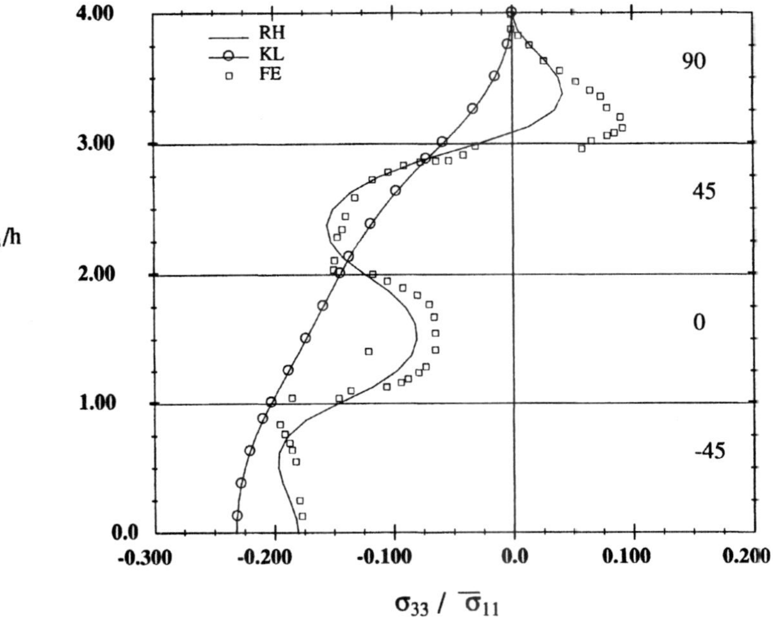

FIGURE 8.38 Interlaminar Stresses σ_{33} in a $[90/45/0/-45]_s$ Laminate

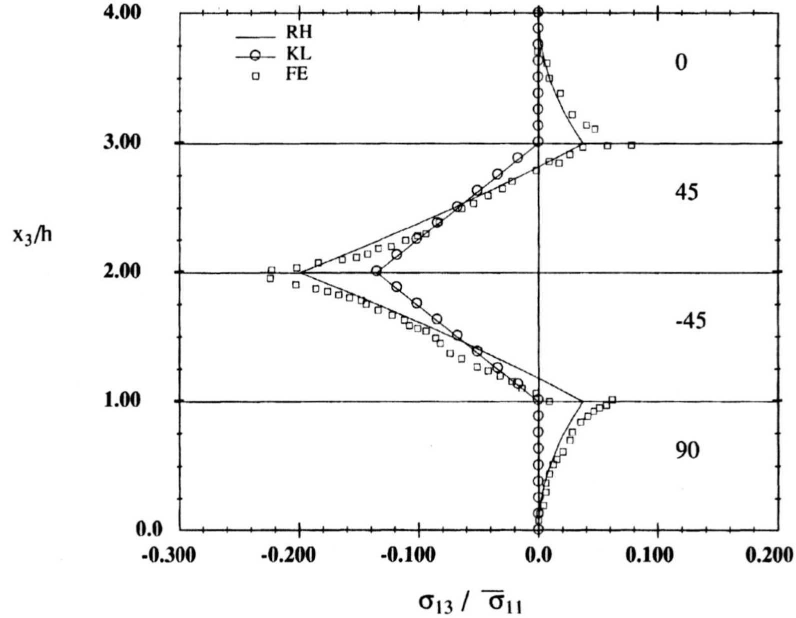

FIGURE 8.39 Interlaminar Stresses σ_{13} in a $[0/\pm45/90]_s$ Laminate

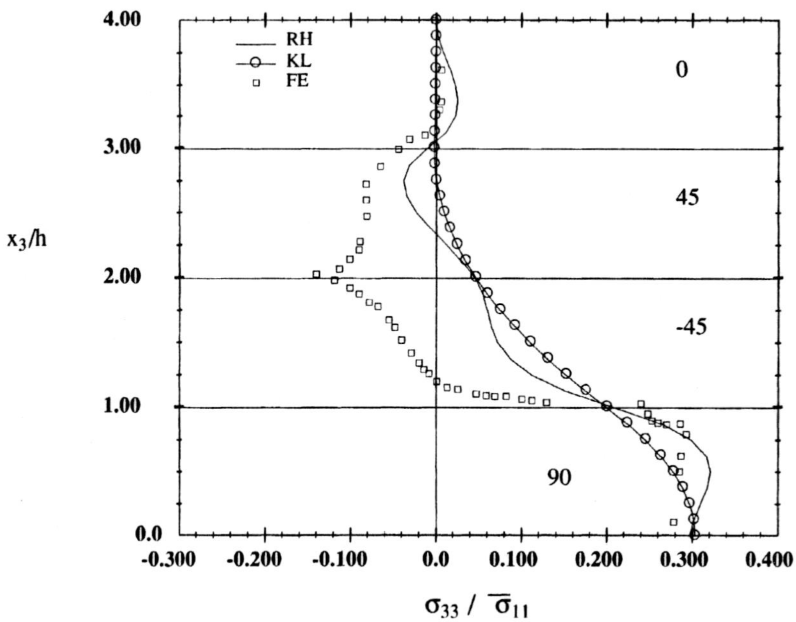

FIGURE 8.40 Interlaminar Stresses σ_{33} in a $[0/\pm45/90]_s$ Laminate

- The maximum interlaminar shear stresses σ_{23} are significantly lower than σ_{13} and σ_{33}.
- Stacking sequence plays an important role in the through-thickness distributions.
- Peak values of the interlaminar shear stress σ_{13} always occur at layer interfaces, with the largest values at the +45/–45 interface of the $[0/45/-45/90]_s$ laminate.
- Peak values of tensile interlaminar normal stress σ_{33} always occur *within* the 90° layer whereas peak compressive values occur in +45° or –45° layers.

$[45/90/0/-45]_s$ Laminate

Figures 8.34–8.36 show comparisons of the three methods of analysis under consideration for the through-thickness distributions of the interlaminar stresses near the edge of a $[45/90/0/-45]_s$ laminate. As the figures indicate, all three methods give similar results, with the finite-element and RH solutions being generally more alike. Both components of interlaminar shear stress exhibit peak values at layer interfaces, with the maximum σ_{13} at both the 45/90 and 0/–45 interfaces. The maximum σ_{23} occurs at the 45/90 interface. The maximum interlaminar normal stress occurs in the center of the 90° layer.

$[90/45/0/-45]_s$ Laminate

The results in Figs. 8.37 and 8.38 show through-thickness comparisons of all three methods for the interlaminar shear stress σ_{13} and the interlaminar normal stress σ_{33} for a $[90/45/0/-45]_s$ laminate. The results are similar to those for the previous laminate with the maximum interlaminar shear stress at the interface between the 0/±45° layers and the maximum interlaminar normal stress within the 90° layer for maximum tensile stress and within the –45° layer for the overall maximum magnitude, which is compressive. The distribution of σ_{33} is highly nonlinear. The RH and finite-element solutions show good agreement with one another.

$[0/\pm45/90]_s$ Laminate

The final laminate considered has adjacent ±45° layers. The maximum interlaminar shear stress σ_{13} (Fig. 8.39) occurs at the ±45 interface. The interlaminar normal stress distribution (Fig. 8.40) is highly nonlinear, with the maximum stress occurring in the 90° layer. While the RH solution provides better agreement for both components of stress, the agreement with the finite-element results for the interlaminar normal stress is less than desired.

8.6 Summary

Edge effects in finite-width tensile coupons have been shown to be significant, with some components of interlaminar stress exhibiting singular behavior at the free edge. The laminate stacking sequence and layer thicknesses have been shown to be very important to the sign and magnitude of interlaminar stresses. Interlaminar stresses are confined to a boundary layer region near the free edge.

References

Bogy, D. B. (1968), "Edge-Bonded Dissimilar Orthogonal Elastic Wedges under Normal and Shear Loading," *J. Appl. Mech.*, vol. 35, pp. 460–466.

Buczek, M. B., Gregory, M. A., and Herakovich, C. T. (1983), *CLFE2D—A Generalized Plain Strain Finite Element Program for Laminated Composites Subjected to Mechanical and Hygrothermal Loading*, VPI-E-83-40, Virginia Tech, Blacksburg.

Czarnek, R., Post, D., and Herakovich, C. T. (1983), "Edge Effects in Composites by Moiré Interferometry," *Exp. Tech.*, vol. 7, no. 1, pp. 18–21.

Dana, J. R. (1974), "Three Dimensional Finite Element Analysis of Thick Laminated Composites—Including Interlaminar and Boundary Effects Near Circular Holes," Ph.D. dissertation, Virginia Tech, Blacksburg.

Foye, R. L., and Baker, D. J. (1971), *Design/Analysis Methods for Advanced Composite Structures*, AFML-TR-7-299, vol. 1, Wright-Patterson Air Force Base, Ohio.

Griffin, O. H., Jr., Kamat, M. P., and Herakovich, C. T. (1981), "Three Dimensional Inelastic Finite Element Analysis of Laminated Composites," *J. Compos. Mater.*, vol. 15, pp. 336–348.

Hayashi, T. (1967), "Analytical Study of Interlaminar Shear Stresses in a Laminated Composite Plate," *Trans. Jpn. Soc. Aeronaut. Space Sci.*, vol. 10, no. 17, pp. 43–48.

Herakovich, C. T., Post, D., Buczek, M. B., and Czarnek, R. (1984), "Free Edge Strain Concentrations in Real Composite Laminates: Experimental-Theoretical Correlation," *J. Appl. Mech.*, vol. 52, no. 4, pp. 787–793.

Herakovich, C. T., Renieri, G. D., and Brinson, H. F. (1976), "Finite Element Analysis of Mechanical and Thermal Edge Effects in Composite Laminates," Proceeding of the Army Symposium on *Composite Materials: The Influence of Mechanics of Failure on Design*, Cape Cod, MA, pp. 237–248.

Kassapoglou, C., and Lagace, P. A. (1986), "An Efficient Method for the Calculation of Interlaminar Stresses in Composite Materials," *J. Appl. Mech.*, vol. 53, pp. 744–750.

Norwood, D. S., Shuart, M. J., and Herakovich, C. T. (1990), *An Analysis of Interlaminar Stresses in Unsymmetrically Laminated Plates*, CCMS-90-11, Virginia Tech, Blacksburg.

Oplinger, D. W., Parker, B. S., and Chiang, F. P. (1974), "Edge-Effect Studies in Fiber-Reinforced Laminates", *Exp. Mech.*, vol. 14, no. 9, pp. 347–354.

Pagano, N. J. (1978), "Stress Fields in Composite Laminates," *Int. J. Solids Struct.*, vol. 14, pp. 385–400.

Pipes, R. B., and Daniel, I. M. (1971), "Moire Analysis of the Interlaminar Shear Edge Effect in Laminated Composites," *J. Compos. Mater.*, vol. 5, pp. 255–259.

Pipes, R. B., and Pagano, N. J. (1970), "Interlaminar Stresses in Composite Laminates under Uniform Axial Extension," *J. Compos. Mater.*, vol. 4, pp. 538–548.

Renieri, G. D., and Herakovich, C. T. (1976), *Nonlinear Analysis of Laminated Fibrous Composites*, NASA CR-148317, VPI-E-76-10, Virginia Tech, Blacksburg.

Rose, C. A., and Herakovich, C. T. (1991), "An Approximate Analytical Solution for Interlaminar Stresses in Angle-Ply Laminates," in *Composites: Design, Manufacture, and Application* (S. W. Tsai and G. S. Springer, eds.), ICCM VIII, SAMPE, Covina, CA, pp. 28-W-1–28-W-13.

Rose, C. A., and Herakovich, C. T. (1992), "An Approximate Solution for Interlaminar Stresses in Laminated Composite Subjected to Bending and Extension," in *Proceedings, American Society of Composites Seventh Technical Conference*, Technomic Publishing Co., Stamford, CT, pp. 1125–1135.

Rose, C. A., and Herakovich, C. T. (1993), "An Approximate Solution for Interlaminar Stresses in Composite Laminates," *Compos. Eng.*, vol. 3, no. 3, pp. 271–285.

Rybicki, E. F. (1971), "Approximate Three-Dimensional Solutions for Symmetric Laminates Under In-Plane Loading," *J. Compos. Mater.*, vol. 5, pp. 354–360.

Tang, S. (1975), "A Boundary Layer Theory: Part I—Laminated Composites in Plane Stress," *J. Compos. Mater.*, vol. 9, pp. 33–41.

Wang, S. S., and Choi, I. (1982), "Boundary-Layer Effects in Composite Laminates: Part I—Free-Edge Stress Singularities," *J. Appl. Mech.*, vol. 49, pp. 541–548.

Exercises

8.1 A finite-width, symmetric laminate consisting of unit-thickness layers is subjected to axial loading N_x. The fiber orientations and material properties result in the σ_y stresses indicated in the figure below. Determine the location and magnitude of all extreme values of the through-thickness distribution of the interlaminar moment M_z in the upper half of the laminate, and sketch the through-thickness distribution. (*Hint:* It may be helpful to work with a local z-coordinate from the bottom of each layer.)

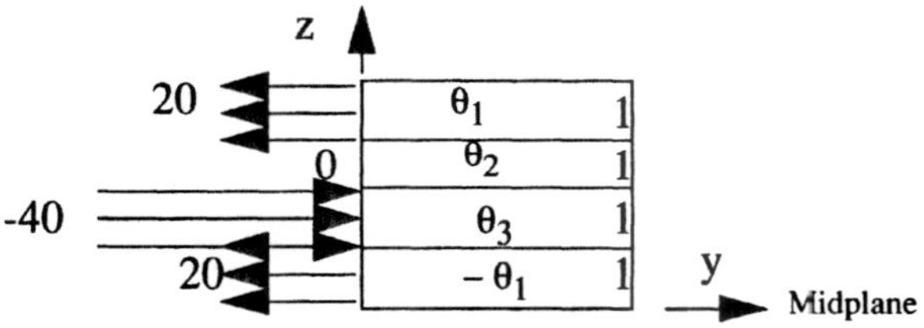

8.2 A symmetric laminate consisting of layers with fiber orientations of 0°, +60°, and –60° is subjected to axial loading N_x. Each layer has a nondimensional thickness of 2. Laminate analysis indicates the following nondimensional stresses:

Fiber Orientation	σ_y	τ_{xy}
0	–2	0
60	1	7
–60	1	–7

a) What stacking sequence and interface results in the largest interlaminar moment M_z?

b) For the laminate selected, plot the through-thickness distribution of the interlaminar force F_{zx}.

8.3 As an engineer experienced with composites, you are called upon to study the influence of stacking sequence on the interlaminar forces and moments in a finite-width tensile coupon. The fiber orientations of the individual layers of a symmetric, eight-layer laminate have been selected based upon stiffness considerations. All layers have a thickness of 2, and laminate analysis provides the stresses and orientations indicated in the table on the following page.

θ	σ_x	σ_y	τ_{xy}
4 @ 0	40	−1	0
2 @ +15	33	1	8
2 @ −15	33	1	−8

a) Plot the through-thickness distribution (upper half of laminate only) of the interlaminar forces F_{yz} and F_{zx} and the interlaminar moment M_z for a $[0/15/0/-15]_s$ laminate.

b) Consider the three laminates $[0/15/0/-15]_s$, $[0/0/15/-15]_s$, and $[15/-15/0/0]_s$. Which stacking sequence results in the worst case as it relates to the interlaminar moment M_z?

8.4 Show that (8.20) represents the most general displacement field for a midplane symmetric laminate that is subjected to axial strain loading.

8.5 Show that (8.21) represents the most general displacement field for a laminate that exhibits symmetry about both the x-y and x-z planes and is subjected to axial strain loading.

8.6 Consider a symmetric laminate that is long in the x-direction, of finite width $2b$ in the y-direction, and subjected to the loading $\{N\} = \{0, 0, 0\}$ and $\{M\} = \{M_x, 0, 0\}$. Develop expressions for the interlaminar shear force F_{yz}, the interlaminar shear force F_{zx}, and the interlaminar moment M_z, all as a function of z. Define all terms introduced. Sketch the distribution of each of the interlaminar quantities through the thickness of a four-layer $[0/90]_s$ laminate if the distribution of σ_y and τ_{xy} for this loading is as given in the figure below.

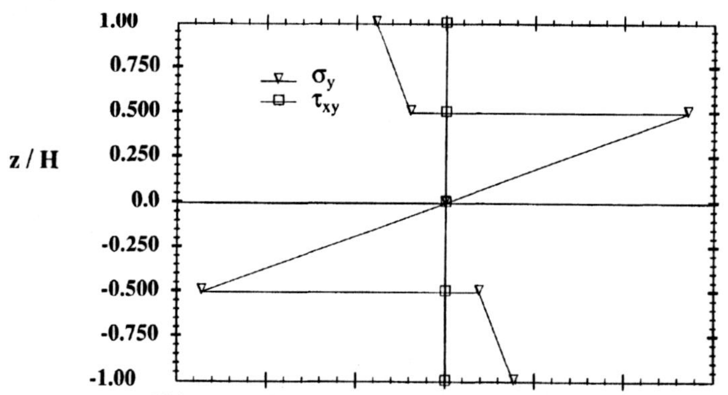

CHAPTER 9 FAILURE AND DAMAGE

"It ain't over til it's over."
Yogi Berra

9.1 Introduction

Failure is often an ill-defined term in reference to composite materials and composite structures. These heterogeneous, laminated materials typically exhibit many local failures prior to rupture into two or more distinct pieces. Thus "first failure" does not necessarily correspond to "final failure." The local failures are referred to as "damage," and the development of additional local failures with increasing load or time is called "damage accumulation." Indeed, the term *damage mechanics* has been coined to describe the study of the initiation and accumulation of damage up to and including rupture.

In this chapter we will discuss local failure mechanisms at the fiber/matrix (micromechanics) level, present several theories for predicting failure of laminae and laminates, and conclude with the presentation of a damage theory. It is emphasized that the failure theories are only *theories*. While significant progress has been made in this area of study, there currently is no single theory that accurately predicts failure at all levels of analysis, for all loading conditions, and for all types of composite materials. While some failure theories have a physical basis, most theories represent attempts to provide mathematical expressions which give a "best fit" of the available experimental data, recognizing the practical limits of data collection and the limits of mathematical representations that are practical from a designer's point of view.

From the standpoint of the structural designer, it is desirable to have failure criteria which are applicable at the level of the lamina, the laminate, and the structural component. Failure at these levels is often the consequence of an accumulation of micro-level failure events. Thus, it is also important to have a fundamental understanding of failure events at the fiber/matrix level. It is *necessary* to have a fundamental understanding of micro-level failure mechanisms in order to develop higher-strength materials. The ultimate goal is to have a failure theory that the designer can use with confidence under the most general structural configuration and loading conditions and that the developer of new materials can use to design and fabricate materials to meet specific needs.

9.2 Failure Mechanisms

Fibrous composite materials fail in a variety of mechanisms at the fiber/matrix (micro) level (Fig. 9.1). Micro-level failure mechanisms include *fiber fracture*, *fiber buckling (kinking)*, *fiber splitting*, *fiber pullout*, *fiber/matrix debonding*, *matrix cracking*, and *radial cracks*. At the laminate level, micro-level mechanisms manifest themselves as *lamina failures* in the form of *transverse cracks* in planes parallel to the fibers, *fiber-dominated failures* in planes perpendicular to the fibers, and *delaminations* between layers of the laminate.

Transverse fiber fracture (breaking a continuous fiber into two or more distinct segments) is the most catastrophic of failure mechanisms, as the fibers are typically the primary load-carrying

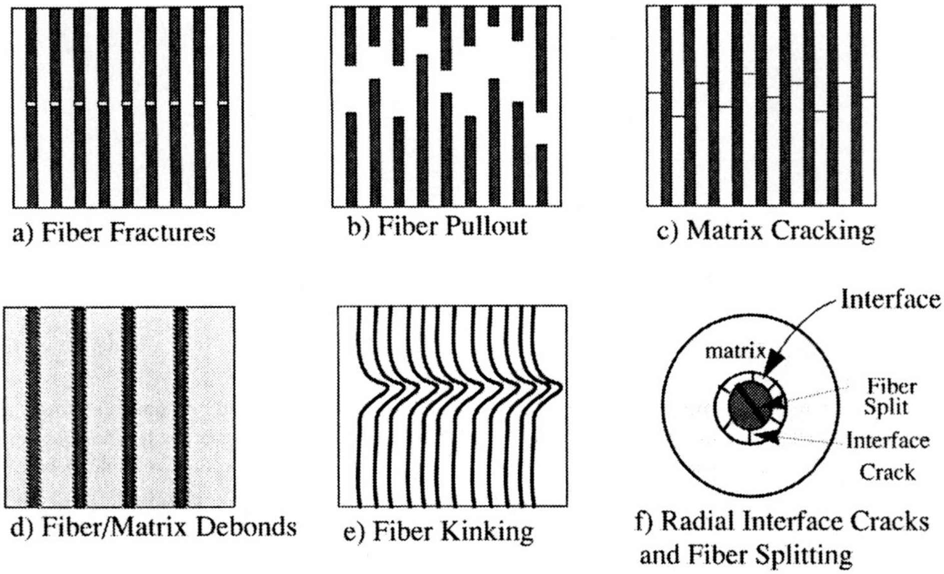

FIGURE 9.1 Micro-Level Failure Mechanisms

component. Fiber failure may be the result of tensile or compressive stresses. Fiber fracture (Fig. 9.1a) occurs under tensile load when the maximum allowable axial tensile stress (or strain) of the fiber is exceeded. Fiber pullout (Fig. 9.1b) occurs when the fiber fractures and is accompanied by fiber/matrix debonding (Fig. 9.1d). Matrix cracking (Fig. 9.1c) occurs when the strength of the matrix is exceeded. Fiber kinking (Fig. 9.1e and Fig. 6.21) occurs when the axial compressive stress causes the fiber to buckle. The critical bucking stress for a fiber embedded in a matrix is a function of the properties of the fiber and the matrix (which provides lateral support to the fiber). Fiber splitting and radial interface cracks (Fig. 9.1f) occur when the transverse or hoop stresses in the fiber or interphase region between the fiber and the matrix reaches its ultimate value.

9.2.1 Axial Tensile Strength

The axial tensile strength of a unidirectional lamina is typically controlled by the fiber ultimate strain (stress). Kelly and Davies (1965) provided the analysis for predicting axial tensile strength as a function of fiber and matrix strengths, and the constituent volume fractions. Assuming that the fibers are identical, continuous, aligned, and uniformly spaced, equilibrium in the axial direction requires that the *composite ultimate average stress,* σ_c^{ult}, be expressed in terms of the *matrix stress at composite failure,* σ_m', the *stress in the fiber at composite failure,* σ_f', and the respective *volume fractions,* V_m and V_f, in the form

$$\sigma_c^{ult} = \sigma_f' V_f + \sigma_m' V_m \tag{9.1}$$

We note that the sum of the fiber and matrix volume fractions must equal 1.0, i.e.,

$$V_f + V_m = 1 \tag{9.2}$$

Since the fiber ultimate strain is typically less than the matrix ultimate strain, the matrix may not have attained its ultimate stress, σ_m^{ult}, when the fibers fail. There are two questions associated with the failure of the fibers: (a) Will the matrix also fail? (b) Does the composite failure stress represent an increase in strength over the bulk matrix strength? There is a minimum fiber volume fraction, V_f^{min}, below which the fibers do not provide additional strength; i.e., the composite does not fail when the fibers fail because the strength of the remaining matrix is sufficient to carry the load. There is also a critical fiber volume fraction, V_f^{crit}, below which the fibers do not provide additional strength because the matrix at $V_m = 1$ would carry more load.

These two important fiber volume fractions can be expressed in terms of constituent strengths and volume fractions in the following manner. For a composite that fails when the fibers attain the *fiber ultimate stress,* σ_f^{ult}, (9.1) and (9.2) give the *fiber-dominated composite failure stress,* σ_c^{fult}, as

$$\sigma_c^{fult} = \sigma_f^{ult} V_f + \sigma'_m (1 - V_f) \tag{9.3}$$

For this composite ultimate stress to represent a reinforcement due to the presence of the fibers, it must be greater than the stress which can be carried by the matrix portion of the composite volume after all fibers have failed. We shall designate the *failure stress of the matrix portion of the volume* as σ_c^{mult}. Substituting this matrix stress in (9.1) with zero fiber stress gives

$$\sigma_c^{mult} = \sigma_m^{ult} V_m = \sigma_m^{ult}(1 - V_f) \tag{9.4}$$

Equations (9.3) and (9.4) are both linear in the fiber volume fraction, V_f. As indicated in Fig. 9.2, they represent two intersecting straight lines on a plot of composite ultimate stress versus fiber volume fraction. The minimum fiber volume fraction, V_f^{min}, necessary for the fibers to actually represent a reinforcement corresponds to the intersection of the two straight lines. The critical fiber volume fraction, V_f^{crit}, required for the fibers to provide additional strength is also shown in the figure. It corresponds to the fiber-dominated composite ultimate (9.3) being equal to the matrix ultimate, i.e.,

$$\sigma_c^{fult} = \sigma_m^{ult} \tag{9.5}$$

For the fiber-dominated composite ultimate, σ_c^{fult}, to be equal to or greater than the fiberless composite ultimate, σ_c^{mult}, from (9.3) and (9.4) we must have

$$\sigma_f^{ult} V_f + \sigma'_m (1 - V_f) \geq \sigma_m^{ult}(1 - V_f) \tag{9.6}$$

The equality condition in (9.6) corresponds to the *minimum fiber volume fraction,* V_f^{min}, *that must be exceeded for the fibers to provide actual reinforcement.* Solving for the fiber volume fraction from (9.6) as an equality gives

$$V_f^{min} = \frac{\sigma_m^{ult} - \sigma'_m}{\sigma_f^{ult} + \sigma_m^{ult} - \sigma'_m} \tag{9.7}$$

Likewise, from (9.3) and (9.5), the V_f^{crit} for the composite to be stronger than the bulk matrix requires that

$$\sigma_f^{ult} V_f + \sigma'_m (1 - V_f) \geq \sigma_m^{ult} \tag{9.8}$$

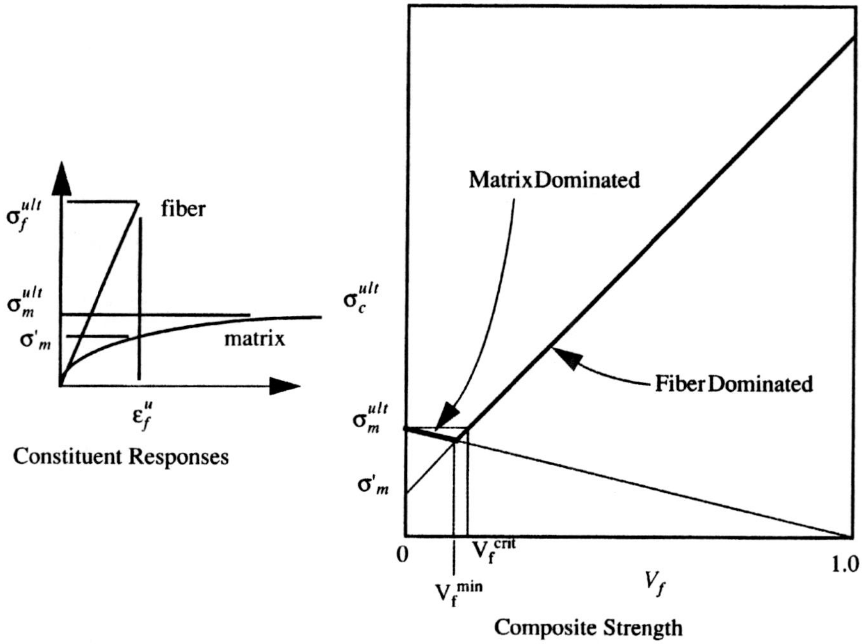

FIGURE 9.2 Composite Strength and Fiber Volume Fraction

The equality condition of (9.8) gives the critical fiber volume fraction for the composite strength to be greater than the bulk matrix strength.

$$V_{crit} = \frac{\sigma_m^{ult} - \sigma'_m}{\sigma_f^{ult} - \sigma'_m} \tag{9.9}$$

Careful study of (9.7) and (9.9) indicates that V_{crit} is always greater than V_{min} since all stresses are positive for tensile loading.

9.2.2 Axial Compressive Strength

Compressive failure of unidirectional composites loaded in the fiber direction is usually considered to be a microbuckling problem (sometimes called shear crippling or fiber kinking). As such, it is influenced by many factors, including fiber size and shape, fiber waviness, fiber/matrix bond strength, fiber and matrix stiffness, and, of course, fiber and matrix compressive strength. Correlation of compression failure models with experimental results is further complicated by the difficulties associated with compression testing, as discussed in Chapter 6. Composites using small-diameter fibers such as carbon, glass, and Kevlar are more susceptible to fiber buckling than those made with larger-diameter fibers such as boron and SCS-6 silicon-carbide.

Hahn and Williams (1986) presented a shear crippling model which includes initial fiber imperfection (fiber curvature). For an elastic–perfectly plastic composite with strong fibers and shear yield stress, τ_y, they proposed the following equation for the compressive strength X_c:

$$X_c = V_f\left(\frac{G_{12}\tau_y}{\tau_y + \dfrac{\pi f_o G_{12}}{L}}\right) \tag{9.10}$$

where V_f is the fiber volume fraction, G_{12} is the in-plane effective composite axial shear modulus, and f_o/L is a fiber curvature parameter. The fiber curvature parameter is determined by "backing it out" from a set of typical known experimental results which include the compressive strength. As an example of the use of this model, Fox et al. (1987) presented the results in Table 9.1 comparing theory and experiment for the compressive strength of T300/934 carbon/epoxy at cryogenic, room, and elevated temperatures for nonirradiated material and material that had been exposed to electron irradiation. The fiber curvature parameter was "backed out" from the nonirradiated cryogenic results and was found to have a value $f_o/L = 0.0041$. As indicated in the table, the comparison of theory and experiment was very good for all cases except the elevated-temperature irradiated condition. A possible explanation for the significantly lower failure loads at elevated temperature for the irradiated specimens is that they failed due to an entirely different mechanism, as it is known that electron irradiation lowers the glass transition temperature of epoxy (Milkovich et al., 1986). Lowering the glass transition temperature has the effect of degrading the matrix properties, thereby reducing its ability to provide lateral support to the fibers.

Condition (V_f)	Temperature, °C (°F)	G_{12}, GPa (Msi)	τ_y, MPa (ksi)	X_c, MPa (ksi)	
				Predicted	Measured
Nonirradiated	−157 (−250)	8.1 (1.17)	50.6 (7.34)	1662 (241)	1662 (241)
($V_f = 0.63$)	24 (75)	4.74 (0.69)	52.1 (7.56)	1372 (199)	1331 (193)
	121 (250)	3.86 (0.56)	27.2 (3.94)	862 (125)	869 (126)
Irradiated	−157 (−250)	7.72 (1.12)	50.0 (7.25)	1572 (228)	1572 (228)
($V_f = 0.61$)	24 (75)	5.38 (0.78)	47.1 (6.83)	1324 (192)	1296 (188)
	121 (250)	2.76 (0.40)	16.1 (2.33)	520 (75.5)	333 (48.3)

TABLE 9.1 Compressive Strength of T300/934 Carbon/Epoxy

9.3 Macroscopic Failure Theories

In the following presentation of macroscopic failure theories we assume that the individual laminae are homogeneous, orthotropic materials with known, measured strengths for one-dimensional states of stress in the principal material directions. We also note that shear strength in the plane of the fibers is independent of the sign of the shear stress in the principal material coordinate directions because, as indicated in Fig. 9.3, positive and negative shear stresses are identical in terms of the stress parallel and perpendicular to the fibers. Shear strength is also independent of sign in the transverse 2-3 plane for transversely isotropic composites.

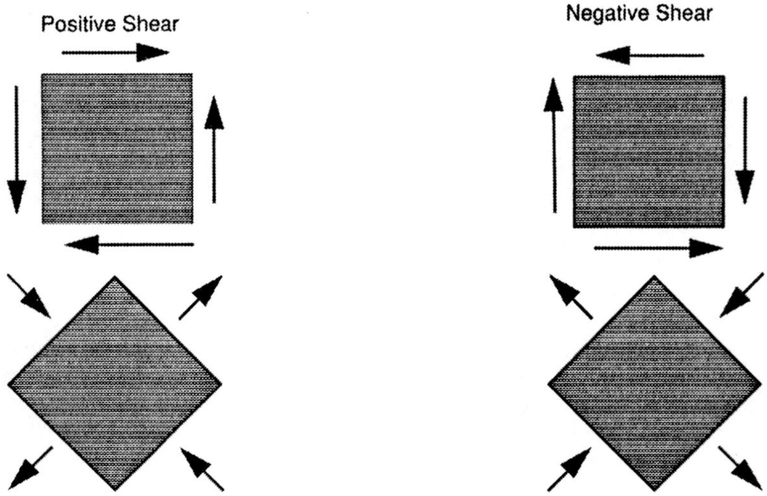

FIGURE 9.3 Positive and Negative Shear Stresses for $\theta = 0°$

The sign independence of shear strength does not hold for coordinate systems other than the principal material coordinates. If the state of plane stress is pure shear for a fiber orientation other than 0° or 90°, the sign of the shear stress can have a significant influence on the stress in the principal material coordinates. As an example, stress states in global x-y and principal material 1-2 coordinates are shown in Fig. 9.4 for the case of $\theta = 45°$. For positive shear, the fibers are in axial tension and transverse compression, whereas for negative shear the opposite is true: the fibers are in axial compression and transverse tension. Because of vastly different strengths in tension and compression for most fibrous composites, the positive and negative shear strength will also vary significantly for off-axis fiber orientations.

It is also evident from this figure that pure macro-level shear stress can result in a variety of micro-level failure mechanisms. The composite may fail due to fiber breakage ($\sigma_1 > 0$), fiber bucking ($\sigma_1 < 0$), fiber/matrix debonding or matrix failure ($\sigma_2 > 0$), or matrix compression failure ($\sigma_2 < 0$). Therefore, while we formulate the failure theories in terms of macroscopic stresses, we should be cognizant of the micro-level failure mechanisms.

Application of the macroscopic failure theories to individual layers of a laminate corresponds to a "first ply" failure theory. That is, it is assumed that the laminate fails when the first ply fails. This, of course, is not the typical case. In general, multiple transverse cracks form in layers at an angle to the primary load direction well in advance of complete failure. Such crack development is referred to as damage accumulation and is discussed later in this chapter.

Numerous macroscopic failure theories have been proposed for fibrous composites. In the following we present several theories selected on the basis of their relationship to physical intuition for one-dimensional states of stress or the correlation of theory and experiment for multidimensional states of stress.

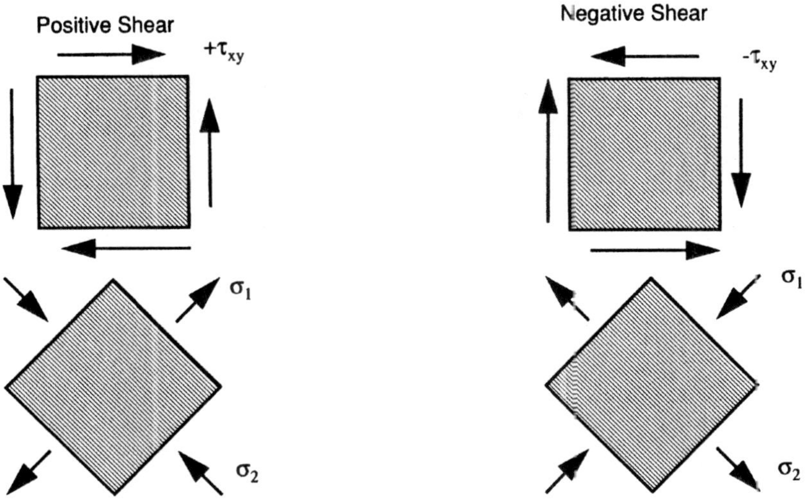

FIGURE 9.4 Positive and Negative Shear Stresses for θ = 45°

9.3.1 Maximum Stress Theory

The maximum stress failure criterion assumes that failure occurs whenever any one component of stress attains its limiting value, independent of the values of all other components of stress. The allowable values must be determined from a series of tests in which the specimen is under a uniform, uniaxial state of stress, as depicted in Fig. 9.5. The notation used for the strength values includes subscripts T and C for tension and compression, respectively; X indicates the ultimate normal stress magnitude in the fiber direction, Y and Z the ultimate normal stress magnitudes in the two transverse directions, and Q, R, and S the ultimate shear stresses.

The maximum stress failure criterion then requires that all of the individual stress components be less than their respective limiting values if failure is not to occur. Thus the "safe" condition for the maximum stress failure criterion can be written mathematically in the form

$$\begin{aligned} X_C &< \sigma_1 < X_T \\ Y_C &< \sigma_2 < Y_T \\ Z_C &< \sigma_3 < Z_T \\ |\tau_{23}| &< Q \\ |\tau_{31}| &< R \\ |\tau_{12}| &< S \end{aligned} \tag{9.11}$$

The initiation of failure according to the maximum stress criterion corresponds to one or more of the inequalities in (9.11) becoming an equality as the load is monotonically increased. For a lamina in plane stress, this failure criterion can be represented as intersecting straight lines in 2-D stress space or intersecting planar surfaces in 3-D stress space. An example of the maximum stress criterion in σ_1-σ_2-τ_{12} stress space is shown in Fig. 9.6. The region interior to the parallelepiped corresponds to the "safe" region. If any component of stress is on or outside one of the limiting planes, the theory predicts failure. The shaded rectangle in Fig. 9.6 corresponds to the safe region for the

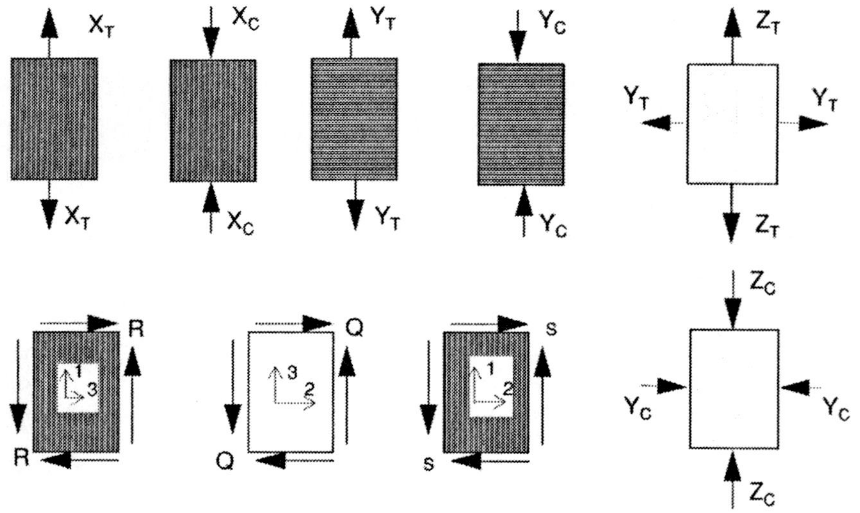

FIGURE 9.5 Lamina Failure Modes

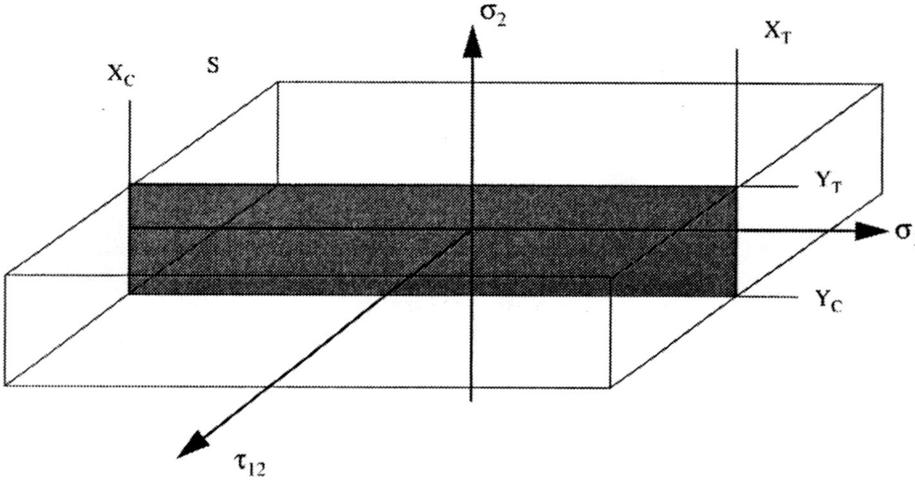

FIGURE 9.6 Maximum Stress Failure in σ_1-σ_2 Stress Space

case of zero shear stress. Any other plane passed through the parallelepiped parallel to the σ_1-σ_2 plane corresponds to the failure criterion for constant but nonzero values of shear stress. Likewise, planes parallel to the other axes correspond to failure surfaces for constant values of the corresponding normal components of stress. For a fully three-dimensional stress state, failure corresponds to a surface in six-dimensional stress space.

9.3.1.1 Off-Axis Tensile Coupon

The maximum stress criterion can be expressed in terms of stresses in global coordinates using the stress transformation equations. As an example, the maximum allowable normal stress σ_x for a uni-

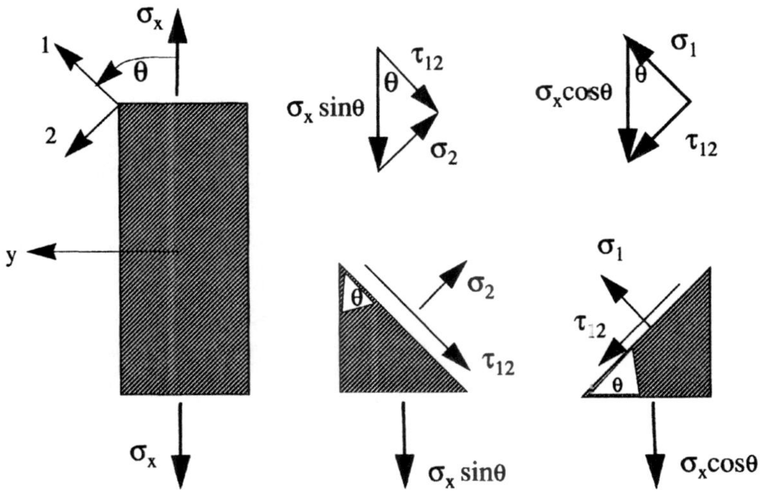

FIGURE 9.7 Off-Axis Tensile Coupon

directional, off-axis coupon under axial stress (Fig. 9.7) can be expressed as a function of the in-plane strength parameters X_T, X_C, Y_T, Y_C, and S and the fiber orientation angle θ.

The plane stress transformation equations (4.23) for transformation from global to principal material directions, for a uniaxial state of stress σ_x, give the stresses

$$\begin{aligned} \sigma_1 &= \sigma_x \cos^2\theta \\ \sigma_2 &= \sigma_x \sin^2\theta \\ \tau_{12} &= -\sigma_x \sin\theta\cos\theta \end{aligned} \tag{9.12}$$

Substitution of these principal material stresses into the failure criterion (9.11) gives the maximum stress failure criterion for a unidirectional, off-axis coupon in terms of the applied stress σ_x and the fiber orientation θ.

$$\begin{aligned} &\frac{X_C}{\cos^2\theta} < \sigma_x < \frac{X_T}{\cos^2\theta} \\ &\frac{Y_C}{\sin^2\theta} < \sigma_x < \frac{Y_T}{\sin^2\theta} \\ &|\sigma_x| < \left|\frac{S}{-\sin\theta\cos\theta}\right| \end{aligned} \tag{9.13}$$

Equations (9.13) are five inequality conditions on the axial stress σ_x. Using the equality conditions gives five equations for the ultimate applied axial stress, σ_x^{ult}, corresponding to five different modes of failure.

The equality conditions can be plotted on a graph of σ_x^{ult} versus fiber orientation θ as indicated in Fig. 9.8, where the three (overlapping) curves for tensile σ_x loading on AS4/3501-6 carbon/epoxy (experimental properties from Beuth and Herakovich, 1987; see Table 9.2) are shown. For any given fiber orientation θ, failure will occur in the mode corresponding to the smallest value of

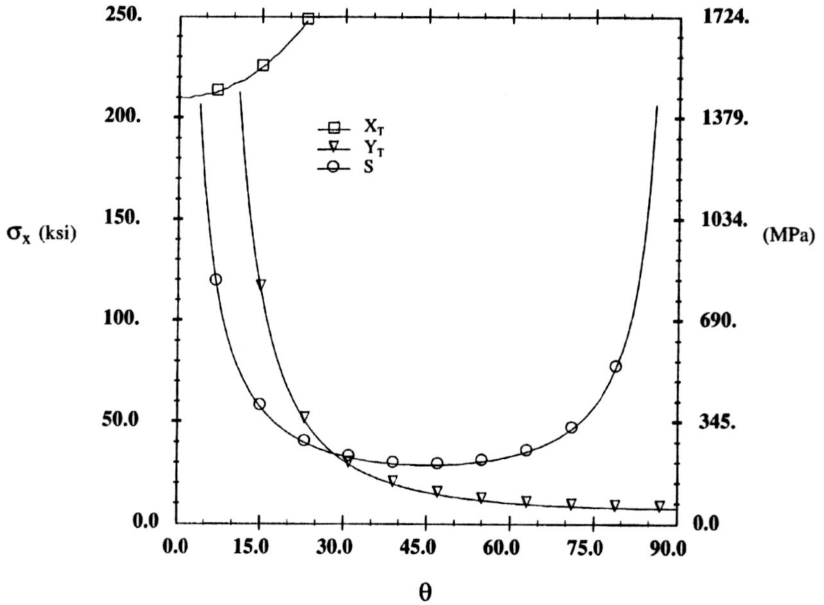

FIGURE 9.8 Maximum Stress Failure Criterion for Axially Loaded Coupon

σ_x. Thus for very small angles (<5°) the mode of failure is fiber tensile failure (X_T), for angles between 5° and 28° the mode of failure is shear (S), and for angles greater than 28° the mode of failure is transverse tension (Y_T). These results are, of course, material dependent. This figure shows that, according to the maximum stress failure criterion, the mode of failure changes in a discontinuous manner as the fiber orientation increases. This is inconsistent with the natural order of real-world failures.

The maximum stress failure criterion has limitations for predicting the failure stress in multi-axial stress states because of its lack of coupling or interaction effects between the various components of stress. It should be intuitively obvious that a multi-axial, far-field average stress state will result in considerably different local micro-level stresses compared with the local stresses associated with each of the individual components of far-field stress.

E_1	E_2	G_{12}	ν_{12}	X_T	Y_T	S
126 GPa (18.3 Msi)	10.50 GPa (1.46 Msi)	5.61GPa (0.81 Msi)	0.30	1450 MPa (210 ksi)	53.4 MPa (7.75 ksi)	99.3 MPa (14.4 ksi)
ε_1^T, %	ε_2^T, %	Γ_{12}, %	F_{12}	X_C	Y_C	
1.15	0.535	2.0	0.0	1296 MPa (−188 ksi)	238 MPa (−34.5 ksi)	

TABLE 9.2 Elastic and Strength Properties of AS4/3501-6 Carbon/Epoxy

9.3.2 Maximum Strain Theory

The maximum strain theory is the strain equivalent of the maximum stress theory. The equations are written in the following form, where the maximum allowable normal strain values have been defined using superscripts T and C to denote tension and compression, respectively, and where Γ_{ij} denotes maximum allowable shear strains.

$$\begin{aligned} \varepsilon_1^C &< \varepsilon_1 < \varepsilon_1^T \\ \varepsilon_2^C &< \varepsilon_2 < \varepsilon_2^T \\ \varepsilon_3^C &< \varepsilon_3 < \varepsilon_3^T \\ |\gamma_{12}| &< \Gamma_{12} \\ |\gamma_{13}| &< \Gamma_{13} \\ |\gamma_{23}| &< \Gamma_{23} \end{aligned} \tag{9.14}$$

The maximum strain failure criterion can be expressed in terms of the principal material stresses for elastic response using the constitutive equations (3.25)–(3.30) for strains in terms of stresses in (9.14). For plane stress, the resulting equations have the form

$$\begin{aligned} \varepsilon_1^C &< \frac{\sigma_1 - \nu_{12}\sigma_2}{E_1} < \varepsilon_1^T \\ \varepsilon_2^C &< \frac{\sigma_2 - \nu_{21}\sigma_1}{E_2} < \varepsilon_2^T \\ \left|\frac{\tau_{12}}{G_{12}}\right| &< |\Gamma_{12}| \end{aligned} \tag{9.15}$$

9.3.2.1 Off-Axis Coupon

For the off-axis coupon under axial stress, (9.12) and (9.15) can be combined to give the maximum strain failure criterion in terms of the applied stress σ_x, the fiber orientation θ, and the elastic properties.

$$\begin{aligned} \varepsilon_1^C &< \frac{(\cos^2\theta - \nu_{12}\sin^2\theta)\sigma_x}{E_1} < \varepsilon_1^T \\ \varepsilon_2^C &< \frac{(\sin^2\theta - \nu_{21}\cos^2\theta)\sigma_x}{E_2} < \varepsilon_2^T \\ \left|\frac{-\sin\theta\cos\theta\sigma_x}{G_{12}}\right| &< |\Gamma_{12}| \end{aligned} \tag{9.16}$$

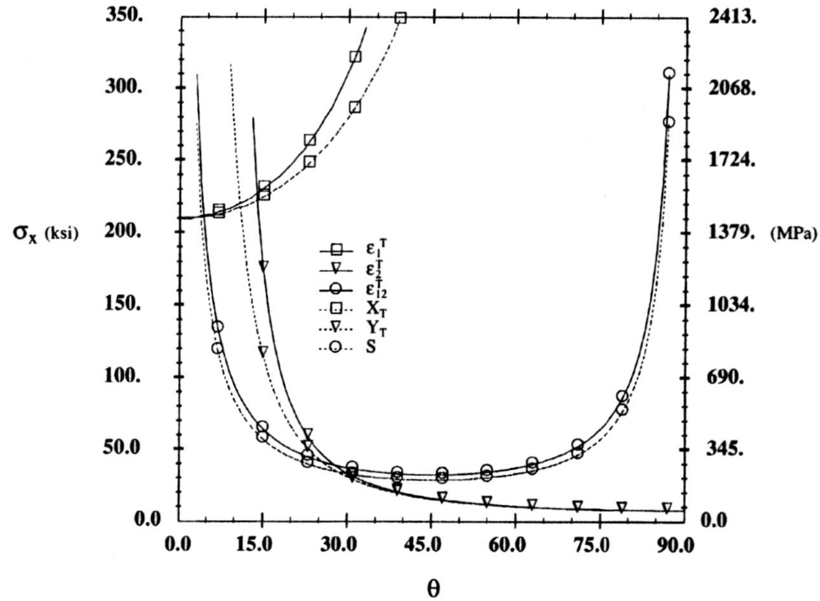

FIGURE 9.9 Maximum Stress and Strain Failure Criteria for Axially Loaded Coupon

Expressing (9.16) explicitly in terms of the axial stress σ_x, we have

$$\frac{E_1\varepsilon_1^C}{\cos^2\theta - \nu_{12}\sin^2\theta} < \sigma_x < \frac{E_1\varepsilon_1^T}{\cos^2\theta - \nu_{12}\sin^2\theta}$$

$$\frac{E_2\varepsilon_2^C}{\sin^2\theta - \nu_{21}\cos^2\theta} < \sigma_x < \frac{E_2\varepsilon_2^T}{\sin^2\theta - \nu_{21}\cos^2\theta} \tag{9.17}$$

$$|\sigma_x| < \left|\frac{G_{12}\Gamma_{12}}{-\sin\theta\cos\theta}\right|$$

Comparison of (9.13) and (9.17) shows that the main difference between the maximum stress and maximum strain failure theories for elastic response of an off-axis coupon is the presence of a Poisson term in the denominator of the maximum strain theory; the numerators are equivalent for elastic response and a one-dimensional state of stress.

The maximum strain and maximum stress failure criteria for AS4/3501-6 carbon/epoxy are compared in Fig. 9.9 for tensile loading of an off-axis coupon. The figure shows that the two theories give very similar predictions for this material system.

9.3.3 Tsai-Hill Theory

Several quadratic failure criteria have been presented in the literature. They represent attempts to provide better correlation between theory and experiment by inclusion of all components of stress in an equation representing the failure criteria. The quadratic criteria are based upon the mathematical premise that a second-order curve has more parameters with which to fit experimental data than does a straight line. These criteria generally are not based upon the physics of the failure mechanisms. While they may provide better correlation between theory and experiment in some situations, they are limited in that the sign of the normal stress components must be known a priori if the positive and negative strengths are different (which is often the case with composites).

The Tsai-Hill theory (Tsai, 1968) represents an attempt to apply Hill's anisotropic plasticity theory (Hill, 1950) to failure of homogeneous, anisotropic materials. The theory assumes a failure surface given by the equation

$$\begin{aligned}(G+H)\sigma_1^2 + (F+H)\sigma_2^2 + (F+G)\sigma_3^2 - 2H\sigma_1\sigma_2 - 2G\sigma_1\sigma_3 \\ - 2F\sigma_2\sigma_3 + 2L\tau_{23}^2 + 2M\tau_{13}^2 + 2N\tau_{12}^2 = 1\end{aligned} \tag{9.18}$$

where F, G, H, L, M, and N are material strength parameters. Stress states inside this surface are "safe," whereas those on or outside the surface correspond to failure.

The strength parameters are expressed in terms of the failure stresses for one-dimensional loading through a series of thought experiments. For pure shear stress loading $\tau_{12} \neq 0$, with the corresponding shear strength S, and all other stresses equal to zero, the failure criterion (9.18) gives

$$2N = \frac{1}{S^2} \tag{9.19}$$

Similar thought experiments for the other two components of shear stress with shear ultimates Q for $\tau_{23} \neq 0$ and R for $\tau_{13} \neq 0$ give

$$\begin{aligned}2L &= \frac{1}{Q^2} \\ 2M &= \frac{1}{R^2}\end{aligned} \tag{9.20}$$

The remaining three strength parameters, F, G, and H, are determined from the simultaneous set of equations obtained from (9.18) for the three individual thought experiments $\sigma_1 \neq 0$, $\sigma_2 \neq 0$, and $\sigma_3 \neq 0$ (all other $\sigma_{ij} = 0$). The results are

$$2H = \frac{1}{X^2} + \frac{1}{Y^2} - \frac{1}{Z^2} \tag{9.21}$$

$$2G = \frac{1}{X^2} - \frac{1}{Y^2} + \frac{1}{Z^2} \tag{9.22}$$

$$2F = -\frac{1}{X^2} + \frac{1}{Y^2} + \frac{1}{Z^2} \tag{9.23}$$

For the case of plane stress in the 1-2 plane (i.e., $\sigma_3 = \tau_{13} = \tau_{23} = 0$) of a transversely isotropic material ($Y = Z$), the Tsai-Hill failure criterion reduces to

$$\frac{\sigma_1^2}{X^2} - \frac{\sigma_1\sigma_2}{X^2} + \frac{\sigma_2^2}{Y^2} + \frac{\tau_{12}^2}{S^2} = 1 \tag{9.24}$$

9.3.3.1 Off-Axis Tensile Coupon

For the off-axis lamina with $\sigma_1 = \sigma_x \cos^2\theta$, $\sigma_2 = \sigma_x \sin^2\theta$, and $\tau_{12} = -\sigma_x \sin\theta\cos\theta$, the Tsai-Hill failure criterion takes the form

$$\frac{\cos^4\theta}{X^2} + \left(\frac{1}{S^2} - \frac{1}{X^2}\right)\sin^2\theta\cos^2\theta + \frac{\sin^4\theta}{Y^2} = \frac{1}{\sigma_x^2} \tag{9.25}$$

Limitations of the Tsai-Hill criterion are evident in (9.24) and (9.25). All strength parameters appear as second-order terms with no distinction between positive and negative strengths (which are generally different in magnitude). While this causes no difficulty for shear stress in the principal material coordinates, it represents a severe limitation for normal stresses in that the sign of the normal stress must be known a priori and the appropriate strength value then used in the failure criterion.

For the special case $X_T = X_C$ and $Y_T = Y_C$, (9.24) represents a continuous surface in σ_1, σ_2, τ_{12} stress space or a curve in any two-dimensional stress space.

9.3.4 Tensor Polynomial Failure Criterion

Failure criteria based upon polynomials of strength tensors represent an attempt to mathematically overcome one of the shortcomings of the quadratic criteria, namely, to account for the differences in tensile and compressive strengths. They have the additional advantage of being scalar equations of tensor quantities. Hence they are invariant, and transformations between coordinate systems can be effected using the tensor transformation laws of Chapter 2.

Tensor polynomial failure criteria were first considered by Gol'denblat and Kopnov (1965) and Ashkenazi (1965).We consider here the second-order tensor polynomial criterion as proposed by Tsai and Wu (1971). It is a complete quadratic tensor polynomial with the linear terms included. The criterion assumes that there exists a scalar function $f(\sigma_i)$ of the form

$$f(\sigma_i) = F_i\sigma_i + F_{ij}\sigma_i\sigma_j \tag{9.26}$$

where F_i and F_{ij} are tensor quantities of strength parameters. Failure corresponds to the condition

$$f(\sigma_i) \geq 1 \tag{9.27}$$

It can be shown (Tsai and Wu, 1971) that the condition $f(\sigma_i) = 1$ is a closed surface in stress space if

$$F_{ii}F_{jj} - F_{ij}^2 > 0 \qquad (i, j) \text{ no sum} \tag{9.28}$$

Thus, states of stress inside the surface are "safe," and those on or outside the surface correspond to failure.

For a 3-D stress state ($i = 1, 2, ..., 6$), $f(\sigma_i)$ has the form

$$\begin{aligned} f(\sigma_i) = {} & F_1\sigma_1 + F_2\sigma_2 + F_3\sigma_3 + F_4\sigma_4 + F_5\sigma_5 + F_6\sigma_6 \\ & + F_{11}\sigma_1\sigma_1 + F_{12}\sigma_1\sigma_2 + F_{13}\sigma_1\sigma_3 + F_{14}\sigma_1\sigma_4 + F_{15}\sigma_1\sigma_5 + F_{16}\sigma_1\sigma_6 \\ & + F_{21}\sigma_2\sigma_1 + F_{22}\sigma_2\sigma_2 + F_{23}\sigma_2\sigma_3 + F_{24}\sigma_2\sigma_4 + F_{25}\sigma_2\sigma_5 + F_{26}\sigma_2\sigma_6 \\ & + F_{31}\sigma_3\sigma_1 + F_{32}\sigma_3\sigma_2 + F_{33}\sigma_3\sigma_3 + F_{34}\sigma_3\sigma_4 + F_{35}\sigma_3\sigma_5 + F_{36}\sigma_3\sigma_6 \\ & + F_{41}\sigma_4\sigma_1 + F_{42}\sigma_4\sigma_2 + F_{43}\sigma_4\sigma_3 + F_{44}\sigma_4\sigma_4 + F_{45}\sigma_4\sigma_5 + F_{46}\sigma_4\sigma_6 \\ & + F_{51}\sigma_5\sigma_1 + F_{52}\sigma_5\sigma_2 + F_{53}\sigma_5\sigma_3 + F_{54}\sigma_5\sigma_4 + F_{55}\sigma_5\sigma_5 + F_{56}\sigma_5\sigma_6 \\ & + F_{61}\sigma_6\sigma_1 + F_{62}\sigma_6\sigma_2 + F_{63}\sigma_6\sigma_3 + F_{64}\sigma_6\sigma_4 + F_{65}\sigma_6\sigma_5 + F_{66}\sigma_6\sigma_6 \end{aligned} \tag{9.29}$$

We assume that F_i and F_{ij} are symmetric tensors. Other simplifications of (9.29) can be effected through recognition that the shear terms $F_4 = F_5 = F_6 = 0$ and the normal/shear coupling terms $F_{14} = F_{15} = F_{16} = F_{24} = F_{25} = F_{26} = F_{34} = F_{35} = F_{36} = F_{45} = F_{56} = F_{64} = 0$. These simplifications are all a direct result of the fact that the shear strength is independent of sign (in the principal material coordinates). Thus, the reduced form of the scalar function $f(\sigma_i)$ for an orthotropic material is

$$\begin{aligned} f(\sigma_i) = {} & F_1\sigma_1 + F_2\sigma_2 + F_3\sigma_3 + F_{11}\sigma_1^2 + F_{22}\sigma_2^2 + F_{33}\sigma_3^2 + F_{44}\sigma_4^2 + F_{55}\sigma_5^2 + F_{66}\sigma_6^2 \\ & + 2F_{12}\sigma_1\sigma_2 + 2F_{13}\sigma_1\sigma_3 + 2F_{23}\sigma_2\sigma_3 \end{aligned} \tag{9.30}$$

The strength tensors F_i and F_{ij} can be written in terms of the engineering strengths X_T, X_C, Y_T, Y_C, Z_T, Z_C, Q, R, and S through a series of thought experiments with one-dimensional loadings. As an example, consider tensile and compressive failure for axial stress $\sigma_1 \neq 0$ (all other $\sigma_i = 0$).

For tensile failure at the stress $\sigma_1 = X_T$, the failure criterion (9.30) reduces to

$$F_1X_T + F_{11}X_T^2 = 1 \tag{9.31}$$

For compressive failure at the stress $\sigma_1 = X_C$, (9.30) reduces to

$$F_1X_C + F_{11}X_C^2 = 1 \tag{9.32}$$

Simultaneous solution of (9.31) and (9.32) gives F_1 and F_{11} as

$$F_1 = \frac{1}{X_T} + \frac{1}{X_C}; \qquad F_{11} = \frac{-1}{X_TX_C} \tag{9.33}$$

In a similar fashion, one-dimensional stress states with nonzero σ_2, σ_3, τ_{12} (σ_6), τ_{13} (σ_5), or τ_{23} (σ_4) give

$$F_2 = \frac{1}{Y_T} + \frac{1}{Y_C}; \qquad F_{22} = \frac{-1}{Y_TY_C} \tag{9.34}$$

$$F_3 = \frac{1}{Z_T} + \frac{1}{Z_C}; \qquad F_{33} = \frac{-1}{Z_TZ_C} \tag{9.35}$$

$$F_{44} = \frac{1}{Q^2} \tag{9.36}$$

$$F_{55} = \frac{1}{R^2} \tag{9.37}$$

$$F_{66} = \frac{1}{S^2} \tag{9.38}$$

The remaining coefficients in (9.30), F_{12}, F_{13}, and F_{23}, correspond to interaction terms involving normal components of stress. In theory, these coefficients can be determined from three independent tests in which $\sigma_1 = \sigma_2$, $\sigma_1 = \sigma_3$, or $\sigma_2 = \sigma_3$, with all other $\sigma_i = 0$. For example, if we consider the loading $\sigma_1 = \sigma_2 = \sigma$, the tensor polynomial criterion predicts failure when

$$f(\sigma) = F_1\sigma + F_2\sigma + F_{11}\sigma^2 + F_{22}\sigma^2 + 2F_{12}\sigma^2 = 1 \tag{9.39}$$

Equation (9.39) can be solved for F_{12} in terms of the strengths X_T, X_C, Y_T, Y_C, and σ as

$$F_{12} = \frac{1}{2\sigma^2} - \frac{1}{2\sigma}\left(\frac{1}{X_T} + \frac{1}{X_C} + \frac{1}{Y_T} + \frac{1}{Y_C}\right) + \frac{1}{2}\left(\frac{1}{X_T X_C} + \frac{1}{Y_T Y_C}\right) \tag{9.40}$$

It is most difficult to conduct a test in which a specimen fails as a result of an equally biaxial state of stress, e.g., $\sigma_1 = \sigma_2$. All indications are that these interaction coefficients are small, and they are often taken to be zero (Narayanaswami and Adelman, 1977).

For a planar state of stress with $F_{12} = 0$, the tensor polynomial failure criterion (9.30) reduces to

$$f(\sigma_i) = F_1\sigma_1 + F_2\sigma_2 + F_{11}\sigma_1^2 + F_{22}\sigma_2^2 + F_{66}\sigma_6^2 = 1 \tag{9.41}$$

9.3.4.1 Off-Axis Tensile Coupon

For the special case of an off-axis coupon under axial tensile stress σ_x, the tensor polynomial failure criterion (9.41), with the definitions (9.33), (9.34), and (9.38) and the transformation equations (9.12), takes the quadratic form

$$A\sigma_x^2 + B\sigma_x - 1 = 0 \tag{9.42}$$

where A and B are defined as functions of the strength parameters and the fiber orientation θ (using $m = \cos\theta$ and $n = \sin\theta$) in the form

$$A = m^4\left(\frac{-1}{X_T X_C}\right) + n^4\left(\frac{-1}{Y_T Y_C}\right) + \frac{m^2 n^2}{S^2} \tag{9.43}$$

$$B = m^2\left(\frac{1}{X_T} + \frac{1}{X_C}\right) + n^2\left(\frac{1}{Y_T} + \frac{1}{Y_C}\right) \tag{9.44}$$

For a given material with known strengths, (9.42) is a quadratic equation in σ_x that can be solved directly for the failure stress σ_x as a function of θ.

9.3.4.2 Transformation for Tensor Polynomial Failure Criterion

The tensor polynomial failure criterion can be expressed in any arbitrary coordinate system through appropriate transformation of the stress and strength tensors. First it must be noted that we have been using reduced notation, and F_i is actually a second-order tensor and F_{ij} is actually a fourth-order tensor. The transformation equations for second- and fourth-order tensors (see Chapter 2) can be written in terms of the direction cosines a_{ij} as

$$\begin{aligned} F'_{ij} &= a_{ki}a_{lj}F_{kl} \\ F'_{ijkl} &= a_{mi}a_{nj}a_{rk}a_{sl}F_{mnrs} \end{aligned} \tag{9.45}$$

In the arbitrary (primed) coordinate system, failure initiation corresponds to the condition

$$f(\sigma'_i) = F'_i\sigma'_i + F'_{ij}\sigma'_i\sigma'_j = 1 \tag{9.46}$$

The form (9.46) is useful for laminates when it is desired to work in the global x-y coordinate system.

9.3.5 Comparison of Failure Criteria

The Tsai-Hill and tensor polynomial failure criteria are compared for unidirectional AS4/3501-6 carbon/epoxy in Figs. 9.10 and 9.11 for the case of plane stress with $\tau_{12} = 0$. Figure 9.10 shows the criteria to the same scale for both components of stress, whereas Fig. 9.11 shows the transverse scale magnified to more clearly distinguish the differences in the criteria. The Tsai-Hill criterion has been plotted with the appropriate strength parameters in each of the four quadrants. This results in discontinuous gradients for this criterion at the intersections with the axes.

Figure 9.10 clearly demonstrates the large differences in strength in the different loading directions for this unidirectional polymeric composite. The transverse strengths are almost negligible in

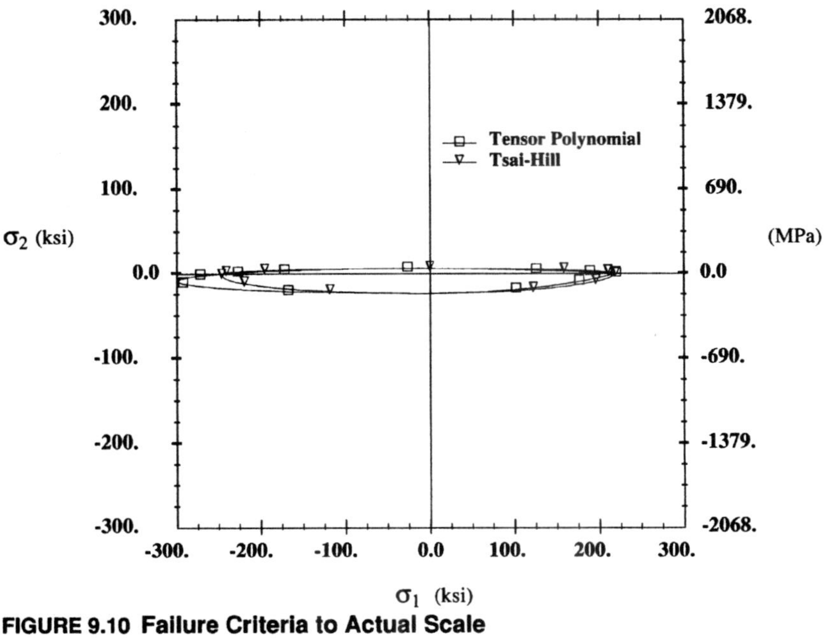

FIGURE 9.10 Failure Criteria to Actual Scale

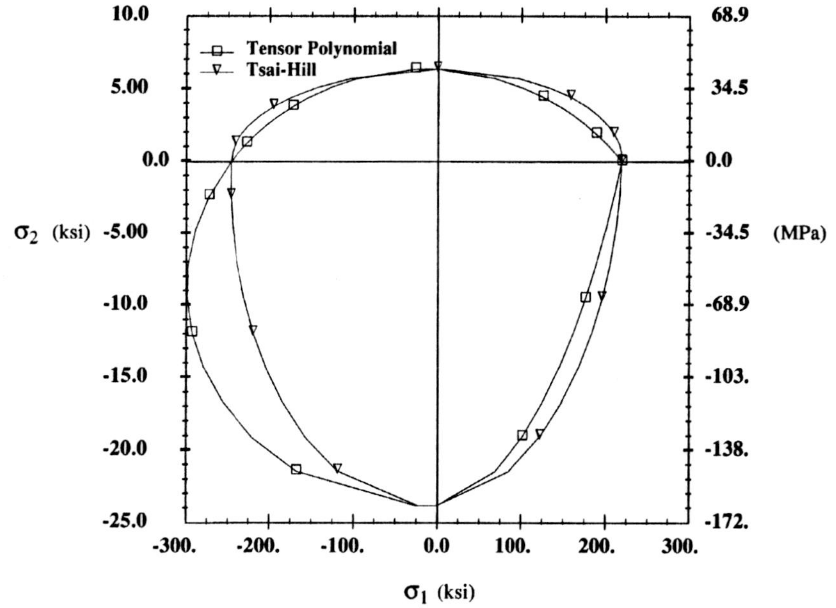

FIGURE 9.11 Failure Criteria with Adjusted Scale

comparison with the axial strengths. Thus it may be said that this material is highly anisotropic (orthotropic) in strength.

As indicated in Fig. 9.11, the predictions are reasonably close in three of the quadrants but exhibit significant differences in the third quadrant. These differences are a direct result of the inclusion of the linear terms in the tensor polynomial criterion.

9.3.6 Comparisons for Off-Axis Laminae

A fundamental evaluation of the various failure criteria can be obtained through a rather simple case study—comparison of theory and experiment for the applied stress, σ_x, for failure of a unidirectional off-axis tensile coupon, as a function of the off-axis fiber orientation θ. For the off-axis lamina, the maximum stress, Tsai-Hill, and tensor polynomial failure criteria have been written in terms of these parameters in equations (9.13), (9.25), and (9.42).

These three failure criteria are compared in Fig. 9.12 for AS4/3501-6 carbon/epoxy. As indicated in the figure, the Tsai-Hill and tensor polynomial predictions are very similar. Predictions based upon the three distinct modes of failure using the maximum stress criterion vary significantly (as discussed previously). For fiber orientations greater than 50° the tensor polynomial, Tsai-Hill, and maximum transverse stress all provide essentially identical predictions. For angles less than 50°, the maximum stress criterion always predicts higher ultimates than the tensor polynomial and Tsai-Hill theories.

Figure 9.13 shows a comparison of the maximum stress and tensor polynomial failure criteria with experimental results for AS4/3501-6 carbon/epoxy (Table 9.2) and Celion 6000/PMR-15 carbon fiber/polyimide matrix (Table 9.3; Pindera and Herakovich, 1981). As indicated in the two tables, the elastic and strength properties of the two materials are very similar. The theoretical predictions in Fig. 9.12 are based upon the properties of AS4/3501-6. As indicated in the figure, there

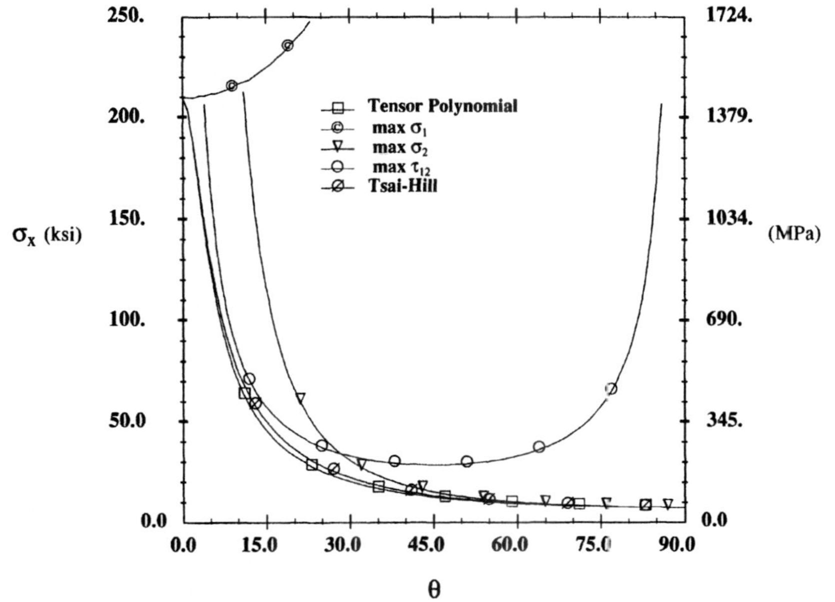

FIGURE 9.12 Comparison of Failure Theories for Off-Axis Coupon

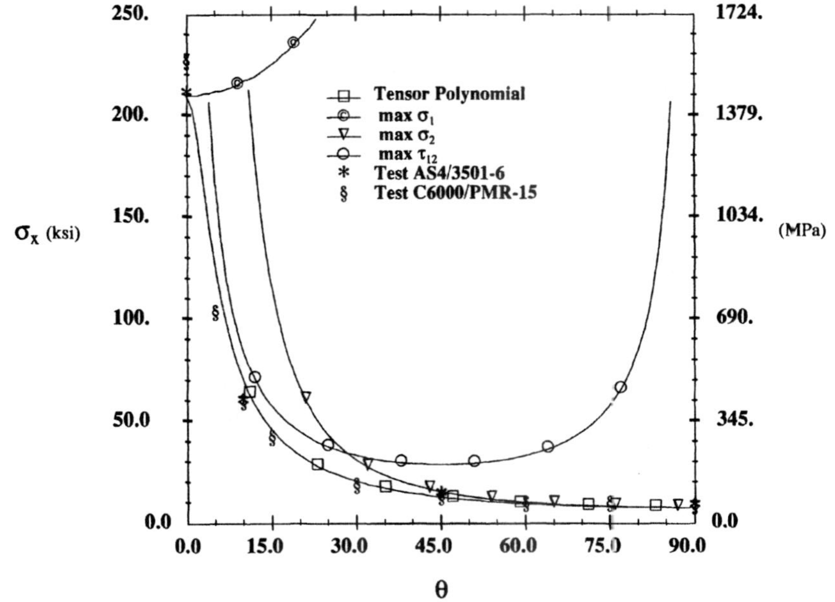

FIGURE 9.13 Off-Axis Failure: Theory and Experiment

E_1	E_2	G_{12}	ν_{12}	X_T	Y_T	S
136 GPa (19.8 Msi)	9.8 GPa (1.42 Msi)	4.96 GPa (0.72 Msi)	0.35	1551 MPa (225 ksi)	51.7 MPa (7.5 Msi)	71.0 (10.3 Msi)
ε_1^T, %	ε_2^T, %	Γ_{12}, %	F_{12}	X_C	Y_C	
1.04	0.505	2.0	0.0	N/A	N/A	

TABLE 9.3 Properties of Celion 6000/PMR-15 Carbon/Polyimide

is excellent correlation between the tensor polynomial theory and experiment for the entire range of fiber orientations. The maximum stress criterion predicts higher strength values than those measured in the laboratory for angles less than 45°. Thus, it is nonconservative for the corresponding stress states.

9.3.6.1 Failure Modes in Off-Axis Coupons

Unidirectional polymeric matrix composites under tensile loading typically fail along lines parallel to the fibers for all fiber orientations. Failed tensile specimens of T300/934 carbon/epoxy (Fig. 9.14; Reed et al., 1986) and Celion 6000 PMR-15 carbon/polyimide (Fig. 9.15; Pindera and Herakovich, 1981) clearly show this over a wide range of fiber orientations. Similar results were obtained for aramid/epoxy by Pindera et al. (1987).

The unidirectional [0] specimens exhibit fiber breakage in addition to the failures parallel to the fibers. The transverse cracks running parallel to the fibers in these specimens are associated with the transverse tensile stresses induced by the constraint of the grips, which restrains the lateral Poisson contraction of the specimen near the grips. The [0] carbon/epoxy specimen (Fig. 9.14) exhibits multiple failures with cracks running the full length of the specimen, whereas the carbon/polyimide specimen (Fig. 9.15) shows only two cracks parallel to the fibers (one in the grip region).

The fact that cracks grow parallel to the fibers for most unidirectional polymeric matrix composites is a clear indication of the high degree of orthotropy in the strength of these materials. Indeed, the transverse strength of carbon/epoxy is typically so low compared with the axial strength

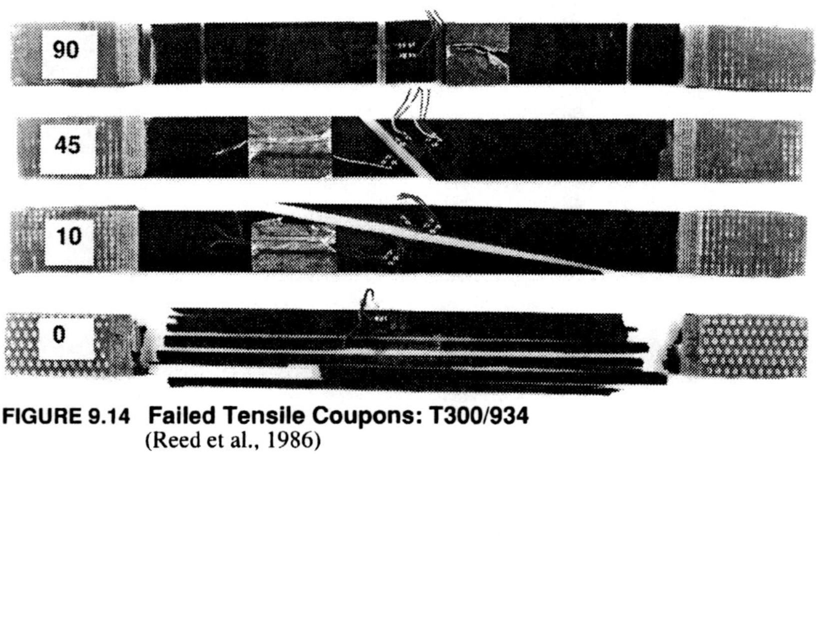

FIGURE 9.14 Failed Tensile Coupons: T300/934
(Reed et al., 1986)

FIGURE 9.15 Failed Celion 6000/PMR-15 Tensile Specimens
(Pindera et al., 1987)

that notched unidirectional carbon/epoxy exhibits crack growth parallel to the fiber direction for all fiber orientations, including 0° (Beuth and Herakovich, 1989). Thus a notched 0° tensile specimen with a centered slit notch fails in an H pattern, with cracks growing from both ends of the slit in both axial directions, as indicated in Fig. 9.16.

Metal matrix composites typically do not exhibit the same high degree of orthotropy in strength as polymeric matrix composites. Failed unidirectional B/Al tensile specimens are shown in Fig. 9.17 (Becker et al., 1987). For fiber angles of 15° and higher, the failure is parallel to the fiber direction and there is no fiber failure. However, for smaller fiber orientations (0° and 10°), fiber failure is evident, with the failure surface perpendicular to the fibers in the case of the 0° specimen. It must be mentioned that transverse tensile failure in metal matrix composites is very dependent on the fiber/matrix bond, and not all metal matrix composites have the good bond exhibited by boron/aluminum.

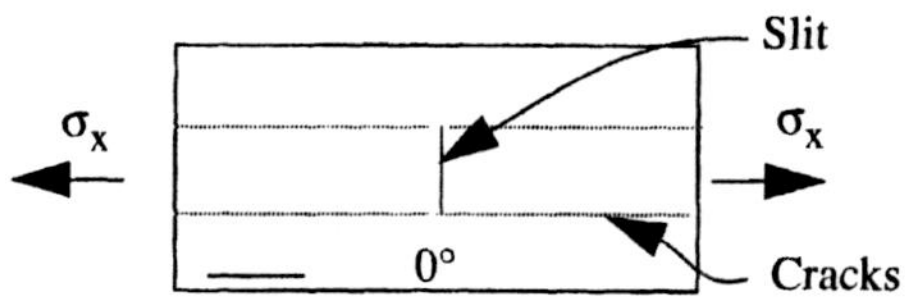

FIGURE 9.16 H Crack in [0] Carbon/Epoxy

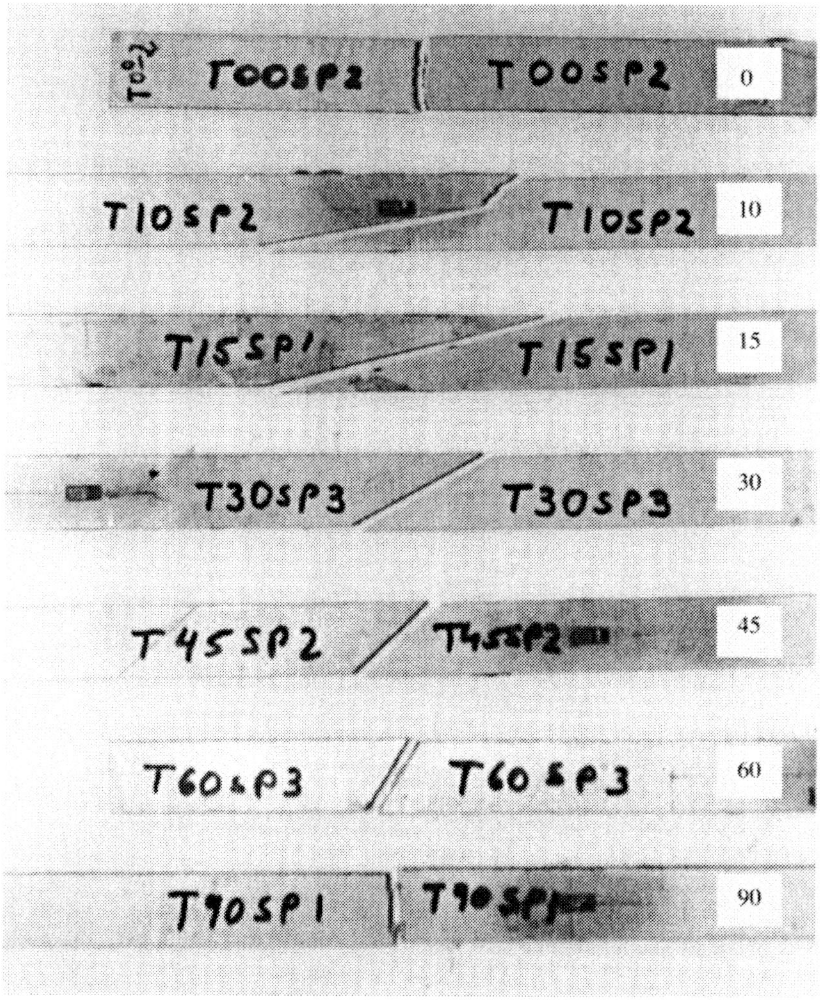

FIGURE 9.17 Failed B/Al Tensile Specimens
(Becker et al., 1987)

9.4 Laminate Failures

Failure of laminated composites is very different from the failure of unidirectional laminae. It is generally true that local failures (that do not lead to catastrophic failure) occur in one or more layers of a laminate prior to the final fracture of the laminate into two or more pieces. These local failures are arrested from catastrophic crack growth by the constraint of adjacent layers. Failure of laminates is also strongly influenced by the interlaminar stresses present in the boundary layer region along free edges. Thus tests on finite-width coupons can reveal strengths that are much different from the strength of a tube of the same stacking sequence. In the following section we discuss laminate failures that are initiated by interlaminar stresses near free edges. This is followed by the presentation of a damage theory for laminated composites.

9.4.1 Free Edge–Initiated Failure in Angle-Ply Laminates

Chapter 8 provided an in-depth discussion of interlaminar stresses in the boundary layer near the free edges of finite-width, polymeric matrix laminated composites. In this section we will present experimental evidence of edge damage in Celion 6000/PMR-15 carbon/polyimide angle-ply laminates and discuss the observed failures in light of the predicted stress distributions and the associated tensor polynomial failure criterion. Angle-ply laminates limit the discussion to the influences of fiber orientation and layer thickness. Examples of the influence of both of these factors on the strength and mode of failure in finite-width coupons will be demonstrated.

The experimental results were obtained by taking edge replicas (Herakovich and Klang, 1984; Herakovich, 1989) at selected intervals during the load history of tensile tests on finite-width coupons. The edge replicas show the type of damage (in the form of cracks) present along the free edge. The finite-element analysis (Nagarkar and Herakovich, 1979) was a nonlinear, temperature-dependent, thermomechanical analysis that included residual curing stresses.

Neglecting the interaction terms F_{ij} $(i \neq j)$ in the expression (9.30) for the tensor polynomial failure criterion, the condition for failure initiation is

$$F_1\sigma_1 + F_2\sigma_2 + F_3\sigma_3 + F_{11}\sigma_1^2 + F_{22}\sigma_2^2 + F_{33}\sigma_3^2 + F_{44}\sigma_4^2 + F_{55}\sigma_5^2 + F_{66}\sigma_6^2 = 1 \qquad (9.47)$$

Figure 9.18 presents results showing the variation of the tensor polynomial function at points along the $+\theta/-\theta$ interface at failure, as determined by the finite-element stress analysis. While the value of the tensor polynomial at the free edge is equal to (or slightly greater than) 1.0 for all laminates, the value away from the edge is well below 1.0 for all laminates investigated other than the $[(\pm 75)]_s$. It is evident from these results that edge effects are critical for laminates with fiber orientations less than 45°. For these laminates 90–95% of the width has a value of the tensor polynomial well below failure, but failure has initiated at the edge. The use of such laminates for axial loading in the presence of free edges would negate any advantage in specific strength gained through the use of the composite.

Table 9.4 (Herakovich et al., 1979) presents results for the individual terms of the tensor polynomial (9.47) at failure, taken from the element that is predicted to fail first. As indicated in the table, for angles of 30° and lower, failure is dominated by the τ_{13} interlaminar shear stress term. As the fiber orientation increases above 30°, edge effects become less dominant but nevertheless are still important, as indicated in Fig. 9.18. For angles greater than 30°, the dominant contribution to the failure polynomial is from the σ_2 transverse normal stress terms.

These results indicate that delaminations, resulting primarily from interlaminar shear stresses, are the expected mode of failure in angle-ply laminates for fiber orientations equal to or less than 30°. For larger fiber angles, the mode of failure is expected to be transverse cracking.

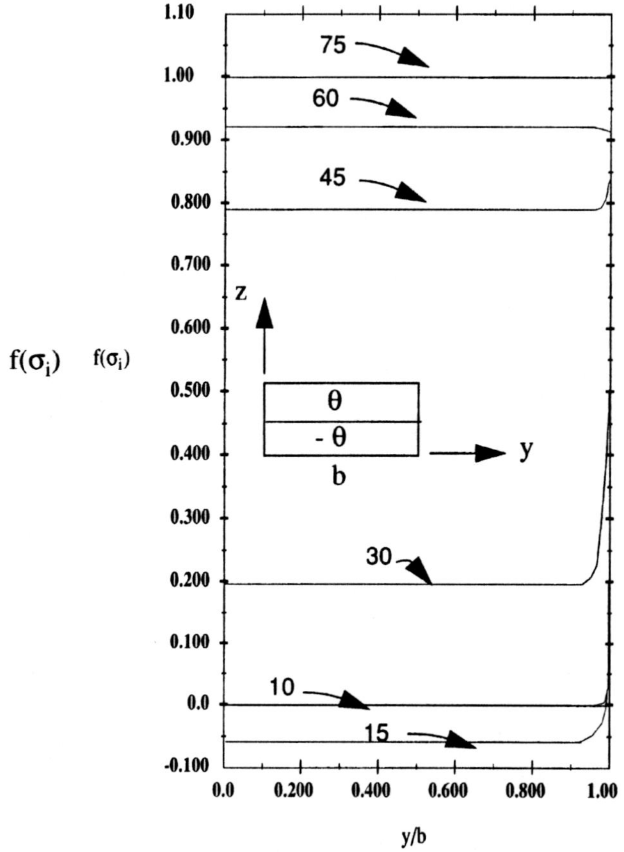

FIGURE 9.18 Tensor Polynomial along Interface of Angle-Ply Laminates

Edge replicas showing the accumulation of damage with increasing tensile load in $[(\pm 10)_2]_s$, $[(\pm 30)_2]_s$, $[(\pm 45)_2]_s$, and $[(\pm 60)_2]_s$ laminates are presented in Figs. 9.19–9.22. The experimental results are in complete agreement with the predictions. Delamination between $+\theta$ and $-\theta$ layers is the mode of failure in the $[(\pm 10)_2]_s$ and $[(\pm 30)_2]_s$ laminates, and transverse cracking is the mode of failure in the $[(\pm 45)_2]_s$ and $[(\pm 60)_2]_s$ laminates.

The results for prediction of delamination in angle-ply laminates are summarized in Fig. 9.23, which shows a comparison of the mismatch in coefficient of mutual influence, the maximum interlaminar shear stress (normalized), and the variation in maximum tensor polynomial for angle-ply laminates, all with respect to fiber orientation. All three curves exhibit strong peaks for $\theta = 15°$, where delaminations are most evident.

Laminate	$F_1\sigma_1$	$F_{11}\sigma_1^2$	$F_2\sigma_2$	$F_{22}\sigma_2^2$	$F_3\sigma_3$	$F_{33}\sigma_3^2$	$F_{44}\tau_{23}^2$	$F_{55}\tau_{13}^2$	$F_{66}\tau_{12}^2$
Angle-Ply Laminates									
$[\pm 10]_s$	0.027	0.020	0.061	0.001	0.056	0.0	0.044	1.040	0 004
$[\pm 15]_s$	0.023	0.015	0.109	0.003	0.112	0.0	0.115	1.070	0.010
$[\pm 30]_s$	0.016	0.012	0.251	0.015	0.375	0.003	0.584	0.867	0.104
$[\pm 37.5]_s$	0.004	0.001	0.511	0.060	0.184	0.0	0.0	0.0	0.240
$[\pm 45]_s$	0.003	0.000	0.588	0.080	0.144	0.0	0.0	0.0	0.185
$[\pm 60]_s$	0.0	0.0	0.724	0.151	0.062	0.0	0.0	0.0	0.092
$[\pm 75]_s$	0.0	0.0	0.810	0.152	0.014	0.0	0.0	0.0	0.025
Cross-Ply Laminates									
$[90/0]_s$	0.0	0.0	0.690	0.110	0.187	0.0	0.0	0.002	0.0
$[0/90]_s$	0.0	0.0	0.717	0.119	0.159	0.0	0.0	0.0	0.0
Quasi-Isotropic Laminates									
$[\pm 45/0/90]_s$	0.0	0.0	0.439	0.045	0.461	0.0	0.0	0.005	0.122
$[90/0/\pm 45]_s$	0.0	0.0	0.621	0.089	0.264	0.0	0.0	0.007	0.238

TABLE 9.4 Tensor Polynomial for Angle-Ply Laminates

$\varepsilon = 0.3\%$ $\varepsilon = 0.5\%$ $\varepsilon = 0.7\%$ $\varepsilon = 0.7\%$

$\varepsilon^{ult} = 0.7\%$

FIGURE 9.19 Edge Damage in a $[(\pm 10)_2]_s$ Laminate

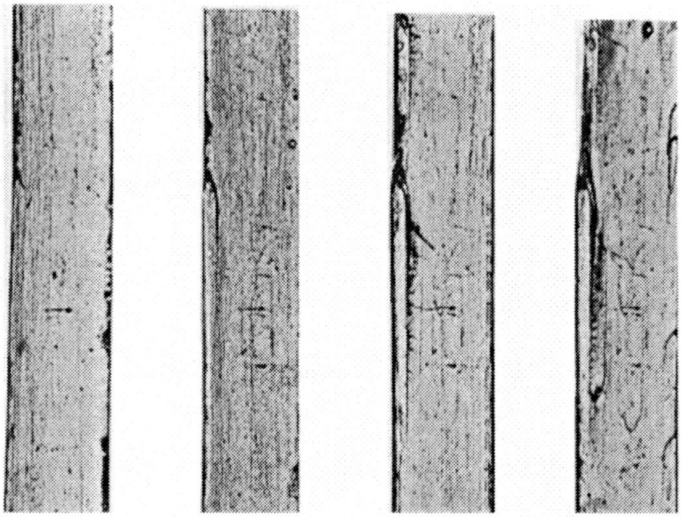

ε = 0.5% ε = 0.8% ε = 1.2% ε = 1.3%

ε^{ult} = 1.4%

FIGURE 9.20 Edge Damage in a $[(\pm 30)_2]_s$ Laminate

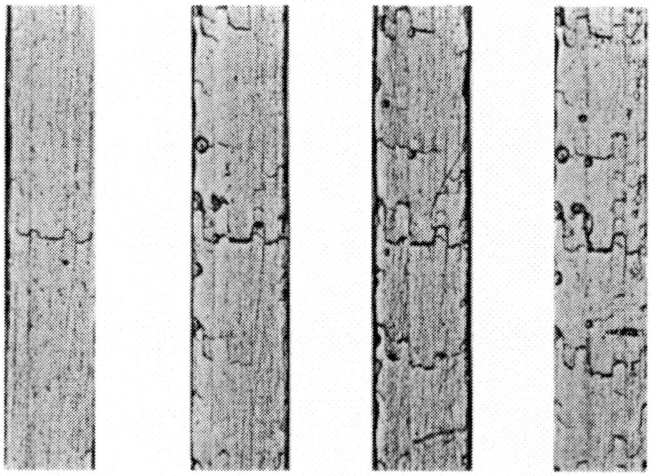

ε = 1.5% ε = 1.7% ε = 1.9% ε = 1.95

ε^{ult} = 1.95%

FIGURE 9.21 Edge Damage in a $[(\pm 45)_2]_s$ Laminate

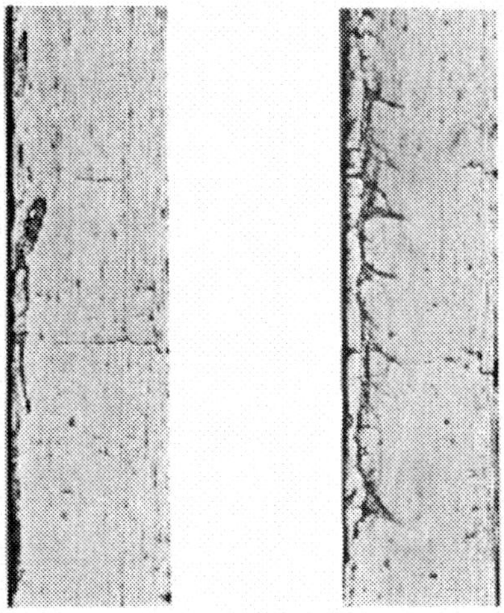

$\varepsilon^{ult} = 0.9\%$

FIGURE 9.22 Edge Damage in a $[(\pm 60)_2]_s$ Laminate

9.4.2 Stacking Sequence–Dependent Angle-Ply Laminate Failure

The strong influence of stacking sequence effects on the strength and mode of failure of laminated composites can be demonstrated by results from tensile tests on angle-ply laminates with the same number of plies at a given fiber orientation, but different layer thicknesses (stacking sequence). Figure 9.24 (Herakovich, 1982) shows failed T300/5208 carbon/epoxy angle-ply specimens of this type. The difference in failure modes for alternating, $[(\pm\theta)_2]_s$, and clustered, $[\theta_2/-\theta_2]_s$, laminates with fiber orientations of 10° and 30° is shown in the figure. It is evident that the mode of failure is completely different depending upon stacking sequence (layer thickness). The laminates with alternating stacking sequence (thinner layers) fail due to a combination of fiber breakage and matrix cracking or fiber/matrix debonding. In contrast, failure of the laminates with clustered stacking sequence (thicker layers) is entirely the result of delamination between the $+\theta$ and $-\theta$ layers and transverse (matrix) failure parallel to the fibers. For these fiber orientations, there is no (or essentially no) fiber failure in the clustered laminates.

The ±30° laminates in Fig. 9.24 provide a clear distinction between the two types of failure. The failure surface for the alternating $[(\pm 30)_2]_s$ laminate is a single fracture plane at 30° to the loading direction. Thus, the fibers are broken in half of the layers whereas the remaining layers fail by transverse cracks parallel to the fibers. In contrast, the failure surface of the clustered $[30_2/-30_2]_s$ laminates consists of delaminations at the $+\theta/-\theta$ interface and transverse failure parallel to the fibers in all layers. Thus, there is no fiber breakage when the $[30_2/-30_2]_s$ laminate fails.

This significant difference in failure modes is the result of the interlaminar stresses. Figure 9.25 shows normalized through-thickness distributions of the interlaminar shear stress τ_{zx} along the free edge of alternating and clustered stacking sequences of the ±10° and ±30° angle-ply laminates, all subjected to the same axial strain (Herakovich, 1982). For both fiber orientations, the maximum interlaminar shear stress of the clustered $[\theta_2/-\theta_2]_s$ configuration is 28% higher than that in the alternating $[(\pm\theta)_2]_s$ stacking sequence for the same applied axial strain. This higher interlaminar shear

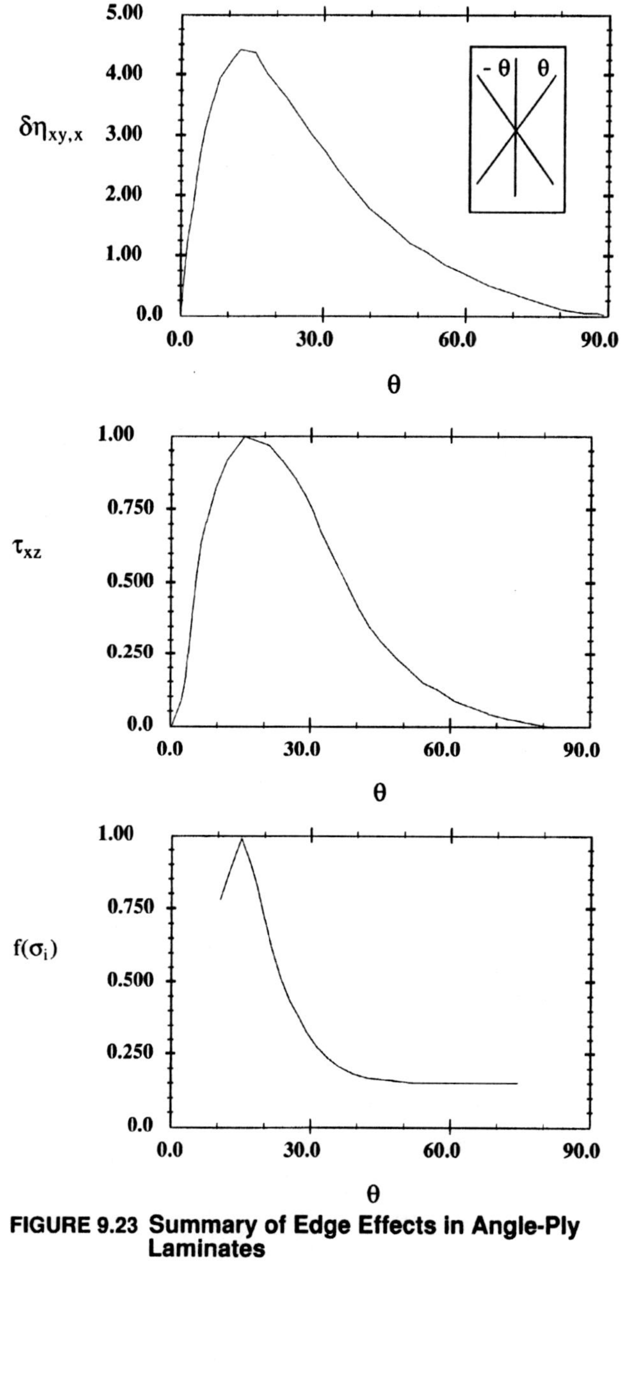

FIGURE 9.23 Summary of Edge Effects in Angle-Ply Laminates

$[(\pm 10)_2]_s$ $[(10_2/-10_2)]_s$ $[(\pm 30)_2]_s$ $[(30_2/-30_2)]_s$

FIGURE 9.24 Failure Modes in Angle-Ply Laminates

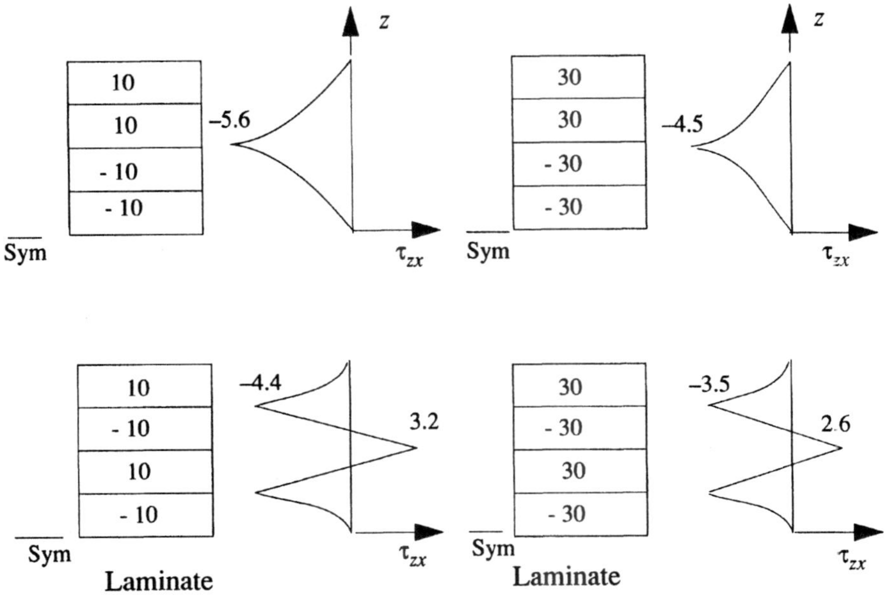

FIGURE 9.25 Maximum Interlaminar Shear Stresses τ_{zx} in Angle-Ply Laminates

stress leads to delamination as the initial mode of failure in the clustered stacking sequence. The ultimate stress of the alternating $[(\pm30)_2]_s$ laminate was 48% higher than that for the clustered $[30_2/-30_2]_s$ laminate, and the strength for the alternating stacking sequence of the ±10° laminate was 28% higher than the strength of the clustered configuration.

9.4.3 Quasi-Isotropic Laminate Failures

The results in Table 9.5 for the tensile strength of quasi-isotropic laminates show that the strength can vary considerably with the stacking sequence. The ultimate strains range from a low of 0.882% to a high of 1.094%, a difference of 24%. The differences in strength are primarily associated with the variable interlaminar stresses due to edge effects, which are stacking sequence dependent as shown in Chapter 8.

Laminate	$[45/90/0/-45]_s$	$[90/45/0/-45]_s$	$[45/90/-45/0]_s$	$[0/\pm45/90]_s$
σ^{ult}, MPa (ksi)	439 (63.7)	505 (73.3)	490 (71.1)	479 (69.5)
ε^{ult}, %	0.882	1.005	0.994	1.094

TABLE 9.5 Strength of Quasi-Isotropic T300/5208 Carbon/Epoxy Laminates

9.5 Damage Mechanics

The goal of damage mechanics is to predict the response of a material in the presence of damage that initiates at some stress state and generally increases with increasing stress up to macroscopic crack initiation or failure; prediction of the conditions for failure is also a fundamental goal of damage mechanics. In general, damage may be thought of as surface discontinuities (microcracks) and volume discontinuities (microvoids). Essentially all materials exhibit cracks and voids at some scale; thus the "damage-free" state is a function of the scale of the phenomena under consideration.

As indicated in Fig. 9.1, the most common types of damage in fibrous composites are fiber breakage, fiber/matrix debonding, matrix cracks, fiber kinking, and, for large-diameter fibers, radial cracks in the fibers. We consider damage that can only increase or remain constant over time; there is no healing. Thus, the evolution of damage results in progressive deterioration of stiffness and nonlinear, inelastic response as depicted in Fig. 9.26, where E^0 is the initial, undamaged modulus and d is a damage parameter representing the percent loss of modulus. As damage occurs, the material loses stiffness and exhibits nonlinear, inelastic response with permanent strains after unloading. The inelastic response is the result of sliding friction at damage sites as well as any inelastic response of the constituent phases. Since the process is irreversible, nonlinear analysis techniques must be employed.

The earliest published works on damage mechanics appear to be those by Kachanov (1958) and Rabotnov (1968) on the application of a continuous damage variable to creep failure of metals. The concept was generalized, within the framework of irreversible thermodynamics, for isotropic materials under multiaxial loads by Lemaitre and Chaboche (1974), Hult (1974), Leckie (1978), and Murakami (1983), among others (see also Krajcinovic, 1989; Lemaitre and Chaboche, 1990; and Bazant and Cedolin, 1991).

In this section we present a damage theory for composites that has its foundation in irreversible thermodynamics. The theory has been shown to be both accurate and versatile for application to a wide variety of fibrous composites.

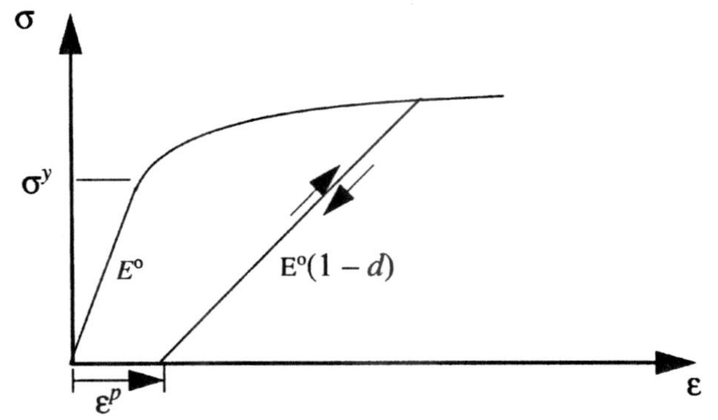

FIGURE 9.26 Idealized Inelastic Response of Damaged Composite

9.5.1 Mesoscale Composite Damage Theory

The general damage theory that serves as the foundation for this model was originally proposed by Ladevèze (1983). He later adapted the general damage theory for application to laminated composites (Ladevèze, 1986). The theory has been shown to be very robust for predicting the damaged response of composites under a wide variety of conditions. It is based upon the method of local state expressed in terms of state variables and the associated thermodynamic forces. According to Lemaitre and Chaboche (1990, p. 57), "The method of local state postulates that the thermodynamic state of a material medium at a given point and instant is completely defined by the knowledge of the values of a certain number of variables at that instant, which depend only upon the (material) point considered. Since the time derivatives of these variables are not involved in the definition of the state, this hypothesis implies that any evolution can be considered as a succession of equilibrium states."

For the most general configuration, Ladevèze's theory considers a composite to be a laminated structure consisting of two elementary constituents: layers of composite and interfaces that separate the composite layers. The interface is considered to be a mechanical surface connecting two adjacent composite layers; it is included in the model only when delamination is of interest. When delamination is not of interest, the model is an assemblage of composite layers.

The theory is called the *mesoscale composite damage theory* because it is based upon the assumption that the damage is uniform through the thickness of individual layers of the composite. A layer is any continuous thickness of plies which have identical fiber orientation. *Mesoscale* is a term indicating that the scale of the analysis is between micromechanics (i.e., the level of the fiber and the matrix) and laminate analysis. The theory is based upon the mean value of the stress in each layer and allows the damage state to vary from layer to layer in a laminate. When delamination is of interest, damage between layers is introduced through consideration of damage to the idealized interfacial layer (Ladevèze et al., 1990; Allix and Ladevèze, 1992). Additional developments of the model were concerned with the computation of the intensities of the different damage mechanisms up to ultimate fracture (Ladevèze, 1992) and extension of the model to variable temperature ranges (Allix et al., 1994).

The mesoscale composite damage theory assumes that the response of a damaged layer, at any instant of time (load state), can be expressed in terms of elastic moduli degradation and inelastic strains due to damage and/or matrix plasticity. Degradation of the elastic moduli is expressed in terms of damage parameters that are functions of the associated thermodynamic forces which serve as damage evolution parameters. The modulus degradation parameters are internal variables, and the thermodynamic forces are the corresponding associated variables (in the thermodynamic sense). The theory includes provisions for coupled damage evolution for the multi-axial states of stress that are typically present in laminates, as well as differences in the effects of damage evolution due to tensile and compressive normal stresses. The latter feature allows for difference in material response associated with crack opening and crack closure.

The form of the damage evolution law generally varies with the type of material, reflecting the dependence on the microstructural damage mechanisms. It is expected that fiber size, microstructure (or morphology) and strength, matrix strength, and fiber/matrix interfacial strength all influence damage evolution. The micro-level damage mechanisms are not identified explicitly in the mesoscale damage model; damage evolution is based on experimental observations of the response of a damaged layer. In this sense, the theory is phenomenological. The theory does allow the damage evolution law to change form during the loading history as new damage mechanisms occur.

The mesoscale composite damage theory has been applied successfully to polymeric, metallic, and ceramic matrix composites made with continuous fibers, short fibers, or woven fabrics. We shall present the theory for the case of the layers of a laminate which are all in a state of plane stress. A more general treatment of the theory and references to other applications can be found in the works of Ladevèze and co-workers (e.g., Ladevèze, 1994).

9.5.1.1 Effective Stress

The starting point for application of continuum damage mechanics to laminated composites is to consider a layer of the laminate in a state of plane stress and define the *effective stresses* $\{\tilde{\sigma}\}$ in the form

$$\{\tilde{\sigma}\} = \begin{Bmatrix} \dfrac{\langle\sigma_{11}\rangle_+}{(1-d_1)} + \langle\sigma_{11}\rangle_- \\ \dfrac{\langle\sigma_{22}\rangle_+}{(1-d_2)} + \langle\sigma_{22}\rangle_- \\ \dfrac{\sigma_{12}}{(1-d_6)} \end{Bmatrix} \tag{9.48}$$

The effective stresses are the stresses acting over the damaged area that effectively resists the forces. The parameters (d_1, d_2, d_6) define the damage state for the three types of stress loadings, with d_i ranging from 0 (the virgin, undamaged state) to 1.0 (corresponding to macroscopic damage or failure). (Note that we have used the reduced notation of Chapter 3 for the d_i, with i = 1, 2, 6 for plane stress, but we otherwise use the full tensor notation in this development of the damage theory.) As the d_i increase (microcracks and microvoids increase in size and/or number), the corresponding effective tensile normal stresses also increase. For compressive normal stresses, microcracks and microvoids close, and hence there is no increase in effective stress. Thus, the bracket notation $\langle\ \rangle$ in (9.48) is defined:

$$\langle a\rangle_+ = a \text{ if } a \geq 0;\ \text{otherwise } \langle a\rangle_+ = 0$$

$$\langle a\rangle_- = a \text{ if } a \leq 0;\ \text{otherwise } \langle a\rangle_- = 0$$

With these definitions, only one (at most) bracketed term associated with each component of the normal stresses, σ_{11} and σ_{22}, is nonzero at any time in the load history. Effective shear stresses are taken to be independent of the sign of the shear stress.

9.5.1.2 Effective Response

The linear elastic constitutive equation for the damaged material is then written according to the *principle of strain equivalence* (Lemaitre and Chaboche, 1990, p. 351), which states that "any deformation behavior, whether uniaxial or multiaxial, of a damaged material is represented by the constitutive laws of the virgin material in which the usual stress is replaced by the effective stress." Thus, the elastic constitutive equations for a damaged orthotropic material in plane stress are

$$
\begin{aligned}
\varepsilon_{11}^{E} &= \frac{\langle\sigma_{11}\rangle_+}{E_1^0(1-d_1)} + \frac{\langle\sigma_{11}\rangle_-}{E_1^0} - \frac{\nu_{12}^0}{E_1^0}\sigma_{22} \\
\varepsilon_{22}^{E} &= \frac{\langle\sigma_{22}\rangle_+}{E_2^0(1-d_2)} + \frac{\langle\sigma_{22}\rangle_-}{E_2^0} - \frac{\nu_{12}^0}{E_1^0}\sigma_{11} \\
\varepsilon_{12}^{E} &= \frac{\sigma_{12}}{2G_{12}^0(1-d_6)}
\end{aligned}
\tag{9.49}
$$

where, based upon experimental results for some materials, we have taken the ratio $\nu_{12}/E_1 = \nu_{12}^0/E_1^0 = \nu_{21}^0/E_2^0$ to remain constant during damage evolution.

From (9.49) we see that the response can be expressed in terms of the elastic moduli E_1, E_2, and G_{12} of the damaged layer, defined in terms of the damage parameters d_1, d_2, and d_6 and the original, undamaged moduli E_1^0, E_2^0, and G_{12}^0 as

$$
\begin{aligned}
E_1 &= E_1^0(1-d_1) \\
E_2 &= E_2^0(1-d_2) \\
G_{12} &= G_{12}^0(1-d_6)
\end{aligned}
\tag{9.50}
$$

From (9.50) we see that the damage parameters, d_i, are internal variables representing the percent degradation of the respective (tensile) modulus.

9.5.1.3 Thermodynamic Forces

Following the method of local state (Lemaitre and Chaboche, 1990), the *thermodynamic forces* Y_1, Y_2, and Y_6 associated with the internal damage variables d_i are related to the density ρ and the mean value (through the layer thickness) of the thermodynamic potential ψ (or, equivalently, the mean value of the strain energy density E_D), at the current states of stress and damage, by the partial derivatives

$$
\begin{aligned}
Y_1 &= \rho\frac{\partial\psi}{\partial d_1}\bigg|_{\tilde{\sigma},\, d_2,\, d_6:\text{cnst}} = \frac{\partial E_D}{\partial d_1}\bigg|_{\tilde{\sigma},\, d_2,\, d_6:\text{cnst}} \\
Y_2 &= \rho\frac{\partial\psi}{\partial d_2}\bigg|_{\tilde{\sigma},\, d_1,\, d_6:\text{cnst}} = \frac{\partial E_D}{\partial d_2}\bigg|_{\tilde{\sigma},\, d_1,\, d_6:\text{cnst}} \\
Y_6 &= \rho\frac{\partial\psi}{\partial d_6}\bigg|_{\tilde{\sigma} d} = \frac{\partial E_D}{\partial d_6}\bigg|_{\tilde{\sigma},\, d_1,\, d_2:\text{cnst}}
\end{aligned}
\tag{9.51}
$$

The mean value of the strain energy density of the damaged layer can be written in terms of the (mean) effective stresses, with the energy associated with tensile normal stress distinguished from that associated with compressive normal stress through the use of the bracket notation.

$$E_D = \frac{1}{2}\left[\frac{\langle\sigma_{11}\rangle_+^2}{E_1^0(1-d_1)} + \frac{\langle\sigma_{11}\rangle_-^2}{E_1^0} - 2\frac{\nu_{12}^0}{E_1^0}\sigma_{11}\sigma_{22} + \frac{\langle\sigma_{22}\rangle_+^2}{E_2^0(1-d_2)} + \frac{\langle\sigma_{22}\rangle_-^2}{E_2^0} + \frac{\sigma_{12}^2}{G_{12}^0(1-d_6)}\right] \tag{9.52}$$

From (9.51) and (9.52), we have the thermodynamic forces in terms of stress components and damage variables:

$$\begin{aligned} Y_1 &= \left.\frac{\partial E_D}{\partial d_1}\right|_{\tilde{\sigma},\, d_2,\, d_6:\text{cnst}} = \frac{\langle\sigma_{11}^2\rangle_+}{2E_1^0(1-d_1)^2} \\ Y_2 &= \left.\frac{\partial E_D}{\partial d_2}\right|_{\tilde{\sigma},\, d_1,\, d_6:\text{cnst}} = \frac{\langle\sigma_{22}^2\rangle_+}{2E_2^0(1-d_2)^2} \\ Y_6 &= \left.\frac{\partial E_D}{\partial d_6}\right|_{\tilde{\sigma},\, d_1,\, d_2:\text{cnst}} = \frac{\sigma_{12}^2}{2G_{12}^0(1-d_6)^2} \end{aligned} \tag{9.53}$$

These associated thermodynamic forces govern damage development in the sense that the previous maximum value of some function of the thermodynamic forces must be exceeded for additional damage to occur.

This is much like the Kaiser effect for acoustic emissions from materials under load (Kaiser, 1950). Acoustic emissions are associated with the release of strain energy due to change (damage) of the material microstructure. The strain energy released from the region of local discontinuity is propagated through the solid in the form of elastic waves (i.e., acoustic emissions). The acoustic emissions persist only as long as the previous maximum load is exceeded. Upon unloading, the acoustic emissions cease, indicating a cessation of energy release (i.e., a cessation of damage development).

9.5.1.4 Damage Evolution

The specific form of the damage evolution laws in the Ladevèze theory is material dependent. In the following discussion we demonstrate the method for a highly orthotropic composite such as carbon/epoxy. The development follows closely that of Ladevèze and Le Dantec (1992).

We first introduce the experimentally verified fact that for carbon/epoxy, and most continuous fiber composites, damage other than fiber breakage does not degrade the fiber direction modulus, E_1; therefore, in the absence of fiber breakage, d_1 is zero throughout the load history, and the only thermodynamic forces of interest (for plane stress) are Y_2 and Y_6.

We define $\underline{Y}_2$ and $\underline{Y}_6$ as the maximum values of the thermodynamic forces attained throughout the load history, $\tau \le t$, for any time τ up to the current time, t.

$$\underline{Y}_2(t) = \max_{\tau \le t} \{Y_2(\tau)\} \tag{9.54}$$

$$\underline{Y}_6(t) = \max_{\tau \le t} \{Y_6(\tau)\} \tag{9.55}$$

Next we define the linear combination of Y_6 and Y_2 as

$$\hat{Y} = (Y_6 + bY_2) \tag{9.56}$$

where b is a material-dependent constant representing coupling between transverse tension and shear effects. The definition of $\hat{Y}$ in (9.56) represents the simplest form of coupling, a linear combination of two effects. Finally we define $\underline{\hat{Y}}$ as the maximum value of $\hat{Y}$ for all previous time τ up to the present time, t, as

$$\underline{\hat{Y}}(t) = \max_{\tau \le t} \{Y_6(\tau) + bY_2(\tau)\} \tag{9.57}$$

With these definitions, experimental results for highly orthotropic composites, such as carbon/epoxy under quasistatic shear loading, indicate that the damage evolution law for d_6 can be written in the form (using the bracket notation defined previously)

$$d_6 = \frac{\langle \sqrt{\underline{\hat{Y}}} - \sqrt{Y_O} \rangle_+}{\sqrt{Y_C}} \quad \text{if } d_6 < 1 \text{ and } \underline{Y}_2 < Y_2^C\text{; otherwise } d_6 = 1 \tag{9.58}$$

where Y_2^C is the brittle-damage threshold corresponding to failure of the fiber/matrix interface due to transverse tension loading, and Y_O and Y_C are damage evolution parameters determined from experimental results.

Experimental results for carbon/epoxy also indicate that the transverse tension damage evolution law for d_2 can be expressed in terms of additional damage parameters Y'_O and Y'_C as

$$d_2 = \frac{\langle \sqrt{\underline{\hat{Y}}} - \sqrt{Y'_O} \rangle_+}{\sqrt{Y'_C}} \quad \text{if } d_2 > 1 \text{ and } \underline{Y}_2 < Y_2^C \text{ ; otherwise } d_2 = 1 \tag{9.59}$$

The shear damage evolution law (9.58) is depicted in Fig. 9.27, and the transverse tension damage evolution law (9.59) is depicted in Fig. 9.28. The figures indicate initiation of damage at $\sqrt{\underline{\hat{Y}}} = \sqrt{Y_O}$ and $\sqrt{\underline{\hat{Y}}} = \sqrt{Y'_O}$ for shear and transverse tension, respectively (as calculated for the specific type of loading), followed by a linear evolution with slopes $1/\sqrt{Y_C}$ for d_6 and $1/\sqrt{Y'_C}$ for d_2.

The parameters Y_O, Y_C, Y'_C, Y'_O, and b are constants that must be determined from appropriate tests, as described in later sections.

9.5.1.5 Inelastic Strains

As depicted in Fig. 9.26, most structural engineering materials exhibit nonlinear response when stressed beyond some elastic (yield) limit stress, σ^y. When the material is completely unloaded from a stress in excess of this limit stress, there are inelastic (permanent) strains ε_{ij}^p. The response during the unloading phase generally exhibits some hysteresis, but the unloading and subsequent reloading often may be approximated as linear elastic, as indicated in Fig. 9.26.

In general, nonlinear material response and permanent strains can be associated with a variety of effects, including plasticity, creep, and damage. In the following, we are concerned with the nonlinear effects in fibrous composites that are associated with plasticity and damage. As most fibers in use today exhibit linear, elastic response, plasticity effects are associated with the matrix. Damage

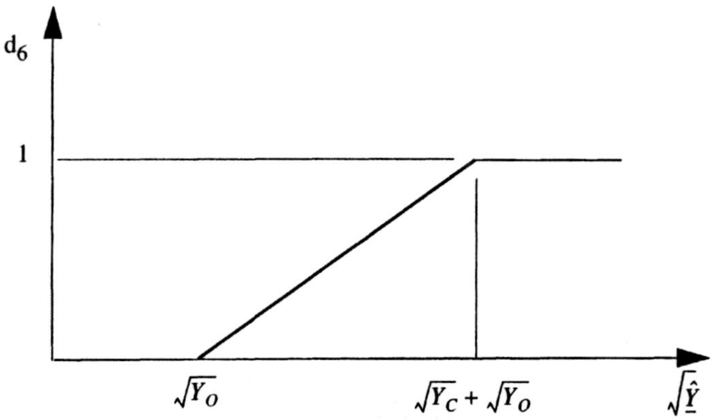

FIGURE 9.27 Typical Shear Damage Evolution

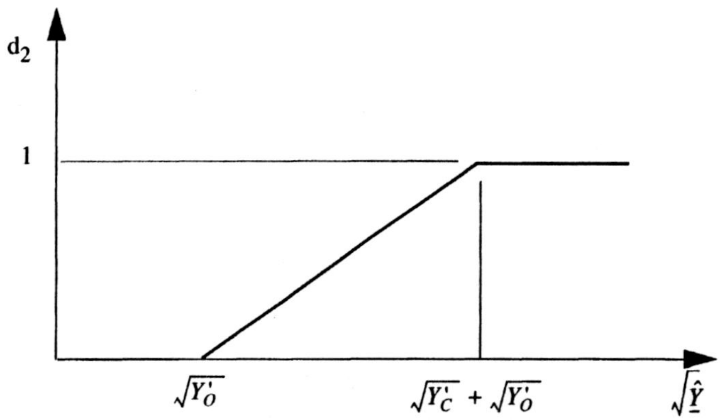

FIGURE 9.28 Typical Transverse Tension Damage Evolution

can be associated with different micromechanisms as discussed previously, including fiber breakage, matrix cracking, and fiber/matrix interfacial debonding.

The formalism for the inelastic strains in the mesoscale composite damage theory follows that of classical plasticity, with an elastic domain function $f(\tilde{\sigma}_{ij}, \tilde{p})$ that is dependent on the current effective stress, $\tilde{\sigma}_{ij}$, and the accumulated effective inelastic strain $\tilde{p}$, for all time up to the present. (The accumulated effective inelastic strain $\tilde{p}$ will be defined specifically in a later section.) Effective inelastic strain increments (rates) $\dot{\tilde{\varepsilon}}_{ij}^{p}$ are normal to this elastic domain function in the same manner that plastic strain rates are normal to the yield surface in plasticity theory (Mises, 1928).

The relationship between the elastic domain function $f(\tilde{\sigma}_{ij}, \tilde{p})$ and the inelastic strain rates is depicted in Fig. 9.29 for loading in the $\tilde{\sigma}_{12}$-$\tilde{\sigma}_{22}$ stress space. The initial elastic domain corresponds to $\tilde{p} = 0$, and subsequent elastic domains correspond to $\tilde{p} > 0$. The hardening (increase in size of the elastic domain) is assumed to be isotropic.

Effective inelastic strain rates, $\dot{\tilde{\varepsilon}}_{ij}^{p}$, are defined such that the rate of work (dissipation of energy) of the effective stresses and the effective inelastic strain rates is identical to that of the actual stress and inelastic strain rates. For composites exhibiting only elastic response in the fiber direction, the

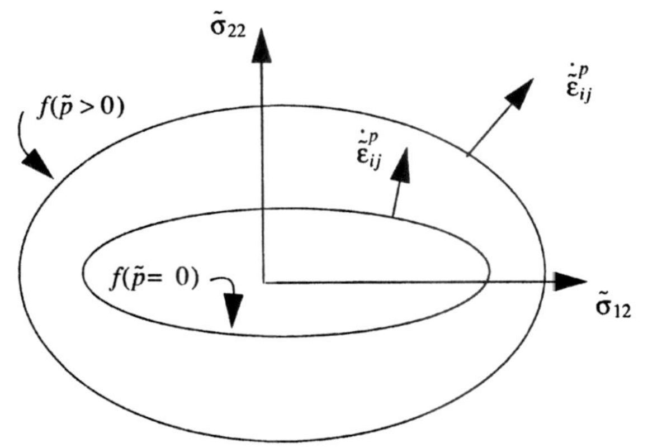

FIGURE 9.29 Normality of Inelastic Strain Rates

effective plastic strain rates are

$$\dot{\tilde{\varepsilon}}_{11}^{p} = 0 \tag{9.60}$$

$$\begin{aligned} \dot{\tilde{\varepsilon}}_{22}^{p} = \dot{\varepsilon}_{22}^{p}(1-d_2) \qquad & \text{if } (\sigma_{22} > 0) \\ \dot{\tilde{\varepsilon}}_{22}^{p} = \dot{\varepsilon}_{22}^{p} \qquad & \text{if } (\sigma_{22} < 0) \end{aligned} \tag{9.61}$$

$$\dot{\tilde{\varepsilon}}_{22}^{p} = \dot{\varepsilon}_{12}^{p}(1-d_6) \tag{9.62}$$

The inelastic dissipation, Φ_p, is then

$$\Phi_p = \left(\tilde{\sigma}_{ij}\dot{\tilde{\varepsilon}}_{ij}^{p}\right) = (\sigma_{ij}\dot{\varepsilon}_{ij}^{p}) = \sigma_{11}\dot{\varepsilon}_{11}^{p} + \sigma_{22}\dot{\varepsilon}_{22}^{p} + 2\sigma_{12}\dot{\varepsilon}_{22}^{p} \tag{9.63}$$

For a state of uniform, plane stress in a unidirectional layer and no inelastic strains in the fiber direction, the elastic domain function f is written in terms of effective stresses $\tilde{\sigma}_{ij}$ and effective inelastic strain $\tilde{p}$ as

$$f = \sqrt{\tilde{\sigma}_{12}^2 + a^2\tilde{\sigma}_{22}^2} - R(\tilde{p}) - R_o \tag{9.64}$$

where $R(\tilde{p})$ is a function of the accumulated effective inelastic strain, R_o is the initial threshold value for $R(\tilde{p} = 0)$, and a^2 is a material constant. The condition $f < 0$ corresponds to the state of stress being inside the elastic domain of Fig. 9.29; the response is elastic for this condition. Inelastic response corresponds to $f = 0$ and $\dot{f} = 0$; the state of stress is on the elastic domain boundary, and the current increment of load is such that the stress states remains on the (hardening) elastic domain boundary.

The normality condition requires that the inelastic strains be normal to the inelastic domain surface. This requirement can be expressed in terms of a parameter $\dot{\lambda}_p$ (where $\dot{\lambda}_p \geq 0$) in the form

$$\dot{\tilde{p}} = -\dot{\lambda}_p\frac{\partial f}{\partial R} = \dot{\lambda}_p \tag{9.65}$$

$$\dot{\varepsilon}_{12}^{p} = \frac{\dot{\lambda}_p}{2}\frac{\partial f}{\partial \tilde{\sigma}_{12}} = \frac{\dot{\tilde{p}}}{2}\frac{\tilde{\sigma}_{12}}{[R(\tilde{p}) + R_o]} \tag{9.66}$$

$$\dot{\varepsilon}_{22}^{p} = \dot{\lambda}_p\frac{\partial f}{\partial \tilde{\sigma}_{22}} = \dot{\tilde{p}}\frac{a^2\tilde{\sigma}_{22}}{[R(\tilde{p}) + R_o]} \tag{9.67}$$

From (9.64) with $f = 0$ and using (9.66) and (9.67), we obtain

$$\dot{\tilde{p}} = \sqrt{4\left(\dot{\varepsilon}_{12}^{p}\right)^2 + \frac{1}{a^2}\left(\dot{\varepsilon}_{22}^{p}\right)^2} \tag{9.68}$$

which gives the effective cumulative inelastic strain:

$$\tilde{p} = \int_0^t \sqrt{4\left(\dot{\varepsilon}_{12}^{p}\right)^2 + \frac{1}{a^2}\left(\dot{\varepsilon}_{22}^{p}\right)^2}\, dt \tag{9.69}$$

Using (9.61) and (9.62), this can be written

$$\tilde{p} = \left(\int_0^t \sqrt{4(\dot{\varepsilon}_{12}^{p})^2(1-d_6)^2 + \frac{1}{a^2}(\dot{\varepsilon}_{22}^{p})^2(1-d_2)}\, dt\right) \tag{9.70}$$

Alternatively, the accumulated permanent strain can be expressed in terms of the applied axial strain ε_x in the form

$$\tilde{p} = \left(\int_0^{\varepsilon_x} \sqrt{4\left(\frac{d\varepsilon_{12}^{p}}{d\varepsilon_x}\right)^2(1-d_6)^2 + \frac{1}{a^2}\left(\frac{d\varepsilon_{22}^{p}}{d\varepsilon_x}\right)^2(1-d_2)}\, d\varepsilon_x\right) \tag{9.71}$$

In both (9.70) and (9.71), d_2 is taken to be zero if $\sigma_{22} < 0$ since there is no effect of damage for compressive transverse stresses.

In practice, a trapezoidal rule can be used to estimate $d\varepsilon_{12}^{p}/d\varepsilon_x$ and $d\varepsilon_{22}^{p}/d\varepsilon_x$ for a given ε_x from the curves of $d\varepsilon_{12}^{p}$ versus $d\varepsilon_x$ and $d\varepsilon_{22}^{p}$ versus $d\varepsilon_x$. The increment of accumulated effective permanent strain $\Delta\tilde{p}$ can then be determined from a tabulation of the values for d_2, d_6, $\Delta\varepsilon_{12}^{p}/\Delta\varepsilon_x$, $\Delta\varepsilon_{22}^{p}/\Delta\varepsilon_x$, and $\Delta\varepsilon_x$ for each value of ε_x. Thus $\tilde{p}(\varepsilon_x)$ is determined.

Points on the strain hardening curve $R(\tilde{p}) + R_o$ versus $\tilde{p}(\varepsilon_x)$ are now determined for every $\tilde{p}(\varepsilon_x)$ and the corresponding $R(\tilde{p}) + R_o$ obtained from (9.64) with $f = 0$:

$$R(\tilde{p}) + R_o = \sqrt{\frac{\sigma_{12}^2}{(1-d_6)^2} + \frac{a^2\sigma_{22}^2}{(1-d_2)^2}} \tag{9.72}$$

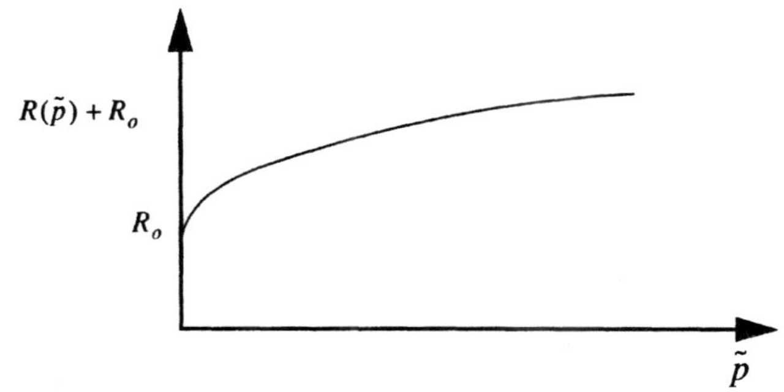

FIGURE 9.30 Typical Strain Hardening Curve for Composites

A typical result for carbon/epoxy is shown in Fig. 9.30; this curve suggests the representation

$$R(\tilde{p}) = \beta \tilde{p}^{\alpha} \tag{9.73}$$

where α and β are material parameters determined from a least squares fit of the data.

With $R(\tilde{p})$ known, the accumulated effective permanent strain rate $\dot{\tilde{p}}$ is determined from (9.64) using $f = 0$ and $\dot{f} = 0$, with the result

$$\dot{\tilde{p}} = \frac{\tilde{\sigma}_{12}\dot{\tilde{\sigma}}_{12} + a^2\tilde{\sigma}_{22}\dot{\tilde{\sigma}}_{22}}{(R(\tilde{p}) + R_o)\dfrac{\partial}{\partial \tilde{p}}R(\tilde{p})} \tag{9.74}$$

The parameter a^2 is determined by eliminating $R(\tilde{p}) + R_o$ from (9.66) and (9.67). After substitution of the definitions for effective stress and strain, the result is

$$a^2 = \frac{\dot{\varepsilon}^p_{22}\sigma_{12}(1 - d_2)^2}{2\dot{\varepsilon}^p_{12}\sigma_{22}(1 - d_6)^2} \tag{9.75}$$

Values for a^2 are now determined from the experimental result from a test which has nonzero transverse and shear stresses. In general, a^2 may be taken to be a constant.

9.5.1.6 Total Strain Rates

The total strain rates, $\dot{\varepsilon}_{ij}$, are the sum of elastic or recoverable strain rates, $\dot{\varepsilon}^E_{ij}$, and the inelastic or permanent strain rates, $\dot{\varepsilon}^p_{ij}$:

$$\dot{\varepsilon}_{ij} = \dot{\varepsilon}^E_{ij} + \dot{\varepsilon}^p_{ij} \tag{9.76}$$

9.5.1.7 Stress Rates for Inelastic Response

The elastic constitutive equations (4.9) for plane stress and (9.76) combine to give the stress rates in the form

$$\dot{\sigma}_{ij} = [\bar{Q}](\dot{\varepsilon}_{ij} - \dot{\varepsilon}^p_{ij}) \tag{9.77}$$

Thus the complete constitutive equation for the response of an inelastic material with anisotropic damage is determined.

9.5.2 Mesoscale Damage Parameter Determination

The damage parameters for a given material can be determined from a series of tests on selected laminates. In addition to the elastic moduli of the undamaged material and the fiber direction strength, the following parameters must be determined for the mesoscale composite damage theory when applied to a material exhibiting the type of damage evolution described in the preceding paragraphs:

- Damage evolution parameters Y_O, Y_O', Y_C, Y_C', and Y_2^C
- Strain hardening parameters R_o and $R = R(\tilde{p}) = \beta\tilde{p}^{\alpha}$
- Coupling constants a^2 and b

For highly orthotropic composites, these damage parameters can be determined from cyclic tensile tests on appropriately chosen angle-ply laminates. For other material systems, it may be necessary to use tests on other laminae or laminates in order to determine the damage parameters.

9.5.2.1 Angle-Ply Tensile Test

As developed in Chapter 5 for elastic response, the stresses in principal material coordinates for any angle-ply laminate under axial load can be expressed in terms of the applied average stress, $\bar{\sigma}_x$, and a coefficient, B, that is a function of the fiber orientation angle, θ, and the lamina elastic properties in the form of (5.130), which is repeated here:

$$\begin{aligned} \sigma_{11} &= B\bar{\sigma}_x \\ \sigma_{22} &= (1-B)\bar{\sigma}_x \\ \sigma_{12} &= \frac{-1}{2mn}(B(1-2m^2)+m^2)\bar{\sigma}_x \end{aligned} \tag{9.78}$$

where

$$B = \left[\frac{m^2(2m^2-1)+4m^2n^2\frac{G_{12}}{E_2}\left(\frac{E_2}{E_1}\nu_{12}+1\right)}{4m^2n^2\frac{G_{12}}{E_2}\left(\frac{E_2}{E_1}+2\frac{E_2}{E_1}\nu_{12}+1\right)+(2m^2-1)(m^2-n^2)}\right] \tag{9.79}$$

For loading in the inelastic range, equations (9.78) can be used in incremental form, where the current engineering properties in the coefficient B are a function of the strain history and defined by (9.50).

For a typical carbon/epoxy with $E_1 >> E_2$, B may be approximated by B_a, where

$$B_a = \left[\frac{m^2(2m^2-1)+4m^2n^2\frac{G_{12}}{E_2}}{4m^2n^2\frac{G_{12}}{E_2}+(2m^2-1)(m^2-n^2)}\right] \tag{9.80}$$

Since both G_{12} and E_2 are degraded as a result of damage, it is assumed that the ratio G_{12}/E_2 remains constant throughout the load history, and thus the stresses are always given by (9.78), with $B = B_a$ as in (9.80).

$[\pm 45]_s$ Tensile Test

For the special case $\theta = 45°$ in a highly orthotropic material, (9.80) reduces to

$$B_a = 1 \tag{9.81}$$

Therefore, a *close approximation* for the stresses in principal material coordinates of a $[\pm 45]_s$ laminate of highly orthotropic material that is subjected to the axial stress $\bar{\sigma}_x$ is

$$\begin{aligned} \sigma_{11} &= \bar{\sigma}_x \\ \sigma_{22} &= 0 \\ \sigma_{12} &= \frac{-\bar{\sigma}_x}{2} \end{aligned} \tag{9.82}$$

where the shear stress σ_{12} is exact, independent of the material properties. This relationship between stresses in principal material coordinates and applied stress $\bar{\sigma}_x$ is valid in the elastic as well as the inelastic region.

Since damage does not affect the fiber direction response of composites such as carbon/epoxy, we see from (9.82) that (to a close approximation) damage evolution in a $[\pm 45]_s$ carbon/epoxy laminate is dominated by the shear stress σ_{12}. Thus the form of the damage evolution law (9.58) can be determined from a cyclic tensile test on a $[\pm 45]_s$ laminate, where the associated force $Y_2 \approx 0$.

Figure 9.31 shows the cyclic shear stress-strain response obtained from a tensile test on a $[\pm 45]_s$ carbon/epoxy laminate. The shear stress σ_{12} is determined from (9.82), and the shear strain ε_{12} is determined from strain gage measurements and appropriate strain transformation equations. As the figure indicates, the response is highly nonlinear, with large permanent strains upon unloading. There is substantial hysteresis during unloading and reloading until the previous maximum stress is attained. The shear modulus degrades with each successively higher maximum stress.

Points for the shear damage evolution curve are determined from results like those in Fig. 9.31. A value of d_6 is obtained from the unloading/reloading modulus corresponding to each maximum stress (e.g., in Fig. 9.31, $G_{12}'' = G_{12}^0(1 - d_6)$ at $\sigma_{12} = \sigma_{12}''$); these values are then used to calculate the corresponding associated force Y_6 using (9.53) and $\underline{\hat{Y}}$ from (9.56), with $Y_2 = 0$. Values of d_6 and $\sqrt{\underline{\hat{Y}}}$ are then plotted and a best fit used to determine the form of the evolution law. Typical results of this process for three tests on IM6/914 carbon/epoxy are shown in Fig. 9.32, where the shear damage parameter d_6 attains an overall maximum value above 0.6 prior to ultimate failure.

The shear damage evolution in Fig. 9.32 can be represented by a straight line with damage initiating at $\sqrt{\underline{\hat{Y}}} = \sqrt{Y_O}$; the slope of the damage evolution line is $1/\sqrt{Y_C}$. The damage evolution parameters Y_O and Y_C are determined from a least squares fit of experimental data. The shear damage evolution law can then be written in the form (9.58), which is repeated here:

$$d_6 = \frac{\langle \sqrt{\underline{\hat{Y}}} - \sqrt{Y_O} \rangle_+}{\sqrt{Y_C}} \qquad \text{if } d_6 < 1 \text{ and } Y_2 < Y_2^C\text{; otherwise } d_6 = 1 \tag{9.83}$$

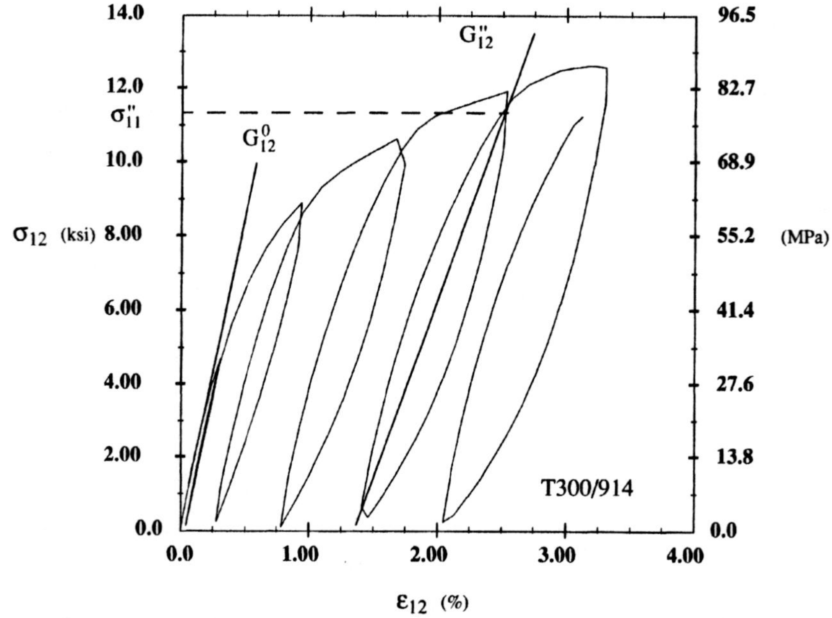

FIGURE 9.31 Cyclic Shear Stress-Strain Response of Carbon/Epoxy
(Ladevèze and Le Dantec, 1991)

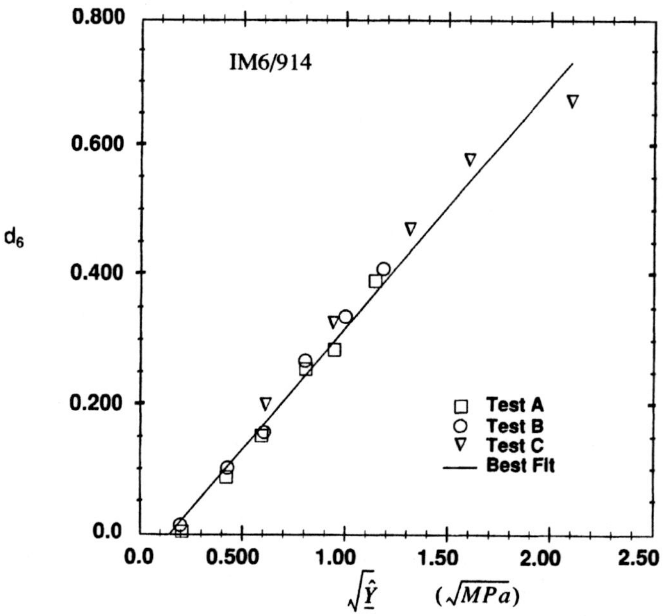

FIGURE 9.32 Shear Damage Evolution in Carbon/Epoxy
(Ladevèze and Le Dantec, 1991)

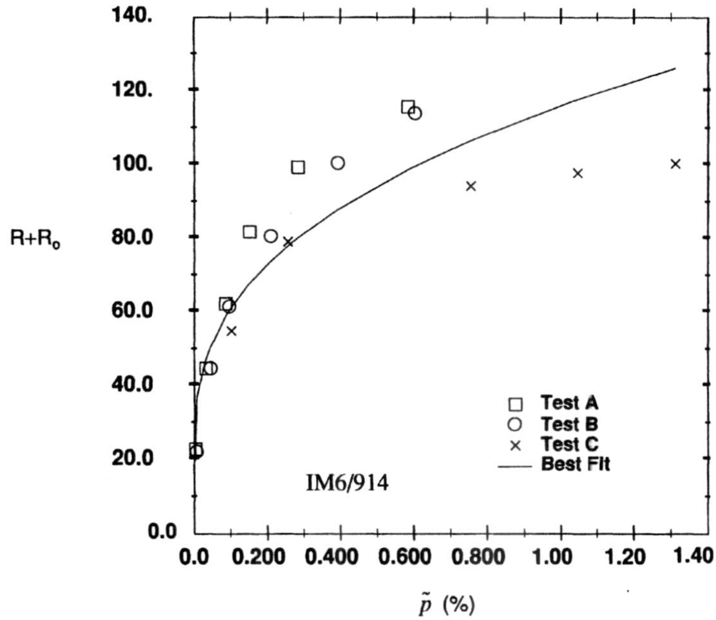

FIGURE 9.33 Strain-Hardening Curve for Carbon/Epoxy
(Ladevèze and Le Dantec, 1991)

The strain-hardening curve $R(\tilde{p}) + R_o$, as a function of the accumulated permanent strain $\tilde{p}$, is also determined from the $[\pm45]_s$ experimental results using (9.71) and (9.72), with ε_{22} and σ_{22} being zero for this test. Representative results for carbon/epoxy are shown in Fig. 9.33.

From the $[\pm45]_s$ test we have determined the parameters Y_O and Y_C for shear damage evolution and the strain-hardening curve $R(\tilde{p}) + R_o$. It still remains to determine the brittle damage threshold in transverse tension, Y_2^C, the coupling coefficients b and a^2, and the damage evolution parameters Y_O' and Y_C'. These parameters are determined from a $[\pm67.5]_s$ laminate as described in the following section.

Before leaving the discussion of the $[\pm45]_s$ tension test, we note that, as the results in Table 9.6 indicate, the exact B in (9.78) differs considerably from the approximate B_a for a metal matrix composite, where the ratio E_2/E_1 is much larger than for carbon/epoxy. Indeed, the exact calculation shows that the $[\pm45]_s$ test is not shear dominated for metal matrix materials; it is closer to equal-magnitude transverse and shear stresses. Thus, a different specimen must be used to obtain the shear response for composites that do not exhibit a high degree of orthotropy.

9.5.2.2 $[\pm67.5]_s$ Tension Test

There are two reasons that the $[\pm67.5]_s$ laminate is chosen for the remaining characterization of highly orthotropic materials. Interlaminar stresses are sufficiently low in this laminate (see Chapter 8) that they do not cause initiation of damage or failure, and the state of stress in the principal material coordinates is biaxial, providing an opportunity to study coupling effects.

Material	Exact				Approximate			
	B	$\frac{\sigma_{11}}{\sigma_x}$	$\frac{\sigma_{22}}{\sigma_x}$	$\frac{\sigma_{12}}{\sigma_x}$	B_a	$\frac{\sigma_{11}}{\sigma_x}$	$\frac{\sigma_{22}}{\sigma_x}$	$\frac{\sigma_{12}}{\sigma_x}$
	±45°							
T300/5208	0.910	0.910	0.090	−0.5	1.0	1.0	0.0	−0.5
SCS-6/Ti-15-3	0.585	0.585	0.415	−0.5	1.0	1.0	0.0	−0.5
	±67.5°							
T300/5208	0.207	0.207	0.793	−0.414	0.209	0.209	0.791	−0.415
SCS-6/Ti-15-3	0.132	0.132	0.868	−0.339	0.118	0.118	0.882	−0.325

TABLE 9.6 Stress Comparisons for Axially Loaded Angle-Ply Laminates

From Table 9.6, the coefficients B and B_a for T300/5208 with $\theta = 67.5°$ are

$$\begin{aligned} B &= 0.207 \\ B_a &= 0.209 \end{aligned} \tag{9.84}$$

Recalling (9.78), the principal material stresses in a $[\pm 67.5]_s$ carbon/epoxy laminate are

$$\begin{aligned} \sigma_{11} &= B\bar{\sigma}_x \\ \sigma_{22} &= (1-B)\bar{\sigma}_x \\ \sigma_{12} &= -(B(0.9998)+0.2065)\bar{\sigma}_x \end{aligned} \tag{9.85}$$

Using the exact value $B = 0.207$, the stresses are

$$\begin{aligned} \sigma_{11} &= 0.207\bar{\sigma}_x \\ \sigma_{22} &= 0.793\bar{\sigma}_x \\ \sigma_{12} &= -0.414\bar{\sigma}_x \end{aligned} \tag{9.86}$$

and with the approximate value $B_a = 0.209$, the stresses are

$$\begin{aligned} \sigma_{11} &= 0.209\bar{\sigma}_x \\ \sigma_{22} &= 0.791\bar{\sigma}_x \\ \sigma_{12} &= -0.415\bar{\sigma}_x \end{aligned} \tag{9.87}$$

It is evident from (9.86) and (9.87) (and Table 9.6) that the differences in stresses between the exact and approximate solutions are very small for ±67.5° carbon/epoxy. The coefficients in (9.87) also can be taken as constants because the variation in the ratio G_{12}/E_2 is small during the entire loading history. It is also interesting to note that, unlike the 45° case, the exact and approximate values are very similar for a metal matrix composite for fiber orientations of ±67.5° (Table 9.6).

From (9.86) we see that a $[\pm 67.5]_s$ carbon/epoxy laminate under axial load exhibits significant transverse normal stress, σ_{22}, and in-plane shear stress, σ_{12}. Thus the desired biaxiality is present. With the stresses given by (9.86) and the strains in principal material coordinates available after transformation of the strain gage measurements, a curve of the transverse response can be determined from this biaxial test. The strains in principal material coordinates can be expressed in terms of the measured axial strain, ε_x, and transverse strain, ε_y ($\varepsilon_{xy} = 0$ for tensile loaded angle-ply laminates) using the transformation equations (4.24) as

$$\begin{aligned} \varepsilon_{11} &= m^2\varepsilon_x + n^2\varepsilon_y \\ \varepsilon_{22} &= n^2\varepsilon_x + m^2\varepsilon_y \\ \varepsilon_{12} &= -nm(\varepsilon_x - \varepsilon_y) \end{aligned} \tag{9.88}$$

Representative results for the shear and transverse response of carbon/epoxy (under a biaxial stress state) obtained from a $[\pm 67.5]_s$ laminate are shown in Figs. 9.34 and 9.35, respectively. It is emphasized that these curves provide the responses under a biaxial state of stress. These figures show that the degree of nonlinearity and the magnitude of permanent strains are much smaller for this loading case than they are for the essentially pure shear results of Fig. 9.31. Nevertheless, d_6 and d_2 can be determined as functions of $\underline{\hat{Y}}$ for this biaxial stress state. In principal material coordinates, the shear response is uncoupled from the normal response and d_6 is determined as a function of σ_{12} directly from Fig. 9.34.

The results for d_6 as a function of $\underline{\hat{Y}}$ under biaxial stress from the $[\pm 67.5]_s$ test can be used with the results for d_6 as a function of Y_6 only under pure shear to determine the coupling coefficient b.

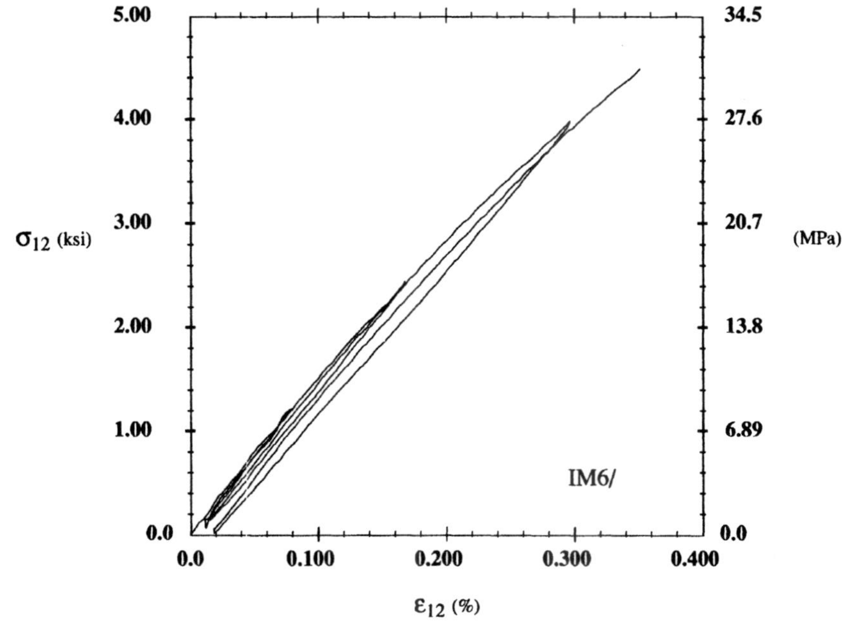

FIGURE 9.34 Shear Response, $[\pm 67.5]_s$ Laminate
(Ladevèze and Le Dantec, 1991)

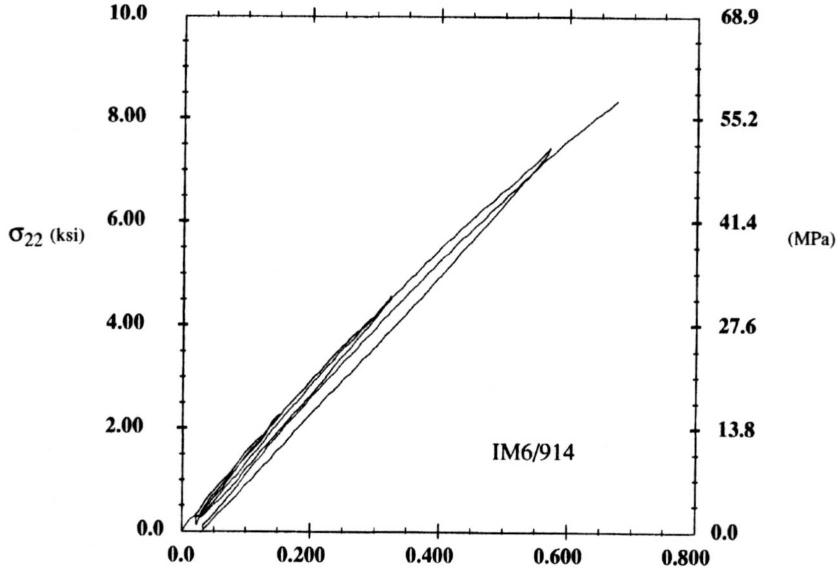

FIGURE 9.35 Transverse Response, [±67.5]$_s$ Laminate
(Ladevèze and Le Dantec, 1991)

From (9.56) and (9.58) we can write

$$b = \frac{(d_6\sqrt{Y_C} + \sqrt{Y_O})^2 - Y_6}{Y_2} \tag{9.89}$$

The constant b can now be determined from a best fit of all available experimental data for this particular material.

With b known, values of $\underline{\hat{Y}}$ can be calculated from (9.56) for a biaxial state of stress using σ_{12} and d_6 from Fig. 9.34 and σ_{22} and d_2 from Fig. 9.35. Thus, it is possible to plot the evolution of the damage parameter d_2 as a function of $\sqrt{\underline{\hat{Y}}}$. Typical results for the transverse tension damage evolution in carbon/epoxy are shown in Fig. 9.36.

As indicated in the figure, the results are similar to those in Fig. 9.32. Thus they can be represented in the form

$$d_2 = \frac{\langle \sqrt{\underline{\hat{Y}}} - \sqrt{Y'_O} \rangle_+}{\sqrt{Y'_C}} \tag{9.90}$$

where the transverse tension damage parameters Y'_C and Y'_O are determined from a least squares best fit of the available experimental data.

Two parameters remain to be determined: the coupling term a^2 in the elastic domain function (9.64) and the brittle damage threshold in transverse tension Y_2^C. The coupling constant a^2 is deter-

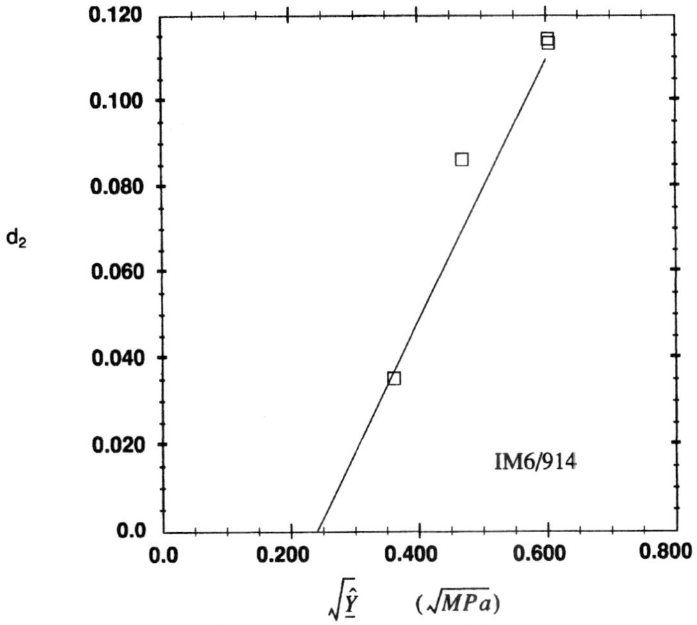

FIGURE 9.36 Transverse Tension Damage Evolution
(Ladevèze and Le Dantec, 1991)

mined from a least squares best fit of values determined from (9.75) at selected stress levels in the $[\pm 67.5]_s$ test, where the strain rates are replaced by increments as in the determination of $\tilde{p}(\varepsilon_x)$.

The brittle damage threshold in transverse tension Y_2^C is determined from the known failure stress $\bar{\sigma}_x$ of the $[\pm 67.5]_s$ tensile test and use of the damage model to predict the response of this laminate. The value of Y_2 at the failure stress is the threshold value of interest. Letting the superscript C refer to the critical value at failure and using (9.53), we can write

$$Y_2^C = \frac{\langle \sigma_{22}^2 \rangle_+^C}{2E_2^0(1 - d_2^C)^2} \tag{9.91}$$

where d_2^C must be determined by exercising the damage evolution model. Note that the $[\pm 67.5]_s$ tensile test provides larger transverse failure strains than the all-90° lamina tensile test, and thus it is preferred for determining the transverse ultimate; the all-90° specimen is overly sensitive to flaws.

Finally, it should be noted that the extent to which the parameters b and a^2 are constant is material dependent, and more sophisticated approaches may be required depending upon the results obtained from the material being studied.

9.5.3 Failure Prediction

In the mesoscale damage theory, failure is predicted to occur by one of three mechanisms: fiber failure, transverse tensile failure defined by Y_2^C, or attainment of an instability condition defined as a zero (or negative) slope of the applied stress versus strain diagram. For the materials studied to date, failure predictions using these three mechanisms have exhibited good correlation with experimental results.

9.5.4 Mesoscale Composite Damage Theory Predictions

The following sections discuss the application of the mesoscale composite damage theory for tensile loading of symmetric laminates of highly orthotropic materials such as carbon/epoxy. As developed in Chapter 5, the strains are uniform through the thickness of the laminate, and the state of stress is planar with the stresses constant in each layer. For this discussion, the axial direction will be denoted by the subscript L and the transverse direction will be denoted by T (Fig. 9.37).

The average stresses applied to the laminate are

$$\{\bar{\sigma}\} = \begin{Bmatrix} \sigma_L \\ 0 \\ 0 \end{Bmatrix} \tag{9.92}$$

and the associated laminate strains are

$$\{\bar{\varepsilon}\} = \begin{Bmatrix} \varepsilon_L \\ \varepsilon_T \\ 0 \end{Bmatrix} \tag{9.93}$$

9.5.4.1 Cross-Ply Laminates

For cross-ply laminates the layer strains in principal material coordinates can be written directly in terms of the laminate strains. For the 0° layer,

$$^{0}\{\varepsilon\}_1 = \begin{Bmatrix} \varepsilon_L \\ \varepsilon_T \\ 0 \end{Bmatrix} \tag{9.94}$$

and for the 90° layer,

$$^{90}\{\varepsilon\}_1 = \begin{Bmatrix} \varepsilon_T \\ \varepsilon_L \\ 0 \end{Bmatrix} \tag{9.95}$$

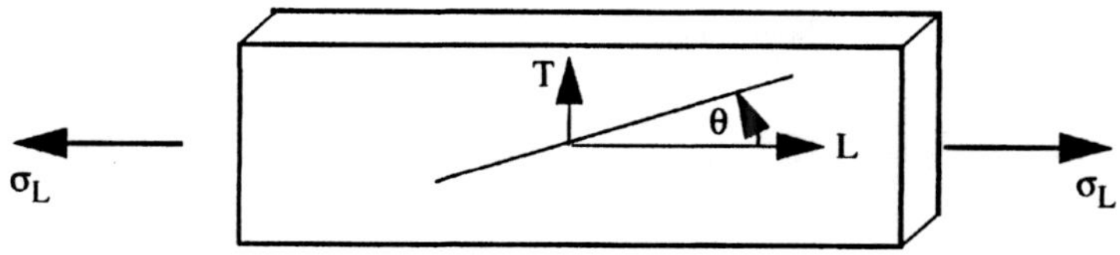

FIGURE 9.37 Tensile Coupon

Substitution of these strains into the plane stress constitutive equations (4.11)—using (4.16) to introduce the engineering material properties, neglecting $(\nu_{12}^0)^2 E_2^0/E_1^0$ in comparison to 1.0, introducing the damage parameters with the definitions (9.50), and assuming that the inelastic strains are negligible in both layers (a valid assumption for the materials under consideration)—provides the layer constitutive equations:

$$\begin{Bmatrix} \sigma_{11} \\ \sigma_{22} \\ \sigma_{12} \end{Bmatrix} = \begin{bmatrix} E_1^0 & \nu_{12}^0 E_2^0(1-d_2) & 0 \\ \nu_{12}^0 E_2^0(1-d_2) & E_2^0(1-d_2) & 0 \\ 0 & 0 & 2G_{12}^0(1-d_6) \end{bmatrix} \begin{Bmatrix} \varepsilon_{11} \\ \varepsilon_{22} \\ \varepsilon_{12} \end{Bmatrix} \tag{9.96}$$

Combining (9.94) and (9.96) gives the stresses in the 0° layer in terms of the laminate strains:

$${}^{0}\begin{Bmatrix} \sigma_{11} \\ \sigma_{22} \\ \sigma_{12} \end{Bmatrix} = \begin{Bmatrix} E_1^0 \varepsilon_L + \nu_{12}^0 E_2^0(1-d_2)\varepsilon_T \\ \nu_{12}^0 E_2^0(1-d_2)\varepsilon_L + E_2^0(1-d_2)\varepsilon_T \\ 0 \end{Bmatrix} \tag{9.97}$$

Likewise, combining (9.95) and (9.96) gives the stresses in the 90° layer:

$${}^{90}\begin{Bmatrix} \sigma_{11} \\ \sigma_{22} \\ \sigma_{12} \end{Bmatrix} = \begin{Bmatrix} E_1^0 \varepsilon_T + \nu_{12}^0 E_2^0(1-d_2)\varepsilon_L \\ \nu_{12}^0 E_2^0(1-d_2)\varepsilon_T + E_2^0(1-d_2)\varepsilon_L \\ 0 \end{Bmatrix} \tag{9.98}$$

With the stresses in each layer now known in terms of the laminate strains, and using equilibrium, we can write the stress-strain relationship for a damaged *laminate* in the form

$$\begin{Bmatrix} \varepsilon_L \\ \varepsilon_T \\ \varepsilon_{LT} \end{Bmatrix} = \begin{bmatrix} \dfrac{2}{E_1^0} & \dfrac{-2E_2^0\nu_{12}^0 D}{(E_1^0)^2} & 0 \\ \dfrac{-2E_2^0\nu_{12}^0 D}{(E_1^0)^2} & \dfrac{2}{E_1^0} & 0 \\ 0 & 0 & \dfrac{2}{G_{12}^0 D} \end{bmatrix} \begin{Bmatrix} \sigma_L \\ 0 \\ 0 \end{Bmatrix} \tag{9.99}$$

where $D \equiv (2 - {}^{0}d_2 - {}^{90}d_2)$.

From (9.99) we see that the axial stiffness of the laminate is constant with a value $E_1^0/2$, and the laminate Poisson's ratio is

$$\nu_{LT} = \frac{2E_2^0\nu_{12}^0(2 - {}^{0}d_2 - {}^{90}d_2)}{E_1^0} \tag{9.100}$$

Thus, unlike the axial modulus, the laminate Poisson's ratio decreases with increase in damage. That these results are representative of real materials is shown by experimental results for the response of a $[0/90]_s$ laminate in Fig. 9.38 and the comparison of theory and experiment for Poisson's ratio in Fig. 9.39. The evolution of damage in each layer of a cross-ply laminate can be expressed in terms of the applied stress σ_L by combining (9.99) with (9.97) and (9.98) to obtain explicit expressions for the nonzero principal material stresses in each layer. The results are

$$^{0}\sigma_{11} = 2\sigma_L \tag{9.101}$$

$$^{0}\sigma_{22} = \frac{2E_2^0\nu_{12}^0(1-{}^{0}d_2)}{E_1^0}\sigma_L \tag{9.102}$$

$$^{90}\sigma_{11} = \frac{-2E_2^0\nu_{12}^0(1-{}^{0}d_2)}{E_1^0}\sigma_L \tag{9.103}$$

$$^{90}\sigma_{22} = \frac{2E_2^0(1-{}^{90}d_2)}{E_1^0}\sigma_L \tag{9.104}$$

Using these stress equations in the damage evolution laws (9.53), (9.57), and (9.59) defines the damage evolution in each layer.

For the 0° layer:

$$^{0}\underline{\hat{Y}} = \frac{b}{2E_2^0}\frac{\langle {}^{0}\sigma_{22}\rangle_+^2}{(1-{}^{0}d_2)^2} = 2bE_2^0\left(\frac{\nu_{12}^0}{E_1^0}\right)^2\sigma_L^2 \tag{9.105}$$

$$^{0}Y_2 = \frac{\langle {}^{0}\sigma_{22}\rangle_+^2}{2E_2^0(1-{}^{0}d_2)^2} = 2E_2^0\left(\frac{\nu_{12}^0}{E_1^0}\right)^2\sigma_L^2 \tag{9.106}$$

$$^{0}d_2 = \frac{\langle\sqrt{{}^{0}\underline{\hat{Y}}}-\sqrt{Y'_O}\rangle_+}{\sqrt{Y'_C}} \qquad \text{if } {}^{0}\underline{Y}_2 < Y_2^C\text{; otherwise } {}^{0}d_2 = 1 \tag{9.107}$$

For the 90° layer:

$$^{90}\underline{\hat{Y}} = \frac{b}{2E_2^0}\frac{\langle {}^{90}\sigma_{22}\rangle_+^2}{(1-{}^{90}d_2)^2} = 2bE_2^0\left(\frac{1}{E_1^0}\right)^2\sigma_L^2 \tag{9.108}$$

$$^{90}\underline{Y}_2 = \frac{\langle {}^{90}\sigma_{22}\rangle_+^2}{2E_2^0(1-{}^{90}d_2)^2} = 2E_2^0\left(\frac{1}{E_1^0}\right)^2\sigma_L^2 \tag{9.109}$$

$$^{90}d_2 = \frac{\langle\sqrt{{}^{90}\underline{\hat{Y}}}-\sqrt{Y'_O}\rangle_+}{\sqrt{Y'_C}} \qquad \text{if } {}^{90}\underline{Y}_2 < Y_2^C\text{, otherwise } {}^{90}d_2 = 1 \tag{9.110}$$

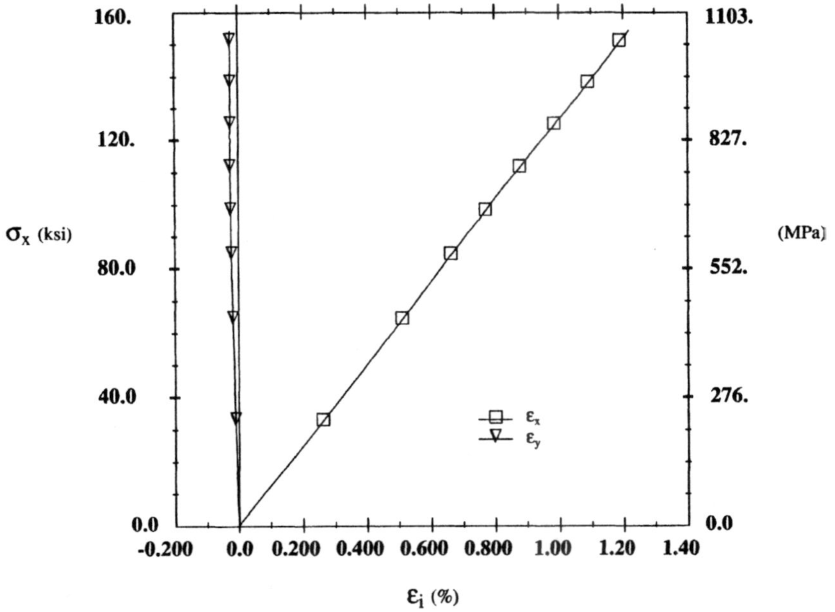

FIGURE 9.38 $[0/90]_s$ Laminate Response for IM6/914 Carbon/Epoxy (Ladevèze & Le Dantec, 1991)

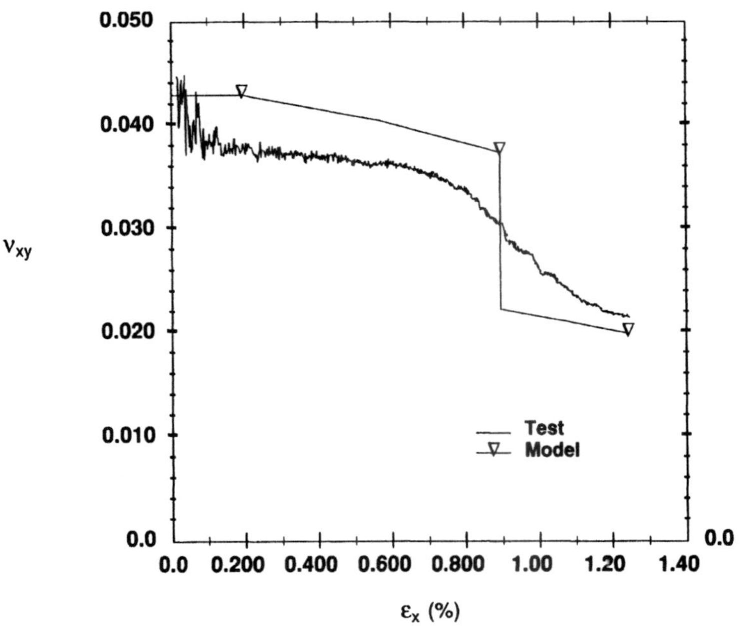

FIGURE 9.39 Poisson's Ratio for a $[0/90]_s$ Laminate (IM6/914) (Ladevèze and Le Dantec, 1991)

Substitution of (9.107) and (9.110) into the laminate stress-strain relationship (9.99) indicates that the strain response of a highly orthotropic cross-ply laminate is a quadratic function of applied stress after damage has initiated.

For IM6/914 carbon/epoxy the identified model parameters at room temperature are given in Table 9.7.

From (9.105) and (9.108) it is apparent that damage initiates in the 90° layer before it does in the 0° layer. And from Table 9.7 and (9.110) with $d_2 = 0$, we see that the mesoscale composite damage theory predicts that damage initiates first in the 90° layer at $\hat{Y} = Y_O = 0.0576$ MPa or $\sigma_L = 516$ MPa. This is consistent with the experimental results of Fig. 9.38, which exhibits nonlinearity in the transverse strain near this stress level. At higher stress levels the damage is higher in the 90° layer than in the 0° layer. As an example, at 800 MPa, ${}^{90}d_2 = 0.22$, whereas ${}^{0}d_2 = 0.035$ for the IM6/914 carbon/epoxy.

Final failure, or rupture, of the $[0/90]_s$ IM6/914 laminate corresponds to the ultimate fiber strain (Fig. 9.38) at approximately 1100 MPa. The corresponding values of the damage parameters are ${}^{90}d_2 = 0.335$ and ${}^{0}d_2 = 0.072$. Clearly, there is substantially more damage in the 90° layer at final failure.

9.5.4.2 Angle-Ply Laminates

Comparison of the experimental results in Figs. 9.38 and 9.40 indicates that tensile loading of $[\pm 45]_s$ carbon/epoxy laminate exhibits response characteristics very different from those of the cross-ply laminate. As discussed previously, the response of a $[\pm 45]_s$ laminate is dominated by shear effects, whereas shear is zero in the cross-ply laminate. The shear dominance in the $[\pm 45]_s$ laminate results in substantial damage and large inelastic strains prior to final failure. And the final failure is an instability condition rather than fiber failure. Thus, aspects of the mesoscale composite damage theory not required for the cross-ply laminate must be included in predicting the tensile response of a $[\pm 45]_s$ laminate.

Since the $[\pm 45]_s$ laminate is dominated by shear strain and $|\sigma_{22}| \ll |\sigma_{12}|$, all other strains are neglected. From (9.76), the total shear strain rate $\dot{\varepsilon}_{12}$ is the sum of the elastic shear strain rate $\dot{\varepsilon}^E_{12}$ and the inelastic (permanent) shear strain rate $\dot{\varepsilon}^p_{12}$:

$$\dot{\varepsilon}_{12} = \dot{\varepsilon}^E_{12} + \dot{\varepsilon}^p_{12} \tag{9.111}$$

The condition for inelastic strains is that the elastic domain function f of (9.64) satisfy the condition $f = 0$. Thus, for a pure shear state of stress we have

$$\frac{\sigma_{12}}{1 - d_6} = R(p) + R_o \tag{9.112}$$

E_1^0, GPa	E_2^0, GPa	G_{12}^0, GPa	ν_{12}	b
170	10.8	5.8	0.34	2.5
Y'_C, MPa	Y'_O, MPa	Y_C, MPa	Y_O, MPa	Y_2^C, MPa
14.29	0.0576	7.673	0.0225	0.49

TABLE 9.7 IM6/914 Damage Parameters

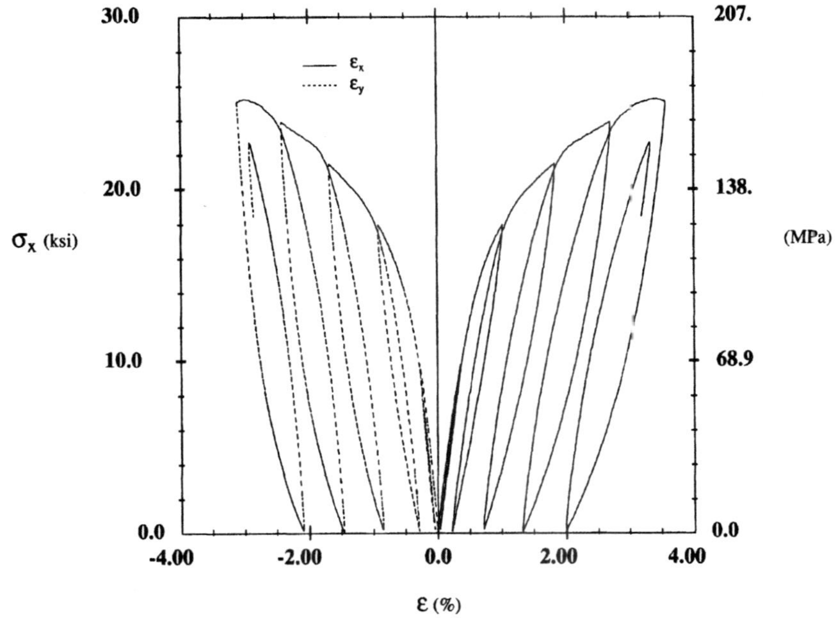

FIGURE 9.40 Tensile Response of [±45]$_s$ T300/914 Carbon/Epoxy
(Ladevèze and Le Dantec, 1991)

From (9.66) the inelastic shear strain rate is then

$$2\dot{\varepsilon}_{12}^{p}(1-d_6) = \dot{\tilde{p}}\frac{\sigma_{12}}{2(1-d_6)[R(\tilde{p})+R_o]} \tag{9.113}$$

and the accumulated permanent strain rate is

$$\dot{\tilde{p}} = \frac{\sigma_{12}\dot{\tilde{\sigma}}_{12}}{(1-d_6)[R(\tilde{p})+R_o]\frac{dR}{dp}} \tag{9.114}$$

Combining (9.113) and (9.114) gives the inelastic shear strain rate:

$$2\dot{\varepsilon}_{12}^{p} = \frac{\dot{\tilde{\sigma}}_{12}}{2(1-d_6)\frac{dR}{dp}} \tag{9.115}$$

The elastic strain rates are

$$\dot{\varepsilon}_{12}^{E} = \frac{\dot{\tilde{\sigma}}_{12}}{2G_{12}^{0}} \tag{9.116}$$

where the shear effective stress rate is

$$\dot{\tilde{\sigma}}_{12} = \frac{\dot{\sigma}_{12}}{(1-d_6)} + \dot{d}\frac{\sigma_{12}}{(1-d_6)^2} \tag{9.117}$$

Recalling the evolution law for shear damage from (9.83),

$$d_6 = \frac{\langle\sqrt{Y_6} - \sqrt{Y_O}\rangle_+}{\sqrt{Y_C}} \tag{9.118}$$

we can write

$$\dot{d}_6 = \frac{\dot{Y}_6}{2\sqrt{Y_C}\sqrt{Y_6}} \tag{9.119}$$

Now, from (9.53),

$$Y_6 = \frac{\sigma_{12}^2}{2G_{12}^0(1-d_6)^2} \tag{9.120}$$

and hence we can write

$$\dot{Y}_6 = \frac{1}{G_{12}^0(1-d_6)}\left(\frac{\sigma_{12}}{(1-d_6)}\dot{\sigma}_{12} + \frac{\sigma_{12}^2}{(1-d_6)^2}\dot{d}_6\right) \tag{9.121}$$

Combining (9.119) through (9.121) gives

$$\dot{d}_6 = \frac{\dot{\sigma}_{12}}{\sqrt{Y_C}\sqrt{2G_{12}^0}(1-d_6)F} \tag{9.122}$$

where

$$F = 1 - \frac{\sigma_{12}}{\sqrt{Y_C}\sqrt{2G_{12}^0}(1-d_6)^2} \tag{9.123}$$

Combining (9.111), (9.115)–(9.117), and (9.122) gives the desired total strain rate relationship:

$$\dot{\varepsilon}_{12} = \left(\frac{1}{G_{12}^0} + \frac{1}{2(1-d_6)\dfrac{\partial R}{\partial p}}\right)\frac{\dot{\sigma}_{12}}{2(1-d_6)F} \tag{9.124}$$

Failure is predicted at the instability condition when the slope of the load-versus-strain curve is zero. This corresponds to the shear stress attaining a maximum value. Thus the stress rate, $\dot{\sigma}_{12}$, is zero and the strain rate, $\dot{\varepsilon}_{12}$, is nonzero. From the last term in (9.124), this corresponds to $d_6 = 1$ or

$F = 0$. The condition $d_6 = 1$ corresponds to shear failure, whereas the condition $F = 0$ gives the limiting shear damage, d_6^c, at the instability condition. From (9.123) and (9.118) with $F = 0$, we have

$$d_6^c = \frac{1}{2}\left(1 - \frac{\sqrt{Y_O}}{\sqrt{Y_C}}\right) \tag{9.125}$$

For carbon/epoxy, the term $\sqrt{Y_O}/\sqrt{Y_C}$ is generally small and d_6^c approaches 0.5 at failure. This value is in relatively good agreement with the experimental results in Fig. 9.27. The shear stress, σ_{12}^f, at failure is determined from (9.123) and (9.118) with $F = 0$. The result is

$$\sigma_{12}^f = \frac{1}{4}(\sqrt{Y_C}\sqrt{2G_{12}^0})\left(1 + \frac{\sqrt{Y_O}}{\sqrt{Y_C}}\right)^2 \tag{9.126}$$

Using the values in Table 9.7 for IM6/914, (9.123) predicts a failure shear stress of 78.6 MPa, which corresponds well with the experimental value of 82 MPa (Ladevèze and Le Dantec, 1992). Equally good results were reported for T300/914, with the predicted failure shear stress of 88.5 MPa and the experimental value 92 MPa. Comparison of the predicted and experimental axial response for two tests on $[\pm 45]_s$ T300/914 laminates, including the failure prediction, exhibits very good correlation, as shown in Fig. 9.41.

9.5.4.3 General Laminates

Equally good comparisons between theory and experiment have been obtained for other laminates. Figures 9.42 and 9.43 show the comparisons for $[-12/78]_s$ and $[67.5/22.5]_s$ laminates (Ladevèze

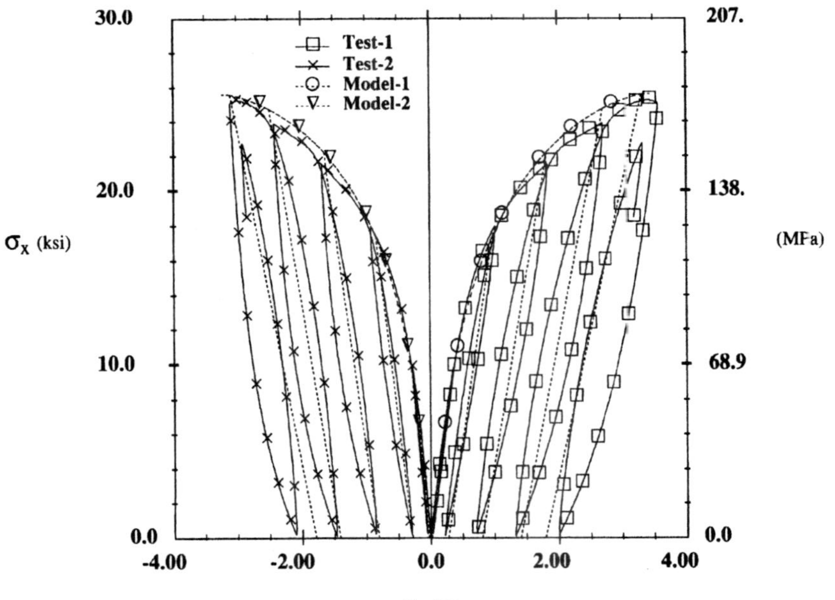

FIGURE 9.41 Theory and Experiment: Axial Response of $[\pm 45]_s$ T300/914 (Ladevèze and Le Dantec, 1991)

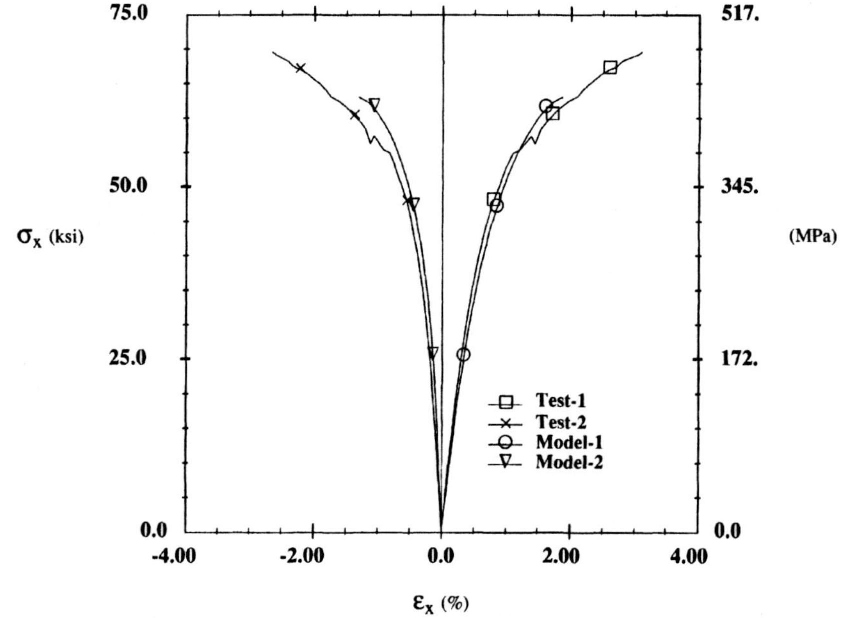

FIGURE 9.42 Theory and Experiment: Axial Response of $[-12/78]_{2s}$ T300/914 (Ladevèze and Le Dantec, 1991)

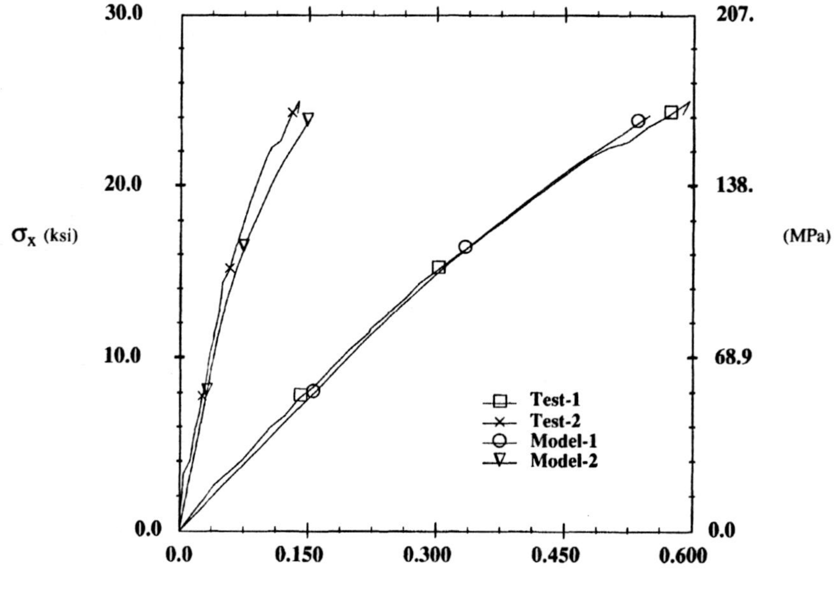

FIGURE 9.43 Theory and Experiment: Axial Response of $[67.5/22.5]_{2s}$ T300/914 (Ladevèze and Le Dantec, 1991)

and Le Dantec, 1992). As indicated in the figures, both the axial and Poisson responses exhibit excellent correlation between theory and experiment for both laminates.

9.6 Summary

Failure theories and a model for damage evolution in composites have been presented. The failure theories have been applied for tensile loading of unidirectional off-axis coupons and angle-ply laminates. The failure theories predict first failure of the composite, and at least one theory, the tensor polynomial theory, provides good predictions for the strength of off-axis laminae. The tensor polynomial failure criterion also was shown to accurately capture the initial mode of failure along the free edge of angle-ply laminates. Laminates exhibit damage evolution, and hence the failure theories generally do not predict their ultimate strength accurately. However, the mesoscale damage theory was shown to be quite accurate for predicting the nonlinear, inelastic response and final failure of carbon/epoxy laminates in the presence of damage.

References

Allix, O., Bahlouli, N., Ladevèze, P., and Perret, L. (1994), "Damage Modelling of Composite Laminates at Various Temperatures," in *Damage Mechanics in Composites,* ASME, AMD vol. 185 (D. H. Allen and J. W. Ju, eds.), American Society of Mechanical Engineers, New York, pp. 21–35.

Allix, O., and Ladevèze, P. (1992), "Interlaminar Interface Modelling for the Prediction of Delamination," *Comp. Struct.,* vol. 22, pp. 235–242.

Ashkenazi, E. K. (1965), "Problems of the Anisotropy of Strength," *Mekhanika Polimerov,* vol. 1, p. 79; English translation *Polymer Mechanics,* vol. 1 (1966), Faraday Press, p. 60.

Bazant, Z. P., and Cedolin, L. (1991), *Stability of Structures,* Oxford University Press, New York, Chapter 13.

Becker, W., Pindera, M. -J., and Herakovich, C. T. (1987), *Mechanical Response of Unidirectional Boron/Aluminum Under Combined Loading,* CCMS-87-06, Virginia Tech, Blacksburg.

Beuth, J. L., Jr., and Herakovich, C. T. (1989), "Analysis of Crack Extension in Anisotropic Materials Based on Local Normal Stress," *Theo. Appl. Fract. Mech.,* vol. 11, pp. 27–46.

Fox, D. J., Sykes, G. F., Jr., and Herakovich, C. T. (1987), *Space Environmental Effects on Graphite-Epoxy Compressive Properties and Epoxy Tensile Properties,* CCMS-87-11, Virginia Tech, Blacksburg.

Gol'denblat, I. I. and Kopnov, V. A. (1965), "Strength of Glass-Reinforced Plastics in the Complex Stress State," *Mekhanika Polimerov,* vol. 1, p. 70; English translation *Polymer Mechanics,* vol. 1 (1966), Faraday Press, p. 54.

Hahn, H. T., and Williams, J. G. (1986), "Compression Failure Mechanisms in Unidirectional Composites," in *Composite Materials: Testing and Design (Seventh Conference),* ASTM STP 893, (J. M. Whitney, ed.), American Society for Testing and Materials, Philadelphia, pp. 115–139.

Herakovich, C. T. (1982), "Influence of Layer Thickness on the Strength of Angle-Ply Laminates," *J. Compos. Mater.,* vol. 16, pp. 216–227.

Herakovich, C. T. (1989), "Edge Effects and Delamination Failures," *J. Strain Anal.,* vol. 24, no. 4, pp. 245–252.

Herakovich, C. T. (1989), "Free Edge Effects in Laminated Composites," in *Handbook of Composites, Vol. 2: Structure and Design* (C. T. Herakovich and Y. M. Tarnopol'skii, eds.), Elsevier Science Publishers B. V., Amsterdam, Chapter 4.

Herakovich, C. T., and Klang, E. C. (1984), "Theoretical/Experimental Correlation for Edge Cracking in Graphite/Polyimide Laminates," International Symposium on Composites Materials and Engineering, University of Delaware, Wilmington.

Herakovich, C. T., Nagarkar, A. P., and O'Brien, D. A. (1979), "Failure Analysis of Composite Laminates with Free Edges," in *Modern Developments in Composite Materials and Structures,* (J. R. Vinson, ed.), American Society of Mechanical Engineers, New York.

Hill, R. (1950), *The Mathematical Theory of Plasticity,* Oxford University Press, London.

Hult, J. (1974), *On Topics in Applied Continuum Mechanics* (J. L. Zeman and F. Ziegler, eds.), Springer, New York.

Kachanov, L. M. (1958), "Time of the Rupture Process Creep Conditions," *Izv. Akad. Nauk SSR Otd. Tekh. Nauk,* 8, pp. 26–31.

Kaiser, J. (1950), "Untersuchungen über das Auftreten Gerauschen Biem Zugversuch," Ph.D. thesis, Technisch Hochschule, Munich.

Kelly, A., and Davies, G. J. (1965), "The Principle of the Fibre Reinforcement of Metals," *Metall. Rev.,* vol. 10, pp. 1–78

Krajcinovic, D. (1989), "Damage Mechanics," *Mech. Mater.,* vol. 8, pp. 117–197.

Ladevèze, P. (1983), *Sur Une Théorie de l'Endommagement Anisotrope,* Rapport Interne no. 34, Laboratoire de Mécanique et Technologie, Cachan, France.

Ladevèze, P. (1986), "Sur la Mécanique de l'Endommagement des Composites," in *Compte-Rendus des JNC* 5 (C. Bathias and D. Menkès, eds.), Pluralis, Paris, pp. 667–683.

Ladevèze, P. (1992), "A Damage Computational Method for Composite Structures," *Comput. Struct.,* vol. 44, pp. 79–87.

Ladevèze, P. (1994), "Inelastic Strains and Damage," in *Damage Mechanics of Composite Materials* (R. Talreja, ed.), Elsevier, Amsterdam.

Ladevèze, P., Allix, O., and Daudeville, L. (1990), "Mesomodeling of Damage for Laminate Composites: Application to Delamination," in *Inelastic Deformation of Composite Materials* (D. V. Dvorak, Ed.), Springer-Verlag, New York, pp. 607–622.

Ladevèze, P., and Le Dantec, E. (1992), "Damage Modelling of the Elementary Ply for Laminated Composites," *Compos. Sci. Technol.,* vol. 43, pp. 257–267.

Leckie, F. A. (1978), "The Constitutive Equations of Continuum Creep Damage Mechanics," *Philos. Trans. R. Soc. London,* vol. A288, p. 27.

Lemaitre, J., and Chaboche, J. -L. (1974), "A Nonlinear Model of Creep-Fatigue Damage Cumulation and Interaction," *Proc. IUTAM Symp. on Mechanics of Visco-elastic Media and Bodies,* Springer, Gothenburg.

Lemaitre, J., and Chaboche, J. -L. (1990), *Mechanics of Solid Materials,* Cambridge University Press, Cambridge (originally *Mécanique des Matériaux Solides,* Dunod, Paris, 1985), p. 57.

Milkovich, S. M., Sykes, G. F., Jr., and Herakovich, C. T. (1986), "Space Radiation Effects on the Thermomechanical Behavior of Graphite-Epoxy Composites," *J. Compos. Mater.,* vol. 20, no. 6, pp. 579–593.

Mises, R. von (1928), "Mechanik der Plastischen Formänderung von Kristallen," *Z. Angew. Math. Mech.,* vol. 8, pp. 161–185.

Murakami, S. (1983), "Notion of Continuum Damage Mechanics and Its Application to Anisotropic Creep Damage Theory," *J. Eng. Mater. Technol.,* vol. 105, pp. 99–105.

Nagarkar, A. P., and Herakovich, C. T. (1979), *Nonlinear Temperature Dependent Failure Analysis of Finite Width Composite Laminates,* VPI-E-79-36, Virginia Tech, Blacksburg.

Narayanaswami, R., and Adelman, H. M. (1977), "Evaluation of the Tensor Polynomial and Hoffman Strength Theories for Composite Materials," *J. Compos. Mater.*, vol. 11, pp. 366–377.

Pindera, M. -J., Gurdal, Z., Herakovich, C. T., Hidde, J. S., and Starbuck, J. M. (1989), "Mechanical Response of Aramid-Epoxy under Tensile, Compressive and Shear Loading," *J. Reinf. Plast. Compos.*, vol 6, no. 3, pp. 234–252.

Pindera, M. -J., and Herakovich, C. T. (1981), *An Endochronic Theory for Transversely Isotropic Fibrous Composites*, VPI-E-81–27, Virginia Tech, Blacksburg.

Rabotnov, Y. N. (1968), "Creep Rupture," in *Proc. XII Int. Congress on Applied Mechanics*, Stanford, Springer, Berlin.

Reed, S. M., Herakovich, C. T., and Sykes, G. F., Jr. (1986), *The Effects of Space Radiation on a Chemically Modified Graphite-Epoxy Composite Material*, CCMS-86–06, Virginia Tech, Blacksburg.

Tsai, S. W. (1968), "Strength Theories of Filamentary Structures," in *Fundamental Aspects of Fiber Reinforced Plastic Composites* (R. T. Schwartz and H. S. Schwartz, eds.), Wiley Interscience, New York, pp. 3–11.

Tsai, S. W., and Wu, E. M. (1971), "A General Theory of Strength for Anisotropic Materials," *J. Compos. Mater.*, vol. 5, pp. 58–80.

Exercises

9.1 Show that equations (9.17) are true for the maximum strain failure criterion applied to an off-axis tensile coupon.

9.2 Show that equations (9.20)–(9.23) are correct for the Tsai-Hill failure criterion.

9.3 Show that the shear strength being independent of sign results in the shear terms F_4, F_5, and F_6 and the normal/shear coupling terms F_{14}, F_{15}, F_{16}, F_{24}, F_{25}, F_{26}, F_{34}, F_{35}, F_{36}, F_{45}, F_{56}, and F_{64} all being zero for the tensor polynomial failure criterion.

9.4 The tensor polynomial failure criterion may be written

$$F_i\sigma_i + F_{ij}\sigma_i\sigma_j = 1$$

Consider a symmetric laminate under the loading $N_x = k_1N$, $N_y = k_2N$, $N_{xy} = 0$, $\{M\} = 0$. Show through equations and a sketch how the tensor polynomial criterion would be used to plot the failure surface of each layer of the laminate in N_x-N_y load space for a generic symmetric laminate. How would you predict first failure of the laminate as a function of k_1, k_2, and N?

CHAPTER 10 LAMINATED TUBES

"No theory is good except on condition that one uses it to go beyond."
André Gide

10.1 Introduction

A common structural element is the hollow cylinder or tube. Tubes are used in a wide variety of fabricated applications including hoses, piping systems, drive shafts, structural members, and space trusses. They are also present in living organisms in functions as diverse as blood vessels and elephant trunks. Fibrous composites can be fabricated accurately and efficiently to specifications as laminated (or layered) tubes using filament winding, pultrusion, or other fabrication methods. And the opportunity for an optimized configuration in terms of materials selection, fiber orientations, and layer thicknesses is ever present. The tubular configurations of living organisms are invariably laminated optimally to meet the demands of the specific use. In the following sections we present the elasticity solution for a long, laminated circular cylinder subjected to uniform axial force and torque at its ends, internal and external pressure uniformly applied along its length, and uniform temperature change throughout the tube (Fig. 10.1). The elasticity solution is valid away from the ends where the axial load and torque are applied.

It will be shown that laminated tubes can exhibit axial-torsional coupling when off-axis layers are present. Specifically, it will be shown that axial force, uniform pressure, and uniform thermal loadings can produce rotation of the tube, and torque can produce axial strain. It will also be shown that, in general, stresses vary through the thickness of the layers, and all coupling coefficients and stresses are a function of the tube radius and wall thickness. It is also noted that the solution for a laminated, hollow cylinder can be applied to a laminated, solid bar with modification of only one boundary condition in the solution for the hollow cylinder.

The laminated circular tube is one of a very few composite structural configurations for which an exact elasticity solution is available. The analytical solution for mechanical loading is based upon the works of Lekhnitskii (1950), Scherrer (1967), Reissner (1970), Pagano (1971), Reissner and Tsai (1974), and Wilson and Orgill (1986). Thermal stresses were included by Hyer et al. (1986), Hyer and Cooper (1986), and Hyer and Rousseau (1987). Orgill and Wilson (1986) included material and geometric nonlinearities. The availability of an analytical solution permits direct and efficient in-depth study of the problem parameters.

We consider a long tube of N layers that is loaded axisymmetrically at its ends and uniformly and axisymmetrically along its length (Fig. 10.1). The inside radius of the tube is R_I and the outside radius R_O. Each layer is considered to be a homogeneous, orthotropic material in the layer material principal coordinates (and a monoclinic material in the global coordinates for off-axis fiber orientations). It is assumed that there is perfect bonding between the layers. We number the layers starting from the inside of the tube.

The elasticity solution for the laminated circular tube is presented first for a single layer, and that solution is extended to the laminated tube by invoking the stress and displacement continuity requirements at the perfectly bonded interfaces between layers.

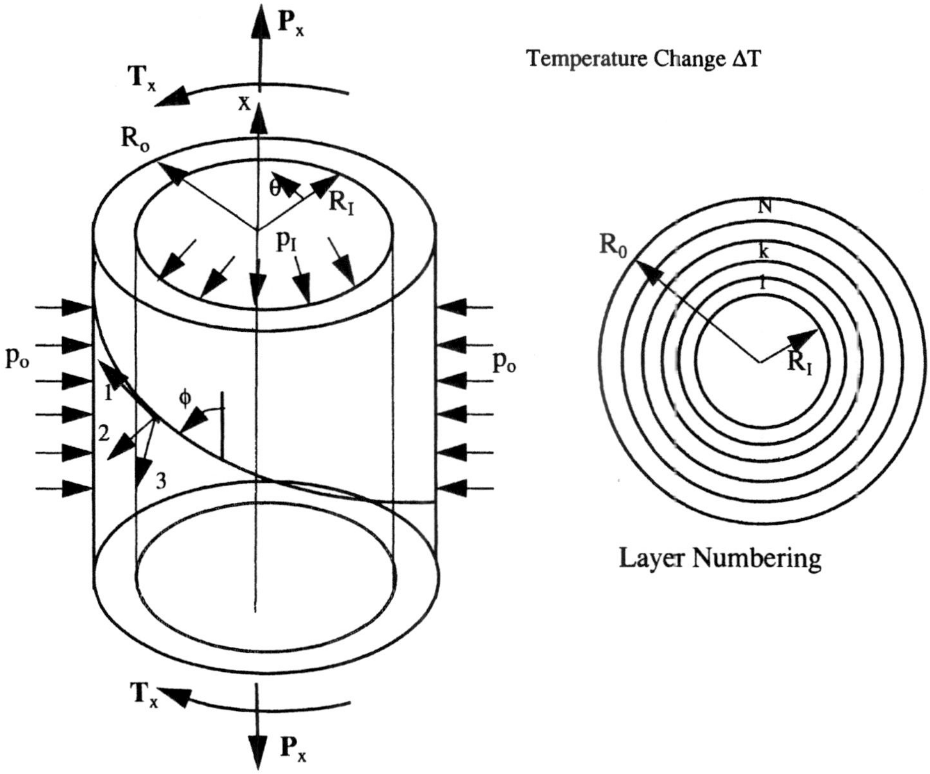

FIGURE 10.1 Laminated Composite Tube

10.2 Single-Layer Elasticity Solution

The single-layer solution is the exact elasticity solution for a long, open-ended, circular tube made from a homogeneous, monoclinic layer and subjected to axisymmetric thermomechanical loading. The mechanical loads are applied axisymmetrically at the ends and uniformly and axisymmetrically along the length. The thermal load is a uniform temperature change. The exact elasticity solution satisfies equilibrium, strain-displacement, compatibility, and boundary conditions for the stated constitutive equations and loading.

10.2.1 Strains in Cylindrical Coordinates

It is simplest to work in cylindrical coordinates (x, θ, r) with x along the axis of the tube (Fig. 10.1). The most general displacement field can then be expressed in terms of the *axial displacement*, $u(x, \theta, r)$; the *tangential displacement*, $v(x, \theta, r)$; and the *radial displacement*, $w(x, \theta, r)$. For the displacement field

$$
\begin{aligned}
u &= u(x, \theta, r) \\
v &= v(x, \theta, r) \qquad (10.1) \\
w &= w(x, \theta, r)
\end{aligned}
$$

the strain-displacement relationships in cylindrical coordinates are (e.g., Fung, 1965)

$$\varepsilon_x = \frac{\partial u}{\partial x} \qquad \varepsilon_\theta = \frac{1}{r}\left(\frac{\partial v}{\partial \theta} + w\right) \qquad \varepsilon_r = \frac{\partial w}{\partial r}$$

$$\gamma_{\theta r} = \frac{1}{r}\left(\frac{\partial w}{\partial \theta} - v + r\frac{\partial}{\partial r}(v)\right) \qquad \gamma_{xr} = \frac{\partial u}{\partial r} + \frac{\partial w}{\partial x} \qquad \gamma_{\theta x} = \frac{\partial v}{\partial x} + \frac{1}{r}\frac{\partial u}{\partial \theta} \tag{10.2}$$

For the axisymmetric tube under consideration, all displacements, strains, and stresses are independent of θ. In a long tube, the radial displacements, w, are also independent of the axial coordinate x away from the ends where the axial force and torque are applied. Thus, the most general displacements for the problem under consideration are

$$\begin{aligned} u &= u(x, r) \\ v &= v(x, r) \\ w &= w(r) \end{aligned} \tag{10.3}$$

Combining (10.2) and (10.3) gives the reduced strain-displacement equations for an axisymmetric tube loaded uniformly along its length:

$$\varepsilon_x = \frac{\partial u}{\partial x} \qquad \varepsilon_\theta = \frac{w}{r} \qquad \varepsilon_r = \frac{\partial w}{\partial r}$$

$$\gamma_{\theta r} = \frac{\partial v}{\partial r} - \frac{v}{r} \qquad \gamma_{xr} = \frac{\partial u}{\partial r} \qquad \gamma_{x\theta} = \frac{\partial v}{\partial x} \tag{10.4}$$

10.2.2 Compatibility

For the strains (10.4), three of the equations of compatibility (2.36) are satisfied identically. The remaining three (in cylindrical coordinates) are

$$\frac{d^2\varepsilon_x}{dr^2} = 0 \tag{10.5}$$

$$\frac{1}{r}\frac{d\varepsilon_x}{dr} = 0 \tag{10.6}$$

$$\frac{1}{2}\frac{d}{dr}\left[\frac{1}{r}\frac{d}{dr}(r\gamma_{x\theta})\right] = 0 \tag{10.7}$$

where total derivatives are used since the strains are a function of the single variable r only. Integration of (10.5) and (10.6) shows that the axial strain is constant in the layer, i.e.,

$$\varepsilon_x = \varepsilon_x^0 \tag{10.8}$$

10.2.3 Constitutive Equations

For an orthotropic layer, the thermo-elastic constitutive equations in principal material (1, 2, 3) coordinates (Fig. 10.1) are

$$\begin{bmatrix} \sigma_1 \\ \sigma_2 \\ \sigma_3 \\ \tau_{23} \\ \tau_{31} \\ \tau_{12} \end{bmatrix} = \begin{bmatrix} C_{11} & C_{12} & C_{13} & 0 & 0 & 0 \\ C_{12} & C_{22} & C_{23} & 0 & 0 & 0 \\ C_{13} & C_{23} & C_{33} & 0 & 0 & 0 \\ 0 & 0 & 0 & C_{44} & 0 & 0 \\ 0 & 0 & 0 & 0 & C_{55} & 0 \\ 0 & 0 & 0 & 0 & 0 & C_{66} \end{bmatrix} \begin{bmatrix} \varepsilon_1 - \varepsilon_1^T \\ \varepsilon_2 - \varepsilon_2^T \\ \varepsilon_3 - \varepsilon_3^T \\ \gamma_{23} \\ \gamma_{31} \\ \gamma_{12} \end{bmatrix} \tag{10.9}$$

where the thermal strains are $\varepsilon_i^T = \alpha_i \Delta T$. The thermo-elastic constitutive equations in the global cylindrical (x, θ, r) coordinates for this orthotropic layer at a fiber angle ϕ to the axial x-direction (Fig. 10.1) are determined by a transformation about the radial axis through the angle ϕ. The resulting constitutive equations have the form of a monoclinic material in cylindrical (x, θ, r) coordinates:

$$\begin{bmatrix} \sigma_x \\ \sigma_\theta \\ \sigma_r \\ \tau_{\theta r} \\ \tau_{xr} \\ \tau_{x\theta} \end{bmatrix} = \begin{bmatrix} \bar{C}_{11} & \bar{C}_{12} & \bar{C}_{13} & 0 & 0 & \bar{C}_{16} \\ \bar{C}_{12} & \bar{C}_{22} & \bar{C}_{23} & 0 & 0 & \bar{C}_{26} \\ \bar{C}_{13} & \bar{C}_{23} & \bar{C}_{33} & 0 & 0 & \bar{C}_{36} \\ 0 & 0 & 0 & \bar{C}_{44} & \bar{C}_{45} & 0 \\ 0 & 0 & 0 & \bar{C}_{45} & \bar{C}_{55} & 0 \\ \bar{C}_{16} & \bar{C}_{26} & \bar{C}_{36} & 0 & 0 & \bar{C}_{66} \end{bmatrix} \begin{bmatrix} \varepsilon_x - \varepsilon_x^T \\ \varepsilon_\theta - \varepsilon_\theta^T \\ \varepsilon_r - \varepsilon_r^T \\ \gamma_{\theta r} \\ \gamma_{xr} \\ \gamma_{x\theta} - \gamma_{x\theta}^T \end{bmatrix} \tag{10.10}$$

where $\bar{C}_{ij}$ are given by the transformation equations (3.53)–(3.65), written in terms of the fiber angle ϕ. Inverting (10.10) gives the layer constitutive equations for strains in terms of stresses and appropriately defined compliance coefficients $\bar{S}_{ij}$:

$$\begin{bmatrix} \varepsilon_x - \varepsilon_x^T \\ \varepsilon_\theta - \varepsilon_\theta^T \\ \varepsilon_r - \varepsilon_r^T \\ \gamma_{\theta r} \\ \gamma_{xr} \\ \gamma_{x\theta} - \gamma_{x\theta}^T \end{bmatrix} = \begin{bmatrix} \bar{S}_{11} & \bar{S}_{12} & \bar{S}_{13} & 0 & 0 & \bar{S}_{16} \\ \bar{S}_{12} & \bar{S}_{22} & \bar{S}_{23} & 0 & 0 & \bar{S}_{26} \\ \bar{S}_{13} & \bar{S}_{23} & \bar{S}_{33} & 0 & 0 & \bar{S}_{36} \\ 0 & 0 & 0 & \bar{S}_{44} & \bar{S}_{45} & 0 \\ 0 & 0 & 0 & \bar{S}_{45} & \bar{S}_{55} & 0 \\ \bar{S}_{16} & \bar{S}_{26} & \bar{S}_{36} & 0 & 0 & \bar{S}_{66} \end{bmatrix} \begin{bmatrix} \sigma_x \\ \sigma_\theta \\ \sigma_r \\ \tau_{\theta r} \\ \tau_{xr} \\ \tau_{x\theta} \end{bmatrix} \tag{10.11}$$

10.2.4 Equilibrium

The equilibrium equations in cylindrical coordinates (with body forces neglected) are

$$\begin{aligned}
\frac{\partial\sigma_r}{\partial r} + \frac{1}{r}(\sigma_r - \sigma_\theta) + \frac{1}{r}\frac{\partial\tau_{\theta r}}{\partial\theta} + \frac{\partial\tau_{xr}}{\partial x} &= 0 \\
\frac{\partial\tau_{\theta r}}{\partial r} + \frac{1}{r}\frac{\partial\sigma_\theta}{\partial\theta} + \frac{\partial\tau_{x\theta}}{\partial x} + \frac{2}{r}\tau_{\theta r} &= 0 \\
\frac{\partial\tau_{xr}}{\partial r} + \frac{1}{r}\frac{\partial\tau_{x\theta}}{\partial\theta} + \frac{\partial\sigma_x}{\partial x} + \frac{1}{r}\tau_{xr} &= 0
\end{aligned} \tag{10.12}$$

For a long axisymmetric tube under the prescribed loading, the stresses are independent of x and θ. Hence, the partial differential equilibrium equations (10.12) reduce to ordinary differential equations in r only:

$$\begin{aligned}
\frac{d\sigma_r}{dr} + \frac{\sigma_r - \sigma_\theta}{r} &= 0 \\
\frac{d\tau_{\theta r}}{dr} + \frac{2\tau_{\theta r}}{r} &= 0 \\
\frac{d\tau_{xr}}{dr} + \frac{\tau_{xr}}{r} &= 0
\end{aligned} \tag{10.13}$$

Integration of last two equations of (10.13) gives the shear stresses in the form

$$\begin{aligned}
\tau_{\theta r} &= \frac{E}{r^2} \\
\tau_{xr} &= \frac{F}{r}
\end{aligned} \tag{10.14}$$

where E and F are constants of integration.

10.2.5 Displacements

Using (10.14) with the constitutive equation for γ_{xr} from (10.11), the strain-displacement equation (10.4) gives the shear strain:

$$\gamma_{xr} = \frac{\partial u}{\partial r} = \bar{S}_{45}\frac{E}{r^2} + \bar{S}_{55}\frac{F}{r} \tag{10.15}$$

Integration of (10.15) with respect to r gives the axial displacement:

$$u(x, r) = -\bar{S}_{45}\frac{E}{r} + \bar{S}_{55}F\ln r + f(x) \tag{10.16}$$

where $f(x)$ is an arbitrary function.

Substitution of (10.8) and (10.16) into the first of the strain-displacement equations (10.4) shows that $f(x)$ must be equal to ε_x^0 plus a constant, say F_1, that represents rigid body displacement. Thus, the final form for the axial displacement is

$$u(x, r) = \varepsilon_x^0 x - \bar{S}_{45}\frac{E}{r} + \bar{S}_{55}F \ln r + F_1 \tag{10.17}$$

The tangential displacements, v, are determined by integrating the third compatibility equation (10.7) to give the shear strain:

$$\gamma_{x\theta} = K_1 r + \frac{K_2}{r} \tag{10.18}$$

where K_1 and K_2 are constants of integration. Combining (10.18) with the strain-displacement relation (10.4) for $\gamma_{x\theta}$ gives the tangential displacement in the form

$$v = \left(K_1 r + \frac{K_2}{r}\right)x + g(r) \tag{10.19}$$

where $g(r)$ is an arbitrary function.

From (10.4), we also have

$$\gamma_{\theta r} = \frac{\partial v}{\partial r} - \frac{v}{r} = g'(r) - \frac{2K_2 x}{r^2} - \frac{g(r)}{r} \tag{10.20}$$

Combining the constitutive equation (10.11) and the results (10.14) from equilibrium, the shear strain $\gamma_{\theta r}$ is

$$\gamma_{\theta r} = \bar{S}_{44}\frac{E}{r^2} + \bar{S}_{45}\frac{F}{r} \tag{10.21}$$

Equating (10.20) and (10.21) for $\gamma_{\theta r}$ shows that $K_2 = 0$ and that $g(r)$ must satisfy the ordinary differential equation

$$\frac{d}{dr}g(r) - \frac{g(r)}{r} = \bar{S}_{44}\frac{E}{r^2} + \bar{S}_{45}\frac{F}{r} \tag{10.22}$$

The solution to (10.22) is

$$g(r) = -\frac{\bar{S}_{44}E}{2r} - \bar{S}_{45}F + G_1 r \tag{10.23}$$

where G_1 is a constant of integration associated with rotational rigid body motion.

Substituting (10.23) into (10.19) (with $K_2 = 0$) and defining $K_1 = \gamma^0$, the angle of twist (in radians) per unit length, gives the final form for $v(x, r)$:

$$v(x, r) = \gamma^0 x r - \bar{S}_{44}\frac{E}{2r} - \bar{S}_{45}F + G_1 r \tag{10.24}$$

Combining the first equilibrium equation of (10.13) with the thermo-elastic constitutive equation (10.10), the strain-displacement relations (10.4), and the displacement fields (10.17) and (10.24) gives a second-order, ordinary differential equation (ODE) for the displacement, w, as a function of r, the uniform axial strain ε_x^0, the angle of twist per unit length γ^0, and the thermal strains ε_i^T, i.e.,

$$\frac{d^2w}{dr^2} + \frac{1}{r}\frac{dw}{dr} - \frac{\bar{C}_{22}w}{\bar{C}_{33}r^2} = \frac{1}{\bar{C}_{33}}\left[\frac{(\bar{C}_{12} - \bar{C}_{13})\varepsilon_x^0 + \Sigma}{r} + (\bar{C}_{26} - 2\bar{C}_{36})\gamma^0\right] \tag{10.25}$$

where Σ is defined as

$$\Sigma = (\bar{C}_{13} - \bar{C}_{12})\varepsilon_x^T + (\bar{C}_{23} - \bar{C}_{22})\varepsilon_\theta^T + (\bar{C}_{33} - \bar{C}_{32})\varepsilon_r^T + (\bar{C}_{63} - \bar{C}_{62})\gamma_{x\theta}^T \tag{10.26}$$

or more simply, using the summation convention of repeated subscripts,

$$\Sigma = (\bar{C}_{i3} - \bar{C}_{i2})\varepsilon_i^T \qquad (i = 1, 2, 3, 6) \tag{10.27}$$

For a uniform temperature change ΔT, the thermal strains in (10.27) can be written in terms of the coefficients of thermal expansion in the form

$$\varepsilon_i^T = \alpha_i \Delta T \tag{10.28}$$

where the coefficients of thermal expansion are

$$\alpha_i = \begin{bmatrix} \alpha_x \\ \alpha_\theta \\ \alpha_r \\ 0 \\ 0 \\ \alpha_{x\theta} \end{bmatrix} \tag{10.29}$$

Defining

$$\tilde{\Sigma} = (\bar{C}_{i3} - \bar{C}_{i2})\alpha_i \qquad (i = 1, 2, 3, 6) \tag{10.30}$$

we can write the thermal term (10.27) as

$$\Sigma = \tilde{\Sigma}\Delta T \tag{10.31}$$

The ODE (10.25) then has the form

$$\frac{d^2w}{dr^2} + \frac{1}{r}\frac{dw}{dr} - \frac{\bar{C}_{22}w}{\bar{C}_{33}r^2} = \frac{1}{\bar{C}_{33}}\left(\frac{(\bar{C}_{12} - \bar{C}_{13})\varepsilon_x^0 + \tilde{\Sigma}\Delta T}{r} + (\bar{C}_{26} - 2\bar{C}_{36})\gamma^0\right) \tag{10.32}$$

The solution to the ODE (10.32) is

$$w(r) = A_1 r^{\lambda} + A_2 r^{-\lambda} + \left(\frac{\bar{C}_{12} - \bar{C}_{13}}{\bar{C}_{33} - \bar{C}_{22}}\right)\varepsilon_x^0 r + \left(\frac{\bar{C}_{26} - 2\bar{C}_{36}}{4\bar{C}_{33} - \bar{C}_{22}}\right)\gamma^0 r^2 + \left(\frac{\tilde{\Sigma}}{\bar{C}_{33} - \bar{C}_{22}}\right) r\Delta T \qquad (10.33)$$

where

$$\lambda = \sqrt{\frac{\bar{C}_{22}}{\bar{C}_{33}}} \qquad (10.34)$$

If we introduce the definitions

$$\left(\frac{\bar{C}_{12} - \bar{C}_{13}}{\bar{C}_{33} - \bar{C}_{22}}\right) = \Gamma \qquad (10.35)$$

$$\left(\frac{\bar{C}_{26} - 2\bar{C}_{36}}{4\bar{C}_{33} - \bar{C}_{22}}\right) = \Omega \qquad (10.36)$$

$$\left(\frac{\tilde{\Sigma}}{\bar{C}_{33} - \bar{C}_{22}}\right) = \Psi \qquad (10.37)$$

and set the rigid body motion constants F_1 in (10.17) and G_1 in (10.24) equal to zero, the displacement for a single-layer tube subjected to axisymmetric thermomechanical loading can be summarized as

$$u(x, r) = \varepsilon_x^0 x - \bar{S}_{45}\frac{E}{r} + \bar{S}_{55} F \ln r \qquad (10.38)$$

$$v(x, r) = \gamma^0 x r - \bar{S}_{44}\frac{E}{2r} - \bar{S}_{45} F \qquad (10.39)$$

$$w(r) = A_1 r^{\lambda} + A_2 r^{-\lambda} + \Gamma \varepsilon_x^0 r + \Omega \gamma^0 r^2 + \Psi r \Delta T \qquad (10.40)$$

If the layer is isotropic or transversely isotropic $\bar{C}_{22} = \bar{C}_{33}$, $\bar{C}_{12} = \bar{C}_{13}$, $\bar{C}_{i6} = 0$, and $\alpha_\theta = \alpha_r$. Thus the ODE (10.32) takes the form

$$\frac{d^2 w}{dr^2} + \frac{1}{r}\frac{dw}{dr} - \frac{w}{r^2} = 0 \qquad (10.41)$$

which has the solution

$$w(r) = A_1 r + A_2 r^{-1} \qquad (10.42)$$

We note that (10.42) is equivalent to (10.40) with $\Gamma = \Omega = \Psi = 0$ and $\lambda = 1$. Therefore, the remainder of the development will be formulated using only (10.40), with the caveat that appropriate terms must be eliminated or modified for transversely isotropic and isotropic layers.

For a given ΔT, there are, in general, six unknown constants in the displacement field specified by (10.38)–(10.40): $\overset{0}{\varepsilon}_x$, E, F, $\overset{0}{\gamma}$, A_1, and A_2. These constants are determined (or known) from the specified boundary conditions and loading conditions.

10.2.6 Stress Boundary Conditions

For a hollow cylinder subjected to normal stresses $\sigma_r = -p_I$ (pressure) on its internal surface, $r = R_I$, and pressure $\sigma_r = -p_o$ on the external surface $r = R_o$, the boundary conditions are

Interior Surface $r = R_I$:

$$\begin{aligned} \sigma_r(R_I) &= -p_I \\ \tau_{\theta r}(R_I) &= 0 \\ \tau_{xr}(R_I) &= 0 \end{aligned} \tag{10.43}$$

External Surface $r = R_o$:

$$\begin{aligned} \sigma_r(R_o) &= -p_o \\ \tau_{\theta r}(R_o) &= 0 \\ \tau_{xr}(R_o) &= 0 \end{aligned} \tag{10.44}$$

The latter two of (10.43) or (10.44) along with the shear relations (10.14) require that the constants E and F in (10.14) are zero, i.e.,

$$E = F = 0 \tag{10.45}$$

10.2.7 Simplified Displacements

Using the conditions (10.45), the displacements (10.38)–(10.40) in a single-layer tube subjected to axisymmetric thermomechanical loading with only normal stresses acting on its surfaces take the simplified forms

$$u(x, r) = \overset{0}{\varepsilon}_x x \tag{10.46}$$

$$v(x, r) = \gamma^0 x r \tag{10.47}$$

$$w(r) = A_1 r^{\lambda} + A_2 r^{-\lambda} + \Gamma \overset{0}{\varepsilon}_x r + \Omega \gamma^0 r^2 + \Psi r \Delta T \tag{10.48}$$

For a given thermal loading ΔT, the number of unknowns has been reduced to four: $\overset{0}{\varepsilon}_x$, $\overset{0}{\gamma}$, A_1, and A_2 (assuming that neither $\overset{0}{\varepsilon}_x$ nor $\overset{0}{\gamma}$ is specified).

10.2.8 Strains

The layer strains can be determined from the strain-displacement relations (10.4) and the expressions (10.38)–(10.40) for displacements. The results are

$$\begin{aligned}
\varepsilon_r &= \frac{\partial w}{\partial r} = \lambda A_1 r^{\lambda-1} - \lambda A_2 r^{-\lambda-1} + \Gamma\varepsilon_x^0 + 2\Omega\gamma^0 r + \Psi\Delta T \\
\varepsilon_\theta &= \frac{\partial v}{\partial \theta} + \frac{w}{r} = A_1 r^{\lambda-1} + A_2 r^{-\lambda-1} + \Gamma\varepsilon_x^0 + \Omega\gamma^0 r + \Psi\Delta T \\
\varepsilon_x &= \frac{\partial u}{\partial x} = \varepsilon_x^0 \\
\gamma_{\theta r} &= \frac{1}{r}\frac{\partial w}{\partial \theta} + \frac{\partial v}{\partial r} - \frac{v}{r} = \gamma^0 x - \gamma^0 x = 0 \\
\gamma_{rx} &= \frac{\partial w}{\partial x} + \frac{\partial u}{\partial r} = 0 + 0 = 0 \\
\gamma_{x\theta} &= \frac{\partial v}{\partial x} + \frac{1}{r}\frac{\partial u}{\partial \theta} = \gamma^0 r
\end{aligned} \tag{10.49}$$

From (10.49) we see that the shear strains $\gamma_{\theta r}$ and γ_{rx} are identically zero and the axial strain ε_x^0 is constant. Further, the shear strain $\gamma_{\theta x}$ varies linearly with r. The normal strains ε_r and ε_θ vary as powers of r.

10.2.9 Stresses

The layer stresses can now be determined directly from the constitutive equation (10.10) and the strains (10.49). The resulting shear stresses are

$$\tau_{r\theta} = \tau_{xr} = 0 \tag{10.50}$$

and

$$\tau_{x\theta} = \bar{C}_{16}\varepsilon_x^0 + \bar{C}_{26}\varepsilon_\theta + \bar{C}_{36}\varepsilon_r + \bar{C}_{66}\gamma^0 r - \bar{C}_{i6}\alpha_i\Delta T \qquad (i \text{ sum, } i = 1, \ldots, 6) \tag{10.51}$$

Combining (10.49) and (10.51) gives the nonzero shear stress in the form

$$\begin{aligned}
\tau_{x\theta} = {} & \{\bar{C}_{16} + (\bar{C}_{26} + \bar{C}_{36})\Gamma\}\varepsilon_x^0 + \{\bar{C}_{66} + (\bar{C}_{26} + 2\bar{C}_{36})\Omega\}\gamma^0 r + [(\bar{C}_{26} + \bar{C}_{36})\Psi - \bar{C}_{i6}\alpha_i]\Delta T \\
& + (\bar{C}_{26} + \lambda\bar{C}_{36})A_1 r^{\lambda-1} + (\bar{C}_{26} - \lambda\bar{C}_{36})A_2 r^{-\lambda-1} \qquad (i \text{ sum, } i = 1, 2, 3, 6)
\end{aligned} \tag{10.52}$$

The normal stresses are determined by the following subset of the constitutive equations (10.10):

$$\begin{Bmatrix} \sigma_x \\ \sigma_\theta \\ \sigma_r \end{Bmatrix} = \begin{bmatrix} \bar{C}_{11} & \bar{C}_{12} & \bar{C}_{13} & \bar{C}_{16} \\ \bar{C}_{12} & \bar{C}_{22} & \bar{C}_{23} & \bar{C}_{26} \\ \bar{C}_{13} & \bar{C}_{23} & \bar{C}_{33} & \bar{C}_{36} \end{bmatrix} \begin{Bmatrix} \varepsilon_x - \alpha_x\Delta T \\ \varepsilon_\theta - \alpha_\theta\Delta T \\ \varepsilon_r - \alpha_r\Delta T \\ \gamma_{x\theta} - \alpha_{x\theta}\Delta T \end{Bmatrix} \tag{10.53}$$

where the strains are given by (10.49).

The resulting explicit expressions for normal stresses are

$$\begin{aligned}
\sigma_x = {} & \{\bar{C}_{11} + (\bar{C}_{13} + \bar{C}_{12})\Gamma\}\varepsilon_x^0 + [(\bar{C}_{12} + 2\bar{C}_{13})\Omega + \bar{C}_{16}]\gamma^0 r + [(\bar{C}_{12} + \bar{C}_{13})\Psi - \bar{C}_{i1}\alpha_i]\Delta T \\
& + (\bar{C}_{12} + \lambda\bar{C}_{13})A_1 r^{\lambda-1} + (\bar{C}_{12} - \lambda\bar{C}_{13})A_2 r^{-\lambda-1} \qquad (i \text{ sum, } i = 1, 2, 3, 6)
\end{aligned} \tag{10.54}$$

$$\sigma_\theta = \{\bar{C}_{12} + (\bar{C}_{22} + \bar{C}_{23})\Gamma\}\varepsilon_x^0 + [(\bar{C}_{22} + 2\bar{C}_{23})\Omega + \bar{C}_{26}]\gamma^0 r + [(\bar{C}_{22} + \bar{C}_{23})\Psi - \bar{C}_{i2}\alpha_i]\Delta T \\ + (\bar{C}_{22} + \lambda\bar{C}_{23})A_1 r^{\lambda - 1} + (\bar{C}_{22} - \lambda\bar{C}_{23})A_2 r^{-\lambda - 1} \quad (i \text{ sum}, i = 1, 2, 3, 6) \tag{10.55}$$

$$\sigma_r = \{\bar{C}_{13} + (\bar{C}_{23} + \bar{C}_{33})\Gamma\}\varepsilon_x^0 + [(\bar{C}_{23} + 2\bar{C}_{33})\Omega + \bar{C}_{36}]\gamma^0 r + [(\bar{C}_{23} + \bar{C}_{33})\Psi - \bar{C}_{i3}\alpha_i]\Delta T \\ + (\bar{C}_{23} + \lambda\bar{C}_{33})A_1 r^{\lambda - 1} + (\bar{C}_{23} - \lambda\bar{C}_{33})A_2 r^{-\lambda - 1} \quad (i \text{ sum}, i = 1, 2, 3, 6) \tag{10.56}$$

It remains to determine the four unknowns ε_x^0, γ^0, A_1, and A_2. Thus four simultaneous equations are required. If ε_x^0 or γ^0 is given, there are two associated force variables, the axial force P and the torque T, to be determined. The four equations used to determine the unknowns are the axial force equilibrium equation, the torque equilibrium equation, and the normal stress boundary conditions on the inner and outer surfaces.

10.2.10 Normal Stress Boundary Conditions

From the normal stress boundary conditions in (10.43) and (10.44), we have

$$\sigma_r(R_I) = -p_I \tag{10.57}$$

and

$$\sigma_r(R_o) = -p_o \tag{10.58}$$

Using (10.56) in the boundary conditions provides two of the required four simultaneous equations for ε_x^0, γ^0, A_1, and A_2:

$$-p_I = \{\bar{C}_{13} + (\bar{C}_{23} + \bar{C}_{33})\Gamma\}\varepsilon_x^0 + [(\bar{C}_{23} + 2\bar{C}_{33})\Omega + \bar{C}_{36}]\gamma^0 r_I + [(\bar{C}_{23} + \bar{C}_{33})\Psi - \bar{C}_{i3}\alpha_i]\Delta T \\ + (\bar{C}_{23} + \lambda\bar{C}_{33})A_1 r_I^{\lambda - 1} + (\bar{C}_{23} - \lambda\bar{C}_{33})A_2 r_I^{-\lambda - 1} \quad (i \text{ sum}, i = 1, 2, 3, 6) \tag{10.59}$$

$$-p_o = \{\bar{C}_{13} + (\bar{C}_{23} + \bar{C}_{33})\Gamma\}\varepsilon_x^0 + [(\bar{C}_{23} + 2\bar{C}_{33})\Omega + \bar{C}_{36}]\gamma^0 r_o + [(\bar{C}_{23} + \bar{C}_{33})\Psi - \bar{C}_{i3}\alpha_i]\Delta T \\ + (\bar{C}_{23} + \lambda\bar{C}_{33})A_1 r_o^{\lambda - 1} + (\bar{C}_{23} - \lambda\bar{C}_{33})A_2 r_o^{-\lambda - 1} \quad (i \text{ sum}, i = 1, 2, 3, 6) \tag{10.60}$$

10.2.11 Axial Force

The axial force P_x required at the ends of the tube is determined by integrating the axial stress σ_x over the cross-sectional area of the tube. For a tube with inside radius R_I and outside radius R_o, the expression for P_x is

$$P_x = \int_{R_I}^{R_o} 2\pi\sigma_x r \, dr \tag{10.61}$$

Using the axial stress (10.54) in (10.61), we obtain a third simultaneous equation.

$$P_x = 2\pi\Bigg\{[\{\bar{C}_{11} + (\bar{C}_{13} + \bar{C}_{12})\Gamma\}\varepsilon_x^0 + \{(\bar{C}_{12} + \bar{C}_{13})\Psi - \bar{C}_{i1}\alpha_i\}\Delta T]\left(\frac{R_o^2 - R_I^2}{2}\right)$$

$$+\left\{\bar{C}_{16} + (\bar{C}_{12} + 2\bar{C}_{13})\Omega\right\}\gamma^0\left(\frac{R_o^3 - R_I^3}{3}\right) + \frac{(\bar{C}_{12} + \lambda\bar{C}_{13})}{\lambda + 1}A_1(R_o^{\lambda+1} - R_I^{\lambda+1}) \tag{10.62}$$

$$+\frac{(\bar{C}_{12} - \lambda\bar{C}_{13})}{-\lambda + 1}A_2\left(R_o^{-\lambda+1} - R_I^{-\lambda+1}\right)\Bigg\} \quad (i \text{ sum}, i = 1, 2, 3, 6)$$

10.2.12 Torque

The torque T_x is determined by integrating the moment of the shear stress $\tau_{x\theta}$ from (10.52) over the cross-sectional area of the tube.

$$T_x = 2\pi\int_{R_I}^{R_o} \tau_{x\theta} r^2\, dr \tag{10.63}$$

Substituting (10.52) into (10.63) and integrating gives the last of the required simultaneous equations:

$$T_x = 2\pi\Bigg\{[\{\bar{C}_{16} + (\bar{C}_{26} + \bar{C}_{36})\Gamma\}\varepsilon_x^0 + \{(\bar{C}_{26} + \bar{C}_{36})\Psi - \bar{C}_{i6}\alpha_i\}\Delta T]\left(\frac{R_o^3 - R_I^3}{3}\right)$$

$$+\left\{\bar{C}_{66} + (\bar{C}_{26} + 2\bar{C}_{36})\Omega\right\}\gamma^0\left(\frac{R_o^4 - R_I^4}{4}\right) + (\bar{C}_{26} + \lambda\bar{C}_{36})\frac{A_1}{\lambda + 2}r^{\lambda+2} \tag{10.64}$$

$$+(\bar{C}_{26} - \lambda\bar{C}_{36})\frac{A_2}{2 - \lambda}r^{2-\lambda}\Bigg\} \quad (i \text{ sum}, i = 1, 2, 3, 6)$$

10.2.13 Thermomechanical Loading

Equations (10.59), (10.60), (10.62), and (10.64) constitute the four simultaneous equations that provide the solution to the thermomechanical loading problem. For a given ΔT, the four unknowns are A_1, A_2, ε_x^0 or P_x, and γ^0 or T_x.

10.2.14 Pure Thermal Loading

Pure thermal loading corresponds to a nonzero temperature change, ΔT, with p_o, p_I, P_x, and T_x all set to zero. The four unknowns are then A_1, A_2, ε_x^0, and γ^0. The axial strain ε_x^0 per unit temperature change corresponds to the axial coefficient of thermal expansion α_x, and the γ^0 per unit temperature change corresponds to a coefficient of thermal twist, which we shall designate as α_γ.

10.2.15 Special Fiber Orientations

For the special cases of orthotropic and transversely isotropic layers there are simplifications in the expressions for radial displacements, w. The expressions for the axial displacements, u (10.38), and the tangential displacements, v (10.39), remain unchanged. With the changes in w, the expressions for strains and stresses change accordingly. In the following two sections we present the modifications in the expressions for the radial displacements, w. It is straightforward to take the appropriate partial derivatives to obtain the strains and then substitute them into the constitutive equations to obtain the stresses. These last two steps are left to the reader.

10.2.15.1 Orthotropic Layers

If a layer is orthotropic, such as a 90° fiber orientation (Fig. 10.2), the stiffness coefficients $\bar{C}_{16}$, $\bar{C}_{26}$, and $\bar{C}_{36}$ for that layer are zero, as is Ω in (10.36). The solution to the modified ODE (10.32) then has the form

$$w(r) = A_1 r^{\lambda} + A_2 r^{-\lambda} + \Gamma \varepsilon_x^0 r + \Psi r \Delta T \tag{10.65}$$

As indicated from equations (10.54) through (10.56), all coupling between normal stress components and twist, γ^0, disappears for orthotropic layers. And from (10.65) we see that the radial displacements, w, are uncoupled from the twist, γ^0.

10.2.15.2 Transversely Isotropic Layers

As discussed previously, if a layer is transversely isotropic in the r-θ plane, as for a 0° fiber orientation (Fig. 10.2), additional simplifications occur. For the transverse isotropy layer $\bar{C}_{16} = \bar{C}_{26} = \bar{C}_{36} = 0$, $\bar{C}_{12} = \bar{C}_{13}$, $\bar{C}_{22} = \bar{C}_{33}$, $\lambda = 1$, $\alpha_r = \alpha_\theta$, and $\Gamma = \Omega = \Sigma = 0$. The governing ODE (10.32) for radial displacement, w, reduces to

$$\frac{d^2 w}{dr^2} + \frac{1}{r}\frac{dw}{dr} - \frac{w}{r^2} = 0 \tag{10.66}$$

which has the solution

$$w(r) = A_1 r + A_2 r^{-1} \tag{10.67}$$

This is, of course, also the solution for an isotropic layer.

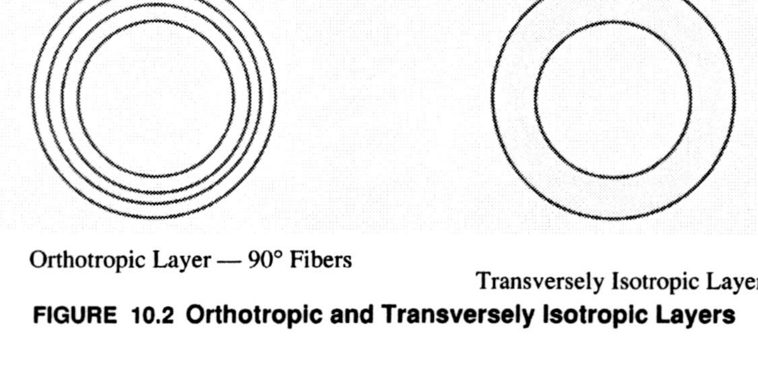

Orthotropic Layer — 90° Fibers

Transversely Isotropic Layer — 0° Fibers

FIGURE 10.2 Orthotropic and Transversely Isotropic Layers

10.3 Laminated Tube

The solution for a laminated, multilayered tube follows directly from the preceding development for a single-layer tube with the additional features that the material properties vary from layer to layer and the requirement that stress and displacement continuity conditions be satisfied at the layer interfaces. For an N-layer laminated tube there are, in general, $6N$ unknowns—ε_x^0, E, F, γ^0, A_1, and A_2—for each layer. However, as will be shown, for the loading cases under consideration with only normal stress (pressure) loading on the internal and external surfaces (i.e., zero shear stresses on these surfaces), the constants E and F are zero in all layers and continuity of displacements between layers requires that ε_x^0 and γ^0 be constants throughout the tube. This reduces the number of unknowns to $2N + 2$. The unknowns are: ε_x^0, γ^0, $A_1^{(k)}$, and $A_2^{(k)}$ $(k = 1,\ldots, N)$. In the following development the generic layer is denoted by k.

10.3.1 Layer Displacements

The most general displacements in the kth layer of a laminated tube subjected to axisymmetric loading follow from (10.38) to (10.40) as

$$u^{(k)}(x, r) = \varepsilon_x^{0^{(k)}} x - \bar{S}_{45}^{(k)} \frac{E^{(k)}}{r} + \bar{S}_{55}^{(k)} F^{(k)} \ln r \tag{10.68}$$

$$v^{(k)}(x, r) = \gamma^{0^{(k)}} xr - \bar{S}_{44}^{(k)} \frac{E^{(k)}}{2r} - \bar{S}_{45}^{(k)} F^{(k)} \tag{10.69}$$

$$w^{(k)}(r) = A_1^{(k)} r^{\lambda} + A_2^{(k)} r^{-\lambda} + \Gamma^{(k)} \varepsilon^{0^{(k)}} r + \Omega^{(k)} \gamma^{0^{(k)}} r^2 + \Psi^{(k)} r \Delta T \tag{10.70}$$

10.3.2 Traction Continuity Conditions

Traction continuity between layers requires that the two interlaminar shear stresses $\tau_{\theta r}$ and τ_{xr} be continuous from layer to layer. Writing these continuity conditions using (10.14) and starting from either the inside or the outside surface where the shear stresses are zero shows that the constants $E^{(k)}$ and $F^{(k)}$ are equal to 0 in all layers, i.e.,

$$E^{(k)} = F^{(k)} = 0 \qquad (k = 1,\ldots, N) \tag{10.71}$$

Thus, the total number of unknowns in the problem is reduced from $6N$ to $4N$.

From (10.14) and (10.71) we have the interesting result that the interlaminar shear stresses $\tau_{\theta r}$ and τ_{xr} are zero throughout a long, laminated tube subjected to axisymmetric, uniform pressure loading along its length, i.e.,

$$\tau_{\theta r}^{(k)} = \tau_{rx}^{(k)} = 0 \qquad (k = 1,\ldots, N) \tag{10.72}$$

10.3.3 Displacement Continuity Conditions

For perfectly bonded layers, all displacements, including the axial displacements, u, must be continuous from layer to layer. Thus the general expression (10.38) for the axial displacement u (with $E^{(k)} = F^{(k)} = 0$ from (10.71)) requires that *the axial strain be constant throughout all layers*, i.e.,

$$\varepsilon^{0^{(k)}} = \varepsilon_x^0 \qquad (k = 1,\ldots, N) \tag{10.73}$$

The tangential displacements, v, must also be continuous from layer to layer. Thus the general expression (10.39) for v (with $E^k = F^k = 0$ from (10.71)) requires that *the angle of twist per unit length be constant for all layers,* i.e.,

$$\gamma^{0(k)} = \gamma^0 \qquad (k = 1, \ldots, N) \tag{10.74}$$

Hence, the total number of unknowns is now reduced to $2N + 2$; they are ε_x^0, γ^0, $NA_1^{(k)}$, and $NA_2^{(k)}$. The $2N + 2$ simultaneous equations required to solve for the unknowns are obtained from the axial force and torque equilibrium equations, the interior and exterior normal stress boundary conditions, and $2(N-1)$ layer interfacial continuity conditions.

10.3.4 Reduced Displacements

With $E^{(k)}$ and $F^{(k)}$ equal to 0 in all layers from (10.71) and ε_x^0 and γ^0 constant throughout the tube from (10.73) and (10.74), the individual layer displacements in the kth layer, (10.38)–(10.40), can now be written in the reduced form

$$u^{(k)}(x, r) = \varepsilon_x^0 x \tag{10.75}$$

$$v^{(k)}(x, r) = \gamma^0 xr \tag{10.76}$$

$$w^{(k)}(r) = A_1^{(k)} r^{\lambda} + A_2^{(k)} r^{-\lambda} + \Gamma^{(k)} \varepsilon_x^0 r + \Omega^{(k)} \gamma^0 r^2 + \Psi^{(k)} r \Delta T \tag{10.77}$$

From the above equations we see that, at a given x, the axial displacements are uniform throughout the thickness of the tube and the tangential displacements vary linearly with r. The radial displacements vary in a more complicated manner, but as a function of r only.

10.3.5 Layer Strains

The individual layer displacements (10.75)–(10.77) for a laminated tube are identical to those of (10.38)–(10.40) for a single-layer tube when the layer material properties and unknown constants A_1 and A_2 appropriate to the layer are used. Therefore, the strain relations (10.49) are appropriate for the layers of a laminated tube, when modified to indicate layerwise properties and constants. The strains can be written

$$\begin{aligned}
\varepsilon_r^{(k)} &= \lambda^{(k)} A_1^{(k)} r^{\lambda^{(k)}-1} - \lambda^{(k)} A_2^{(k)} r^{-\lambda^{(k)}-1} + \Gamma^{(k)} \varepsilon^0 + 2\Omega^{(k)} \gamma^0 r + \Psi^{(k)} \Delta T \\
\varepsilon_\theta^{(k)} &= A_1^{(k)} r^{\lambda^{(k)}-1} + A_2^{(k)} r^{-\lambda^{(k)}-1} + \Gamma^{(k)} \varepsilon^0 + \Omega^{(k)} \gamma^0 r + \Psi^{(k)} \Delta T \\
\varepsilon_x^{(k)} &= \varepsilon_x^0 \\
\gamma_{\theta r}^{(k)} &= 0 \\
\gamma_{rx}^{(k)} &= 0 \\
\gamma_{x\theta}^{(k)} &= \gamma^0 r
\end{aligned} \tag{10.78}$$

10.3.6 Layer Stresses

It was noted in (10.72) that the interlaminar shear stresses are zero in all layers of a laminated tube when the surface tractions on the interior and exterior surfaces of the tube are limited to normal stresses. The remaining individual layer stresses follow from the constitutive equations (10.10) and the strains (10.78). The results are the same as those given in (10.52) and (10.54)–(10.56), with appropriate modification for layerwise properties and constants. Thus,

$$\begin{aligned}\tau_{x\theta} &= \{\bar{C}_{16} + (\bar{C}_{26} + \bar{C}_{36})\Gamma\}\varepsilon_x^0 + \{\bar{C}_{66} + (\bar{C}_{26} + 2\bar{C}_{36})\Omega\}\gamma^0 r + [(\bar{C}_{26} + \bar{C}_{36})\Psi - \bar{C}_{i6}\alpha_i]\Delta T \\ &\quad + (\bar{C}_{26} + \lambda\bar{C}_{36})A_1 r^{\lambda-1} + (\bar{C}_{26} - \lambda\bar{C}_{36})A_2 r^{-\lambda-1} \quad (i \text{ sum}, i = 1, 2, 3, 6)\end{aligned} \tag{10.79}$$

$$\begin{aligned}\sigma_x &= \{\bar{C}_{11} + (\bar{C}_{13} + \bar{C}_{12})\Gamma\}\varepsilon_x^0 + [(\bar{C}_{12} + 2\bar{C}_{13})\Omega + \bar{C}_{16}]\gamma^0 r + [(\bar{C}_{12} + \bar{C}_{13})\Psi - \bar{C}_{i1}\alpha_i]\Delta T \\ &\quad + (\bar{C}_{12} + \lambda\bar{C}_{13})A_1 r^{\lambda-1} + (\bar{C}_{12} - \lambda\bar{C}_{13})A_2 r^{-\lambda-1} \quad (i \text{ sum}, i = 1, 2, 3, 6)\end{aligned} \tag{10.80}$$

$$\begin{aligned}\sigma_\theta &= \{\bar{C}_{12} + (\bar{C}_{22} + \bar{C}_{23})\Gamma\}\varepsilon_x^0 + [(\bar{C}_{22} + 2\bar{C}_{23})\Omega + \bar{C}_{26}]\gamma^0 r + [(\bar{C}_{22} + \bar{C}_{23})\Psi - \bar{C}_{i2}\alpha_i]\Delta T \\ &\quad + (\bar{C}_{22} + \lambda\bar{C}_{23})A_1 r^{\lambda-1} + (\bar{C}_{22} - \lambda\bar{C}_{23})A_2 r^{-\lambda-1} \quad (i \text{ sum}, i = 1, 2, 3, 6)\end{aligned} \tag{10.81}$$

$$\begin{aligned}\sigma_r &= \{\bar{C}_{13} + (\bar{C}_{23} + \bar{C}_{33})\Gamma\}\varepsilon_x^0 + [(\bar{C}_{23} + 2\bar{C}_{33})\Omega + \bar{C}_{36}]\gamma^0 r + [(\bar{C}_{23} + \bar{C}_{33})\Psi - \bar{C}_{i3}\alpha_i]\Delta T \\ &\quad + (\bar{C}_{23} + \lambda\bar{C}_{33})A_1 r^{\lambda-1} + (\bar{C}_{23} - \lambda\bar{C}_{33})A_2 r^{-\lambda-1} \quad (i \text{ sum}, i = 1, 2, 3, 6)\end{aligned} \tag{10.82}$$

10.3.7 Axial Force

The axial force P_x required at the ends of the tube is determined by integrating the axial stress σ_x over the area of the cross section. For an N-layer tube with inside radius R_I and outside radius R_o, the expression for P_x is

$$P_x = \int_{R_I}^{R_o} 2\pi\sigma_x r\, dr = 2\pi\sum_{k=1}^{N}\int_{r_{k-1}}^{r_k} \sigma_x^{(k)}(r) r\, dr \tag{10.83}$$

This provides one of the required simultaneous equations for determining the problem unknowns.

10.3.8 Torque

The torque T_x is determined in a similar manner by integrating the moment of the shear stress $\tau_{x\theta}$ over the cross-sectional area of the tube. Thus,

$$T_x = 2\pi\int_{R_I}^{R_o} \tau_{x\theta} r^2\, dr = 2\pi\sum_{k=1}^{N}\int_{r_{k-1}}^{r_k} \tau_{x\theta}^{(k)}(r) r^2\, dr \tag{10.84}$$

This provides the second of the required simultaneous equations for determining the problem unknowns.

10.3.9 Boundary Conditions

The boundary conditions for a laminated cylinder with only pressure loading on its internal and external surfaces are identical to those for the single layer, with appropriate modification for layer properties and layer numbers. On the internal surface $r = R_I$ ($k = 1$), the boundary condition on σ_r is

$$\begin{aligned} -p_I = {} & [\bar{C}_{13}^{(1)} + (\bar{C}_{23}^{(1)} + \bar{C}_{33}^{(1)})\Gamma^{(1)}]\varepsilon_x^0 + [(\bar{C}_{23}^{(1)} + 2\bar{C}_{33}^{(1)})\Omega^{(1)} + \bar{C}_{36}^{(1)}]\gamma^0 R_I \\ & + [(\bar{C}_{23}^{(1)} + \bar{C}_{33}^{(1)})\Psi^{(1)} - \bar{C}_{i3}^{(1)}\alpha_i^{(1)}]\Delta T \\ & + (\bar{C}_{23}^{(1)} + \lambda^{(1)}\bar{C}_{33}^{(1)})A_1^{(1)} R_I^{\lambda^{(1)}-1} \\ & + (\bar{C}_{23}^{(1)} - \lambda^{(1)}\bar{C}_{33}^{(1)})A_2^{(1)} R_I^{-\lambda^{(1)}-1} \quad (i \text{ sum}, i = 1, 2, 3, 6) \end{aligned} \tag{10.85}$$

A similar equation describes the boundary condition on the outer surface $r = R_o$ ($k = N$):

$$\begin{aligned} -p_o = {} & \{\bar{C}_{13}^{(N)} + (\bar{C}_{23}^{(N)} + \bar{C}_{33}^{(N)})\Gamma^{(N)}\}\varepsilon_x^0 + [(\bar{C}_{23}^{(N)} + 2\bar{C}_{33}^{(N)})\Omega^{(N)} + \bar{C}_{36}^{(N)}]\gamma^0 R_o \\ & + [(\bar{C}_{23}^{(N)} + \bar{C}_{33}^{(N)})\Psi^{N} - \bar{C}_{i3}^{(N)}\alpha_i^{(N)}]\Delta T \\ & + (\bar{C}_{23}^{(N)} + \lambda^{(N)}\bar{C}_{33}^{(N)})A_1^{(N)} R_0^{\lambda^{(N)}-1} \\ & + (\bar{C}_{23}^{(N)} - \lambda^{(N)}\bar{C}_{33}^{(N)})A_2^{(N)} R_0^{-\lambda^{(N)}-1} \quad (i \text{ sum}, i = 1, 2, 3, 6) \end{aligned} \tag{10.86}$$

Thus, we have two more of the required simultaneous equations needed for the solution to our problem.

10.3.10 Traction and Displacement Continuity Requirements

There are two remaining continuity requirements that must be satisfied at each interface: continuity of the radial stresses, σ_r, and continuity of the radial displacements, w. For an N-layer laminate with $N - 1$ interfaces these conditions represent the final $2(N - 1)$ equations required for the complete set of simultaneous equations. We can write these continuity conditions in the form

$$w^{(k)} = w^{(k+1)} \qquad \text{(interfaces } k = 1, N-1) \tag{10.87}$$

$$\sigma_r^{(k)} = \sigma_r^{(k+1)} \qquad \text{(interfaces } k = 1, N-1) \tag{10.88}$$

where the radial displacements are given by (10.40) and the radial stresses are given by (10.82).

10.3.11 Simultaneous Equations for a Laminated Tube

The unknowns in the laminated tube problem described above include NA_1, NA_2, ε_x^0, and γ^0 for a total of $2N + 2$ unknowns. The corresponding equations are the $2(N - 1)$ continuity equations (10.87) and (10.88), the axial force equilibrium equation (10.83), the torque equilibrium equation (10.84), and the two surface boundary conditions (10.85) and (10.86). Simultaneous solution of these equations provides the desired unknowns. If ε_x^0 and/or γ^0 is specified, the number of unknowns is reduced accordingly, and the required axial force and torque for the specified ε_x^0 and γ^0 are determined as a postsolution step from (10.83) and (10.84), respectively.

It is also possible to treat ΔT as an unknown for specified conditions on ε_x^0 or P_x and γ^0 or T_x. For example, the problem can be to determine the ΔT for a given laminated tube to have a specified axial strain, ε_x^0, if the tube is completely free of mechanical loads.

10.3.11.1 Local/Global Stiffness Solution Method

An efficient method for solving the system of simultaneous equations which often occur in the solution to problems in laminated composites was presented by Pindera (1991). The solution procedure uses the interfacial continuity requirements on tractions and displacements to effectively reduce the number of unknowns in the problem. The corresponding reduction in computer resources can be significant for laminated composites with a large number of layers. This is especially true if material or geometric nonlinearities require an incremental solution.

The approach is to eliminate the unknown coefficients $A_1^{(k)}$ and $A_2^{(k)}$ from the problem by expressing them in terms of the unknown radial displacements w_k^- and w_k^+ on the inner and outer radii of the kth layer, respectively (Fig. 10.3).

Using the general expression (10.40) for the radial displacements in the kth layer, we can write

$$w_k^- = A_1^{(k)} r_k^{\lambda} + A_2^{(k)} r_k^{-\lambda} + \Gamma^{(k)} \varepsilon_x^0 r_k + \Omega^{(k)} \gamma^0 r_k^2 + \Psi^{(k)} r_k \Delta T \tag{10.89}$$

$$w_k^+ = A_1^{(k)} r_{k+1}^{\lambda} + A_2^{(k)} r_{k+1}^{-\lambda} + \Gamma^{(k)} \varepsilon_x^0 r_{k+1} + \Omega^{(k)} \gamma^0 r_{k+1}^2 + \Psi^{(k)} r_{k+1} \Delta T \tag{10.90}$$

These two equations can be solved for $A_1^{(k)}$ and $A_2^{(k)}$ in terms of $w_k^- = w(r_k)$, $w_k^+ = w(r_{k+1})$ and other known and unknown quantities, including ε_x^0, γ^0, and ΔT. The solution for the generic kth layer can be written in the following form:

$$\begin{pmatrix} A_1 \\ A_2 \end{pmatrix}^{(k)} = \begin{bmatrix} k_{11} & k_{12} & k_{13} & k_{14} & k_{15} \\ k_{21} & k_{22} & k_{23} & k_{24} & k_{25} \end{bmatrix}^{(k)} \begin{bmatrix} w^+ \\ w^- \\ \varepsilon_x^0 \\ \gamma^0 \\ \Delta T \end{bmatrix}^{(k)} \tag{10.91}$$

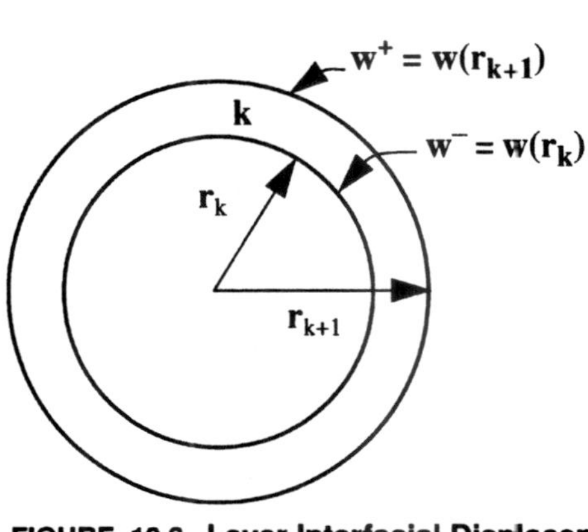

FIGURE 10.3 Layer Interfacial Displacements

where k_{ij} are appropriately defined quantities for the generic layer. These results for $A_1^{(k)}$ and $A_2^{(k)}$ can now be substituted into the radial stress continuity conditions (10.88) and the stress boundary conditions (10.85) and (10.86) as well as the axial force and torque equilibrium equations (10.83) and (10.84). The set of simultaneous equations is now in terms of the $(N-1)$ interfacial radial displacements, w_k, the radial displacements at the inner and outer radii, w_1 and w_{N+1}, and the unknowns ε_x^0 and γ^0, for a total of $N-1+4=N+3$ unknowns. This can be a significant reduction in the number of simultaneous equations, approaching 50% as the number of layers becomes very large.

The result of the above reformulation is a system of $(N-1)$ interfacial continuity equations plus two boundary conditions, for a total of $N+1$ equations. If ε_x^0, γ^0, and ΔT are given, there are $N+1$ unknowns: the radial displacements, w, at the $N-1$ interfaces and at the two external surfaces. If P_x or T_x is specified, then the additional equations necessary to determine ε_x^0 or γ^0 are the force and torque equilibrium equations (10.83) and (10.84).

10.4 Tube Response

With all of the unknowns of a given tube problem determinable using the approach detailed in the previous section, it is now possible to determine effective elastic and thermal properties of tubes. In this section we describe methods and show representative results for effective axial modulus, shear modulus, axial and radial coefficients of thermal expansion, and several coupling coefficients. Results are presented as a function of the tube aspect ratio R_I/h, where h is the tube thickness.

10.4.1 Effective Axial Modulus

The *effective axial modulus*, $\bar{E}_x$, of a tube can be written in the form

$$\bar{E}_x = \frac{P_x}{A\varepsilon_x^0} \tag{10.92}$$

where P_x is the applied axial force, A is the cross sectional area, and ε_x^0 is the uniform axial strain obtained from the solution to a system of simultaneous equations as described in Section 10.3. The effective moduli of tubes consisting of unidirectional off-axis and angle-ply laminates, as a function of the fiber orientation ϕ and aspect ratio R_I/h, are presented in Fig. 10.4 for T300/5208 carbon/epoxy. As indicated in the figure, the modulus of the angle-ply tubes is larger than that of the unidirectional off-axis lamina up to an angle of approximately 60°, after which the moduli are essentially identical. We see from the figure that the tube aspect ratio has only a small influence on the effective modulus of the angle-ply tubes, and then only for the smallest aspect ratio. The results in Fig. 10.4 for aspect ratios of 5 and 75 are identical to those in Fig. 5.9 for flat laminates. As expected, the curvature of the tube has no effect on the effective axial modulus for tubes with large aspect ratios.

10.4.2 Effective Poisson's Ratio

We define the *effective Poisson's ratio* $\bar{\nu}_{x\theta}$ as the negative ratio of hoop strain at the inner radius $r = R_I$ to the uniform axial strain ε_x^0 for axial loading P_x. Using (10.4),

$$\bar{\nu}_{x\theta} = \left.\frac{-\varepsilon_\theta}{\varepsilon_x^0}\right|_{R_I} = \frac{-w(R_I)}{\varepsilon_x^0 R_I} \tag{10.93}$$

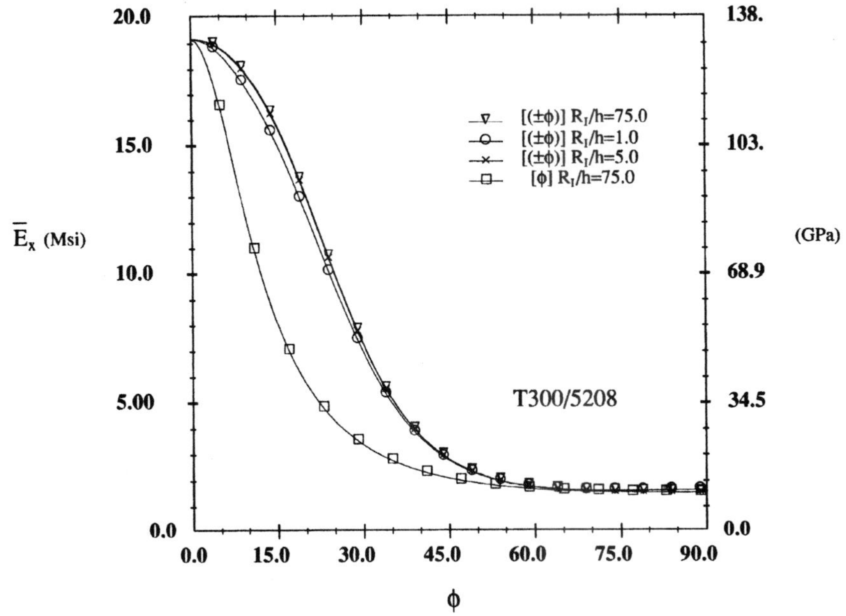

FIGURE 10.4 Effective Axial Modulus for Carbon/Epoxy Tubes

Thus the effective Poisson's ratio $\bar{\nu}_{x\theta}$ gives the change in inner radius. We note from (10.4) and (10.78) that, in general, Poisson's ratio $\nu_{x\theta}$ varies with radial location.

Predictions for the effective Poisson's ratio $\bar{\nu}_{x\theta}$ at the inner radius of unidirectional off-axis and [±ϕ] tubes are presented in Fig. 10.5. The results for angle-ply tubes are presented for three aspect ratios, 1, 5, and 75. The results in the figure show that large changes in Poisson's ratio (and the inner radius) can occur in angle-ply laminated tubes. The maximum Poisson's ratio ranges from 1.25 for an aspect ratio of 75 to more than 1.85 for an aspect ratio of 1.0. Further, for small aspect ratios (<5) and large fiber angles the Poisson's ratio is negative. These results can be used to advantage for gripping circular members under axial loading where it is desired to clamp down on the member.

As is generally true with composites, laminate configurations, fiber orientations, and layer thicknesses can be optimized for a particular application. That is, tubes with high, low, or negative Poisson's ratio can be configured. The tube aspect ratio has a more pronounced influence on Poisson's ratio, with only the results for $R_I/h = 75$ being identical to those for flat laminates in Fig. 5.12.

10.4.3 Effective Axial Shear Modulus

The *effective axial shear modulus*, $\bar{G}_{x\theta}$, of a laminated composite tube can be defined in much the same way as for isotropic tubes. Recalling the fundamental relationship for a homogeneous tube ($\Phi = TL/JG$), the relationship between angle of twist per unit length γ^0, the applied torque T_x, the polar moment of inertia J, the length L, the angle of twist Φ, and the effective axial shear modulus $\bar{G}_{x\theta}$ is

$$\gamma^0 = \frac{\Phi}{L} = \frac{T_x}{J\bar{G}_{x\theta}} \tag{10.94}$$

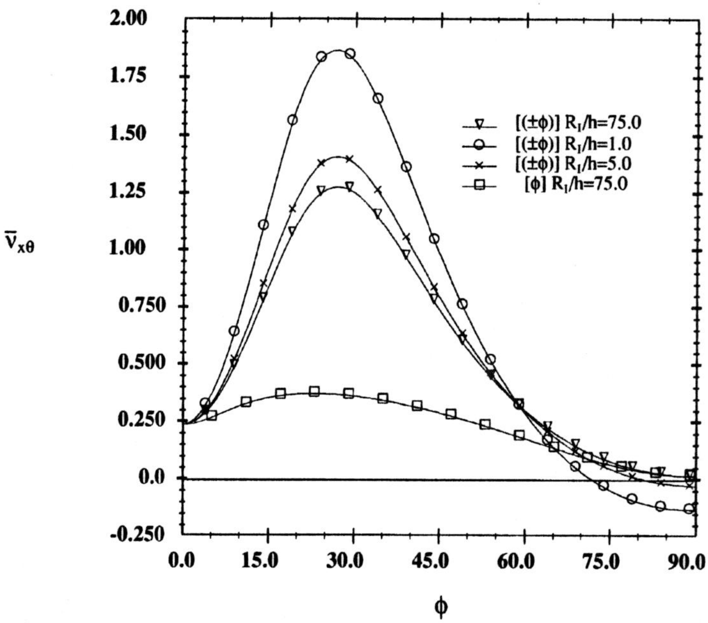

FIGURE 10.5 Effective Poisson's Ratios $\bar{\nu}_{x\theta}$ for Carbon/Epoxy Tubes

From the preceding equation, the effective shear modulus can be written

$$\bar{G}_{x\theta} = \frac{T_x}{J\gamma^0} \tag{10.95}$$

Typical results for the effective axial shear modulus of unidirectional off-axis and angle-ply T300/5208 tubes are shown in Fig. 10.6. As indicated in the figure, the effective shear modulus, at a given angle, is larger for the angle-ply tube, with both unidirectional and laminated tubes having their maximum modulus at a fiber orientation $\phi = 45°$. Comparison of the tube results in Fig. 10.6 with those for laminated plates in Fig. 5.13 indicates that the results are identical for large aspect ratios.

10.4.4 Effective Coefficients of Thermal Expansion

The *effective axial coefficient of thermal expansion*, $\bar{\alpha}_x$, is

$$\bar{\alpha}_x = \frac{\varepsilon_x^0}{\Delta T} \tag{10.96}$$

where ε_x^0 is determined from the system of simultaneous equations for the loading $\Delta T \neq 0$, $P_x = T_x = p_I = p_o = 0$.

The results in Fig. 10.7 for T300/5208 carbon/epoxy show that for all but the smallest aspect ratio $\bar{\alpha}_x$ is larger for unidirectional off-axis tubes for all fiber angles. The results for the angle-ply tubes are very interesting in that $\bar{\alpha}_x$ initially decreases further from its small negative value at $\theta = 0°$, attaining a maximum negative value at $\theta = 30°$ and then passing through 0.0 and coin-

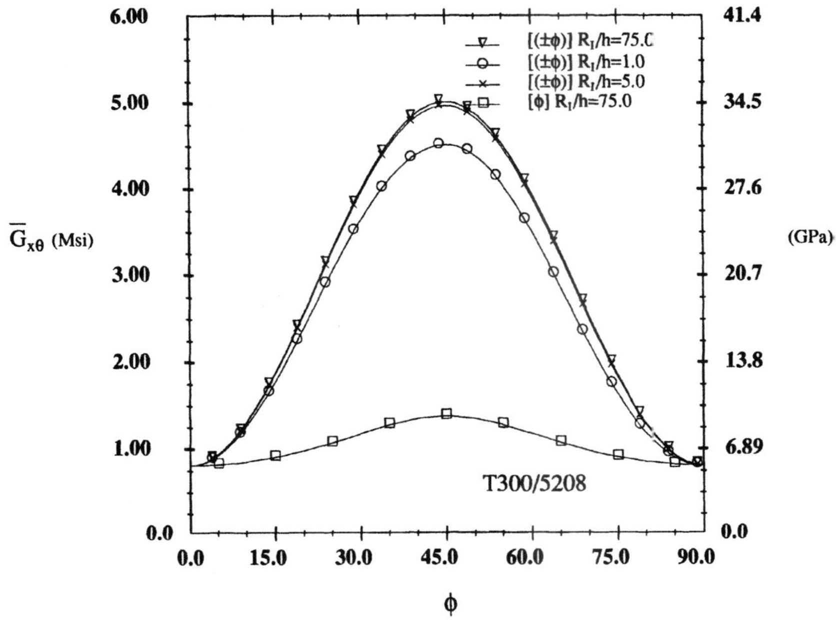

FIGURE 10.6 Effective Axial Shear Stiffness for Carbon/Epoxy Tubes

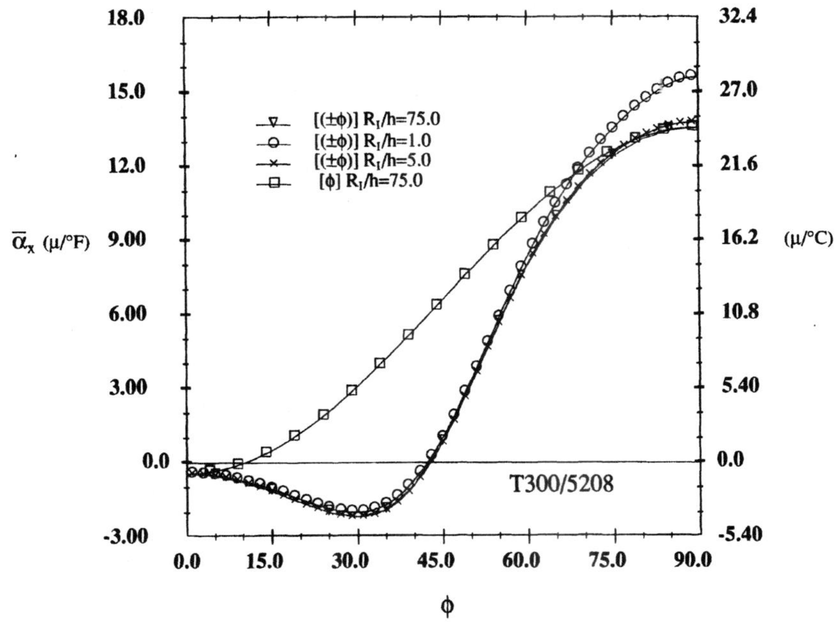

FIGURE 10.7 Axial Thermal Expansion Coefficients for Tubes

ciding with the unidirectional lamina at $\theta = 90°$ (except for very small aspect ratios). The effective axial CTEs for the large-aspect-ratio tubes are identical to those for flat laminates shown in Fig. 5.28. We also note that for large fiber angles the axial CTE of very small aspect ratio angle-ply tubes actually exceeds the transverse CTE α_2.

Possibly a more interesting property of tubes is the change in radius associated with thermal loading. We shall call this quantity the *effective radial coefficient of thermal expansion* (CTE), $\bar{\alpha}_r$, and define it as the change in radius per unit length of inner radius, R_I, per unit temperature change ΔT. Thus it has the dimensions of a coefficient of thermal expansion, i.e., strain per unit temperature change. If R_I^f is the final inner radius after the temperature change ΔT, the radial CTE, $\bar{\alpha}_r$, is

$$\bar{\alpha}_r = \frac{R_I^f - R_I}{R_I \Delta T} \tag{10.97}$$

The curves in Fig. 10.8 show the variation in radial coefficients of thermal expansion for unidirectional and angle-ply tubes. We note a most interesting fact when we compare the results in Fig. 10.8 for a tube to those in Fig. 5.28 for a flat laminate. The effective radial CTE of a large-aspect-ratio tube is identical to the effective transverse CTE, α_y, of a flat laminate, and, like the laminates, the effective radial CTE of a tube (Fig. 10.8) is 90° out of phase with the effective axial CTE of a tube (Fig. 10.7). There is, however, a major difference between the axial and radial CTE of angle-ply tubes as the aspect ratio is decreased. For small aspect ratios the radial CTE becomes more negative as the aspect ratio decreases, attaining large negative values at $\phi = 55°$ for an aspect ratio of 1.0; the radius exhibits large shrinkage for an increase in temperature for this case.

10.4.5 Axial-Rotational Coupling

Coupling effects between normal and shear response as well as coupling between in-plane and bending response were discussed in Chapters 4 and 5 in reference to flat laminae and laminates.

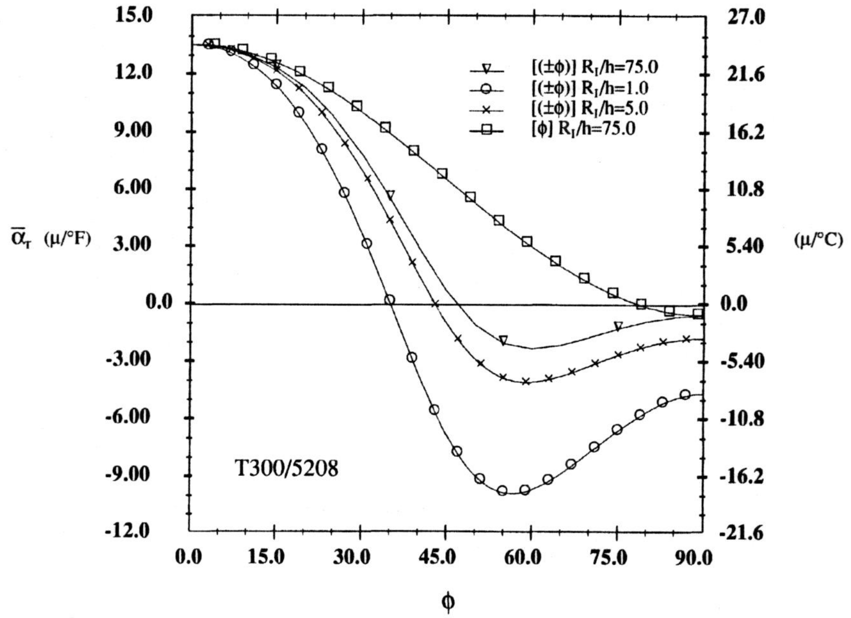

FIGURE 10.8 Radial Coefficients of Thermal Expansion for Tubes

It was shown in Chapter 5 that laminates with equal-thickness $+\theta$ and $-\theta$ layers do not exhibit coupling. The same cannot be said for tubes. Tubes exhibit coupling between axial and shear responses not found in flat laminates. The fundamental reason for this additional coupling in tubes is that the individual layers of a laminated tube are not at the same distance (radius) from the axis of the tube.

Specific examples showing the type and magnitude of coupling for a variety of tube and loading configurations are presented in Table 10.1 for a tube aspect ratio (ratio of inner radius, R_I, to wall thickness, h) of 300. As indicated in the table there is no axial-shear coupling for the orthogonal 0° and 90° tubes. Neither the unidirectional tubes nor the cross-ply tube exhibit coupling. Axial strain does not produce twist, and torsional loading does not produce axial strain. However, for all tubes with off-axis layers there is coupling between axial and shear response. This is true for axial, torsional, internal pressure, and thermal loading. It is noted that the degree of coupling for the angle-ply tubes is much less than that of the unidirectional off-axis tube. It follows that for a tube with a large number of $+\theta$ and $-\theta$ layers, an alternating stacking sequence of the $(\pm\theta)_n$ type will result in smaller coupling than a clustered stacking sequence of the $(\theta_n/-\theta_n)$ type. It is also noted that reversing the stacking sequence of the angle-ply tube also reverses the sign of the coupled response.

Another interesting result in Table 10.1 is noted from a comparison of the radial displacements w_I of the inside surface of the tubes for axial loading. Tubes of ±45 laminates exhibit large, negative radial displacements that are approximately three times as large as those in the unidirectional 0° and 45° tubes and many times larger than those in all other tubes. This is a direct result of the large Poisson's ratio for ±45 laminates. Thus the ±45 tube configuration can be used to provide a large clamping force for a tube under tensile loading.

In general, tube coupling coefficients ζ_{ij} can be defined as the ratio of two strains associated with a particular type of loading in the same manner that Poisson's ratio and coefficients of mutual influence are defined. Coefficients to quantify coupling effects in tubes can be defined in terms of the uniform axial strain, ε_x^0, and the shear strain, $\gamma^0 R_I$, on the inside surface of the tube. The cou-

Loading		Laminate					
		[0]	[90]	[45]	$[0/90]_s$	$[45/-45]_s$	$[-45/45]_s$
Axial $\varepsilon_x^0 = 0.1\%$	$\gamma^0 (\times 10^{-5})$	0.0	0.0	−2.077	0.0	-1.469×10^{-6}	$+1.469\times10^{-6}$
	$w_I (\times 10^{-3})$	−7.20	−0.56	−8.76	−1.063	−22.12	−22.12
Torque $\gamma^0 = 0.1\%$	$\varepsilon_x^0 (\times 10^{-2})$	0.0	0.0	−1.232	0.0	-2.341×10^{-6}	$+2.341\times10^{-6}$
	w_I	0.0	0.0	−0.371	0.0	-7.065×10^{-8}	$+7.065\times10^{-8}$
Internal pressure p_I = 10 psi	$\gamma^0 (\times 10^{-5})$	0.0	0.0	−2.945	0.0	-1.554×10^{-7}	$+1.554\times10^{-7}$
	$w_I (\times 10^{-2})$	5.79	0.470	4.26	0.87	3.167	3.167
	$\varepsilon_x^0 (\times 10^{-4})$	0.374	0.359	−4.127	−0.102	−7.756	−7.756
Thermal ΔT = 100°F	$\varepsilon_x^0 (\times 10^{-4})$	−0.43	13.6	6.5848		0.8319	0.8319
	$\gamma^0 (\times 10^{-5})$	0.0	0.0	−92.909	0.0	0.53118	−0.53118
	$w_I (\times 10^{-4})$	20.4	0.785	9.8067		1.06152	−1.06152

w_I = radial displacement in inches at the inside radius R_I

TABLE 10.1 Coupled Response of T300/5208 Tubes (R/h = 300)

pling coefficients are a function of material properties, fiber orientation and stacking sequence, and the tube geometry; they must be defined for each specific type of loading. We define coupling coefficients for axial force, torque, internal pressure, and thermal loading in the following paragraphs. Figures showing the variation of the coefficients with fiber orientation ϕ, stacking sequence, and tube aspect ratio (R_I/h) are presented and discussed. The figures show plots for two aspect ratios ($R_I/h = 5$ and $R_I/h = 75$) of unidirectional tubes and one aspect ratio ($R_I/h = 75$) for angle-ply tubes. All results are for T300/5208 carbon/epoxy tubes.

10.4.5.1 Axial Force Coupling Coefficient

The *axial force coupling coefficient*, $\zeta_{P\gamma}$, is defined as the ratio of the shear strain (at $r = R_I$) to axial strain for applied axial force. Thus,

$$\zeta_{P\gamma} = \frac{\gamma^0 R_I}{\varepsilon_x^0} \tag{10.98}$$

The curves in Fig. 10.9 show the variation of $\zeta_{P\gamma}$ as a function of fiber orientation. It is evident that coupling is essentially zero in the angle-ply tubes but attains high values in the unidirectional off-axis tubes. There is only a small dependence on R_I/h. The maximum coupling in the unidirectional tubes occurs at a fiber orientation of 12° (for the material considered), where the shear strain is approximately twice the axial strain.

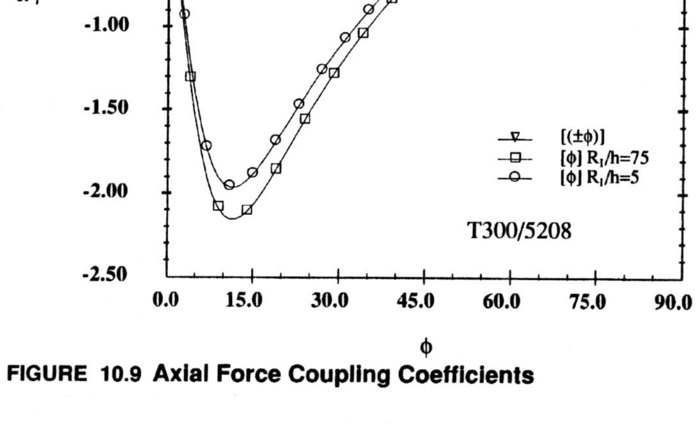

FIGURE 10.9 Axial Force Coupling Coefficients

10.4.5.2 Torque Coupling Coefficient

The *torque coupling coefficient,* $\zeta_{T\varepsilon}$, is defined as the ratio of the axial strain to shear strain (at $r = R_I$) for pure torque loading. Thus,

$$\zeta_{T\varepsilon} = \frac{\varepsilon_x^0}{\gamma^0 R_I} \tag{10.99}$$

As indicated in Fig. 10.10, the coupling in angle-ply tubes is again essentially zero, but the unidirectional off-axis tubes exhibit significant coupling, with the maximum at 37° where the axial strain is approximately 50% of the shear strain. Tube aspect ratio has a small effect on the values.

10.4.5.3 Thermal Coupling Coefficient

We define the *thermal coupling coefficient,* $\zeta_{\Delta T}$, as the ratio of shear strain (at $r = R_I$) to axial strain for uniform thermal loading. Thus,

$$\zeta_{\Delta T} = \frac{\gamma^0 R_I}{\varepsilon_x^0} \tag{10.100}$$

As indicated in Fig. 10.11, the thermal coupling coefficient exhibits quite different behavior from the two previous coefficients. The angle-ply tubes again exhibit essentially no coupling; however, the unidirectional off-axis tubes exhibit much larger coupling with a critical fiber orientation of 10°, where the axial strain passes through zero as the axial expansion changes from negative to positive, with a resulting singularity in $\zeta_{\Delta T}$. This effect is a direct result of the fact that the coefficient of thermal expansion for the carbon/epoxy is negative for 0° fiber orientation and then increases to positive values as the fiber orientation increases up to 90°. There is also much less dependence on the aspect ratio for unidirectional tubes.

10.4.5.4 Internal Pressure Coupling Coefficient

The *internal pressure coupling coefficient,* ζ_{PI}, is defined as the ratio of shear strain (at $r = R_I$) to axial strain for uniform internal pressure loading. Thus,

$$\zeta_{PI} = \frac{\gamma^0 R_I}{\varepsilon_x^0} \tag{10.101}$$

As the results in Fig. 10.12 indicate, the behavior of this coefficient is similar to the thermal coupling coefficient in some aspects, but quite different in others. Like the thermal coupling coefficient, ζ_{PI} exhibits a singular point. However, for this coefficient the singularity occurs approximately at a fiber orientation of 80° and is present only for the smaller-aspect-ratio tube. The singularity occurs because the axial strain passes through zero as it changes sign with an increase in the fiber orientation. For the angle-ply tubes, the coefficient is essentially zero.

10.4.5.5 Coupling Coefficient Summary

As indicated in the previous paragraphs, axial-torsional coupling is present whenever a laminated tube has one or more off-axis layers. The coupling is much larger for unidirectional off-axis tubes but is present even for angle-ply tubes. While the coupling may be small for laminated tubes, it can be important, particularly if fatigue loading is involved.

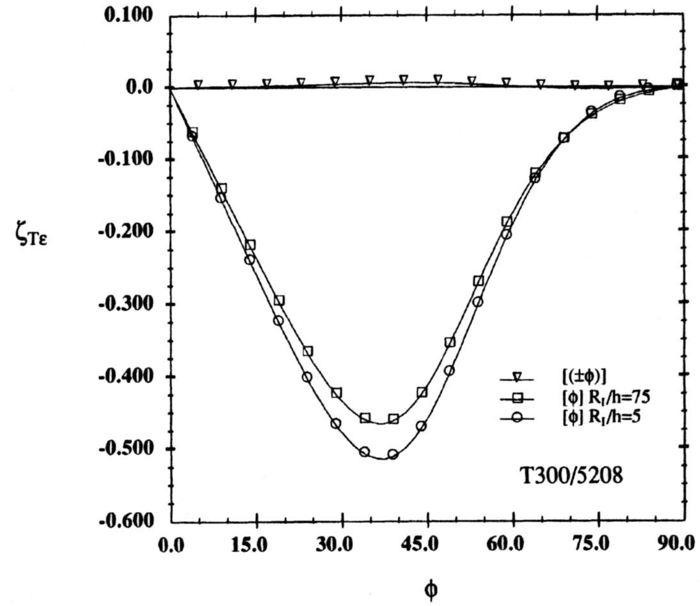

FIGURE 10.10 Torque Coupling Coefficients

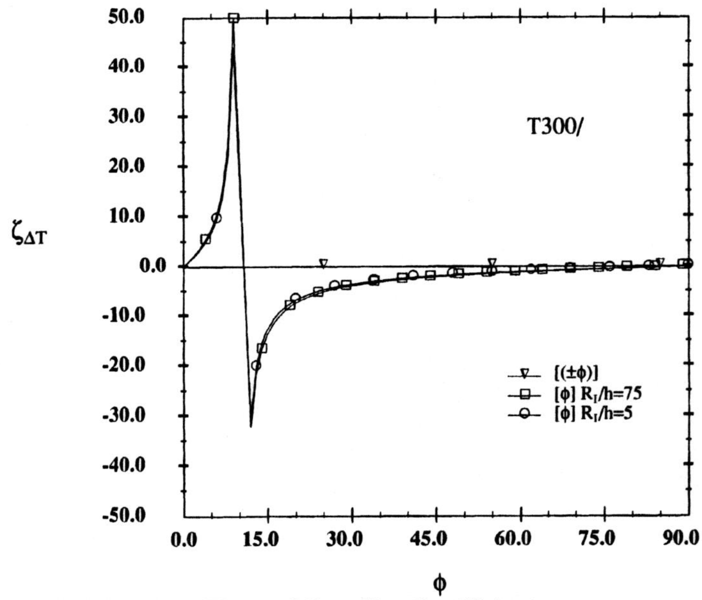

FIGURE 10.11 Thermal Coupling Coefficients

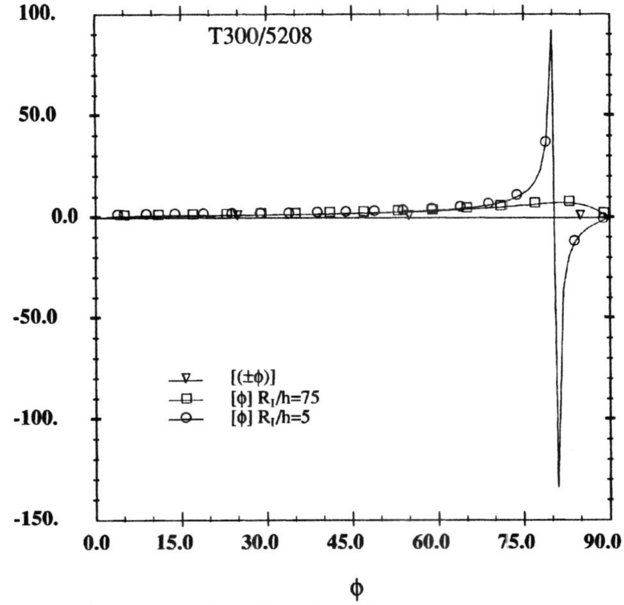

FIGURE 10.12 Pressure Coupling Coefficients

10.5 Stress Distributions

It is instructive to assess the through-thickness distribution of the stresses in tubes, as stresses are of fundamental importance in the prediction of failure. It is desired to demonstrate the linearity or nonlinearity of the distributions as well as the influence of load type and specimen aspect ratio. A [+45/–45] T300/5208 carbon/epoxy tube has been chosen to demonstrate the various influences. Through-thickness stress distributions are presented in Figs. 10.13 through 10.28 for four different loading conditions: axial tensile load P_x, torsion T_x, uniform temperature change ΔT, and internal pressure p_I. The +45° layer is on the inside of the tube. Results are presented for two different tube aspect ratios to demonstrate "thickness" effects. The large aspect ratio ($R_I/h = 75$) corresponds to a thin tube, and the small aspect ratio ($R_I/h = 5$) corresponds to a "thick" tube. It is noted that the distribution of stresses in thin tubes approaches lamination theory stresses for a symmetric laminate subjected to the equivalent loading.

The results presented here are normalized to more clearly demonstrate the dependence on the radial coordinate and the tube aspect ratio. The radial coordinate, r, is normalized as $(r - R_I)/h$. The stresses are normalized by the following stress quantities:

- Axial loading: average axial stress σ_o, where

$$\sigma_o = \frac{P_x}{\pi(R_o^2 - R_I^2)} \tag{10.102}$$

- Torsion: average laminate shear stress τ_o, where

$$\tau_o = \frac{3T_x}{2\pi(R_o^3 - R_I^3)} \tag{10.103}$$

- Internal pressure: applied internal pressure p_I
- Temperature change: equivalent thermal stress $\sigma_T = \bar{E}_x\bar{\alpha}_x\Delta T$, where σ_T is the magnitude of the average axial stress in a tube subjected to a temperature change ΔT when the ends of the tube are constrained against axial displacement, but free to displace radially, and $\bar{E}_x$ and $\bar{\alpha}_x$ are the tube effective axial modulus and effective coefficient of axial thermal expansion, respectively

The figures are grouped according to the component of stress under consideration. Figures 10.13 to 10.16 are for axial stress σ_x. Figures 10.17 to 10.20 are for hoop stress σ_θ. Figures 10.21 to 10.24 are for radial stress σ_r, and Figs. 10.25 to 10.28 are for shear stress $\tau_{x\theta}$.

While these figures present typical stress distributions in laminated tubes, it is emphasized that the results are for the specific case analyzed. Stress distributions in laminated tubes are a function of the material properties, fiber orientations, stacking sequence, layer thicknesses, and tube aspect ratio.

As indicated in the figures, the dependence of the stress magnitude on the tube aspect ratio varies depending upon the component of stress and the type of loading. In some cases, the stress magnitude is independent of the tube aspect ratio, whereas in other cases it varies with the aspect ratio to the 1, −1, or −2 power. The influence of tube aspect ratio is summarized in Table 10.2 for all four loading cases and all four nonzero components of stress.

10.5.1 Axial Stresses σ_x

The axial stress distributions are slightly nonlinear through the thickness of each layer and exhibit discontinuities at the layer interfaces. The distributions approach linearity in each layer with increase in aspect ratio. For axial (Fig. 10.13) and torsional (Fig. 10.14) loadings the axial stress also approaches uniformity within each layer for the larger tube aspect ratio. For thermal (Fig. 10.15) and internal pressure (Fig. 10.16) loadings, the axial stress exhibits a sign change within each layer. In all but the internal pressure loading case, the maximum stress occurs at the ±45 interface. For internal pressure loading the maximum axial stress occurs at the inside surface of the tube.

Stress	P_x	T_x	p_I	ΔT
σ_x	1	$(R_I/h)^{-1}$	1	$(R_I/h)^{-1}$
σ_θ	$(R_I/h)^{-1}$	1	$(R_I/h)^1$	$(R_I/h)^{-1}$
σ_r	$(R_I/h)^{-2}$	$(R_I/h)^{-1}$	1	$(R_I/h)^{-2}$
$\tau_x\theta$	1	1	$(R_I/h)^1$	1

TABLE 10.2 Dependence of Stress Components on Tube Aspect Ratio

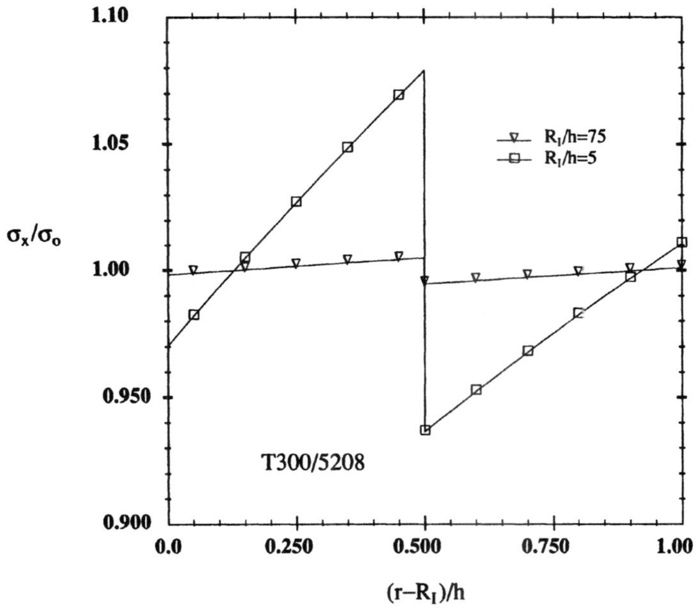

FIGURE 10.13 Axial Stresses in [+45/–45] Tubes: Axial Loading

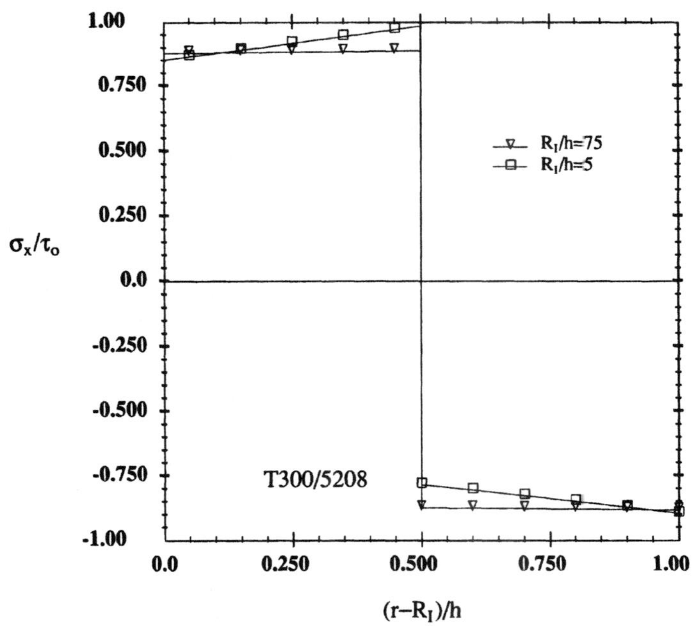

FIGURE 10.14 Axial Stresses in [+45/–45] Tubes: Torque Loading

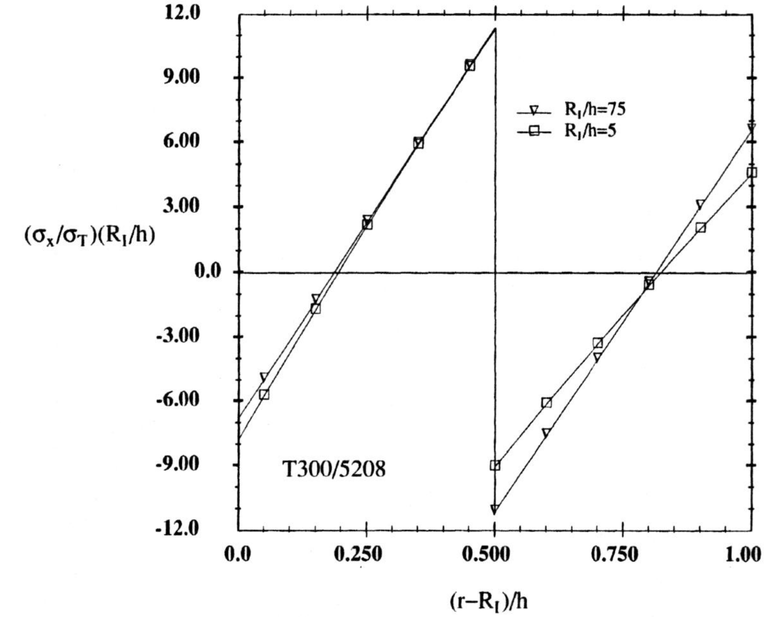

FIGURE 10.15 Axial Stresses in [+45/–45] Tubes: Thermal Loading

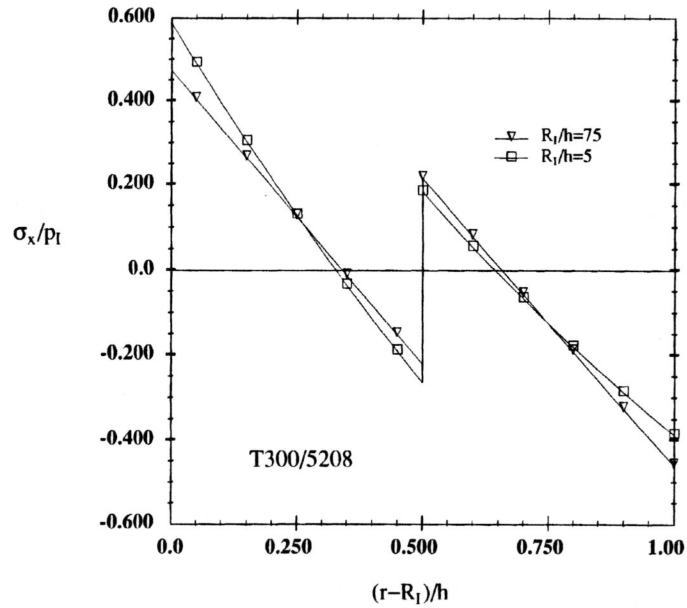

FIGURE 10.16 Axial Stresses in [+45/–45] Tubes: Internal Pressure

10.5.2 Hoop Stresses σ_θ

The hoop stress distributions are similar to those for the axial stress, being nearly linear in each layer and becoming more linear with increasing aspect ratio. Discontinuities are present at the layer interfaces for all four loading cases. There is a sign change through the thickness of each layer for axial and thermal loadings. The maximum hoop stress occurs at the ±45 interface for axial (Fig. 10.17) and torque (Fig. 10.18) loadings. For thermal (Fig. 10.19) and internal pressure (Fig. 10.20) loading, the maximum hoop stress occurs at the inside surface of the tube.

10.5.3 Radial Stresses σ_r

The radial stress distributions are highly nonlinear through the thickness of individual layers for axial (Fig. 10.21) and thermal (Fig. 10.23) loadings. The distributions are linear for torque loading (Fig. 10.22) and approach linearity with increasing aspect ratio for internal pressure loading (Fig. 10.24). The location of the maximum magnitude of the radial stress varies considerably with the type of loading. For axial and thermal loading the maximum radial stress is at midlayer of the inner layer, whereas for torque loading it is at the ±45 interface.

10.5.4 Shear Stresses $\tau_{x\theta}$

The shear stress distributions (Figs. 10.25 to 10.28) are similar to those for the axial and hoop stresses, exhibiting nearly linear distributions within individual layers and discontinuities at the layer interface, with the discontinuities being larger for the smaller-aspect-ratio tubes. The location of the maximum shear stress varies with the type of loading but is never internal to a layer.

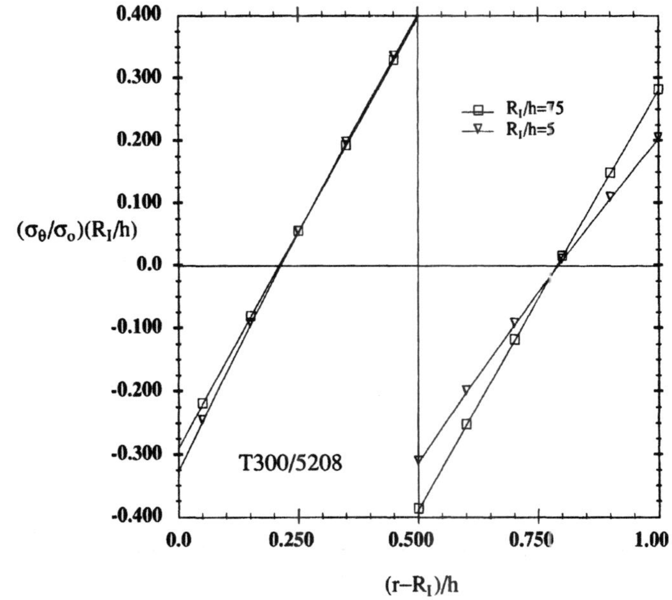

FIGURE 10.17 Hoop Stresses in [+45/–45] Tubes: Axial Loading

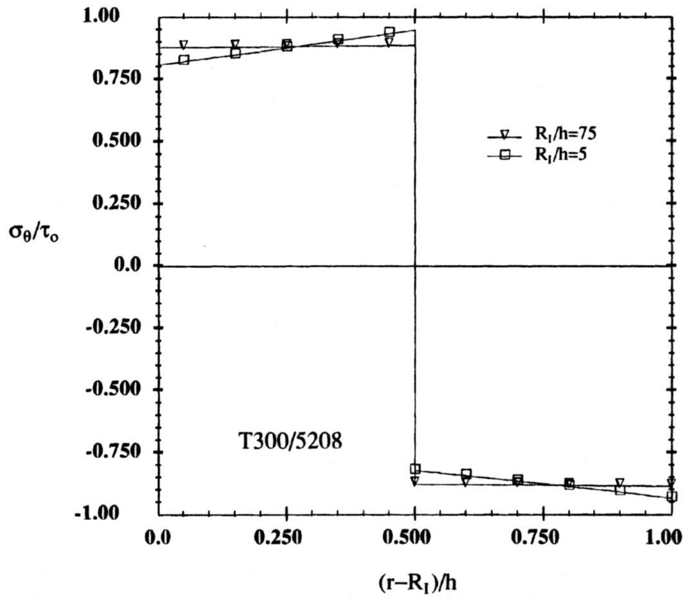

FIGURE 10.18 Hoop Stresses in [+45/−45] Tubes: Torque Loading

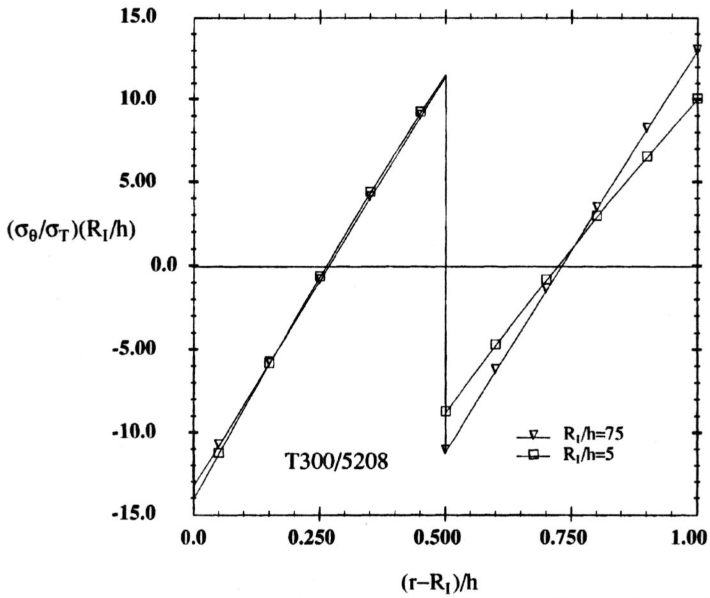

FIGURE 10.19 Hoop Stresses in [+45/−45] Tubes: Thermal Loading

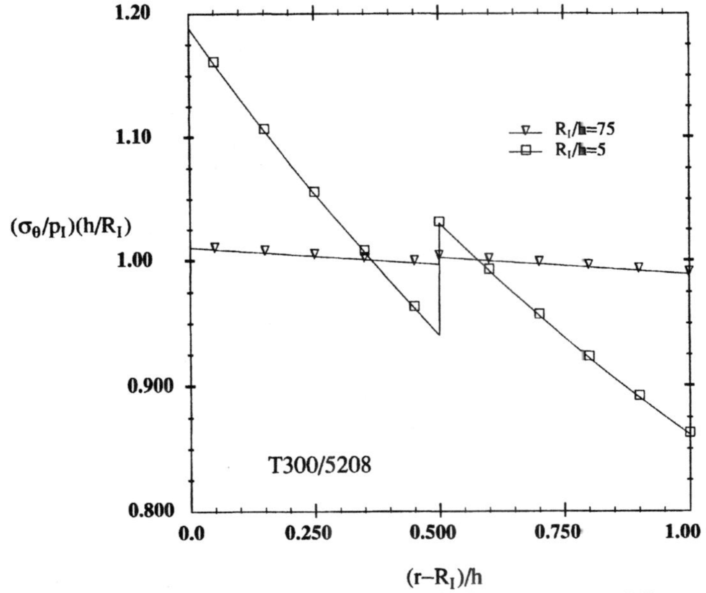

FIGURE 10.20 Hoop Stresses in [+45/–45] Tubes: Internal Pressure

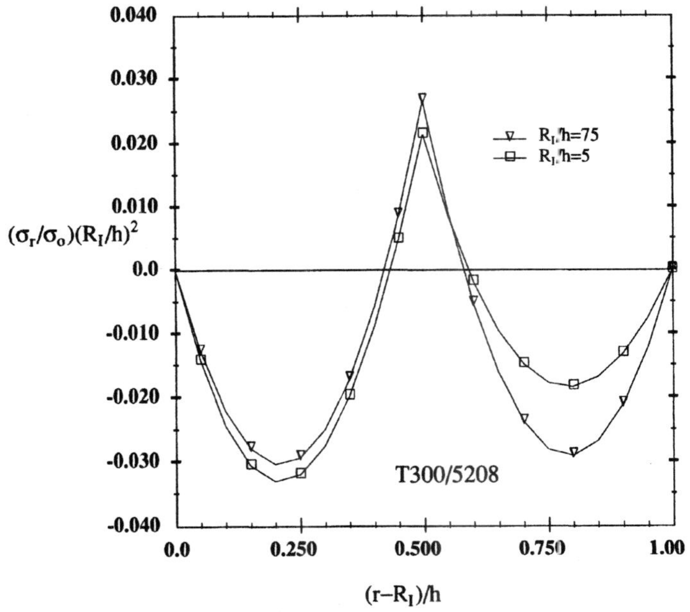

FIGURE 10.21 Radial Stresses in [+45/–45] Tubes: Axial Loading

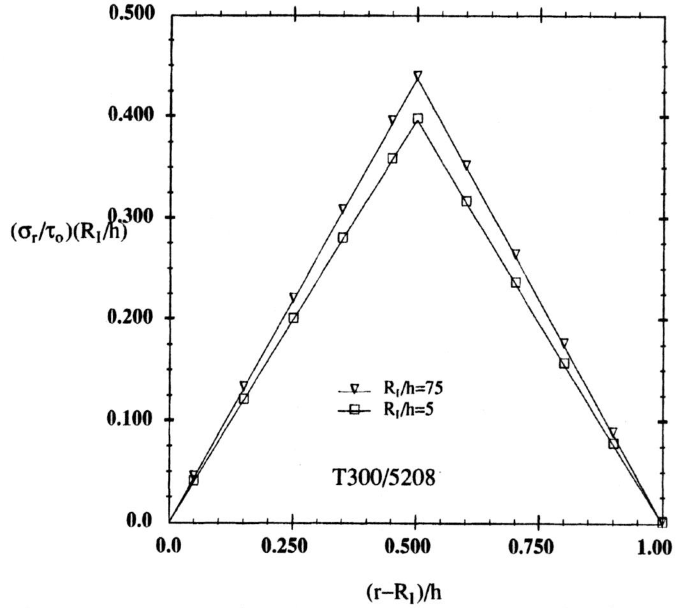

FIGURE 10.22 Radial Stresses in [+45/–45] Tubes: Torque Loading

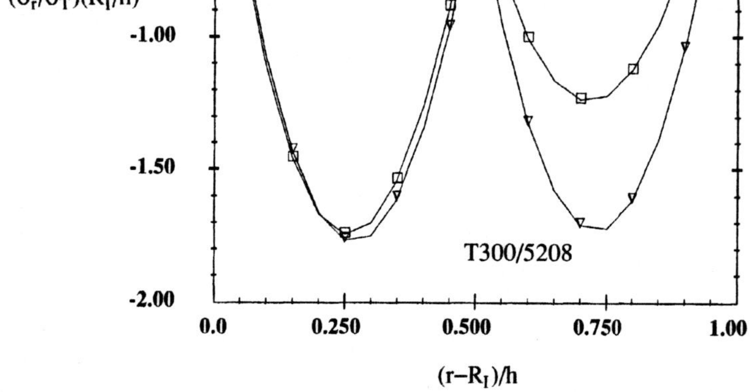

FIGURE 10.23 Radial Stresses in [+45/–45] Tubes: Thermal Loading

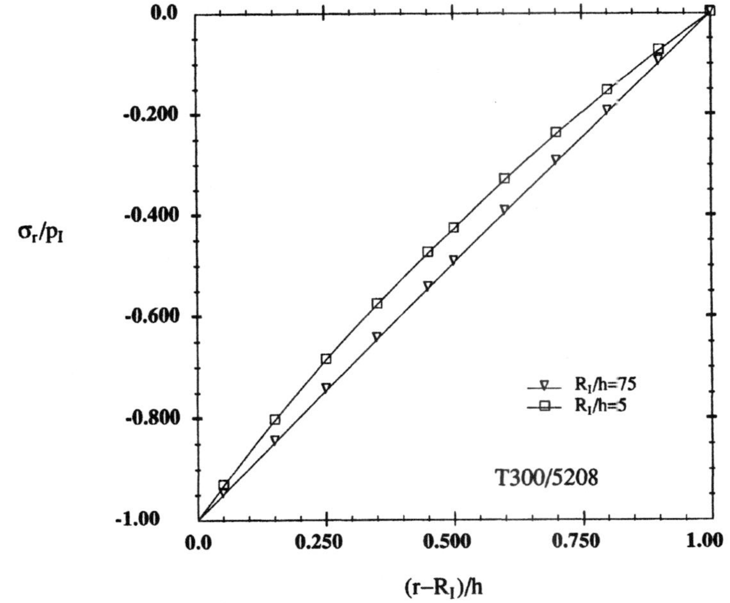

FIGURE 10.24 Radial Stresses in [+45/−45] Tubes: Internal Pressure

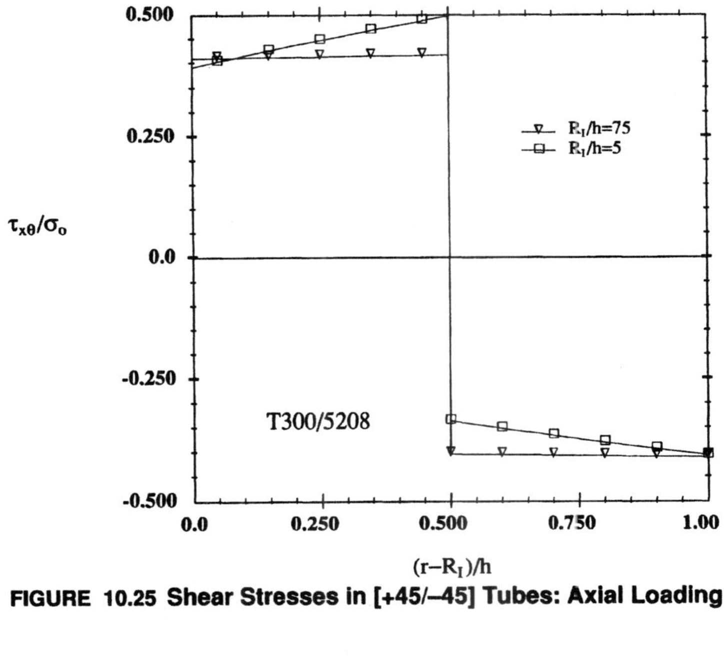

FIGURE 10.25 Shear Stresses in [+45/−45] Tubes: Axial Loading

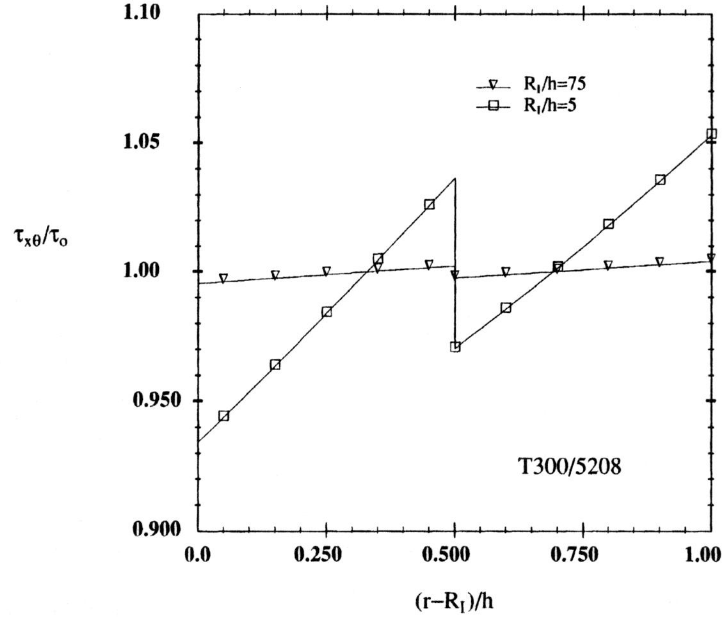

FIGURE 10.26 Shear Stresses in [+45/–45] Tubes: Torque Loading

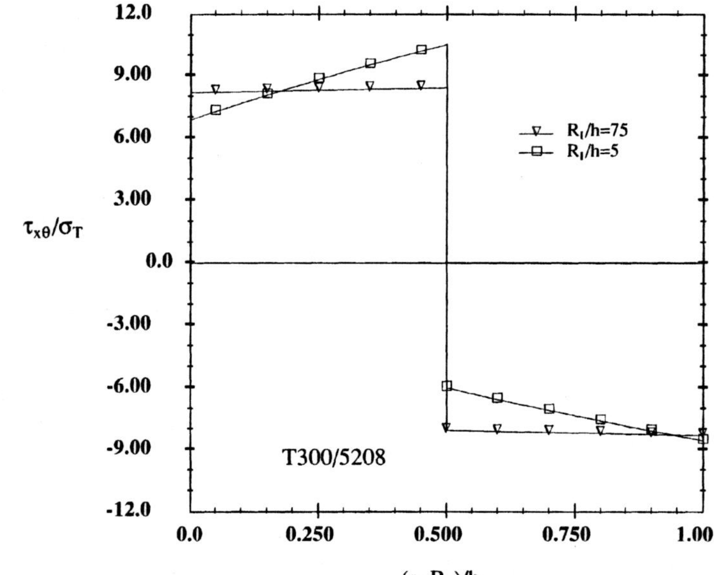

FIGURE 10.27 Shear Stresses in [+45/–45] Tubes: Thermal Loading

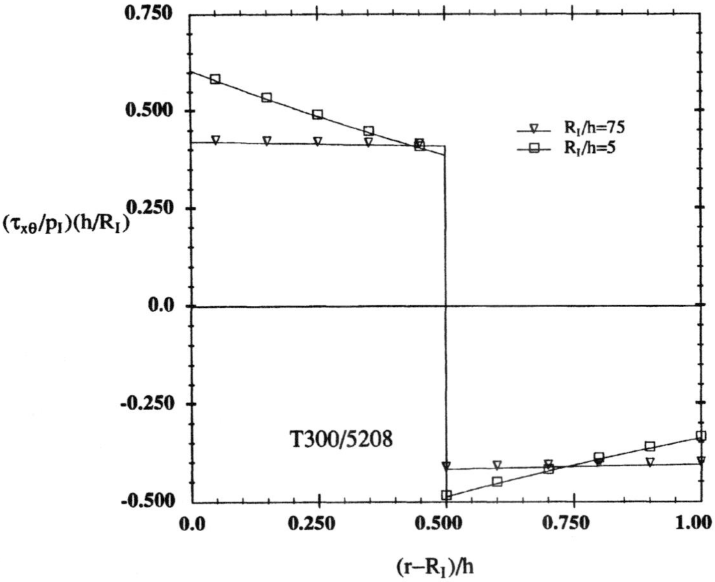

FIGURE 10.28 Shear Stresses in [+45/–45] Tubes: Internal Pressure

10.6 Nonlinear Response and Dynamic Loading

Several studies of laminated tubes have been conducted that were concerned with effects other than linear elasticity and static loading. The nonlinear, thermomechanical response of laminated tubes was treated by Derstine and Pindera (1989). The dynamic response of pulse pressure–loaded laminated tubes was considered by Larom et al. (1991).

10.7 Solid Cylinders

The preceding solution can be applied to solid cylinders if the traction boundary condition on the inside surface is replaced with the requirement that the radial displacement, w, be zero at $r = 0$. From (10.78) we have that $A_2^{(1)} = 0$ in order to eliminate the (otherwise) singular displacement at $r = 0$. The remainder of the solution follows in the same manner as described for the hollow tube.

The solution for layered, solid cylinders was first applied to the micromechanics problem of a fiber embedded in a matrix to predict the effective properties of a composite (Hashin and Rosen, 1964). Other applications of the solid, laminated cylinder solution at the micromechanics level for the prediction of stress distributions in and effective properties of fibrous composites have been presented by Avery and Herakovich (1986), Knott and Herakovich (1988, 1991), Hashin (1990), and Pindera et al. (1993).

10.8 Summary

The analytical solution for an infinite-length laminated cylindrical tube subjected to combined axial, torsional, pressure, and thermal loading has been presented for the case of uniform loading along the length. Effective axial and shear moduli, coefficients of thermal expansion, and coupling coefficients have been presented for unidirectional off-axis laminae and angle-ply laminates. The chapter closes with through-thickness stress distributions for ±45° angle-ply tubes as a function of the tube aspect ratio and the type of loading.

References

Avery, W. B., and Herakovich, C. T. (1986), "Effect of Fiber Anisotropy on Thermal Stresses in Fibrous Composites," *J. Appl. Mech.*, vol. 53, pp. 751–756.

Derstine, M. S., and Pindera, M. -J. (1989), "Nonlinear Response of Composite Tubes under Combined Thermomechanical Loading," in *Composites and Other New Materials for PVP: Design and Analysis Considerations*, PVP vol. 174, American Society of Mechanical Engineers, New York.

Fung, Y. C. (1965), *Foundations of Solid Mechanics*, Prentice-Hall, Englewood Cliffs, NJ, p. 114.

Hashin, Z. (1990), "Thermoelastic Properties and Conductivity of Carbon/Carbon Fiber Composites," *Mech. Mater.*, vol. 8, pp. 293–308.

Hashin, Z., and Rosen, B. W. (1964), "The Elastic Moduli of Fiber-Reinforced Materials," *J. Appl. Mech.*, vol. 31, pp. 223–232.

Hyer, M. W., and Cooper, D. E. (1986), "Stresses and Deformations in Composite Tubes Due to a Circumferential Temperature Gradient," *J. Appl. Mech.*, vol. 53, pp. 757–764.

Hyer, M. W., Cooper, D. E., and Cohen, D. (1986), "Stresses and Deformations in Cross-Ply Tubes Subjected to a Uniform Temperature Change," *J. Therm. Stresses*, vol. 9, pp. 97–117.

Hyer, M. W., and Rousseau, C. Q. (1987), "Thermally Induced Stresses and Deformations in Angle-Ply Composite Tubes," *J. Comp. Mater.*, vol. 21, no. 5, pp. 454–480.

Knott, T. W., and Herakovich, C. T. (1988), *Effect of Fiber Morphology on Composite Properties*, CCMS-88-09 and VPI-E-88-16, Virginia Tech, Blacksburg.

Knott, T. W., and Herakovich, C. T. (1991), "Effect of Fiber Orthotropy on Effective Composite Properties," *J. Compos. Mater.*, vol. 25, no. 6, pp. 732–759.

Larom, D., Herakovich, C. T., and Aboudi, J. (1991), "Dynamic Response of Pulse Loaded Laminated Composite Cylinders," *Int. J. Impact Eng.*, vol. 11, no. 2., pp. 233–248.

Lekhnitskii, S. G. (1950), *Teoriia Uprugosti Anisotropnovo Tela*, Government Publishing House for Technical-Theoretical Works, Moscow; translated as *Theory of Elasticity of an Anisotropic Elastic Body*, Holden-Day, Inc., San Francisco (1963), pp. 243–273.

Orgill, G., and Wilson, J. F. (1986), "Finite Deformations of Nonlinear, Orthotropic Cylindrical Shells," *J. Appl. Mech.*, vol. 53, pp. 257–265.

Pagano, N. J. (1971), "Stress Gradients in Laminated Composite Cylinders," *J. Compos. Mater.*, vol. 5, pp. 260–265.

Pindera, M. -J. (1991), "Local/Global Stiffness Matrix Formulation for Composite Materials and Structures," *Compos. Eng.*, vol. 1, no. 2, pp. 69–83.

Pindera, M. -J., Freed, A. D., and Arnold, S. M. (1993), "Effects of Fiber and Interfacial Layer Morphologies on the Thermoplastic Response of Metal Matrix Composites," *Int. J. Solids Struct.*, vol. 30, no. 9, pp. 1213–1238.

Reissner, E. (1970), "On Uniform Stress and Strain in Axially Homogeneous Cylindrical Shells," *Int. J. Solids Struct.*, vol. 6, pp. 133–138.

Reissner, E., and Tsai, W. T. (1974), "Pure Bending, Stretching, and Twisting of Anisotropic Cylindrical Shells," *J. Appl. Mech.*, vol. 4, pp. 168–174.

Scherrer, R. E. (1967), "Filament-Wound Cylinders with Axial-Symmetric Loads," *J. Compos. Mater.*, vol. 1, pp. 344–355.

Wilson, J. F., and Orgill, G. (1986), "Linear Analysis of Uniformly Stressed, Orthotropic Cylindrical Shells," *J. Appl. Mech.*, vol. 53, pp. 249–256.

Exercises

10.1 Determine expressions for the stresses in an orthotropic layer of a tube.

10.2 Show that $\Gamma = \Omega = \Psi = 0$ for a transversely isotropic layer.

10.3 Determine expressions for the stresses in a transversely isotropic layer of a tube.

CHAPTER 11 MICROMECHANICS

"The end is where we start from."
T. S. Eliot, "Little Gidding," 1943

11.1 Introduction

Fibrous composites are heterogeneous materials consisting of two or more phases. The two primary phases are the fiber and the matrix. Additional phases may include coatings applied to the fiber or reaction product zones that develop during the fabrication process (Fig. 11.1).

The study of composites at the fiber and matrix level is referred to as *micromechanics.* It is desired to predict the overall effective (or average) thermo-elastic properties and inelastic response of a composite based upon the properties, arrangement, and volume fraction of the phases. A wide variety of methods are now available for predicting effective thermo-elastic properties of composites. The earliest works are those of Paul (1960), Hashin (1962), Hill (1964), Hashin and Rosen (1964), and the monographs by Hashin (1972) and Achenbach (1975). More recently, the book by Aboudi (1991) surveyed the methods available for determining the effective elastic properties and discussed the application of cell methods for elastic and inelastic response.

11.2 Effective Properties

As indicated in Chapter 3, a unidirectional, continuous fiber composite generally is *orthotropic* with *nine independent elastic constants.* However, for a unidirectional composite which exhibits *isotropic* properties in a plane transverse to the fibers (same properties in all directions in the x_2-x_3 plane, Fig. 11.2), the effective response is *transversely isotropic.* In this case there are only five independent elastic constants. Layers of unidirectional composites with a large number of fibers through the layer thickness generally are considered to be transversely isotropic.

When the full tensor notation is used for the stresses, σ_{ij}, and the strains, ε_{ij}, the average or effective constitutive equations for a transversely isotropic material have the form

$$\begin{Bmatrix} \sigma_{11} \\ \sigma_{22} \\ \sigma_{33} \\ \sigma_{23} \\ \sigma_{31} \\ \sigma_{12} \end{Bmatrix} = \begin{bmatrix} C_{11}^* & C_{12}^* & C_{12}^* & 0 & 0 & 0 \\ C_{12}^* & C_{22}^* & C_{23}^* & 0 & 0 & 0 \\ C_{12}^* & C_{23}^* & C_{22}^* & 0 & 0 & 0 \\ 0 & 0 & 0 & \dfrac{C_{22}^* - C_{23}^*}{2} & 0 & 0 \\ 0 & 0 & 0 & 0 & C_{66}^* & 0 \\ 0 & 0 & 0 & 0 & 0 & C_{66}^* \end{bmatrix} \begin{Bmatrix} \varepsilon_{11} \\ \varepsilon_{22} \\ \varepsilon_{33} \\ 2\varepsilon_{23} \\ 2\varepsilon_{31} \\ 2\varepsilon_{12} \end{Bmatrix} \tag{11.1}$$

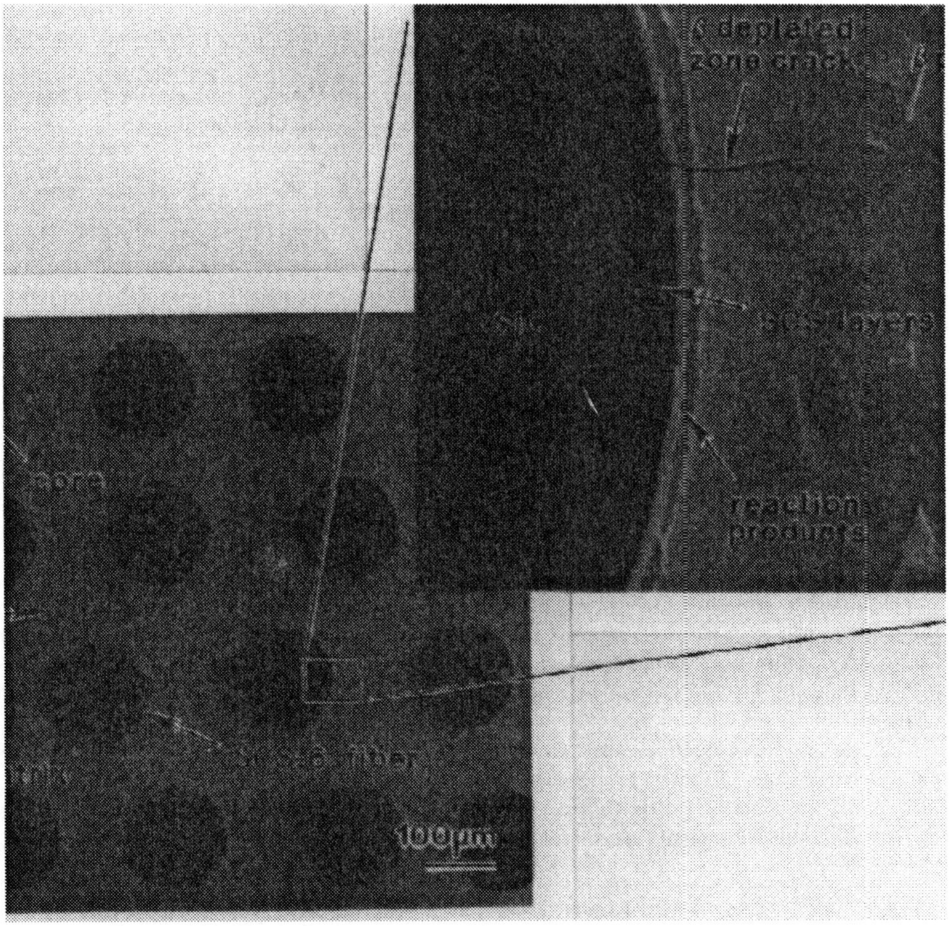

FIGURE 11.1 SCS-6 Silicon-Carbide Fiber in Titanium Matrix
(Photo courtesy of D. Sypeck and H. N. G. Wadley, Univ. of Virginia)

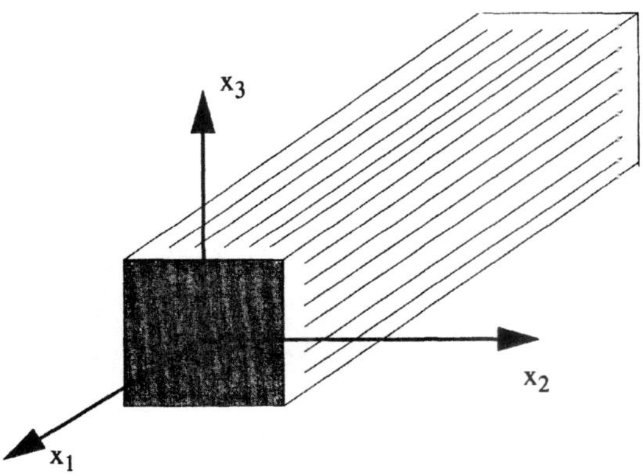

FIGURE 11.2 Transversely Isotropic Composite

where the five elastic coefficients C^*_{11}, C^*_{22}, C^*_{12}, C^*_{23}, and C^*_{66} are the *effective stiffness coefficients of the equivalent homogeneous material.* They can be expressed in terms of the effective engineering properties as indicated in (3.39). The goal of micromechanics, as far as elastic response is concerned, is to determine the *effective (or average) stiffness coefficients,* C^*_{ij}, in terms of the fiber and matrix properties, the fiber volume fraction, and the arrangement of fibers in the matrix.

11.3 Strength of Materials Approximations

Approximate expressions for four of the five properties of a transversely isotropic composite can be developed using simplified strength of materials approaches. The strength of materials approaches involve assumptions which do not necessarily satisfy the requirements of an exact elasticity solution; thus the resulting equations are only approximate. We use the reduced notation for stresses and strains in this section.

We develop approximate expressions for the four effective properties: axial (fiber direction) modulus E^*_1, transverse (perpendicular to the fibers) modulus $E^*_2 = E^*_3$, axial Poisson's ratio $\nu^*_{12} = \nu^*_{13}$ (loading in the x_1-direction), and axial shear modulus $G^*_{12} = G^*_{13}$ (shear stress parallel to the fibers). The representative volume element under consideration consists of a uniform arrangement of straight, continuous fibers as depicted in Fig. 11.2.

We assume that the fibers and matrix are perfectly bonded with no slip. The effective properties will be expressed in terms of the elastic properties and volume fractions of the fiber and matrix. In the following developments, fiber and matrix properties are denoted by the subscripts f and m, respectively. The reader is reminded that although curves in the following section present theoretical predictions for fiber volume fraction up to 1.0, there is a physical limit to the fiber volume fraction. As shown in Exercise 1.5, the maximum fiber volume for circular fibers in a hexagonal array is 0.907.

All results in this section on strength of materials approximations are based upon the constituent properties given in Table 11.1. Predictions are presented for polymeric (glass/epoxy and carbon/epoxy) and metallic matrix (boron/aluminum) composites. The carbon fiber is taken to be transversely isotropic, and all other materials are assumed to be isotropic.

11.3.1 Effective Axial Modulus

Consider a transversely isotropic fibrous composite under uniform axial strain loading $\bar{\varepsilon}_1$. We *assume* that the axial strain in the composite is uniform such that the axial strains in the fibers and the matrix are identical, i.e.,

$$\bar{\varepsilon}_1 = \varepsilon_f = \varepsilon_m \tag{11.2}$$

	Boron	Carbon	Glass	Aluminum	Epoxy
E^f_A, GPa (Msi)	379 (55.0)	227 (33.0)	86 (12.5)	69 (10.0)	3.4 (0.5)
E^f_T, GPa (Msi)	379 (55.0)	15 (2.2)	86 (12.5)	69 (10.0)	3.4 (0.5)
ν^f_A	0.21	0.28	0.20	0.33	0.35
G^f_A, GPa (Msi)	165 (22.7)	24 (3.5)	34 (5.0)	26 (3.76)	1.28 (0.185)

TABLE 11.1 Constituent Properties for Strength of Materials Predictions

Let E_f be the fiber modulus, E_m the matrix modulus, V_f the fiber volume fraction, and V_m the matrix volume fraction. Axial force equilibrium of the composite requires that the average axial composite stress $\bar{\sigma}_1$ be related to the axial stress in the fiber and matrix through the respective area quantities ($A = A_f + A_m$), i.e.,

$$\bar{\sigma}_1 A = \sigma_f A_f + \sigma_m A_m \tag{11.3}$$

Thus, using (11.2), the *effective axial modulus* E_1^* is

$$E_1^* = \frac{\bar{\sigma}_1}{\bar{\varepsilon}_1} = \frac{\sigma_f}{\varepsilon_f} \cdot \frac{A_f}{A} + \frac{\sigma_m}{\varepsilon_m} \cdot \frac{A_m}{A} \tag{11.4}$$

Invoking Hooke's law for the fiber and matrix ($\sigma = E\varepsilon$) and recognizing that the area and volume relationships are equivalent, we have

$$E_1^* = E_f \cdot V_f + E_m \cdot V_m \tag{11.5}$$

This type of equation is known as a *rule of mixtures*. It is a fundamental relationship expressing the effective axial modulus of a unidirectional fibrous composite in terms of the modulus and the volume fraction of each constituent.

Experimental evidence indicates that the rule of mixtures provides very accurate prediction of the effective axial modulus over a wide rage of fiber volume fractions. Using the total volume relationship $V_f + V_m = 1$, the rule of mixtures may also be written in the form

$$E_1^* = (E_f - E_m)V_f + E_m \tag{11.6}$$

This equation clearly shows that the effective axial modulus is a linear function of the fiber volume fraction. Thus, the fiber volume fraction should always be quoted in reporting the effective properties of a composite.

Rule of mixtures predictions for the axial modulus of three different composite materials are presented in Fig. 11.3. The materials considered are carbon/epoxy (C/Ep), boron/aluminum (B/Al) and glass/epoxy (Gl/Ep). This figure shows the linear variation with fiber volume fraction, the significantly higher axial modulus of the metal matrix composite, and the lower modulus of the glass/epoxy.

11.3.2 Effective Axial Poisson's Ratio

The effective axial Poisson's ratio ν_{12}^* is defined as the negative ratio of lateral strain $\bar{\varepsilon}_2$ to applied axial strain $\bar{\varepsilon}_1$:

$$\nu_{12}^* = \frac{-\bar{\varepsilon}_2}{\bar{\varepsilon}_1} \tag{11.7}$$

for a stress state $\bar{\sigma}_1 \neq 0$, $\bar{\sigma}_2 = \bar{\sigma}_{12} = 0$, and all other stresses identically zero.

For the dimensions shown in Fig. 11.4, the transverse strain $\bar{\varepsilon}_2$ under uniform axial stress, $\bar{\sigma}_x$, can be expressed as the transverse deformation, δH, per unit height, H, in terms of fiber height H_f, matrix height H_m, and the respective Poisson's ratios, ν_f and ν_m, as

$$\bar{\varepsilon}_2 = \frac{\delta H}{H} = \frac{\delta H_f + \delta H_m}{H} = \frac{-(\nu_f \bar{\varepsilon}_1 V_f H + \nu_m \bar{\varepsilon}_1 V_m H)}{H} \tag{11.8}$$

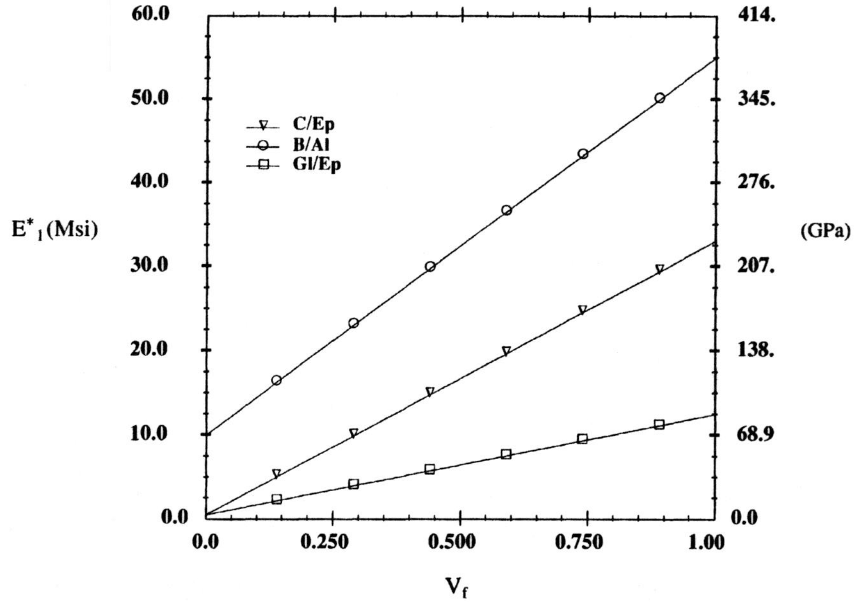

FIGURE 11.3 Rule of Mixtures Predictions for Effective Axial Moduli E_1^*

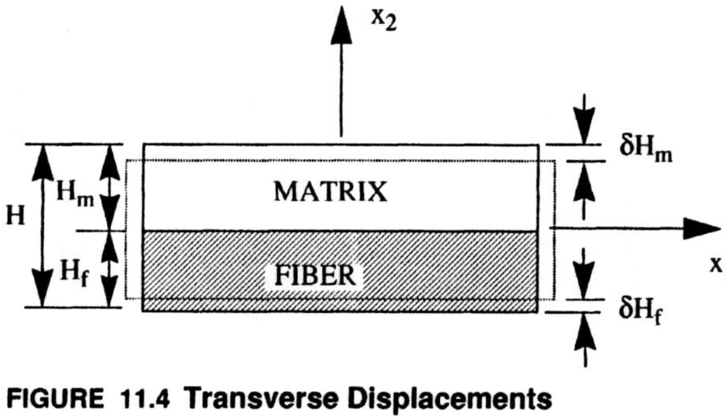

FIGURE 11.4 Transverse Displacements

The effective axial Poisson's ratio, ν_{12}^*, is then

$$\nu_{12}^* = \frac{-\bar{\varepsilon}_2}{\bar{\varepsilon}_1} = \nu_f V_f + \nu_m V_m = V_f(\nu_f - \nu_m) + \nu_m \tag{11.9}$$

Thus the effective axial Poisson's ratio ν_{12}^* is approximated by a *rule of mixtures* expression, as is the axial modulus when the assumption of uniform axial strain is invoked.

Predictions for the axial Poisson's ratio of the three materials under consideration are shown in Fig. 11.5. The predictions are linear, with the largest variation exhibited by the glass/epoxy and the

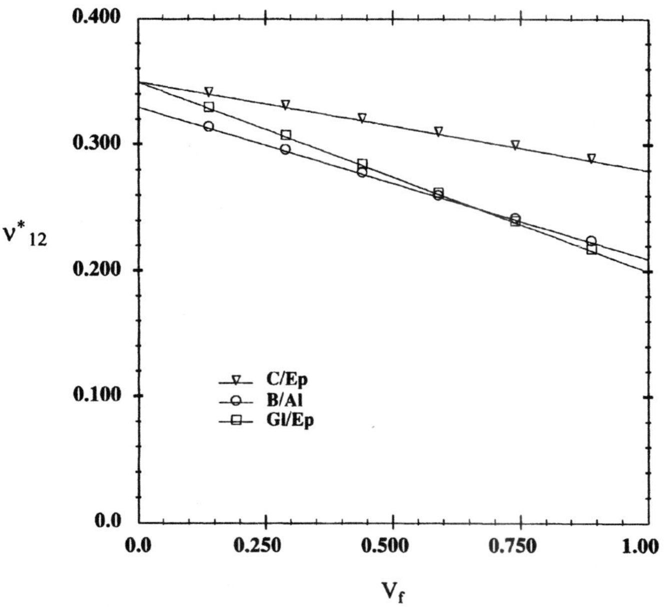

FIGURE 11.5 Rule of Mixtures Predictions for Effective Axial Poisson's Ratio ν_{12}^*

smallest by the carbon/epoxy. Experimental results indicate that the rule of mixtures is a reasonably good predictor of the effective axial Poisson's ratio.

11.3.3 Effective Transverse Modulus

An approximation for the *effective transverse modulus*, E_2^*, can be determined through consideration of a unidirectional composite subjected to the uniform average transverse stress $\bar{\sigma}_2$ as depicted in Fig. 11.6. In this case we *assume* that the transverse stress is constant throughout the composite. The transverse strain in the fiber and matrix is then given by $\varepsilon_f = \bar{\sigma}_2/E_f$ and $\varepsilon_m = \bar{\sigma}_2/E_m$.

The total transverse deformation δ_2 of the composite for a total thickness H is

$$\delta_2 = \bar{\varepsilon}_2 H = \varepsilon_f V_f H + \varepsilon_m V_m H \tag{11.10}$$

Therefore, the average transverse strain in the composite is

$$\bar{\varepsilon}_2 = \varepsilon_f V_f + \varepsilon_m V_m \tag{11.11}$$

i.e., a rule of mixtures expression.

The transverse modulus is determined from (11.11) and Hooke's law for the fiber and matrix.

$$\bar{\varepsilon}_2 = \frac{\bar{\sigma}_2}{E_2^*} = \frac{\bar{\sigma}_2}{E_f} V_f + \frac{\bar{\sigma}_2}{E_m} V_m \tag{11.12}$$

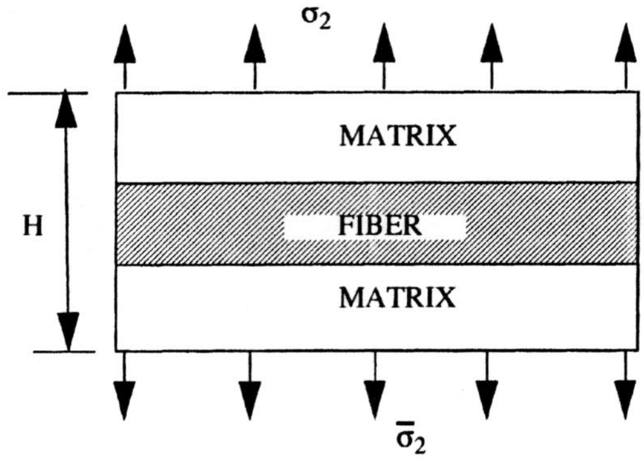

FIGURE 11.6 Composite under Transverse Loading

Eliminating the common stress term results in an expression for the effective transverse modulus E_2^*:

$$\frac{1}{E_2^*} = \frac{V_f}{E_f} + \frac{V_m}{E_m} \tag{11.13}$$

This equation can be written explicitly in terms of the fiber volume fraction as

$$E_2^* = \frac{E_m}{\left[V_f\left(\frac{E_m}{E_f} - 1\right) + 1\right]} \tag{11.14}$$

Clearly, this expression for the effective transverse modulus is nonlinear in the fiber volume fraction. Predictions for the three composites under consideration are shown in Fig. 11.7. The metal matrix composite is predicted to have a much higher transverse modulus than the polymer matrix composites, and both the glass/epoxy and the boron/aluminum exhibit large variations with increase in fiber volume fraction.

11.3.4 Effective Axial Shear Modulus

The *effective axial shear modulus*, G_{12}^*, is

$$G_{12}^* = \frac{\bar{\tau}_{12}}{\bar{\gamma}_{12}} \tag{11.15}$$

for a pure shear loading

$$\bar{\tau}_{12} \neq 0 \tag{11.16}$$

Assuming that the stress state is uniform throughout both the fiber and matrix, the shear strains in the fiber and matrix are

$$\gamma_m = \frac{\bar{\tau}_{12}}{G_m} \tag{11.17}$$

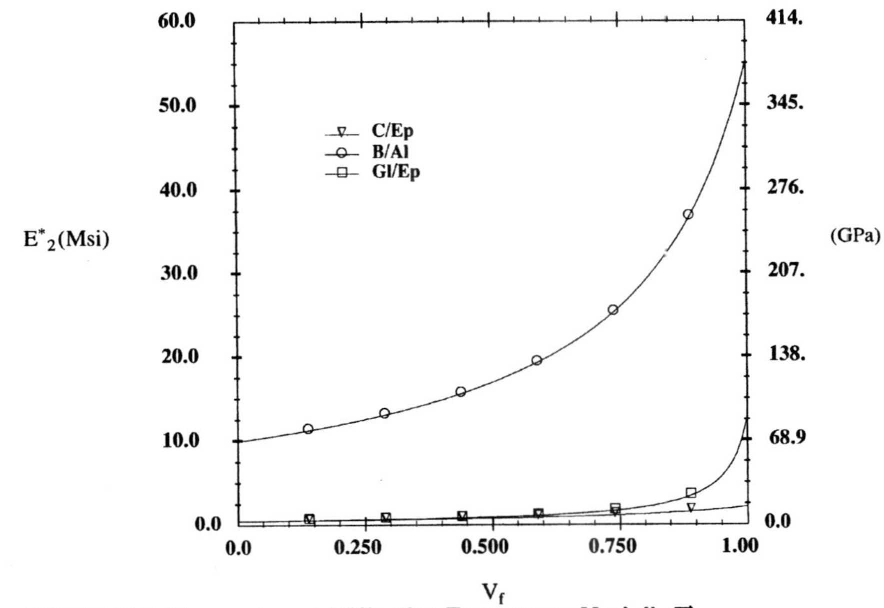

FIGURE 11.7 Comparison of Effective Transverse Moduli E_2^*

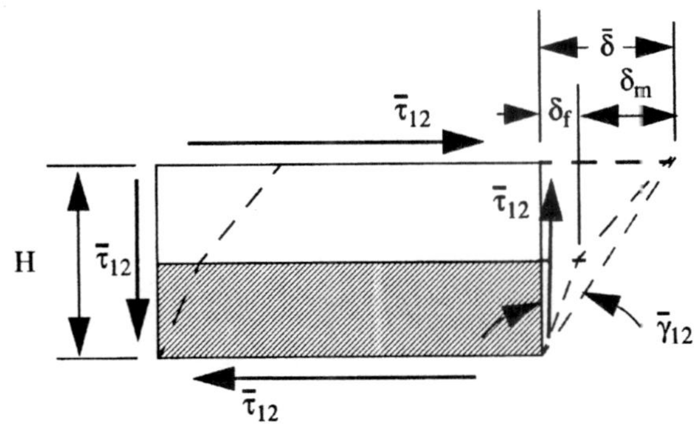

FIGURE 11.8 Shear Deformations

$$\gamma_f = \frac{\bar{\tau}_{12}}{G_f} \tag{11.18}$$

The average composite shear strain $\bar{\gamma}_{12}$ can be expressed as the sum of the changes in right angles in the fiber and matrix of a representative volume element, as depicted in Fig. 11.8. From the figure, we have

$$\tan\bar{\gamma}_{12} = \frac{\bar{\delta}}{H} \tag{11.19}$$

For small angles, the tangent of the angle is approximately equal to the angle, thus

$$\bar{\gamma}_{12} \approx \frac{\bar{\delta}}{H} \tag{11.20}$$

From the distorted geometry of Fig. 11.8, we can write

$$\bar{\delta} = \delta_f + \delta_m \tag{11.21}$$

or

$$\bar{\delta} = \frac{\bar{\tau}_{12}}{G_f} V_f H + \frac{\bar{\tau}_{12}}{G_m} V_m H \tag{11.22}$$

Combining (11.15) with (11.20)–(11.22) gives the result

$$\frac{1}{G_{12}^*} = \frac{\bar{\gamma}_{12}}{\bar{\tau}_{12}} = \frac{V_f}{G_f} + \frac{V_m}{G_m} \tag{11.23}$$

This equation for the effective shear modulus is identical in form to (11.13) for the effective transverse modulus. And, of course, (11.23) can be written in the same form as (11.14) to show the explicit dependence on fiber volume fraction. The results in Fig. 11.9 show that the predictions are highly nonlinear functions of the fiber volume fraction.

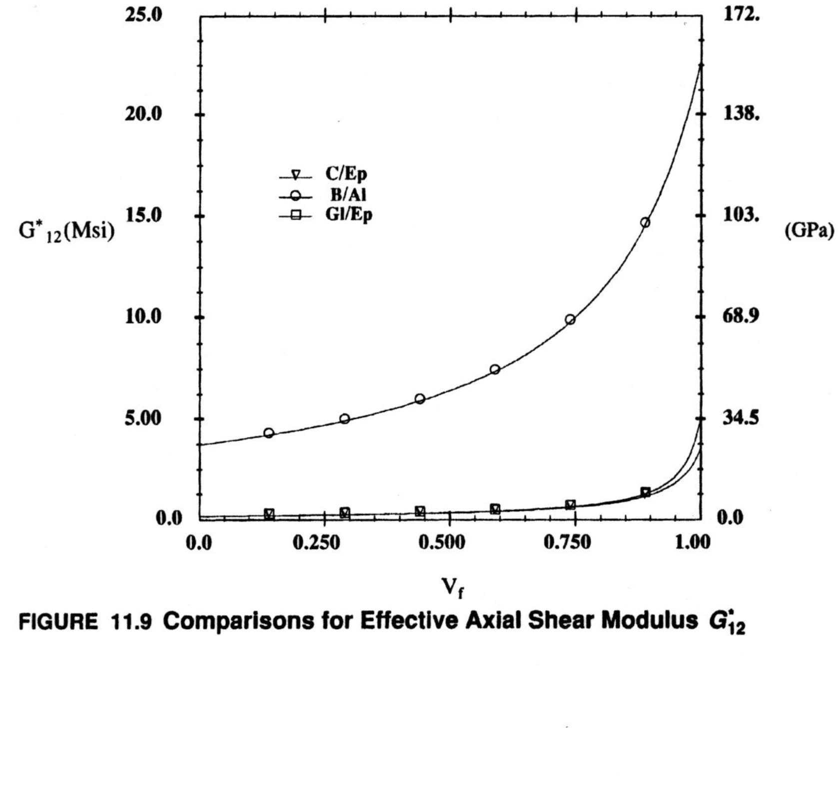

FIGURE 11.9 Comparisons for Effective Axial Shear Modulus G_{12}^*

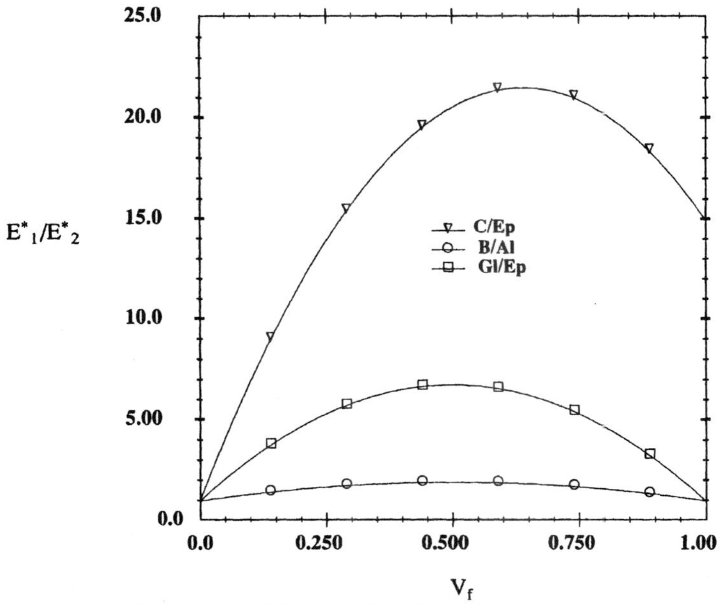

FIGURE 11.10 Degree of Composite Orthotropy

11.3.5 Degree of Orthotropy

One measure of the degree of orthotropy of a material is the ratio of the axial modulus to the transverse modulus. Values of this ratio for the three composite materials under consideration, as predicted using (11.6) and (11.14), are shown in Fig. 11.10 as a function of fiber volume fraction. As indicated in the figure, the carbon/epoxy exhibits the highest degree of orthotropy, with the ratio exceeding 20.0 for fiber volume fraction from 50% to 75%. The maximum degree of orthotropy is for V_f of approximately 63%. In contrast, the maximum degree of orthotropy for the glass/epoxy and boron/aluminum is only approximately 6.5 and 2.0, respectively. We also note that the maximum degree of orthotropy shifts with fiber volume fraction depending upon the constituent properties.

11.3.6 Summary for Strength of Materials Approaches

Simple expressions for four of the five elastic properties of transversely isotropic fibrous composites have been presented. We have noted that two properties, E_1^* and ν_{12}^*, are predicted accurately using these simple strength of materials approaches, whereas two other properties, E_2^* and G_{12}^*, are not predicted as accurately. Although it is true that the properties can be measured in the laboratory, it would not be cost-effective to require experimental results in all cases. Thus there is a need for more accurate micromechanics models.

We have also observed that the degree of orthotropy of a fibrous composite can vary greatly depending upon the properties and volume fraction of the constituents.

11.4 Continuum Approaches

11.4.1 Equivalent Homogeneity

All materials are heterogeneous when evaluated on a sufficiently small scale. However, if the scale of interest is large enough, most materials exhibit statistical homogeneity. For a fibrous composite we can discuss statistical homogeneity in terms of a characteristic dimension of the inhomogeneity. We choose the fiber spacing, λ, as the characteristic dimension (Fig. 11.11). Then there exists a length scale $\delta >> \lambda$ over which the properties can be averaged in a meaningful way. If δ is small compared with the characteristic dimensions of the structure, the material can be idealized as being effectively homogeneous, and the problem can be analyzed using average or effective material properties.

We define a *representative volume element* (RVE) as a volume of material that exhibits statistically homogeneous material properties. A representative volume element and two nonrepresentative volume elements are identified in Fig. 11.11. To be representative, the volume element must include a sufficient number of fibers and surrounding matrix to adequately represent the interaction between the phases. Obviously, a region of all fiber or all matrix is not representative of the effective properties of the composite.

11.4.2 Volumetric Averaging

The effective stress-strain relations are based upon volumetric averaging of the stress and the strain in the composite and then defining the effective elastic properties as the coefficients relating these average quantities. The *composite average stress*, $\bar{\sigma}_{ij}$, is defined as the volume average of all stresses, $\sigma_{ij}(\underset{\sim}{x})$, at position $\underset{\sim}{x}$ in the volume V, i.e.,

$$\bar{\sigma}_{ij} = \frac{1}{V}\int_V \sigma_{ij}(\underset{\sim}{x})dV \tag{11.24}$$

Likewise, the *composite average strain*, $\bar{\varepsilon}_{ij}$, in V is defined

$$\bar{\varepsilon}_{ij} = \frac{1}{V}\int_V \varepsilon_{ij}(\underset{\sim}{x})dV \tag{11.25}$$

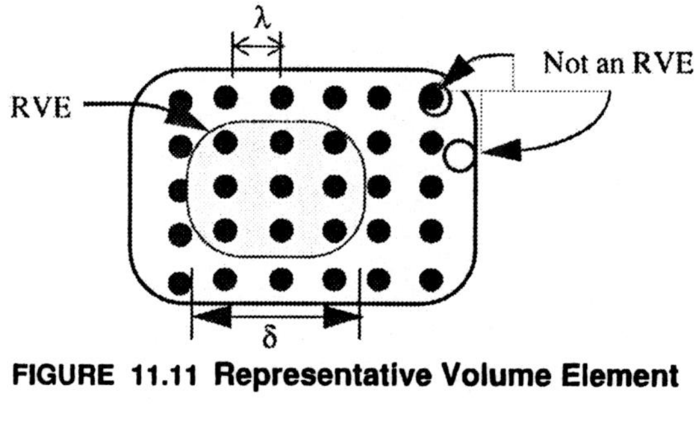

FIGURE 11.11 Representative Volume Element

The *effective stiffness*, C^*_{ijkl}, is then defined

$$\bar{\sigma}_{ij} = C^*_{ijkl}\bar{\varepsilon}_{kl} \tag{11.26}$$

Inverting (11.26) provides the *effective compliance* S^*_{ijkl}, where

$$S^*_{ijkl} = (C^*_{ijkl})^{-1} \tag{11.27}$$

From (11.26) and (11.27) we see that the effective stiffness and compliance coefficients can be determined if the integrals in (11.24) and (11.25) can be evaluated.

In the following discussion, several micromechanical methods for predicting the effective elastic constants of unidirectional fibrous composites are presented. After the methods have been presented, the results will be compared with one another.

11.4.3 Hill's Concentration Factors

Consider a composite consisting of two elastic phases, fiber and matrix, with the properties of each phase exhibiting uniform (homogeneous) properties throughout. From (11.24) the composite average stress can be written in terms of integrals over the *fiber volume*, V_f, and the *matrix volume*, V_m, i.e.,

$$\bar{\sigma}_{ij} = \frac{1}{V}\int_V \sigma_{ij}(\underset{\sim}{x})dV = \frac{1}{V}\int_{V_f} \sigma^{(f)}_{ij}(\underset{\sim}{x})dV + \frac{1}{V}\int_{V_m} \sigma^{(m)}_{ij}(\underset{\sim}{x})dV \tag{11.28}$$

Now we define the *average stress in the fiber*, $\bar{\sigma}^{(f)}_{ij}$, and the *average stress in the matrix*, $\bar{\sigma}^{(m)}_{ij}$, in terms of the respective integrals. Thus,

$$\bar{\sigma}^{(f)}_{ij} = \frac{1}{V_f}\int_{V_f} \sigma^{(f)}_{ij}(\underset{\sim}{x})dV \tag{11.29}$$

$$\bar{\sigma}^{(m)}_{ij} = \frac{1}{V_m}\int_{V_m} \sigma^{(m)}_{ij}(\underset{\sim}{x})dV \tag{11.30}$$

Using (11.29) and (11.30) in (11.28), the average composite stress can be written

$$\bar{\sigma}_{ij} = \frac{V_f}{V}\bar{\sigma}^{(f)}_{ij} + \frac{V_m}{V}\bar{\sigma}^{(m)}_{ij} = V_f\bar{\sigma}^{(f)}_{ij} + V_m\bar{\sigma}^{(m)}_{ij} \tag{11.31}$$

where for a unit volume V, the quantities V_f and V_m are the fiber and matrix volume fractions, respectively, and $V_f + V_m = 1$.

In a similar manner we can write the composite average strains (11.25) in terms of the *average strains in the fiber*, $\bar{\varepsilon}^{(f)}_{ij}$, and the *average strains in the matrix*, $\bar{\varepsilon}^{(m)}_{ij}$, and the respective volume fractions:

$$\bar{\varepsilon}_{ij} = V_f\bar{\varepsilon}^{(f)}_{ij} + V_m\bar{\varepsilon}^{(m)}_{ij} \tag{11.32}$$

The constitutive equations at any point $(\underset{\sim}{x})$ in the fiber or matrix can be written in terms of the respective stiffnesses in the forms

$$\sigma_{ij}^{(f)}(\underset{\sim}{x}) = C_{ijkl}^{(f)}\varepsilon_{kl}^{(f)}(\underset{\sim}{x}) \qquad \sigma_{ij}^{(m)}(\underset{\sim}{x}) = C_{ijkl}^{(m)}\varepsilon_{kl}^{(m)}(\underset{\sim}{x}) \tag{11.33}$$

and in terms of compliances as

$$\varepsilon_{ij}^{(f)}(\underset{\sim}{x}) = S_{ijkl}^{(f)}\sigma_{kl}^{(f)}(\underset{\sim}{x}) \qquad \varepsilon_{ij}^{(m)}(\underset{\sim}{x}) = S_{ijkl}^{(m)}\sigma_{kl}^{(m)}(\underset{\sim}{x}) \tag{11.34}$$

Using (11.33), the *average stress in the fiber*, $\bar{\sigma}_{ij}^{(f)}$, is

$$\bar{\sigma}_{ij}^{(f)} = \int_{V_f} C_{ijkl}^{(f)}\ \varepsilon_{kl}^{(f)}(\underset{\sim}{x})dV_f = C_{ijkl}^{(f)}\bar{\varepsilon}_{kl}^{(f)} \tag{11.35}$$

and the *average stress in the matrix*, $\bar{\sigma}_{ij}^{(m)}$, is

$$\bar{\sigma}_{ij}^{(m)} = \int_{V_m} C_{ijkl}^{(m)}C_{ijkl}\varepsilon_{kl}^{(m)}(\underset{\sim}{x})dV_m = C_{ijkl}^{(m)}\bar{\varepsilon}_{kl}^{(m)} \tag{11.36}$$

Inverting (11.35) and (11.36) gives expressions for the average strains in the constituents in terms of the average stresses and the compliances:

$$\bar{\varepsilon}_{kl}^{(f)} = S_{ijkl}^{(f)}\bar{\sigma}_{kl}^{(f)} \tag{11.37}$$

$$\bar{\varepsilon}_{ij}^{(m)} = S_{ijkl}^{(m)}\bar{\sigma}_{kl}^{(m)} \tag{11.38}$$

Combining (11.31) with (11.35) and (11.36) gives the average composite stress in terms of the average strains in the fiber and matrix, and the respective stiffnesses and volume fractions:

$$\bar{\sigma}_{ij} = V_f C_{ijkl}^{(f)}\bar{\varepsilon}_{kl}^{(f)} + V_m C_{ijkl}^{(m)}\bar{\varepsilon}_{kl}^{(m)} \tag{11.39}$$

Likewise, combining (11.32) with (11.37) and (11.38) gives the average strain in the composite in terms of compliances and average stresses in the constituents.

$$\bar{\varepsilon}_{ij} = V_f S_{ijkl}^{(f)}\bar{\sigma}_{kl}^{(f)} + V_m S_{ijkl}^{(m)}\bar{\sigma}_{kl}^{(m)} \tag{11.40}$$

Now, for an RVE of the composite under homogeneous traction boundary conditions, $T_i = \sigma_{ji}^{o}n_j$, where σ_{ji}^{o} is a constant stress, it can be shown (the proof will follow) that the average stress is given by

$$\bar{\sigma}_{ij} = \sigma_{ij}^{o} \tag{11.41}$$

and for an RVE under homogeneous displacement boundary conditions, $u_i = \varepsilon_{ij}^{o}x_j$, where ε_{ij}^{o} is a constant strain, the average strain in the composite is

$$\bar{\varepsilon}_{ij} = \varepsilon_{ij}^{o} \tag{11.42}$$

It is convenient to express the average stresses and strains in the fiber and matrix in terms of phase-average concentration factors as developed by Hill (1964). The approach is as follows.

The local strains at any point $(\underset{\sim}{x})$ can be written in terms of location-dependent *strain concentration factors* $A_{ijkl}(\underset{\sim}{x})$ and the composite average strains in the form

$$\varepsilon_{ij}^{(f)}(\underset{\sim}{x}) = A_{ijkl}^{(f)}(\underset{\sim}{x})\bar{\varepsilon}_{kl} \qquad \varepsilon_{ij}^{(m)}(\underset{\sim}{x}) = A_{ijkl}^{(m)}(\underset{\sim}{x})\bar{\varepsilon}_{kl} \tag{11.43}$$

Likewise, the local stresses can be written in terms of *stress concentration factors* $B_{ijkl}(\underset{\sim}{x})$ and the composite average stresses as

$$\sigma_{ij}^{(f)}(\underset{\sim}{x}) = B_{ijkl}^{(f)}(\underset{\sim}{x})\bar{\sigma}_{kl} \qquad \sigma_{ij}^{(m)}(\underset{\sim}{x}) = B_{ijkl}^{(m)}(\underset{\sim}{x})\bar{\sigma}_{kl} \tag{11.44}$$

The local strains and stresses in (11.43) and (11.44) can be integrated over their respective volumes to give the average quantities in terms of *phase-average concentration factors* $\bar{A}_{ijkl}^{(\alpha)}$ and $\bar{B}_{ijkl}^{(\alpha)}$, where $\alpha = f$ for the fiber and $\alpha = m$ for matrix. Thus,

$$\bar{\sigma}_{ij}^{(f)} = \bar{B}_{ijkl}^{(f)}\bar{\sigma}_{kl} \qquad \bar{\sigma}_{ij}^{(m)} = \bar{B}_{ijkl}^{(m)}\bar{\sigma}_{kl} \tag{11.45}$$

and

$$\bar{\varepsilon}_{ij}^{(f)} = \bar{A}_{ijkl}^{(f)}\bar{\varepsilon}_{kl} \qquad \bar{\varepsilon}_{ij}^{(m)} = \bar{A}_{ijkl}^{(m)}\bar{\varepsilon}_{kl} \tag{11.46}$$

Combining (11.39) and (11.46) gives the composite average stresses in terms of composite average strains:

$$\bar{\sigma}_{ij} = V_f C_{ijkl}^{(f)}\bar{A}_{klrs}^{(f)}\bar{\varepsilon}_{rs} + V_m C_{ijkl}^{(m)}\bar{A}_{klrs}^{(m)}\bar{\varepsilon}_{rs} \tag{11.47}$$

Rearranging terms, we have

$$\bar{\sigma}_{ij} = [V_f C_{ijkl}^{(f)}\bar{A}_{klrs}^{(f)} + V_m C_{ijkl}^{(m)}\bar{A}_{klrs}^{(m)}]\bar{\varepsilon}_{rs} \tag{11.48}$$

Defining the quantity in brackets as the *effective stiffness* C_{ijrs}^{*}, we have

$$C_{ijrs}^{*} = V_f C_{ijkl}^{(f)}\bar{A}_{klrs}^{(f)} + V_m C_{ijkl}^{(m)}\bar{A}_{klrs}^{(m)} \tag{11.49}$$

and the effective stress-strain relation can be written

$$\bar{\sigma}_{ij} = C_{ijkl}^{*}\bar{\varepsilon}_{kl} \tag{11.50}$$

In a similar fashion, (11.32), (11.37), (11.38), and (11.45) lead to

$$\bar{\varepsilon}_{ij} = S_{ijkl}^{*}\bar{\sigma}_{kl} \tag{11.51}$$

where the *effective compliance* S_{ijkl}^{*} is defined

$$S_{ijkl}^{*} = V_f S_{ijkl}^{(f)}\bar{B}_{klrs}^{(f)} + V_m S_{ijkl}^{(m)}\bar{B}_{klrs}^{(m)} \tag{11.52}$$

From (11.32) and (11.46), we can write

$$\bar{\varepsilon}_{ij} = V_f \bar{A}^{(f)}_{ijkl}\bar{\varepsilon}_{kl} + V_m \bar{A}^{(m)}_{ijkl}\bar{\varepsilon}_{kl} \tag{11.53}$$

which reduces to

$$V_f \bar{A}^{(f)}_{ijkl} + V_m \bar{A}^{(m)}_{ijkl} = I_{ijkl} \tag{11.54}$$

where I_{ijkl} is the fourth-order identity tensor.

Likewise (11.31) and (11.45) give

$$V_f \bar{B}^{(f)}_{ijkl} + V_m \bar{B}^{(m)}_{ijkl} = I_{ijkl} \tag{11.55}$$

Combining (11.49) and (11.54) gives

$$C^*_{ijkl} = C^{(m)}_{ijkl} + V_f (C^{(f)}_{ijrs} - C^{(m)}_{ijrs}) \bar{A}^{(f)}_{rskl} \tag{11.56}$$

and combining (11.52) and (11.55) gives

$$S^*_{ijkl} = S^{(m)}_{ijkl} + V_f (S^{(f)}_{ijrs} - S^{(m)}_{ijrs}) \bar{B}^{(f)}_{rskl} \tag{11.57}$$

From (11.56) and (11.57) we see that the effective stiffness and compliance of the composite can be determined if the strain or stress concentration matrix in one of the phases of the RVE is known.

11.4.4 Voigt Approximation

Voigt (1889) assumed that the strains are constant throughout the composite. Thus the phase-average strain concentration factors are equal, i.e., $\bar{A}^{(f)}_{ijkl} = \bar{A}^{(m)}_{ijkl}$. Using this fact in (11.54) gives

$$\bar{A}^{(f)}_{ijkl} = \bar{A}^{(m)}_{ijkl} = I_{ijkl} \tag{11.58}$$

Combining (11.58) with (11.56) gives the composite effective stiffness:

$$C^*_{ijkl} = C^{(m)}_{ijkl} + V_f (C^{(f)}_{ijkl} - C^{(m)}_{ijkl}) = V_m C^{(m)}_{ijkl} + V_f C^{(f)}_{ijkl} \tag{11.59}$$

Equation (11.59) states that the individual terms of the effective stiffness matrix can be determined from a simple rule of mixtures on the corresponding terms of the fiber and matrix stiffnesses.

11.4.5 Reuss Approximation

Reuss (1929) assumed that the stresses were constant throughout the composite. Thus the phase-average stress concentration factors are equal, i.e., $\bar{B}^{(f)}_{ijkl} = \bar{B}^{(m)}_{ijkl}$. Using this fact in (11.55) gives

$$B^{(f)}_{ijkl} = B^{(m)}_{ijkl} = I_{ijkl} \tag{11.60}$$

Using (11.60) in (11.57) gives the composite effective compliance:

$$S^*_{ijkl} = V_m S^{(m)}_{ijkl} + V_f S^{(f)}_{ijrs} \tag{11.61}$$

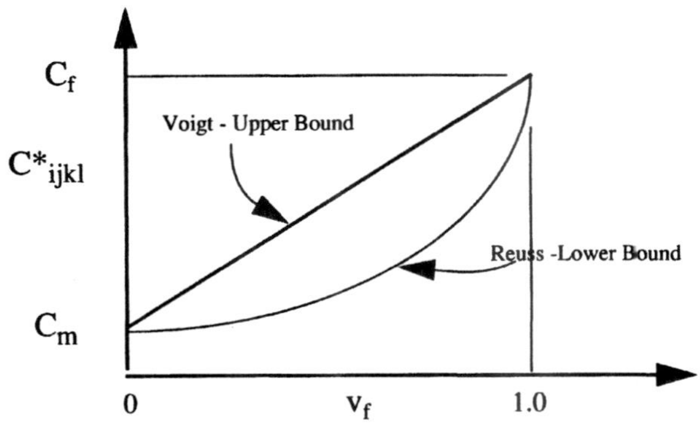

FIGURE 11.12 Bounds on Effective Stiffness

Thus the assumption of uniform stresses through the composite results in a rule of mixtures on the compliances.

Hill has shown that the Voigt approximation is an upper bound and the Reuss approximation is a lower bound on the stiffness coefficients (Fig. 11.12). It is emphasized that the bounds are on stiffness coefficients and not on engineering constants.

11.4.6 Concentric Cylinder Assemblage Model

The concentric cylinder assemblage (CCA) model was proposed by Hashin and Rosen (1964). In this model, a unidirectional, continuous fiber composite is represented by an assemblage of concentric cylinders, each consisting of a fiber core surrounded by a matrix annulus, such that the size of the cylinders varies as needed to fill the entire volume while maintaining the fiber volume fraction constant in all cylinders (Fig. 11.13). Such a material is transversely isotropic. Maintaining constant fiber volume fraction in all cylinders is equivalent to maintaining the ratio of fiber radius to cylinder radius constant. The CCA model is a variation on the concentric spheres model proposed by Hashin (1962). The final forms of the equations for effective properties given in the following discussion were first presented by Hill (1964).

We present the development for the CCA model assuming that the phases are isotropic. If the phases are transversely isotropic, the form of the solution is the same as the isotropic phases case with appropriate transversely isotropic properties used. Solutions for orthotropic phases have been presented by Avery and Herakovich (1986), Knott and Herakovich (1988, 1991), and Hashin (1990). Inelastic response using the CCA model was considered by Pindera et al. (1993).

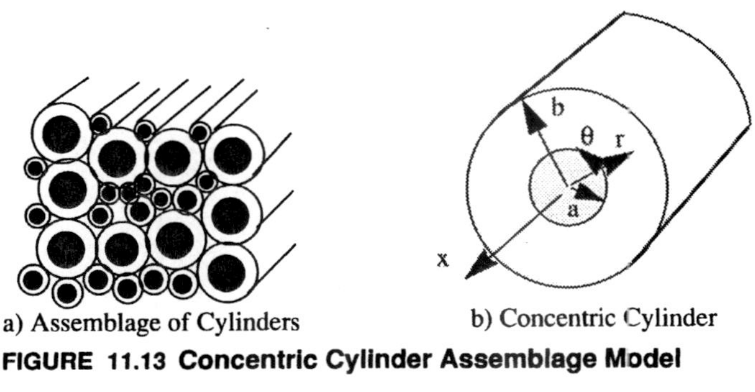

a) Assemblage of Cylinders b) Concentric Cylinder

FIGURE 11.13 Concentric Cylinder Assemblage Model

The elasticity solution for a hollow laminated cylinder (a tube) subjected to uniform loading along its length was discussed in detail in Chapter 10. As noted in Section 10.7, the only difference between the solution for a solid and a hollow cylinder is one of the boundary conditions. For the hollow cylinder, the inside surface is traction free; for the solid cylinder, the radial displacement at $r = 0$ must be finite. The solution for the tube, modified for the $r = 0$ boundary condition, will be used in the following whenever possible.

The value of the CCA model is that four of the five effective elastic properties of a transversely isotropic composite can be determined from the analysis of a single concentric cylinder. The four effective properties are the axial modulus, E_1^*, the axial Poisson's ratio, ν_{12}^*, the axial shear modulus, G_{12}^*, and the plane strain bulk modulus, k_{23}^*. Only one concentric cylinder need be analyzed to determine these four effective properties because, as the theorems of minimum potential and minimum complementary energy (Chapter 2) show, the solution for one cylinder is equivalent to the solution for the representative volume element consisting of the assemblage of concentric cylinders (see Christensen, 1979).

Before proceeding with the development for these effective properties, let us define the plane strain bulk modulus, k_{23}^*, that corresponds to the strain state:

$$\begin{aligned} \varepsilon_{11} &= 0 \\ \varepsilon_{22} &= \varepsilon_{33} = \varepsilon^0 \end{aligned} \tag{11.62}$$

From the constitutive equation (11.1) for the effective homogeneous material, the stresses for this strain state are

$$\begin{aligned} \sigma_{11} &= 2C_{12}^*\varepsilon^0 \\ \sigma_{22} &= (C_{22}^* + C_{23}^*)\varepsilon^0 \equiv \sigma \\ \sigma_{33} &= (C_{22}^* + C_{23}^*)\varepsilon^0 \equiv \sigma \end{aligned} \tag{11.63}$$

If the effective plane strain bulk modulus, k_{23}^*, is defined in terms of the in-plane stresses, σ, and strains, ε^0, acting on the equivalent homogeneous material as

$$\sigma = 2k_{23}^*\varepsilon^0 \tag{11.64}$$

then we see from (11.63) and (11.64) that

$$k_{23}^* = \frac{1}{2}(C_{22}^* + C_{23}^*) \tag{11.65}$$

The remaining four effective engineering constants of interest can be expressed in terms of the effective stiffness coefficients as

$$\begin{aligned} E_1^* &= C_{11}^* - \frac{2(C_{12}^*)^2}{C_{22}^* + C_{23}^*} \\ \nu_{12}^* &= \frac{C_{12}^*}{C_{22}^* + C_{23}^*} \\ G_{12}^* &= C_{66}^* \\ G_{23}^* &= \frac{(C_{22}^* - C_{23}^*)}{2} \end{aligned} \tag{11.66}$$

The proof of (11.66) is left as an exercise.

We also note that the constitutive equations for an isotropic, linear elastic material can be written in the form

$$\sigma_{ij} = 2\mu\varepsilon_{ij} + \lambda\varepsilon_{kk}\delta_{ij} \tag{11.67}$$

where μ is the shear modulus and λ is the Lame constant, with $\lambda = k - \mu$, where k is the plane strain bulk modulus of the isotropic material.

11.4.6.1 Effective Axial Modulus

Consider a concentric cylinder (Fig. 11.13b) subjected to uniform axial strain loading $\varepsilon_x = \varepsilon^0$ and traction-free lateral surfaces. Using the radial displacements from (10.42) for the isotropic phases, the axial displacements (10.38), and the tangential displacements (10.39), the displacement field in the concentric cylinder can be written in terms of unknown constants A_i and B_i in the form

$$w^{(f)}(r) = A_f r \tag{11.68}$$

$$w^{(m)}(r) = A_m r + \frac{B_m}{r} \tag{11.69}$$

$$u^{(f)}(x, r) = u^{(m)}(x, r) = \varepsilon^0 x \tag{11.70}$$

$$v^{(f)}(x, r) = v^{(m)}(x, r) = 0 \tag{11.71}$$

where the constant $B_f = 0$ for finite displacements at $r = 0$, and the subscripts and superscripts f and m refer to the fiber and matrix, respectively.

Using the displacements (11.68)–(11.71) in the strain-displacement equations (10.2) gives the normal strains, in cylindrical coordinates, as

$$\begin{aligned} \varepsilon_r^{(f)} &= \frac{\partial w^{(f)}}{\partial r} = A_f & \varepsilon_r^{(m)} &= \frac{\partial w^{(m)}}{\partial r} = A_m - \frac{B_m}{r^2} \\ \varepsilon_\theta^{(f)} &= \frac{w^{(f)}}{r} = A_f & \varepsilon_\theta^{(m)} &= \frac{w^{(m)}}{r} = A_m + \frac{B_m}{r^2} \\ \varepsilon_x^{(f)} &= \frac{\partial u^{(f)}}{\partial x} = \varepsilon^0 & \varepsilon_r^{(m)} &= \frac{\partial u^{(m)}}{\partial x} = \varepsilon^0 \end{aligned} \tag{11.72}$$

This same set of equations shows that all shear strains are identically zero for the displacements (11.68)–(11.71).

Using the constitutive equations (11.67) and the strains (11.72), the axial stresses in the fiber and matrix are constants given by

$$\begin{aligned} \sigma_x^{(f)} &= (2\mu_f + \lambda_f)\varepsilon^0 + 2\lambda_f A_f \\ \sigma_x^{(m)} &= (2\mu_m + \lambda_m)\varepsilon^0 + 2\lambda_m A_m \end{aligned} \tag{11.73}$$

The radial stresses are constant in the fiber but a function of radial location in the matrix,

$$\begin{aligned} \sigma_r^{(f)} &= \lambda_f \varepsilon^0 + 2k_f A_f \\ \sigma_r^{(m)} &= \lambda_m \varepsilon^0 + 2k_m A_m - \frac{2\mu_m}{r^2} B_m \end{aligned} \tag{11.74}$$

and the hoop stresses are identically equal to the radial stresses:

$$\begin{aligned} \sigma_\theta^{(f)} &= \lambda_f \varepsilon^0 + 2k_f A_f \\ \sigma_\theta^{(m)} &= \lambda_m \varepsilon^0 + 2k_m A_m - \frac{2\mu_m}{r^2} B_m \end{aligned} \tag{11.75}$$

All shear stresses are zero.

The three unknown constants A_f, A_m, and B_m are determined from the three simultaneous equations expressing the stress-free boundary condition at $r = b$ and the interfacial stress and displacement continuity conditions at $r = a$.

Continuity of radial displacements at the fiber/matrix interface, i.e., $w^{(f)}(a) = w^{(m)}(a)$, using (11.68) and (11.69), gives

$$A_f a^2 - A_m a^2 - B_m = 0 \tag{11.76}$$

Continuity of the radial stress at the fiber/matrix interface requires that

$$\sigma_r^{(f)}(a) = \sigma_r^{(m)}(a) \tag{11.77}$$

Using (11.74), this gives

$$2k_f A_f a^2 - 2k_m A_m a^2 + 2\mu_m B_m = a^2 \varepsilon^0 (\lambda_m - \lambda_f) \tag{11.78}$$

The radial stress-free boundary condition $\sigma_r^{(m)}(b) = 0$ at $r = b$ gives

$$2k_m b^2 A_m - 2\mu_m B_m = -b^2 \lambda_m \varepsilon^0 \tag{11.79}$$

Solution of the three simultaneous equations (11.76), (11.78), and (11.79) gives the three unknown coefficients in terms of the applied strain ε^0, the fiber and matrix material properties, and the fiber and matrix radii:

$$A_f = -\frac{1}{2}\varepsilon^0 \frac{[(\lambda_m - \lambda_f)\mu_m a^2 - (k_m \lambda_f + \mu_m \lambda_m) b^2]}{[(k_m - k_f)\mu_m a^2 - (\mu_m + k_f) k_m b^2]} \tag{11.80}$$

$$A_m = -\frac{1}{2}\varepsilon^0 \frac{[(\lambda_m - \lambda_f)\mu_m a^2 - (k_f \lambda_m + \mu_m \lambda_m) b^2]}{[(k_m - k_f)\mu_m a^2 - (\mu_m + k_f) k_m b^2]} \tag{11.81}$$

$$B_m = \frac{1}{2}\varepsilon^0 a^2 b^2 \frac{k_m \lambda_f - \lambda_m k_f}{[(k_m - k_f)\mu_m a^2 - (\mu_m + k_f) k_m b^2]} \tag{11.82}$$

Now the effective axial modulus E_{11}^* is defined in terms of the average axial stress σ_x and the applied axial strain ε^0 as

$$E_1^* = \frac{\bar{\sigma}_x}{\varepsilon^0} = \frac{1}{\pi b^2 \varepsilon^0}\int_0^{2\pi}\left\{\int_0^a \sigma_x^{(f)} r\, dr + \int_a^b \sigma_x^{(m)} r\, dr\right\} d\theta \tag{11.83}$$

Using (11.73) and (11.80)–(11.82), the effective axial modulus can be written

$$E_1^* = V_f E_f + (1 - V_f)E_m + \frac{4V_f(1 - V_f)(\nu_f - \nu_m)^2 \mu_m}{\dfrac{(1 - V_f)\mu_m}{\left(k_f + \dfrac{\mu_f}{3}\right)} + \dfrac{V_f \mu_m}{\left(k_m + \dfrac{\mu_m}{3}\right)} + 1} \tag{11.84}$$

The first two terms in (11.84) correspond to a rule of mixtures, and the last term is typically very small for the composites in use. Thus, the CCA model and the rule of mixtures are essentially equal for the axial modulus of a unidirectional composite.

11.4.6.2 Effective Axial Poisson's Ratio

The preceding problem of an axially loaded cylinder can be used to determine the effective axial Poisson's ratio ν_{12}^*. Defining the lateral strain of the cylinder as the change in radius per unit length associated with an axial strain ε^0 due to applied stress $\bar{\sigma}_x$, we have

$$\nu_{12}^* = -\frac{[w^{(m)}(b)]/b}{\varepsilon^0} \tag{11.85}$$

Now using (11.69) for $w^{(m)}(b)$, we have

$$\nu_{12}^* = \frac{-1}{\varepsilon^0}\left(A_m + \frac{B_m}{b^2}\right) \tag{11.86}$$

Now using (11.81) and (11.82) for A_m and B_m, we have

$$\nu_{12}^* = (1 - V_f)\nu_m + V_f \nu_f + \frac{V_f(1 - V_f)(\nu_f - \nu_m)\left(\dfrac{\mu_m}{\left(k_m + \dfrac{\mu_m}{3}\right)} - \dfrac{\mu_m}{\left(k_f + \dfrac{\mu_f}{3}\right)}\right)}{\dfrac{(1 - V_f)\mu_m}{\left(k_f + \dfrac{\mu_f}{3}\right)} + \dfrac{V_f \mu_m}{\left(k_m + \dfrac{\mu_m}{3}\right)} + 1} \tag{11.87}$$

The first two terms in (11.87) correspond to a rule of mixtures, and the third term is typically very small. Thus, the rule of mixtures is a good approximation for the effective axial Poisson's ratio. However, it is noted that the third term in the expression for effective Poisson's ratio is not as small as that in the expression for effective axial modulus.

11.4.6.3 Effective Plane Strain Bulk Modulus

As discussed previously, the *effective plane strain bulk modulus*, k_{23}^*, corresponds to the strain state

$$\varepsilon_{11} = 0 \qquad \varepsilon_{22} = \varepsilon_{33} = \varepsilon^0 \tag{11.88}$$

The corresponding displacement boundary conditions (in cylindrical coordinates) on the exterior of a cylinder for the strain state (11.88) are

$$\begin{aligned} u(r,\theta) &= 0 \\ v(b,\theta) &= 0 \\ w(b,\theta) &= \varepsilon^0 b \end{aligned} \tag{11.89}$$

This axisymmetric problem has the same form of solution as that given by (11.68) to (11.71). The constants A_f, A_m, and B_m are different from the previous solution because of the different boundary condition at $r = b$.

The continuity conditions (11.76) and (11.78) at the fiber/matrix interface are not altered. The boundary condition $w(b,\theta) = \varepsilon^0 b$ takes the form

$$A_m b^2 + B_m = \varepsilon^0 b^2 \tag{11.90}$$

Simultaneous solution of (11.76), (11.78), and (11.90) gives the unknown constants.

The displacement boundary conditions at $r = b$ are equivalent to pure radial stress tractions $\sigma_{rr}(b)$ on the surface of a homogeneous cylinder. When these tractions are transformed to rectangular coordinates, it is determined that

$$\sigma_{22} = \sigma_{33} = \sigma = \sigma_{rr}(b) \tag{11.91}$$

The effective plane strain bulk modulus of the equivalent homogeneous material is then defined as

$$k_{23}^* = \frac{\sigma}{\varepsilon_{22} + \varepsilon_{33}} = \frac{\sigma_{rr}(b)}{2\varepsilon^0} \tag{11.92}$$

When written in terms of the fiber and matrix properties, the effective plane strain bulk modulus of the equivalent homogeneous material can be written (Hill, 1964)

$$k_{23}^* = k_m + \frac{G_m}{3} + \frac{V_f}{\dfrac{1}{\left(k_f - k_m + \frac{1}{3}(G_f - G_m)\right)} + \dfrac{(1 - V_f)}{\left(k_m + \frac{4}{3}G_m\right)}} \tag{11.93}$$

It is evident from (11.93) that the rule of mixtures is not a good approximation for the plane strain bulk modulus.

11.4.6.4 Effective Axial Shear Modulus

The *effective axial shear modulus*, G_{12}^*, of a unidirectional fibrous composite is obtained from the boundary displacements

$$\begin{aligned} u_1 &= \varepsilon_{12}^0 x_2 \\ u_2 &= \varepsilon_{12}^0 x_1 \\ u_3 &= 0 \end{aligned} \tag{11.94}$$

If this displacement field is applied to a homogeneous, transversely isotropic circular cylinder with axial shear modulus G_{12}^*, the resulting strains are

$$\begin{bmatrix} \varepsilon_{ij} \end{bmatrix} = \begin{bmatrix} 0 & \varepsilon_{12}^0 & 0 \\ \varepsilon_{12}^0 & 0 & 0 \\ 0 & 0 & 0 \end{bmatrix} \tag{11.95}$$

and the stresses are

$$\begin{bmatrix} \sigma_{ij} \end{bmatrix} = \begin{bmatrix} 0 & 2G_{12}^* & 0 \\ 2G_{12}^* & 0 & 0 \\ 0 & 0 & 0 \end{bmatrix} \tag{11.96}$$

Thus the equivalent homogeneous cylinder is in a state of pure axial shear.

Solution of the concentric cylinder elasticity problem for the displacement boundary conditions (11.94) gives the final result

$$\frac{G_{12}^*}{G_m} = \frac{G_f(1+V_f)+G_m(1-V_f)}{G_f(1-V_f)+G_m(1+V_f)} \tag{11.97}$$

From (11.97) we see that the rule of mixtures is not a good approximation for the axial shear modulus. It is interesting to note that the effective shear modulus of the composite is a function of only the fiber and matrix shear moduli and not any other properties of the constituents.

11.4.6.5 Three-Phase Composite Cylinder for Transverse Shear

It has not been possible to determine the *effective transverse shear modulus*, G_{23}^*, using the CCA model as originally proposed. However, a closed-form expression for this property has been proposed by Christensen and Lo (1979). This model is based upon a three-phase cylinder in which the fiber and matrix are embedded in an annulus of the equivalent homogeneous material. The full expression for the transverse modulus using this three-phase model is quite involved (see Christensen and Lo, 1979); however for small fiber volume fractions the expression reduces to

$$\frac{G_{23}^*}{G_m} = 1 + \cfrac{V_f}{\cfrac{G_m}{(G_f - G_m)} + \cfrac{\left(K_m + \frac{7}{3}G_m\right)}{\left(2K_m + \frac{8}{3}G_m\right)}} \tag{11.98}$$

where K_m is the bulk modulus of the matrix relating hydrostatic stress to the change in volume, i.e.,

$$\sigma_{kk} = 3K_m\varepsilon_{kk} \tag{11.99}$$

From (11.98) we see that there is no correlation with the rule of mixtures.

11.4.7 The Method of Cells

The method of cells (MOC) (Aboudi, 1991) and its extension, the generalized method of cells (GMC) (Paley and Aboudi, 1992; Aboudi, 1995), are approximate analytical methods for predicting the elastic as well as inelastic response of fibrous composites. The methods can be used for two-dimensional (e.g., continuous fibers) or three-dimensional (e.g., short fibers or inclusions) analysis. In the following discussion we concentrate on the original method of cells with four subcells in order to present the details of the method. The generalized method of cells considers an arbitrary number of subcells and provides improved predictions, particularly when inelastic effects are under consideration. The extension to the generalized method of cells is discussed in the following section.

The response of a unidirectional, continuous fiber composite can be modeled through consideration of a generic plane section transverse to the fibers. Such a plane section with equally spaced fibers is depicted in Fig. 11.14.

As in most micromechanics models, the analysis is limited to a representative volume element that includes one fiber and the surrounding matrix material. For the method of cells, a repeating volume element, or cell (indicated by the lines in Fig. 11.14), consists of four rectangular subcells as shown in Fig. 11.15. The four subcells are identified by the subscript notation $\beta\gamma$, with β and γ having the range 1 to 2. In Fig. 11.15, subcell $\beta = 1, \gamma = 1$ is fiber, and the remaining three subcells are matrix. The total volume of the repeating element is V, and the volume of any subcell is $V_{\beta\gamma} = h_\beta l_\gamma$. The x_1 axis is parallel to the fiber and x_2 and x_3 are perpendicular to the fibers.

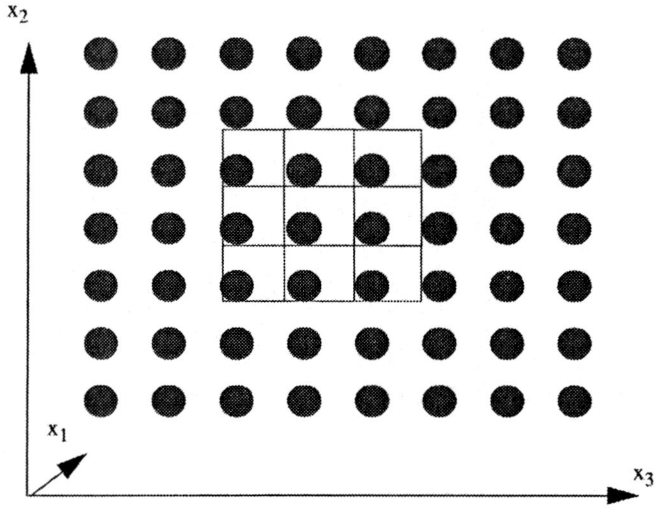

FIGURE 11.14 Cells in Uniformly Spaced Fibrous Composite

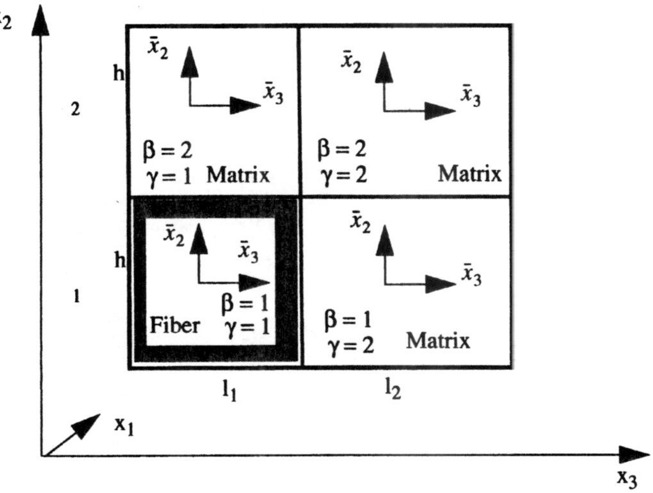

FIGURE 11.15 Cell Representative Volume Element

At this point in the development, it is noted that while the fiber subcell is initially taken to be rectangular (or square) in cross section, the actual shape of the fiber does not affect the final result. The averaging procedures for stress and displacement continuity mask the actual shape of the fiber. The final results for the effective response of the composite are a function of the constituent properties and the fiber volume fraction. The results using this model have been shown to provide excellent correlation with numerical and experimental results for both the effective elastic properties and the effective inelastic response of fibrous composites.

In the following presentation an overbar is used to denote average quantities, and tensors are underscored with a tilde. Thus the *average stress tensor in subcell* $\beta\gamma$ is denoted $\bar{\underset{\sim}{\sigma}}^{(\beta\gamma)}$, and the *average stress tensor in the repeating cell* is

$$\bar{\underset{\sim}{\sigma}} = \frac{1}{V}\sum_{\beta,\gamma} V_{\beta\gamma}\bar{\underset{\sim}{\sigma}}^{(\beta\gamma)} \tag{11.100}$$

For elastic response, Hill's strain concentration tensors $\underset{\sim}{A}^{(\beta\gamma)}$ (Hill, 1954) are now used to express the average subcell strains $\bar{\underset{\sim}{\varepsilon}}^{(\beta\gamma)}$ in terms of average composite (cell) strains $\bar{\underset{\sim}{\varepsilon}}$:

$$\bar{\underset{\sim}{\varepsilon}}^{(\beta\gamma)} = \underset{\sim}{A}^{(\beta\gamma)}\bar{\underset{\sim}{\varepsilon}} \tag{11.101}$$

With six components of strain in each of four subcells we have a total of 24 strain components, $\bar{\underset{\sim}{\varepsilon}}^{(\beta\gamma)}$, and 144 coefficients of the strain concentration tensors, $\underset{\sim}{A}^{(\beta\gamma)}$, to be determined.

Hooke's law for each subcell is known and can be written in terms of the strain concentration tensors in the form

$$\bar{\underset{\sim}{\sigma}}^{(\beta\gamma)} = \underset{\sim}{C}^{(\beta\gamma)}\bar{\underset{\sim}{\varepsilon}}^{(\beta\gamma)} = \underset{\sim}{C}^{(\beta\gamma)}\underset{\sim}{A}^{(\beta\gamma)}\bar{\underset{\sim}{\varepsilon}} \tag{11.102}$$

where $\underset{\sim}{C}^{(\beta\gamma)}$ is the stiffness matrix for subcell $\beta\gamma$.

Combining (11.100) and (11.102) gives Hooke's law for the cell (or repeating volume element):

$$\bar{\underset{\sim}{\sigma}} = \frac{1}{V}\sum_{\beta,\gamma} V_{\beta\gamma}\underset{\sim}{C}^{(\beta\gamma)}\underset{\sim}{A}^{(\beta\gamma)}\bar{\underset{\sim}{\varepsilon}} \tag{11.103}$$

For given average strains, $\bar{\underset{\sim}{\varepsilon}}$, applied to a composite whose constituents and volume fraction are known, all quantities in the effective constitutive equation (11.103) are known except for the strain concentration tensors $A^{(\beta\gamma)}$. Therefore, if we find $A^{(\beta\gamma)}$, the elastic problem is solved. The constitutive equation (11.103) is of the form $\bar{\underset{\sim}{\sigma}} = \underset{\sim}{C}^*\bar{\underset{\sim}{\varepsilon}}$, where the effective stiffness of the composite is

$$\underset{\sim}{C}^* = \frac{1}{V}\sum_{\beta,\gamma} V_{\beta\gamma}\underset{\sim}{C}^{(\beta\gamma)}\underset{\sim}{A}^{(\beta\gamma)} \tag{11.104}$$

11.4.7.1 Displacements and Strains

At this point we introduce specifics of the method of cells in order to determine the strain concentration tensors $A^{(\beta\gamma)}$. First we establish local coordinates $\bar{x}_2, \bar{x}_3$ in each subcell (Fig. 11.15). Next we assume a linear displacement field u_i in each subcell, $\beta\gamma$. Thus

$$u_i^{(\beta\gamma)}(x_1, x_2, x_3; \bar{x}_2^{(\beta)}, \bar{x}_3^{(\gamma)}) = w_i^{(\beta\gamma)}(x_1, x_2, x_3) + \bar{x}_2^{(\beta)}\phi_i^{(\beta\gamma)}(x_1) + \bar{x}_3^{(\gamma)}\psi_i^{(\beta\gamma)}(x_1) \tag{11.105}$$

where $\phi_i^{(\beta\gamma)}(x_1)$ and $\psi_i^{(\beta\gamma)}(x_1)$ are unknowns (called *microvariables*) associated with subcell ($\beta\gamma$), and $w_i^{(\beta\gamma)}(x_1, x_2, x_3)$ are the displacements of the center of subcell ($\beta\gamma$).

The strains in any subcell are determined from the strain-displacement equations written in the form

$$\varepsilon_{ij}^{(\beta\gamma)} = \frac{1}{2}(\partial_j u_i^{(\beta\gamma)} + \partial_i u_j^{(\beta\gamma)}) \tag{11.106}$$

where the partial derivatives are defined $\partial_1 = \partial/(\partial x_1)$, $\partial_2 = \partial/(\partial \bar{x}_2^{(\beta)})$, and $\partial_3 = \partial/(\partial \bar{x}_3^{(\gamma)})$.

Using the displacements (11.105) in the strain-displacement equations (11.106), the subcell strains are, for $i = j = 1$,

$$\varepsilon_{11}^{(\beta\gamma)} = \frac{\partial w_1^{(\beta\gamma)}}{\partial x_1} + \bar{x}_2^{(\beta)}\frac{\partial \phi_1}{\partial x_1} + \bar{x}_3^{(\gamma)}\frac{\partial \psi_1^{(\beta\gamma)}}{\partial x_1} \tag{11.107}$$

Averaging over the subcell, we have the subcell average strain $\bar{\varepsilon}_{11}^{(\beta\gamma)}$:

$$\bar{\varepsilon}_{11}^{(\beta\gamma)} = \frac{\partial w_1^{(\beta\gamma)}}{\partial x_1} \tag{11.108}$$

Likewise, for $i = j = 2$,

$$\varepsilon_{22}^{(\beta\gamma)} = \frac{\partial u_2^{(\beta\gamma)}}{\partial \bar{x}_2^{(\beta)}} = \phi_2^{(\beta\gamma)} \tag{11.109}$$

and averaging over the subcell (with $\phi_2^{(\beta\gamma)} = \text{const}$),

$$\bar{\varepsilon}_{22}^{(\beta\gamma)} = \phi_2^{(\beta\gamma)} \tag{11.110}$$

For $i = 1, j = 2$,

$$2\varepsilon_{12}^{(\beta\gamma)} = \frac{\partial u_1}{\partial \bar{x}_2} + \frac{\partial u_2}{\partial x_1} = \phi_1^{(\beta\gamma)} + \frac{\partial w_2^{(\beta\gamma)}}{\partial x_1} + \bar{x}_2\frac{\partial \phi_2}{\partial x_1} + \bar{x}_3\frac{\partial \Psi_2}{\partial x_1} \tag{11.111}$$

and averaging over the subcell,

$$2\bar{\varepsilon}_{12}^{(\beta\gamma)} = \phi_1^{(\beta\gamma)} + \frac{\partial w_2^{(\beta\gamma)}}{\partial x_1} \tag{11.112}$$

For all other average subcell strain components the same procedure leads to

$$\begin{aligned} 2\bar{\varepsilon}_{13}^{(\beta\gamma)} &= \psi_1^{(\beta\gamma)} + \frac{\partial w_3^{(\beta\gamma)}}{\partial x_1} \\ 2\bar{\varepsilon}_{23}^{(\beta\gamma)} &= \psi_2^{(\beta\gamma)} + \phi_3^{(\beta\gamma)} \\ \bar{\varepsilon}_{33}^{(\beta\gamma)} &= \psi_3^{(\beta\gamma)} \end{aligned} \tag{11.113}$$

11.4.7.2 Displacement Continuity

For perfectly bonded materials, the exact solution of the continuum problem requires pointwise continuity of displacements. For the method of cells with linear displacements specified in each subcell, displacement continuity is satisfied throughout the continuum if we require continuity along the subcell interfaces of Fig. 11.15 as well as along the interfaces between repeating cells (Fig. 11.14). In Fig. 11.16 we show a repeating cell and four adjacent subcells. Satisfaction of displacement continuity along all of the subcell interfaces in this figure provides all of the unique relationships possible for displacement continuity in a material that is in a uniform state of far-field average stress or strain. Thus displacement continuity must be satisfied along the four interface lines, *A-A*, *B-B*, *C-C*, and *D-D*. The global coordinates to the center of generic subcells and interfaces are shown in Fig. 11.16. The superscripts in the coordinate notation indicate either the subcell in question (i.e., γ for x_3 coordinates and β for x_2 coordinates) or the interface referred to (i.e., *A*, *B*, *C*, or *D*). We note that because of the repeating nature of the cells and subcells, all fiber subcells are designated $\beta = 1, \gamma = 1$, with the remaining subcells being matrix. Finally, in the discussion to follow we use a ^ over a symbol to denote quantities associated with an adjacent subcell.

Now continuity along any interface $\bar{x}_2 = \text{const}$ (i.e., along *A-A* and *B-B*) can be written

$$u_i^{(\beta\gamma)}\Big|_{\bar{x}_2^{(\beta)} = h_\beta/2} = u_i^{(\hat{\beta}\gamma)}\Big|_{\bar{x}_2^{(\hat{\beta})} = -h_{\hat{\beta}}/2} \tag{11.114}$$

where along *A-A* we have $\beta = 1, \hat{\beta} = 2$ and along *B-B* $\beta = 2, \hat{\beta} = 1$.

Similarly, continuity along any interface $\bar{x}_3 = \text{const}$ can be written

$$u_i^{(\beta\gamma)}\Big|_{\bar{x}_3^{(\gamma)} = l_\gamma/2} = u_i^{(\beta\hat{\gamma})}\Big|_{\bar{x}_3^{(\hat{\gamma})} = -l_{\hat{\gamma}}/2} \tag{11.115}$$

This must be true along *C-C*, where $\gamma = 1, \hat{\gamma} = 2$ and along *D-D*, where $\gamma = 2, \hat{\gamma} = 1$.

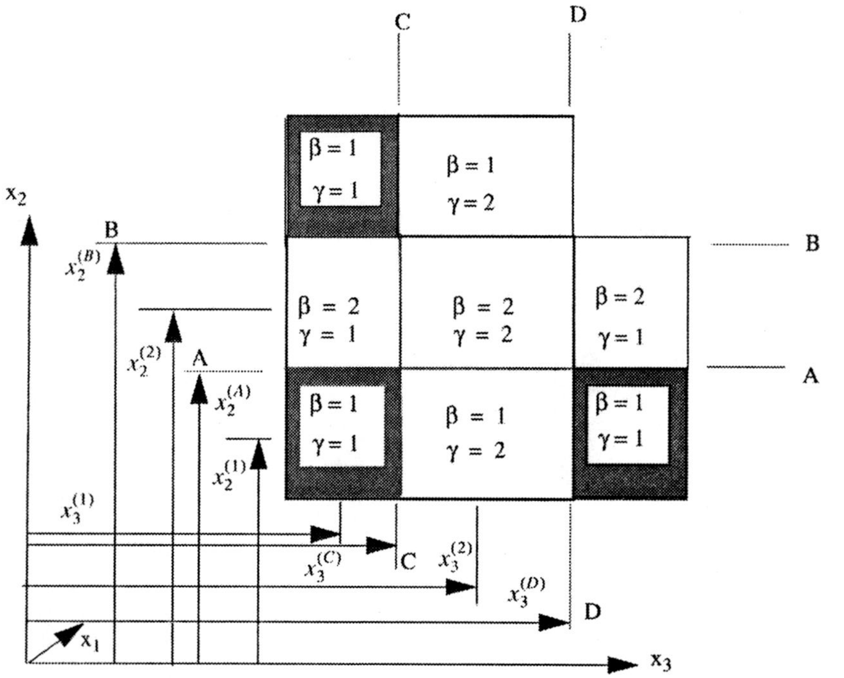

FIGURE 11.16 Subcells and Adjacent Cells

In the method of cells, the pointwise displacement continuity is relaxed somewhat. Displacement continuity is required to be satisfied only *on average* over each subcell interface. Satisfaction of displacement continuity on average along $\bar{x}_2$ = const can be written

$$\int_{-\frac{1}{2}l_\gamma}^{\frac{1}{2}l_\gamma} u_i^{(\beta\gamma)}\Big|_{\bar{x}_2^{(\beta)} = h_\beta/2} d\bar{x}_3^{(\gamma)} = \int_{-\frac{1}{2}l_\gamma}^{\frac{1}{2}l_\gamma} u_i^{(\beta\gamma)}\Big|_{\bar{x}_2^{(\beta)} = -h_{\beta}/2} d\bar{x}_3^{(\gamma)} \tag{11.116}$$

Combining (11.105) and (11.116) gives

$$w_i^{(\beta\gamma)} + \frac{1}{2}h_\beta\phi_i^{(\beta\gamma)} = w_i^{(\hat{\beta}\gamma)} - \frac{1}{2}h_{\hat{\beta}}\phi_i^{(\hat{\beta}\gamma)} \tag{11.117}$$

where the variables $w_i^{(\beta\gamma)}$, $w_i^{(\hat{\beta}\gamma)}$, $\phi_i^{(\beta\gamma)}$, and $\phi_i^{(\hat{\beta}\gamma)}$ are evaluated at the center of the corresponding subcell.

For continuity along the interface $\bar{x}_3$ = const, a similar procedure leads to

$$w_i^{(\beta\gamma)} + \frac{1}{2}l_\gamma\psi_i^{(\beta\gamma)} = w_i^{(\beta\hat{\gamma})} - \frac{1}{2}l_{\hat{\gamma}}\psi_i^{(\beta\hat{\gamma})} \tag{11.118}$$

where, again, all quantities are evaluated at the center of the corresponding subcell.

We now use Taylor series expansion to express displacement continuity in terms of quantities evaluated at the interface, $x_2^{(I)}$, between subcells (i.e., we use a first-order expansion, dropping terms of $O(h^2)$ and higher). Recalling that Taylor series expansion for a function f has the form

$$f(x+h) = f(x) + hf'(x) + O(h^2) \tag{11.119}$$

for our case involving quantities at the center of the subcell and at the interface, where the (directed) distance between these locations is h, we have

$$f(\text{center}) = f(\text{interface}) + hf'(\text{interface}) + O(h^2) \tag{11.120}$$

As a specific example, expansion from the interface A-A at $x_2^{(I)}$ to the center at $x_2^{(\beta)}$ in subcell $(\beta\gamma) = (11)$, the directed distance, h, is $x_2^{(\beta)} - x_2^{(I)} = (-1/2)h_\beta$, and thus (11.120) gives

$$w_i^{(\beta\gamma)}(x_2^{(\beta)}) = w_i^{(\beta\gamma)}(x_2^{(I)}) + \left(\frac{-1}{2}h_\beta\right)f'(x_2^{(I)}) + O(h^2) \tag{11.121}$$

Expansion of the term $(1/2)h_\beta\phi_i^{(\beta\gamma)}$ in (11.117) does not result in any additional terms since we drop all terms $O(h^2)$ and higher. Thus, we can write (11.117) in terms of functions and derivatives of functions evaluated at the interface, $x_2^{(I)}$, in subcells $\beta\gamma$ and $\hat{\beta}\gamma$ as

$$w_i^{(\beta\gamma)} - \frac{1}{2}h_\beta\frac{\partial w_i^{(\beta\gamma)}}{\partial x_2} + \frac{1}{2}h_\beta\phi_i^{(\beta\gamma)} = w_i^{(\hat{\beta}\gamma)} + \frac{1}{2}h_{\hat{\beta}}\frac{\partial w_i^{(\hat{\beta}\gamma)}}{\partial x_2} - \frac{1}{2}h_{\hat{\beta}}\phi_i^{(\hat{\beta}\gamma)} \tag{11.122}$$

Next we define

$$f_i^{(\beta)} = -\frac{1}{2}h_\beta\frac{\partial w_i^{(\beta\gamma)}}{\partial x_2} + \frac{1}{2}h_\beta\phi_i^{(\beta\gamma)} \tag{11.123}$$

and

$$F_i^{(\beta)} = w_i^{(\beta\gamma)} + f_i^{(\beta)} - w_i^{(\hat{\beta}\gamma)} + f_i^{(\hat{\beta})} \tag{11.124}$$

Now combining (11.122) through (11.124) shows that

$$F_i^{(\beta)} = 0 \tag{11.125}$$

Therefore, it follows that

$$\sum_{\beta=1}^{2} F_i^{(\beta)} = 0 \tag{11.126}$$

Further, from (11.124) and (11.126), we have

$$\sum_{\beta=1}^{2} f_i^{(\beta)} = 0 \tag{11.127}$$

Likewise, for continuity along x_3 = const, define

$$G_i^{(\gamma)} = w_i^{(\beta\gamma)} + g_i^{(\gamma)} - w_i^{(\beta\hat{\gamma})} + g_i^{(\hat{\gamma})} \tag{11.128}$$

and

$$g_i^{(\gamma)} = \frac{-1}{2} l_\gamma \frac{\partial w_i^{(\beta\gamma)}}{\partial x_3} + \frac{1}{2} l_\gamma \psi_i^{(\beta\gamma)} \tag{11.129}$$

In the same manner as above, we have

$$G_i^{(\gamma)} = 0 \tag{11.130}$$

$$\sum_{\gamma=1}^{2} G_i^{(\gamma)} = 0 \tag{11.131}$$

It follows from (11.128) and (11.131) that

$$\sum_{\gamma=1}^{2} g_i^{(\gamma)} = 0 \tag{11.132}$$

Now, from (11.123) and using first-order theory with $\partial^2 f/\partial x_2^2 = 0$ and $\phi_i^{(\beta\gamma)}$ = const, we have

$$\frac{\partial f_i^{(\beta)}}{\partial x_2} = 0 \tag{11.133}$$

From (11.124) and (11.125) we can write

$$\frac{\partial F_i^{(\beta)}}{\partial x_2} = 0 = \frac{\partial w_i^{(\beta\gamma)}}{\partial x_2} + \frac{\partial f_i^{(\beta)}}{\partial x_2} - \frac{\partial w_i^{(\hat{\beta}\gamma)}}{\partial x_2} + \frac{\partial f_i^{(\hat{\beta})}}{\partial x_2} \tag{11.134}$$

Now, using (11.133),

$$\frac{\partial w_i^{(\beta\gamma)}}{\partial x_2} = \frac{\partial w_i^{(\hat{\beta}\gamma)}}{\partial x_2} \equiv \frac{\partial w_i}{\partial x_2} \tag{11.135}$$

Likewise, using $\partial g_i^{(\gamma)}/\partial x_3 = 0$ for first-order expansion and $G_i^{(\gamma)} = 0$ from (11.130), we have

$$\frac{\partial w_i^{(\beta\gamma)}}{\partial x_3} = \frac{\partial w_i^{(\beta\hat{\gamma})}}{\partial x_3} \equiv \frac{\partial w_i}{\partial x_3} \tag{11.136}$$

From the definition (11.123), (11.127), and (11.135) we can write

$$\sum_{\beta=1}^{2} h_\beta \phi_i^{(\beta\gamma)} = (h_i + h_2)\frac{\partial w_i}{\partial x_2} \tag{11.137}$$

Likewise, (11.129), (11.132), and (11.136) give

$$\sum_{\gamma=1}^{2} l_r \psi_i^{(\beta\gamma)} = (l_1 + l_2)\frac{\partial w_i}{\partial x_3} \tag{11.138}$$

11.4.7.3 Average Cell Strains

If there is perfect bonding throughout the composite, the average strain in the cell (representative volume element) is defined

$$\bar{\varepsilon}_{ij} = \frac{1}{(h_1 + h_2)(l_1 + l_2)} \sum_{\beta=1}^{2} \sum_{\gamma=1}^{2} h_\beta l_\gamma \bar{\varepsilon}_{ij}^{(\beta\gamma)} \tag{11.139}$$

For $i = 1, j = 1$, and using (11.108) and (11.136), the cell axial strain is

$$\bar{\varepsilon}_{11} = \frac{1}{(h_1 + h_2)(l_1 + l_2)} \sum_{\beta=1}^{2} \sum_{\gamma=1}^{2} h_\beta l_\gamma \frac{\partial w_1^{(\beta\gamma)}}{\partial x_1} = \frac{\partial w_1}{\partial x_1} \tag{11.140}$$

For $i = j = 2$, we recall from (11.110) that the subcell strains are $\bar{\varepsilon}_{22}^{(\beta\gamma)} = \phi_2^{(\beta\gamma)}$. Averaging over all subcells gives the cell average strain:

$$\bar{\varepsilon}_{22} = \frac{1}{(h_1 + h_2)(l_1 + l_2)} \sum_{\beta=1}^{2} \sum_{\gamma=1}^{2} h_\beta l_\gamma \phi_2^{(\beta\gamma)} \tag{11.141}$$

This can be rewritten

$$\bar{\varepsilon}_{22} = \frac{1}{(h_1 + h_2)(l_1 + l_2)} \sum_{\gamma=1}^{2} l_\gamma \left(\sum_{\beta=1}^{2} h_\beta \phi_2^{(\beta\gamma)} \right) \tag{11.142}$$

And using (11.138) gives

$$\bar{\varepsilon}_{22} = \frac{\sum_{\gamma=1}^{2} l_\gamma (h_1 + h_2)\frac{\partial w_2}{\partial x_2}}{(h_1 + h_2)(l_1 + l_2)} \tag{11.143}$$

Expanding, we have

$$\bar{\varepsilon}_{22} = \frac{[l_1(h_1 + h_2) + l_2(h_1 + h_2)]\dfrac{\partial w_2}{\partial x_2}}{(h_1 + h_2)(l_1 + l_2)} \tag{11.144}$$

This reduces to

$$\bar{\varepsilon}_{22} = \frac{\partial w_2}{\partial x_2} \tag{11.145}$$

For $i = 1, j = 2$, the subcell average shear strain, using (11.112), is

$$2\bar{\varepsilon}_{12}^{(\beta\gamma)} = \phi_1^{(\beta\gamma)} + \frac{\partial w_2^{(\beta\gamma)}}{\partial x_1} \tag{11.146}$$

Therefore, the cell average shear strain is

$$2\bar{\varepsilon}_{12} = \frac{1}{(h_1 + h_2)(l_1 + l_2)} \sum_{\beta=1}^{2} \sum_{\gamma=1}^{2} \left(\phi_1^{(\beta\gamma)} + \frac{\partial w_2^{(\beta\gamma)}}{\partial x_1} \right) h_\beta l_\gamma \tag{11.147}$$

Using (11.135), and (11.137), this reduces to

$$2\bar{\varepsilon}_{12} = \frac{\partial w_1}{\partial x_2} + \frac{\partial w_2}{\partial x_1} \tag{11.148}$$

In a similar manner, starting from (11.113), it can be shown that the remaining cell average strains are

$$\bar{\varepsilon}_{33} = \frac{\partial w_3}{\partial x_3} \tag{11.149}$$

$$2\bar{\varepsilon}_{13} = \frac{\partial w_1}{\partial x_3} + \frac{\partial w_3}{\partial x_1} \tag{11.150}$$

$$2\bar{\varepsilon}_{23} = \frac{\partial w_2}{\partial x_3} + \frac{\partial w_3}{\partial x_2} \tag{11.151}$$

We now develop relationships between the individual components of the subcell strains and the corresponding component of the cell average strains. From (11.108) for $i = j = 1$, we have

$$\bar{\varepsilon}_{11}^{(\beta\gamma)} = \frac{\partial w_1^{(\beta\gamma)}}{\partial x_1} \tag{11.152}$$

and from (11.140),

$$\bar{\varepsilon}_{11} = \frac{\partial w_1}{\partial x_1} \tag{11.153}$$

Now, for a unidirectional fibrous composite, the axial strain is uniform throughout; thus the axial strain in each subcell, $\bar{\varepsilon}_{11}^{(\beta\gamma)}$ equals the axial strain in the cell, $\bar{\varepsilon}_{11}$, i.e.,

$$\bar{\varepsilon}_{11} = \bar{\varepsilon}_{11}^{(\beta\gamma)} \tag{11.154}$$

Combining (11.153) and (11.154) gives

$$\frac{\partial w_1}{\partial x_1} = \frac{\partial w_1^{(\beta\gamma)}}{\partial x_1} \tag{11.155}$$

For $i = j = 2$, (11.110), (11.137), and (11.145) combine to give

$$\sum_{\beta=1}^{2} h_\beta \bar{\varepsilon}_{22}^{(\beta\gamma)} = (h_1 + h_2)\frac{\partial w_2}{\partial x_2} = (h_1 + h_2)\bar{\varepsilon}_{22} \tag{11.156}$$

For $i = j = 3$, (11.113), (11.138), and (11.149) combine to give

$$\sum_{\gamma=1}^{2} l_\gamma \bar{\varepsilon}_{33}^{(\beta\gamma)} = (l_1 + l_2)\frac{\partial w_3}{\partial x_3} = (l_1 + l_2)\bar{\varepsilon}_{33} \tag{11.157}$$

For $i = 1, j = 2$, (11.112), (11.137), and (11.148) combine to give

$$\sum_{\beta=1}^{2} h_\beta \bar{\varepsilon}_{12}^{(\beta\gamma)} = (h_1 + h_2)\bar{\varepsilon}_{12} \tag{11.158}$$

For $i = 1, j = 3$, (11.113), (11.138), and (11.150) combine to give

$$\sum_{\gamma=1}^{2} l_\gamma \bar{\varepsilon}_{13}^{(\beta\gamma)} = (l_1 + l_2)\bar{\varepsilon}_{13} \tag{11.159}$$

For $i = 2, j = 3$, (11.113), (11.137), (11.138), and (11.151) combine to give

$$\sum_{\beta=1}^{2}\sum_{\gamma=1}^{2} h_\beta l_\gamma \bar{\varepsilon}_{23}^{(\beta\gamma)} = (h_1 + h_2)(l_1 + l_2)(\bar{\varepsilon}_{23}) \tag{11.160}$$

We note that equations (11.154) through (11.160) represent a total of 13 equations when evaluated for all subcells. This is depicted in the following table.

Equation Number	Number of Equations	Equation Number	Number of Equations
(11.154)	4	(11.158)	2
(11.156)	2	(11.159)	2
(11.157)	2	(11.160)	1

11.4.7.4 Traction Continuity

For perfectly bonded materials, the tractions must be continuous everywhere, including along the interfaces between all subcells (Fig. 11.16). In general, the interfacial traction continuity conditions can be stated in terms of the average cell stresses as follows. Along *A-A* and *B-B* where $\bar{x}_2$ = const,

$$\bar{\sigma}_{2j}^{(\beta\gamma)} = \bar{\sigma}_{2j}^{(\hat{\beta}\gamma)} \tag{11.161}$$

Along *C-C* and *D-D* where $\bar{x}_3$ = const,

$$\bar{\sigma}_{3j}^{(\beta\gamma)} = \bar{\sigma}_{3j}^{(\beta\hat{\gamma})} \tag{11.162}$$

Due to the repeating nature of the subcells, there is a limited number of independent traction continuity conditions. We can write them as follows. Along *A-A* where $\beta = 1, \hat{\beta} = 2$, and $\gamma = 1, 2$, (11.161) gives

$$\bar{\sigma}_{22}^{(\beta\gamma)} = \bar{\sigma}_{22}^{(\hat{\beta}\gamma)} \tag{11.163}$$

$$\bar{\sigma}_{21}^{(\beta\gamma)} = \bar{\sigma}_{21}^{(\hat{\beta}\gamma)} \tag{11.164}$$

$$\bar{\sigma}_{23}^{(\beta\gamma)} = \bar{\sigma}_{23}^{(\hat{\beta}\gamma)} \tag{11.165}$$

Along *C-C* where $\gamma = 1, \hat{\gamma} = 2$, and $\beta = 1, 2$, (11.162) gives

$$\bar{\sigma}_{31}^{(\beta\gamma)} = \bar{\sigma}_{31}^{(\beta\hat{\gamma})} \tag{11.166}$$

$$\bar{\sigma}_{32}^{(\beta\gamma)} = \bar{\sigma}_{32}^{(\beta\hat{\gamma})} \tag{11.167}$$

$$\bar{\sigma}_{33}^{(\beta\gamma)} = \bar{\sigma}_{33}^{(\beta\hat{\gamma})} \tag{11.168}$$

It is left as an exercise to show that the traction continuity conditions along *B-B* and *D-D* do not provide any additional independent equations. Equations (11.163) through (11.168) each correspond to two equations. However, only three of the four from (11.165) and (11.167) are independent. Thus, there are 11 independent traction continuity equations.

Now, in general, the subcell elastic constitutive equations are

$$\bar{\sigma}_{ij}^{(\beta\gamma)} = C_{ijkel}^{(\beta\gamma)} \bar{\varepsilon}_{kl}^{(\beta\gamma)} \tag{11.169}$$

For a transversely isotropic material, the constitutive equations for the individual subcells can be written

$$\begin{Bmatrix} \bar{\sigma}_{11} \\ \bar{\sigma}_{22} \\ \bar{\sigma}_{33} \\ \bar{\sigma}_{23} \\ \bar{\sigma}_{31} \\ \bar{\sigma}_{12} \end{Bmatrix} = \begin{bmatrix} C_{11} & C_{12} & C_{12} & 0 & 0 & 0 \\ C_{12} & C_{22} & C_{23} & 0 & 0 & 0 \\ C_{12} & C_{23} & C_{22} & 0 & 0 & 0 \\ 0 & 0 & 0 & \dfrac{C_{22}-C_{23}}{2} & 0 & 0 \\ 0 & 0 & 0 & 0 & C_{66} & 0 \\ 0 & 0 & 0 & 0 & 0 & C_{66} \end{bmatrix} \begin{Bmatrix} \bar{\varepsilon}_{11} \\ \bar{\varepsilon}_{22} \\ \bar{\varepsilon}_{33} \\ 2\bar{\varepsilon}_{23} \\ 2\bar{\varepsilon}_{31} \\ 2\bar{\varepsilon}_{12} \end{Bmatrix} \tag{11.170}$$

For isotropic phases, the number of independent constants in (11.170) reduces to two, as discussed in Chapter 3. Combining the traction continuity equations (11.163) through (11.168) with the specific form of the constitutive equations (11.170) appropriate for each subcell gives the traction continuity equations in terms of known stiffness coefficients and unknown subcell strains.

We now have the equations necessary to determine the 24 components of subcell strains, $\bar{\underset{\sim}{\varepsilon}}^{(\beta\gamma)}$, and the 144 coefficients of the strain concentration tensors, $\underset{\sim}{A}^{(\beta\gamma)}$.

The 13 independent equations (11.154) through (11.160) corresponding to displacement continuity between subcells can be written in matrix form as

$$[A_G]\{\varepsilon_s\} = [J]\{\bar{\varepsilon}\} \tag{11.171}$$

where $[A_G]$ is a 13×24 matrix of known coefficients involving geometric details, $\{\varepsilon_s\}$ is a 24×1 column vector of unknown subcell strains, $[J]$ is a 13×6 matrix of known coefficients, and $\{\bar{\varepsilon}\}$ is a 6×1 column vector of the average cell strains:

$$\{\varepsilon_s\} = \left\{ \bar{\varepsilon}_{11}^{(11)}\ \bar{\varepsilon}_{22}^{(11)}\ \bar{\varepsilon}_{33}^{(11)}\ \bar{\varepsilon}_{23}^{(11)}\ \bar{\varepsilon}_{31}^{(11)}\ \bar{\varepsilon}_{12}^{(11)}\ \bar{\varepsilon}_{11}^{(12)}\ \ldots\ \bar{\varepsilon}_{12}^{(22)} \right\}^T \tag{11.172}$$

The 11 independent traction continuity equations (11.163) through (11.168) can be written in matrix form as

$$[A_M]\{\varepsilon_s\} = 0 \tag{11.173}$$

where $[A_M]$ is an 11×24 matrix of known coefficients involving material parameters.

Combining (11.171) and (11.173), we have

$$\begin{bmatrix} A_M \\ A_G \end{bmatrix} \{\varepsilon_s\} = \begin{bmatrix} 0 \\ J \end{bmatrix} \{\bar{\varepsilon}\} \tag{11.174}$$

which can be written in terms of the 24×24 matrix of coefficients $[\tilde{A}]$ and the 24×6 coefficient matrix $[K]$ as

$$[\tilde{A}]\{\varepsilon_s\} = [K]\{\bar{\varepsilon}\} \tag{11.175}$$

Solving (11.175) for the subcell strains, ε_s, we have

$$\{\varepsilon_s\} = [\tilde{A}]^{-1}[K]\{\bar{\varepsilon}\} \tag{11.176}$$

or

$$\{\varepsilon_s\} = [A]\{\bar{\varepsilon}\} \tag{11.177}$$

where $[A]$ is the matrix of coefficients for Hill's strain concentration tensor, defined as

$$[A] = [\tilde{A}]^{-1}[K] \tag{11.178}$$

We can now write the subcell strains ε_s in terms of Hill's strain concentration tensor and the average cell strains as

$$\{\varepsilon_s\} = [A]\{\bar{\varepsilon}\} \tag{11.179}$$

and (11.101) is readily obtained by partitions.

In summary, the solution to the elastic micromechanics problem based upon the method of cells is known once the coefficient matrices $[\tilde{A}]$ and $[K]$ in (11.178) have been determined. While the method of cells has proven to be quite useful for predicting the elastic and inelastic response of fibrous composites, it does have some limitations—notably the fact that the effective properties of the unidirectional composite are not transversely isotropic unless an additional step is taken to average the properties about the fiber axis (Aboudi, 1991; Brayshaw, 1994).

11.4.8 The Generalized Method of Cells

The generalized method of cells (GMC) (Paley and Aboudi, 1992) extends the original method of cells to any number of rectangular subcells as depicted in Fig. 11.17. This generalization permits improved modeling of the specific fiber shape as well as the ability to model the arrangement of fibers in the composite. Further, the inclusion of interfacial regions or gradations of properties in the fiber or matrix can be modeled. The generalized method of cells is particularly valuable for improved prediction of the inelastic response of composites. As an example of the efficiency of GMC, Wilt (1995), in a comparison with a finite-element analysis for the viscoplastic, transverse tensile response of silicon-carbide/titanium, showed that in order to have the same degree of accuracy 1088 finite elements were required, compared with only 49 subcells, and the CPU time of the finite-element solution was 3550 times that of GMC.

The method follows that of the original method of cells identically, extending the analysis to $N_\beta \times N_\gamma$ subcells. The details will not be repeated here.

11.4.9 Self-Consistent Method

The self-consistent method is based on the solution to an auxiliary inclusion problem where a single ellipsoidal inclusion is embedded in an infinite medium. The bond between the inclusion and the infinite medium is assumed to be perfect, resulting in displacement and traction continuity across the interface between phases. Uniform stresses or strains are applied at infinity with the objective of determining the stresses and strains in the inclusion. Eshelby (1957) showed that in these types of problems, the stress and strain fields in the inclusion are uniform. Elastic properties

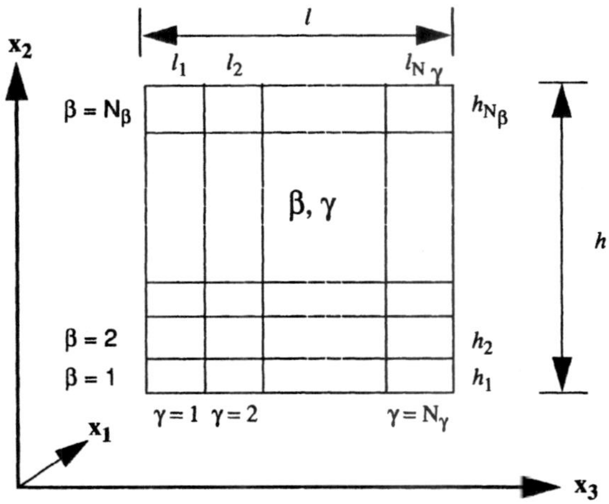

FIGURE 11.17 GMC Representative Repeating Unit Cell

can be determined by finding the relationship between the far-field stresses and strains and the homogeneous stresses and strains in the inclusion, or the stress or strain concentration factors.

The self-consistent model was originally applied to unidirectional composites by Hill (1965) and Budiansky (1965). The approach is to take the average stress and strain fields in the fiber to be equal to those in the inclusion problem. The infinite medium is taken to be homogeneous and to have the properties of the composites. Previously, Hill (1964) had shown that due to the internal structure of the composite only three of the five effective properties are actually independent; the remaining properties are related through so-called universal conditions.

For isotropic phases, the three independent properties are determined by the self-consistent method to be

$$\frac{1}{2G_{12}^*} = \frac{V_f}{G_{12}^* - G_m} + \frac{(1 - V_f)}{G_{12}^* - G_f} \tag{11.180}$$

$$\frac{V_f k_f}{k_f + G_{23}^*} + \frac{(1 - V_f)k_m}{k_m + G_{23}^*} = 2\left(\frac{V_f G_m}{G_m - G_{23}^*} + \frac{(1 - V_f)G_f}{G_f - G_{23}^*}\right) \tag{11.181}$$

$$k^* = \left(\frac{V_f}{k_f + \bar{G}_{23}} + \frac{(1 - V_f)}{k_m + \bar{G}_{23}}\right)^{-1} - G_{23}^* \tag{11.182}$$

The so-called universal connections relating these three constants to the two remaining constants are

$$E_1^* = V_f E_f + (1 - V_f)E_m + \frac{4(\nu_m - \nu_f)^2}{\left(\frac{1}{k_m} - \frac{1}{k_f}\right)^2}\left(\frac{V_f}{k_f} + \frac{(1 - V_f)}{k_m} - \frac{1}{k^*}\right) \tag{11.183}$$

$$\nu_{12}^* = V_f \nu_f + (1 - V_f)\nu_m + \frac{(\nu_m - \nu_f)}{\left(\frac{1}{k_m} - \frac{1}{k_f}\right)}\left(\frac{1}{k^*} - \frac{(1 - V_f)}{k_f} - \frac{V_f}{k_m}\right) \tag{11.184}$$

11.4.10 Mori-Tanaka Method

The original method proposed by Mori and Tanaka (1973) was restated in a more tractable from by Benveniste (1987). The essential assumption is that the average strain in a typical inclusion (fiber) is related to the average strain in the matrix by a fourth-order tensor $\underset{\sim}{T}$ where $\underset{\sim}{T}$ is defined to give the relation between the uniform strain in the inclusion embedded in an all-matrix material subjected to an imposed uniform strain at infinity. The fiber strain concentration factors are found to be

$$A_f = \underset{\sim}{T}[V_f\underset{\sim}{T} + (1 - V_f)\underset{\sim}{I}]^{-1} \tag{11.185}$$

where

$$\underset{\sim}{T} = [\underset{\sim}{S}\underset{\sim}{C}_m^{-1}(\underset{\sim}{C}_f - \underset{\sim}{C}_m) + \underset{\sim}{I}]^{-1} \tag{11.186}$$

$\underset{\sim}{S}$ is Eshelby's tensor, and $\underset{\sim}{C}_f$ and $\underset{\sim}{C}_m$ are the fiber and matrix stiffness tensors, respectively. It is believed that $\underset{\sim}{T}$ accounts for the interaction of adjacent fibers in an approximate manner. Dvorak (1990) presented explicit expressions for the effective properties of unidirectional composites with transversely isotropic constituents. These expression are given in terms of Hill's moduli, which can in turn be related to the effective engineering properties. They are

$$k^* = \frac{k_f k_m + m_m[V_f k_f + V_m k_m]}{V_f k_m + V_m k_f + m_m} = -\left(\frac{1}{G_{23}^*} - \frac{4}{E_2^*} + \frac{4(\nu^*)_{12}^2}{E_1^*}\right)^{-1} \tag{11.187}$$

$$l^* = \frac{V_f l_f(k_m + m_m) + V_m l_m(k_f + m_m)}{V_f(k_m + m_m) + V_m(k_f + m_m)} = 2k^*\nu_{12}^* \tag{11.188}$$

$$n^* = V_f n_f + V_m n_m + (l^* - V_f l_f - V_m l_m)\frac{l_f - l_m}{k_f - k_m} = E_1^* + 4k^*(\nu^*)_{12}^2 = E_1^* + \frac{(l^*)^2}{k^*} \tag{11.189}$$

$$m^* = \frac{m_f m_m(k_m + 2m_m) + k_m m_m(V_f m_f + V_m m_m)}{k_m m_m + (k_m + 2m_m)(V_f m_m + V_m m_f)} = G_{23}^* \tag{11.190}$$

$$p^* = \frac{2V_f p_f p_m + V_m(p_f p_m + p_m^{\;2})}{2V_f p_m + V_m(p_f + p_m)} = G_{12}^* \tag{11.191}$$

11.4.11 Micromechanics Using Finite Elements

Finite-element methods have been used extensively to predict the effective elastic properties and inelastic response of fibrous composites under thermomechanical loading. Finite elements are particularly useful for detailed study of the stress distributions in the fiber and surrounding matrix material as a function of the actual fiber arrangement. The earliest application of finite elements for micromechanics appears to be that by Foye (1968), who studied the effective elastic properties, inelastic response, and stress distributions for unidirectional boron/epoxy. Other early finite-element micromechanics studies were presented by Adams (1970), Lin et al. (1972), and Dvorak et al. (1973). These early studies considered both polymeric matrix and metal matrix composites. Numerous finite-element studies on a variety of issues at the micromechanics level have appeared since the original works.

11.4.12 Model Comparisons

Comparisons between various micromechanics models are presented in this section. Results are presented for effective elastic properties of glass/epoxy and carbon/epoxy composites as a function of fiber volume fraction. The constituent material properties used are given in Table 11.2. The epoxy matrix and the glass fiber are taken to be isotropic, and the carbon fiber is taken to be transversely isotropic. The results presented here follow closely those presented by Lissenden and Herakovich (1992). The predictions of the method of cells used in the following were averaged to provide transversely isotropic composite properties. These predictions are denoted MOC-TI in the figures. When no analytic expression for a property is possible using the CCA model, bounds (Hashin, 1979) are used. In these cases the upper bound is denoted CCA+ and the lower bound is denoted CCA–. The strengths of materials ("St Matls") predictions are also presented in the figures.

Material	E_A, GPa (Msi)	E_T, GPa (Msi)	G_A, GPa (Msi)	G_T, GPa (Msi)	ν_T
Epoxy	5.35 (0.78)	5.35 (0.78)	1.98 (0.29)	1.98 (0.29)	0.350
Glass	113.4 (16.5)	113.4 (16.5)	46.5 (6.7)	46.5 (6.7)	0.220
Carbon	232.0 (33.6)	15.0 (2.2)	5.02 (0.73)	24.0 (3.5)	0.279

TABLE 11.2 Glass, Carbon, and Epoxy Elastic Properties

11.4.12.1 Micromechanics Predictions for Glass/Epoxy

As indicated in Fig. 11.18, all models except the Reuss model predict the same results for the effective axial modulus E_1^*. That is, the effective axial modulus follows a rule of mixtures. The Reuss model assumes that the axial stresses are uniform over the entire composite. Obviously, this is not a good assumption when the fiber and matrix which have different moduli are subjected to the same uniform strain. The predictions for effective axial Poisson's ratio, ν_{12}^* (Fig. 11.19), indicate that with the exception of Reuss, Voigt, and the self-consistent model, the results are essentially the same, with a linear variation in effective Poisson's ratio with fiber volume fraction.

In contrast to these linear predictions, the results for effective transverse modulus, E_2^* (Fig. 11.20), are all highly nonlinear with the exception of the Voigt model, which assumes uniform strains throughout the composite. While all models but the Voigt model predict the same general behavior, there is clearly a significant percentage difference between all of them. However, it may be stated that the MOC-TI, Mori-Tanaka, CCA+, and CCA– models are reasonably close.

The predictions for effective axial modulus, G_{12}^*, are shown in Fig. 11.21. Here again we see that the Voigt model is not acceptable. Also, the self-consistent scheme leaves much to be desired. The MOC-TI, Mori-Tanaka, and CCA models give essentially identical results. The predictions for effective transverse shear modulus, G_{23}^* (Fig. 11.22), are identical to those for the effective axial shear modulus for this composite with isotropic constituents.

The predictions for the effective transverse Poisson's ratio, ν_{23}^*, are shown in Fig. 11.23. It should be noted that the bounds using the CCA model are for the effective stiffness coefficients and not the effective engineering constants. This explains why the CCA– prediction in Fig. 11.23 is greater than the CCA+ prediction.

11.4.12.2 Micromechanics Predictions for Carbon/Epoxy

The predictions for the effective engineering properties for carbon/epoxy are shown in Figs. 11.24–11.29. The carbon/epoxy differs from the glass/epoxy in that the fiber is transversely isotropic, with

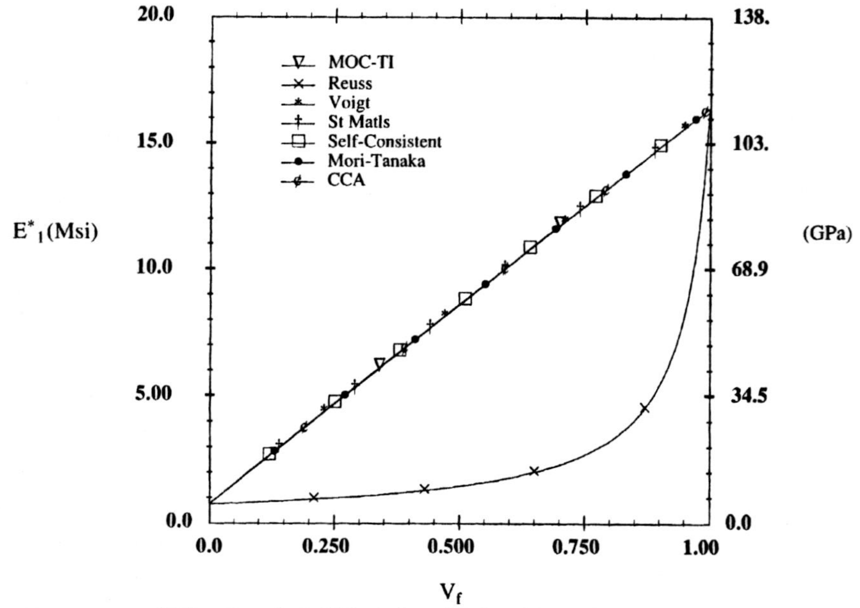

FIGURE 11.18 Effective Axial Modulus E_1^* for Glass/Epoxy

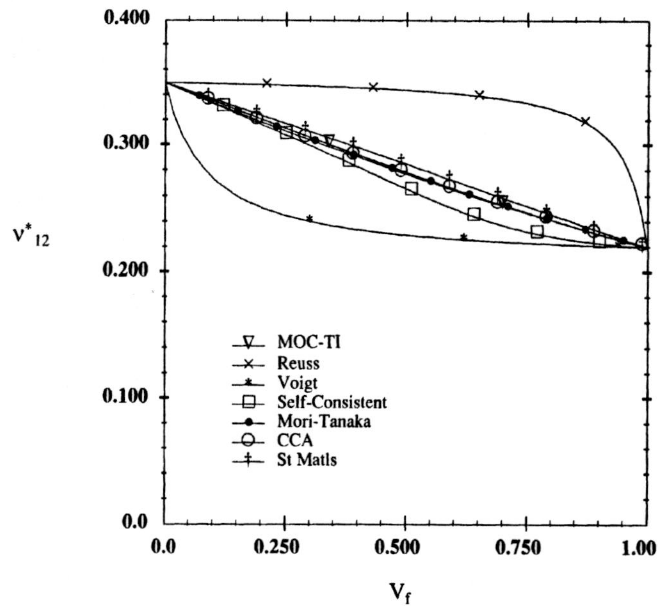

FIGURE 11.19 Effective Axial Poisson's Ratio ν_{12}^* for Glass/Epoxy

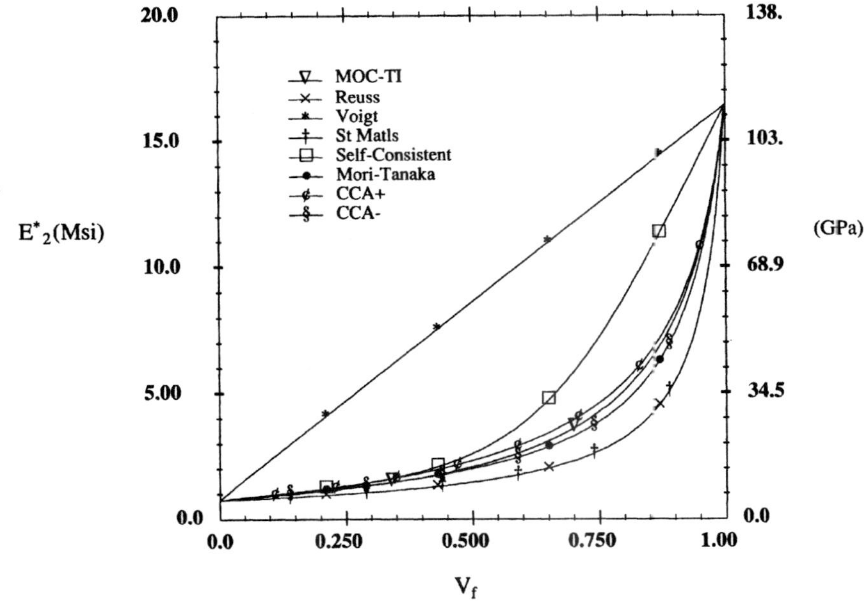

FIGURE 11.20 Effective Transverse Modulus E_2^* for Glass/Epoxy

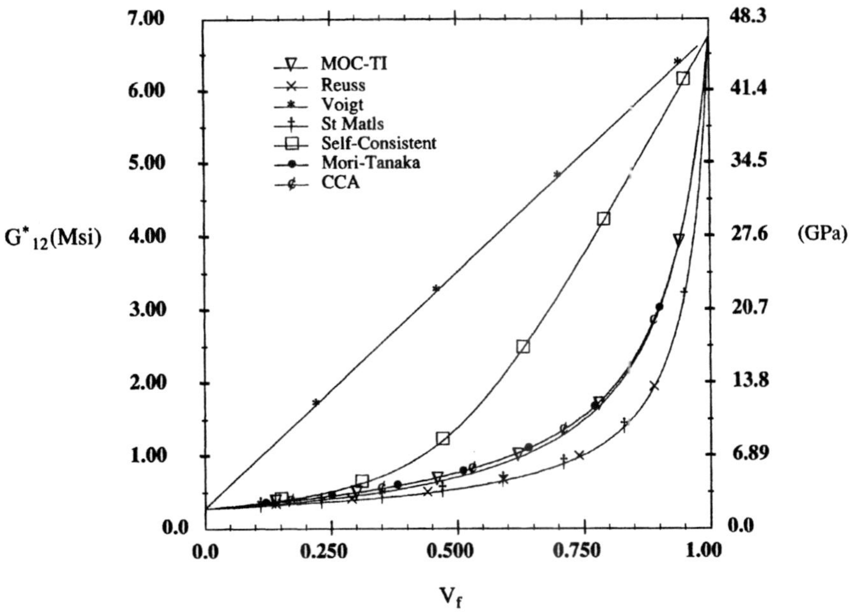

FIGURE 11.21 Effective Axial Shear Modulus G_{12}^* for Glass/Epoxy

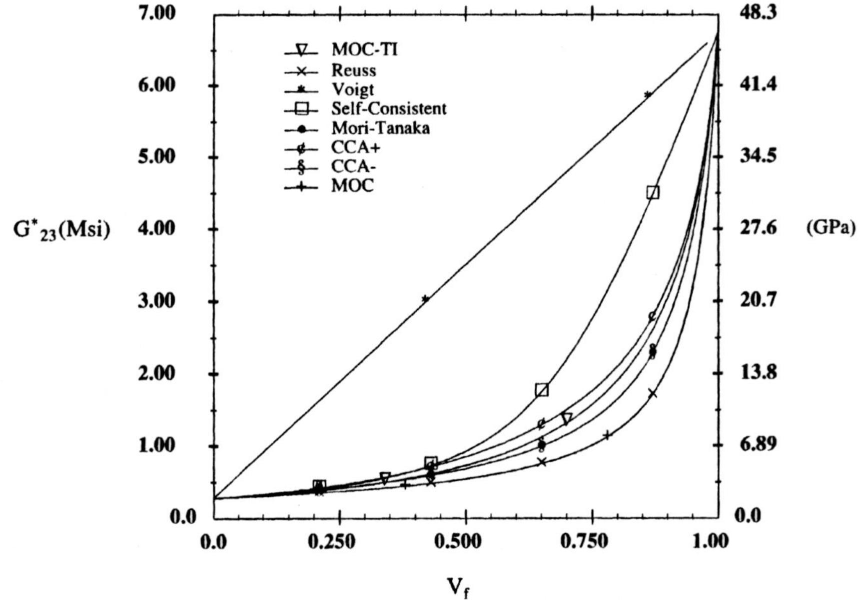

FIGURE 11.22 Effective Transverse Shear Modulus G_{23}^* for Glass/Epoxy

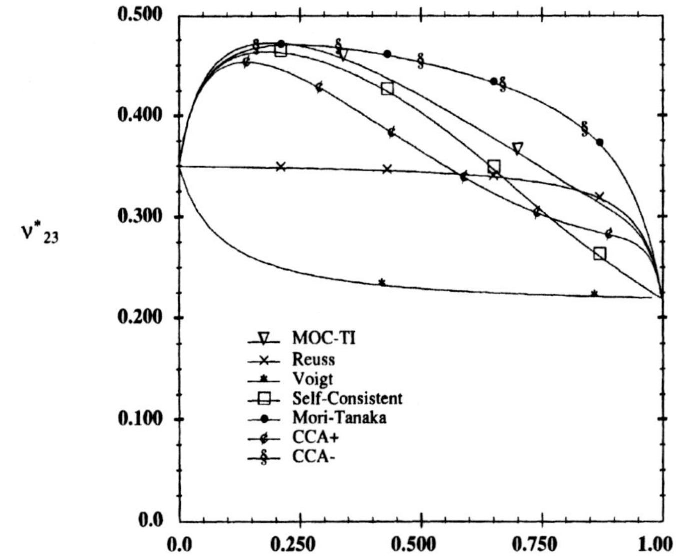

FIGURE 11.23 Effective Transverse Poisson's Ratio ν_{23}^* for Glass/Epoxy

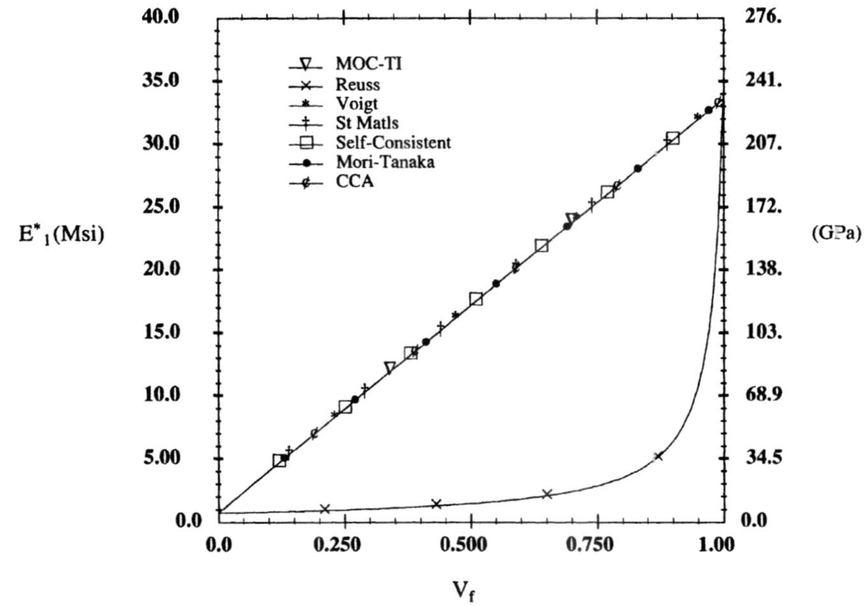

FIGURE 11.24 Effective Axial Modulus E_1^* for Carbon/Epoxy

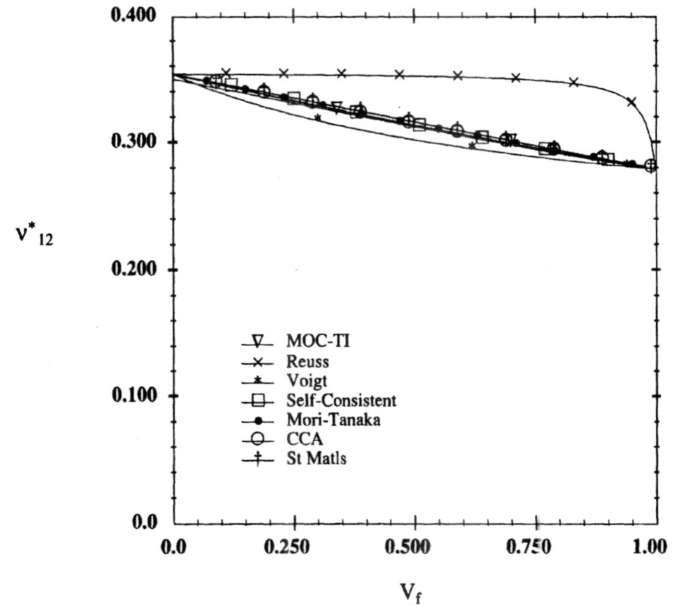

FIGURE 11.25 Effective Axial Poisson's Ratio ν_{12}^* for Carbon/Epoxy

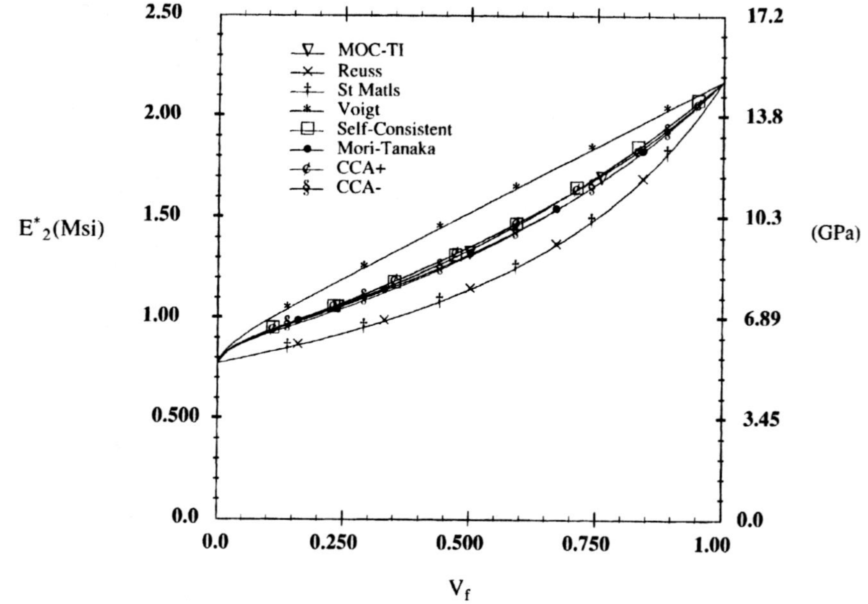

FIGURE 11.26 Effective Transverse Modulus E_2^* for Carbon/Epoxy

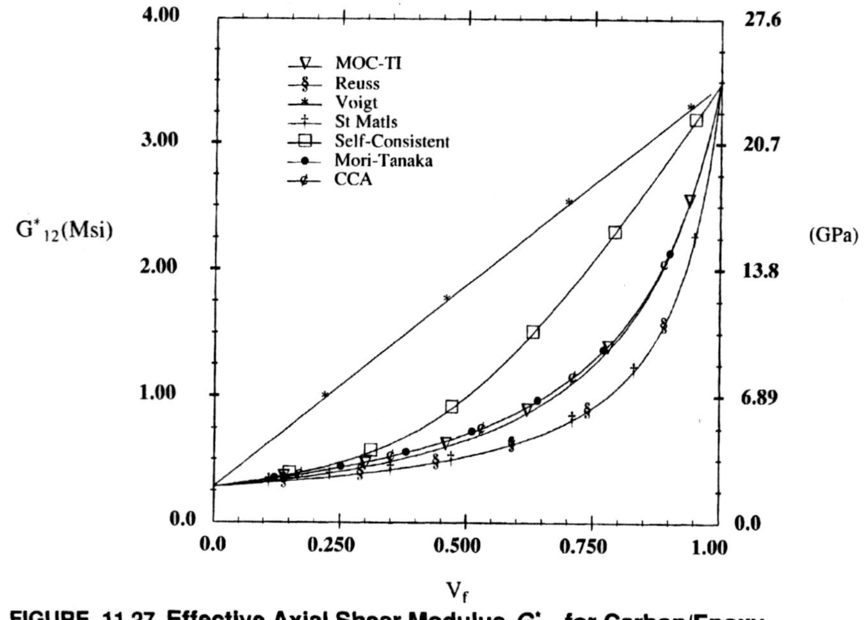

FIGURE 11.27 Effective Axial Shear Modulus G_{12}^* for Carbon/Epoxy

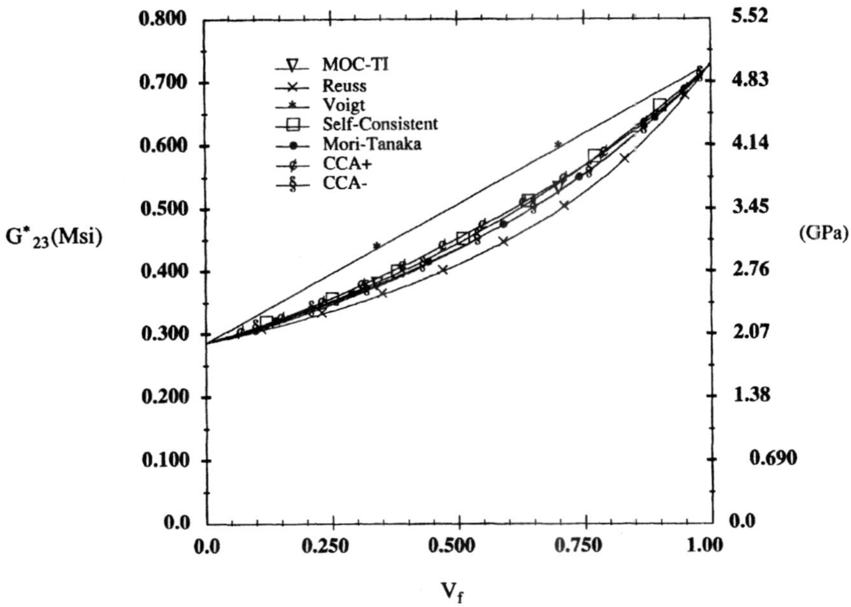

FIGURE 11.28 Effective Transverse Shear Modulus G_{23}^* for Carbon/Epoxy

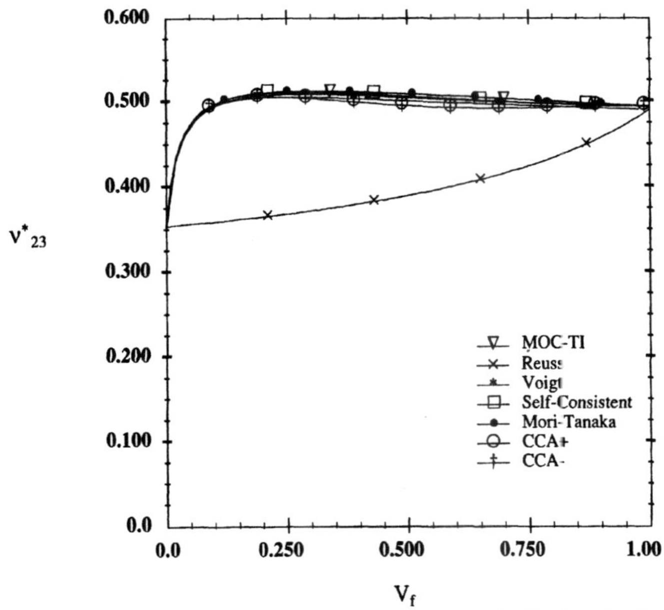

FIGURE 11.29 Effective Transverse Poisson's Ratio ν_{23}^* for Carbon/Epoxy

the axial modulus being much greater than the transverse modulus. The main differences between the results for glass/epoxy and carbon/epoxy are in the effective transverse properties, E_2^*, G_{23}^*, and ν_{23}^*. For the highly orthotropic carbon/epoxy, these properties exhibit much more linear behavior than for the glass/epoxy.

11.5 Summary

We have seen that there are several micromechanics models that provide identical or similar predictions for the effective elastic properties. In general, it may be stated that the method of cells (MOC), the Mori-Tanaka method, and the concentric cylinder assemblage (CCA) give similar predictions in most cases. The Reuss, Voigt, and self-consistent models provide good results only in special cases.

Experimental results for the effective engineering properties of composites as a function of fiber volume fraction are quite scarce due to the difficulties in measuring these properties accurately, if at all. Experimental results and comparisons between theory and experiment can be found in Tsai (1964), Whitney and Riley (1966), Noyes and Jones (1968), Dean and Turner (1973), Aboudi (1991), and Lissenden and Herakovich (1992). The comparisons generally indicate that the experimental results are in line with the method of cell (MOC), the Mori-Tanaka method, and the concentric cylinder assemblage (CCA), as well as with other models for properties where a rule of mixtures suffices.

References

Aboudi, J. (1991), *Mechanics of Composite Materials: A Unified Micromechanical Approach,* Elsevier, Amsterdam.

Aboudi, J. (1995), "Micromechanical Analysis of Thermo-Inelastic Multiphase Short-Fiber Composite," *Compos. Eng.* vol. 5, pp. 839–850.

Achenbach, J. D. (1975), *A Theory of Elasticity with Microstructure for Directionally Reinforced Composites,* CISM Courses and Lectures no. 167, Springer-Verlag, New York.

Adams, D. F. (1970), "Inelastic Analysis of a Unidirectional Composite Subjected to Transverse Normal Loading," *J. Compos. Mater.*, vol. 4, pp. 310–329.

Avery, W. B., and Herakovich, C. T. (1986), "Effect of Fiber Anisotropy on Thermal Stresses in Fibrous Composites," *J. Appl. Mech.*, vol. 53, no. 4, pp. 751–756.

Benveniste, Y. (1987), "A New Approach to the Application of Mori-Tanaka's Theory in Composite Materials," *Mech. Mater.*, vol. 36, pp. 147–157.

Brayshaw, J. B. (1994), "Consistent Formulation of the Method of Cells for Transversely Isotropic Metal Matrix Composites," Ph.D. dissertation, University of Virginia, Charlottesville.

Budiansky, B. (1965), "On the Elastic Moduli of Some Heterogeneous Materials," *J. Mech. Phys. Solids,* vol. 13, pp. 223–227.

Christensen, R. M. (1979), *Mechanics of Composite Materials,* John Wiley & Sons, Inc., New York.

Christensen, R. M., and Lo, K. H. (1979), "Solutions for Effective Shear Properties in Three Phase Sphere and Cylinder Models," *J. Mech. Phys. Solids,* vol. 27, pp. 315–330.

Dean, G. D. and Turner, P. (1973), "The Elastic Properties of Carbon Fibres and Their Composites," *Composites,* vol. 4, no. 4, pp. 174–180.

Dvorak, G. J. (1990), "Plasticity Theories for Fibrous Composite Materials," in *Metal Matrix Composites, Vol. 2: Mechanisms and Properties* (R. K. Everett and R. J. Arsenault, eds.), Treatise on Materials Science and Technology vol. 33, Academic Press, Boston, pp. 1–77.

Dvorak, G. J., Rao, M. S. M., and Tarn, J. Q. (1973), "Yielding in Unidirectional Composites under External Loads and Temperature Changes," *J. Compos. Mater.*, vol. 7, p. 194.

Eshelby, J. D. (1957), "The Determination of the Elastic Field of an Ellipsoidal Inclusion, and Related Problems," *Proc. R. Soc. London*, vol. A241, pp. 376–396.

Foye, R. L. (1968), "Advanced Design Concepts for Advanced Composite Airframes," AFML-TR-68-91, Air Force Materials Laboratory, Wright-Patterson Air Force Base, Ohio.

Hashin, Z. (1962), "The Elastic Moduli of Heterogeneous Materials," *J. Appl. Mech.*, vol. 29, p. 143.

Hashin, Z. (1972), *Theory of Fiber Reinforced Materials*, NASA CR–1974, National Aeronautics and Space Administration, Washington, DC.

Hashin, Z. (1979), "Analysis of Properties of Fiber Composites with Anisotropic Constituents," *J. Appl. Mech.*, vol. 446, pp. 543–550.

Hashin, Z. (1990), "Thermoelastic Properties and Conductivity of Carbon/Carbon Fiber Composites," *Mech. Mater.*, vol. 8, pp. 293–308.

Hashin, Z., and Rosen, B. W. (1964), "The Elastic Moduli of Fiber Reinforced Materials," *J. Appl. Mech.*, vol. 31, pp. 223–232.

Hill, R. (1964), "Theory of Mechanical Properties of Fibre-Strengthened Materials—I. Elastic Behaviour," *J. Mech. Phys. Solids*, vol. 12, pp. 199–212.

Hill, R. (1965), "A Self-Consistent Mechanics of Composite Materials," *J. Mech. Phys. Solids*, vol. 13, no. 4, pp. 213–222.

Knott, T. W., and Herakovich, C. T. (1988), "Effect of Fiber Morphology on Composite Properties," CCMS-88-09, VPI-E-88-16, Virginia Tech, Blacksburg.

Knott, T. W., and Herakovich, C. T. (1991), "Effect of Fiber Orthotropy on Effective Composite Properties," *J. Compos. Mater.*, vol. 25, no.6, pp. 732–759.

Lin, T. H., Salinas, D., and Ito, Y. M. (1972), "Elastic-Plastic Analysis of Unidirectional Composites," *J. Compos. Mater.*, vol. 6, pp. 48–61.

Lissenden, C. J., and Herakovich, C. T. (1992), "Comparison of Micromechanical Models for Elastic Properties," in *Engineering, Construction and Operations in SPACE III* (W. Z. Sadeh, S. Sture, and R. J. Miller, eds.), Space '92, Proc. 3rd Int. Conference, American Society of Civil Engineers, New York, pp. 1309–1322.

Mori, T., and Tanaka, K. (1973), "Average Stress in Matrix and Average Elastic Energy of Materials with Misfitting Inclusions," *Acta Metall.*, vol. 21, pp. 571–574.

Noyes, J. V., and Jones, B. H. (1968), "Analytical Design Procedures for the Strength and Elastic Properties of Multilayer Fiber Composites," Proc. AIAA/ASME 9th Structures, Dynamics and Materials Conf., Paper no. 68-336.

Paley, M., and Aboudi, J. (1992), "Micromechanical Analysis of Composites by the Generalized Cells Method," *Mech. Mater.*, vol. 14, pp. 127–139.

Paul, B. (1960), "Prediction of Elastic Constants of Multiphase Materials," *Trans. AIME*, vol. 218, pp. 36–41.

Pindera, M. -J., Freed, A. D., and Arnold, S. M. (1993), "Effects of Fiber and Interfacial Layer Morphologies on the Thermoplastic Response of Metal Matrix Composites," *Int. J. Solids Struct.*, vol. 30, no. 9, pp. 1213–1238.

Reuss, A. (1929), "Berechnung der Fliessgrenze von Mischkristallen auf Grund der Plastizitätsbedingung für Einkristalle," *Z. Angew. Math. Mech.*, vol. 9, pp. 49–58.

Tsai, S. W. (1964), "Structural Behavior of Composite Materials," NASA CR-71, National Aeronautics and Space Administration, Washington, DC.

Voigt, W. (1889), "Über die Beziehung Zwischen den Beiden Elastizitätskonstanten Isotroper Körper," *Wied. Ann.*, vol. 38, pp. 573–587.

Whitney, J. M. and Riley, M. B. (1966), "Elastic Properties of Fiber Reinforced Composite Materials," *AIAA J.*, vol. 4, p. 1537.

Wilt, T. E. (1995), "On the Finite Element Implementation of the Generalized Method of Cells Micromechanics Constitutive Model," NASA CR 195451, National Aeronautics and Space Administration, Washington, DC.

Exercises

11.1 Prove that equations (11.66) for engineering properties in terms of stiffness coefficients are true.

11.2 Show that the third term in the CCA model prediction for axial modulus, (11.84), is very small for T300/5208 carbon/epoxy and SCS-6/Ti-15-3 silicon-carbide/titanium.

11.3 Show that the rule of mixtures is a better prediction for the effective axial modulus than it is for the effective axial Poisson's ratio by comparing the third term in (11.87) to the third term in (11.84) for T300/5208 carbon/epoxy and SCS-6/Ti-15-3 silicon-carbide/titanium.

11.4 Show that the traction continuity conditions along *B*-*B* and *D*-*D* of the method of cells as depicted in Fig. 11.16 do not provide any additional equations independent from the set (11.163) through (11.168). Also show that only three of the four equations (11.165)–(11.168) are independent.

APPENDIX A MATRIX INVERSION

Given a $n \times n$ square, nonsingular matrix $[A]$, the inverse of this matrix, $[A]^{-1}$, is obtained by replacing each element A_{ij} of the matrix by its cofactor a_{ij}, transposing the resulting matrix, and dividing by the determinant of $[A]$ (Kreyszig, 1962). As an example, for $n = 3$,

$$[A] = \begin{bmatrix} A_{11} & A_{12} & A_{13} \\ A_{21} & A_{22} & A_{23} \\ A_{31} & A_{32} & A_{33} \end{bmatrix} \tag{A.1}$$

with the determinant of $[A]$ denoted $|A| = \Delta$, the inverse of $[A]$ is given by

$$[A]^{-1} = \frac{1}{\Delta}\begin{bmatrix} a_{11} & a_{12} & a_{13} \\ a_{21} & a_{22} & a_{23} \\ a_{31} & a_{32} & a_{33} \end{bmatrix}^T = \frac{1}{\Delta}\begin{bmatrix} a_{11} & a_{21} & a_{31} \\ a_{12} & a_{22} & a_{32} \\ a_{13} & a_{23} & a_{33} \end{bmatrix} \tag{A.2}$$

where the cofactor a_{ij} is determined by suppressing the ith row and jth column from $[A]$, and forming the determinant of the remaining matrix (called the minor of A_{ij}), and then multiplying the minor by $(-1)^{i+j}$.

The determinant Δ is defined as the sum of the products of the elements of any row or column times their respective cofactors. Thus, for the 3×3 matrix $[A]$, the determinant of $[A]$ is

$$\Delta = A_{11}(A_{22}A_{33} - A_{23}A_{32}) - A_{12}(A_{21}A_{33} - A_{31}A_{23}) + A_{13}(A_{21}A_{32} - A_{22}A_{31}) \tag{A.3}$$

and the inverse of $[A]$ is

$$[A]^{-1} = \frac{1}{\Delta}\begin{bmatrix} (A_{22}A_{33} - A_{23}A_{32}) & -(A_{12}A_{33} - A_{13}A_{32}) & (A_{12}A_{23} - A_{13}A_{22}) \\ -(A_{21}A_{33} - A_{23}A_{31}) & (A_{11}A_{33} - A_{13}A_{31}) & -(A_{11}A_{23} - A_{13}A_{21}) \\ (A_{21}A_{32} - A_{22}A_{31}) & -(A_{11}A_{32} - A_{12}A_{31}) & (A_{11}A_{22} - A_{12}A_{21}) \end{bmatrix} \tag{A.4}$$

Reference

Kreyszig, E. (1962), *Advanced Engineering Mathematics,* John Wiley & Sons, Inc., New York, p. 401.

APPENDIX B BOOKS ON COMPOSITES

Aboudi, J. (1991), *Mechanics of Composite Materials: A Unified Micromechanical Approach*, Elsevier, Amsterdam.

Achenbach, J. D. (1975), *A Theory of Elasticity with Microstructure for Directionally Reinforced Composites*, CISM Courses and Lectures no. 167, Springer-Verlag, New York.

Agarwal, B. D., and Broutman, L. J. (1990), *Analysis and Performance of Fiber Composites*, 2nd ed., John Wiley & Sons, Inc., New York.

Ambartsumyan, S. A. (1970), *Theory of Anisotropic Plates*, Technomic Publishing Co., Stamford, CT.

Ashton, J. E., Halpin, J. C., and Petit, P. H. (1969), *Primer on Composite Materials: Analysis*, Technomic, Publishing Co., Stamford, CT.

Ashton, J. E., and Whitney, J. M. (1970), *Theory of Laminated Plates*, Technomic Publishing Co., Stamford, CT.

Broutman, L. J., and Krock, R. H. (eds.) (1974), *Composite Materials*, vols. 1–8, Academic Press, San Diego.

Calcote, L. R. (1969), *The Analysis of Laminated Composite Structures*, Van Nostrand Reinhold, New York.

Chawla, K. K. (1987), *Composite Materials: Science and Engineering*, Springer-Verlag, New York.

Christensen, R. M. (1979), *Mechanics of Composite Materials*, John Wiley & Sons, Inc., New York.

Daniel, I. M., and Ishai, O. (1994), *Engineering Mechanics of Composite Materials*, Oxford University Press, Oxford, England.

Gibson, R. F. (1994), *Principles of Composite Material Mechanics*, McGraw-Hill, Inc., New York.

Halpin, J. C. (1992), *Revised Primer on Composite Materials: Analysis*, 2nd ed., Technomic Publishing Co., Stamford, CT.

Hashin, Z. (1972), *Theory of Fiber Reinforced Materials*, NASA CR-1974, National Aeronautics and Space Administration, Washington, DC.

Hearmon, R. F. S. (1961), *An Introduction to Applied Anisotropic Elasticity*, Oxford University Press, Oxford, England.

Holaway, L. (1993), *Polymer Composites for Civil and Structural Engineering*, Blackie Academic & Professional, Glasgow.

Jones, R. M. (1975), *Mechanics of Composite Materials*, McGraw-Hill, Inc., New York.

Kelly, A. (1994), *Concise Encyclopedia of Composite Materials*, rev. ed., Elsevier Science, Ltd., Amsterdam.

Kelly, A., and Rabotnov, Y. N. (eds.) (1986), *Handbook of Composite*, vols. 1–4, North-Holland, New York.

Lee, S. J. (ed.) (1991), *International Encyclopedia of Composites*, vols. 1–5, VCH Publishers, Inc., New York.

Lekhnitskii, S. G. (1963), *Theory of Elasticity of an Anisotropic Body*, Holden-Day, San Francisco.

Lubin, G. (ed.) (1982), *Handbook of Composites*, Van Nostrand Reinhold, New York.

Obraztsov, I. F., and Vasil'ev, V. V. (eds.) (1982), *Mechanics of Composites*, MIR Publishers, Moscow.

Mallick, P. K. (1988), *Fiber-Reinforced Composites: Materials, Manufacturing and Design*, Marcel Dekker, Inc., New York.

Nemat-Nasser, S., and Hori, M. (1993), *Micromechanics: Overall Properties of Heterogeneous Materials,* Elsevier Science Publishers B. V., Amsterdam.

Ochoa, O. O., and Reddy, J. N. (1992), *Finite Element Analysis of Composite Laminates,* Kluwer Academic Publishers, Dordrecht, The Netherlands.

Spencer, A. J. M. (1972), *Deformations of Fibre-Reinforced Materials,* Clarendon Press, New York.

Springer, G. S. (1988), *Environmental Effects in Composite Materials.* vol. 3, Technomic Publishing Co., Stamford, CT.

Suresh, S., Mortensen, A., and Needleman, A. (1993), *Fundamentals of Metal Matrix Composites,* Butterworth-Heinemann, Stoneham, MA.

Taya, M., and Arsenault, R. J. (1989), *Metal Matrix Composites,* Pergamon Press, Elmsford, NY.

Tsai, S. W. (1985), *Composite Design, 1986,* Think Composites, Dayton, OH.

Tsai, S. W., and Hahn, H. T. (1980), *Introduction to Composite Materials,* Technomic Publishing Co., Stamford, CT.

Vinson, J. R. (1993), *The Behavior of Shells Composed of Isotropic and Composite Materials,* Kluwer Academic Publishers, Dordrecht, The Netherlands.

Vinson, J. R., and Sierakowski, R. L. (1987), *The Behavior of Structures Composed of Composite Materials,* Martinus Nijhoff, Dordrecht, The Netherlands.

Whitney, J. M. (1987), *Structural Analysis of Laminated Anisotropic Plates,* Technomic Publishing Co., Stamford, CT.

AUTHOR INDEX

SUBJECT INDEX

Printed in the United States
31755LVS00002B/1-30